Chloroplasts

Current Research and Future Trends

Edited by

Helmut Kirchhoff

Institute of Biological Chemistry
Washington State University
Pullman, WA
USA

Caister Academic Press

Caister Academic Press
Norfolk, UK

www.caister.com

British Library Cataloguing-in-Publication Data
A catalogue record for this book is available from the British Library

ISBN: 978-1-910190-47-0 (paperback)
ISBN: 978-1-910190-48-7 (ebook)

Cover design adapted from images provided by Helmut Kirchhoff

Contents

Current Books of Interest

Brain-eating Amoebae: Biology and Pathogenesis of *Naegleria fowleri*	2016
Staphylococcus: Genetics and Physiology	2016
Microbial Biodegradation: From Omics to Function and Application	2016
Influenza: Current Research	2016
MALDI-TOF Mass Spectrometry in Microbiology	2016
Aspergillus and *Penicillium* in the Post-genomic Era	2016
Omics in Plant Disease Resistance	2016
Acidophiles: Life in Extremely Acidic Environments	2016
Climate Change and Microbial Ecology: Current Research and Future Trends	2016
Biofilms in Bioremediation: Current Research and Emerging Technologies	2016
Microalgae: Current Research and Applications	2016
Gas Plasma Sterilization in Microbiology: Theory, Applications, Pitfalls and New Perspectives	2016
Virus Evolution: Current Research and Future Directions	2016
Arboviruses: Molecular Biology, Evolution and Control	2016
Shigella: Molecular and Cellular Biology	2016
Aquatic Biofilms: Ecology, Water Quality and Wastewater Treatment	2016
Alphaviruses: Current Biology	2016
Thermophilic Microorganisms	2015
Flow Cytometry in Microbiology: Technology and Applications	2015
Probiotics and Prebiotics: Current Research and Future Trends	2015
Epigenetics: Current Research and Emerging Trends	2015
Corynebacterium glutamicum: From Systems Biology to Biotechnological Applications	2015
Advanced Vaccine Research Methods for the Decade of Vaccines	2015
Antifungals: From Genomics to Resistance and the Development of Novel Agents	2015
Bacteria-Plant Interactions: Advanced Research and Future Trends	2015
Aeromonas	2015
Antibiotics: Current Innovations and Future Trends	2015
Leishmania: Current Biology and Control	2015
Acanthamoeba: Biology and Pathogenesis (2nd edition)	2015
Microarrays: Current Technology, Innovations and Applications	2014
Metagenomics of the Microbial Nitrogen Cycle: Theory, Methods and Applications	2014

Full details at www.caister.com

Contributors

Monica Balsera
Instituto de Recursos Naturales y Agrobiología de Salamanca
Consejo Superior de Investigaciones Científicas
Salamanca
Spain

monica.balsera@csic.es

Bob B. Buchanan
Department of Plant and Microbial Biology
University of California
Berkeley, CA
USA

view@berkeley.edu

Michael Cantrell
Department of Biology
Colorado State University
Fort Collins, CO
USA

michael.cantrell@colostate.edu

Peter Dörmann
Institute of Molecular Physiology and Biotechnology of Plants (IMBIO)
University of Bonn
Bonn
Germany

doermann@uni-bonn.de

Karsten Fischer
Institute for Arctic and Marine Biology
University of Tromsø
Tromsø
Norway

karsten.fischer@uit.no

Nicholas Fisher
MSU-DOE Plant Research Laboratory
Michigan State University
East Lansing, MI
USA

nefisher@msu.edu

Christiane Funk
Department of Chemistry
Umeå University
Umeå
Sweden

christiane.funk@umu.se

Arthur R. Grossman
Department of Plant Biology
Carnegie Institution for Science
Stanford, CA
USA

arthurg@stanford.edu

Miguel A. Hernández-Prieto
ARC Centre of Excellence for Translational Photosynthesis; and
School of Life and Environmental Sciences
The University of Sydney
Sydney
Australia

miguel.hernandez-prieto@sydney.edu.au

Ricarda Höhner
School of Biological Sciences
Washington State University
Pullman, WA
USA

rhoehner@wsu.edu

Georg Hölzl
Institute of Molecular Physiology and Biotechnology of Plants (IMBIO)
University of Bonn
Bonn
Germany

hoelzl@uni-bonn.de

Denis Jallet
Department of Biology
Colorado State University
Fort Collins, CO
USA

denis.jallet@colostate.edu

Felix Kessler
Laboratory of Plant Physiology
University of Neuchâtel
Neuchâtel
Switzerland

felix.kessler@unine.ch

Helmut Kirchhoff
Institute of Biological Chemistry
Washington State University
Pullman, WA
USA

kirchhh@wsu.edu

David M. Kramer
MSU-DOE Plant Research Laboratory
Michigan State University
East Lansing, MI
USA

kramerd8@msu.edu

Hans-Henning Kunz
School of Biological Sciences
Washington State University
Pullman, WA
USA

henning.kunz@wsu.edu

Graham Peers
Department of Biology
Colorado State University
Fort Collins, CO
USA

graham.peers@colostate.edu

Thibaut Pralon
Laboratory of Plant Physiology
University of Neuchâtel
Neuchâtel
Switzerland

thibaut.pralon@unine.ch

Sujith Puthiyaveetil
Institute of Biological Chemistry
Washington State University
Pullman, WA
USA

s.puthiyaveetil@wsu.edu

Shai Saroussi
Department of Plant Biology
Carnegie Institution for Science
Stanford, CA
USA

ssaroussi@carnegiescience.edu

Peter Schürmann
Laboratoire de Biologie Moléculaire et Cellulaire
Université de Neuchâtel
Neuchâtel
Switzerland

peter.schurmann@unine.ch

Trudi A. Semeniuk
ARC Centre of Excellence for Translational Photosynthesis; and
School of Life and Environmental Sciences
The University of Sydney
Sydney
Australia

tasem.editing@gmail.com

Deserah D. Strand
Max-Planck-Institut für Molekulare Pflanzenphysiologie
Potsdam
Germany

strand@mpimp-qolm.mpg.de

Tania Tibiletti
Laboratory of Plant Genetics and Biophysics
Aix-Marseille University
Marseille
France

tania.tibiletti@univ-amu.fr

Alexander N. Tikhonov
Faculty of Physics
M.V. Lomonosov Moscow State University
Moscow
Russia

an_tikhonov@mail.ru

Katharina vom Dorp
Institute of Molecular Physiology and Biotechnology of Plants (IMBIO)
University of Bonn
Bonn
Germany

katharina.vom.dorp@googlemail.com

Andreas P.M. Weber
Institut für Biochemie der Pflanzen
Cluster of Excellence on Plant Sciences
Heinrich-Heine-Universität
Düsseldorf
Germany

andreas.weber@uni-duesseldorf.de

Tyler M. Wittkopp
Department of Plant Biology
Carnegie Institution for Science
Stanford, CA
USA

tylerwittkopp@gmail.com

Yasusi Yamamoto
Graduate School of Natural Science and Technology
Okayama University
Okayama
Japan

yasusiya@cc.okayama-u.ac.jp

Wenqiang Yang
Department of Plant Biology
Carnegie Institution for Science
Stanford, CA
USA

wenqiang_yang@126.com

Miho Yoshioka-Nishimura
Graduate School of Natural Science and Technology
Okayama University
Okayama
Japan

yoshi-m3@cc.okayama-u.ac.jp

Yonghong Zhou
Institute of Molecular Physiology and Biotechnology of Plants (IMBIO)
University of Bonn
Bonn
Germany

yonghong.zhou@outlook.com

Preface

Life must be constantly fuelled with energy to keep the numerous chemical reactions in the cell from thermodynamic equilibrium, which constitutes the energetic driving force for metabolism. The chloroplast organelles in plant leaves have been tuned by more than a billion years of evolution to function as the most efficient and robust powerhouses for life on earth by using energy from the sun. Besides fuelling cell metabolism, the chloroplast also represents a sophisticated anabolic factory for producing a variety of primary and secondary metabolites. This gives the organelle a central role in maintaining life on our planet. Furthermore, in recent decades, multiple endeavours have been initiated to make use of the unique bioenergetic and anabolic potential of chloroplasts to help solve some of the most urgent problems of mankind in the areas of sustainable energy supply, paving the way for a second green revolution, and in the production of valuable chemicals, for example for medical applications.

This book surveys hot topics in chloroplast biology showing that this is a highly active and dynamic area in plant sciences. The chapters in this book document one of the main aspects of chloroplast biology, that is, the plasticity of chloroplast functions in context of a dynamic environment and of different metabolic needs. In recent years, the research community has appreciated that we have to understand how plants in general, and the chloroplast in particular, respond to often unpredictable changes in natural parameters (i.e. light intensity, temperature, etc.). We now have a better understanding of the fact that many gene products have evolved to deal with the environmental and metabolic fluctuations required to ensure plant fitness and survival.

The book gives a state-of-the-art view on eminent areas in chloroplast research. In detail, it starts with the building blocks of the energy-converting thylakoid membranes (lipids, pigments, proteins and membranes); this is complemented by a chapter on plastoglobuli, which are attached to thylakoid membranes. The book continues by presenting current knowledge on electron and proton fluxes and the regulation and repair of the energy-converting machinery. The regulatory aspect is widened to redox regulation, which is followed by a chapter addressing how ions and metabolites are transported across chloroplast membranes. The book closes by presenting system-based approaches for identification and characterization of unique chloroplast-hosted proteins.

The chapters were prepared by internationally, highly acknowledged colleagues who work at the forefront of their research area and, therefore, provide expert insight in their scientific field. I'm deeply thankful for my colleagues' hard work and the fact that they have shared their expert knowledge in excellently written chapters. Last but not least, the realization of this book would not have been possible without the volunteering reviewers. Their thorough job significantly increased the quality of the individual chapters. Their fresh eyes and constructive critique provided valuable input to the contributors. For this essential contribution, I'm highly indebted to the reviewers (alphabetical order): Drs Claire Brehelin, Rikard Fristedt, Jingpeng Gao, Mark Heinnickel, Toru Hisabori, Zhirong Li, Sujith Puthiyaveetil, Mark Aurel Schoettler and Ildiko Szabo.

Helmut Kirchhoff

Chloroplast Lipids

1

Yonghong Zhou, Katharina vom Dorp, Peter Dörmann and Georg Hölzl

Abstract

Chloroplasts are the major site of fatty acid *de novo* synthesis in the cells of plants and algae. Fatty acids are incorporated into glycerolipids in the chloroplast (prokaryotic pathway) or are exported to the endoplasmic reticulum for glycerolipid synthesis following the eukaryotic pathway. The four major glycerolipids of chloroplasts are monogalactosyldiacylglycerol (MGDG), digalactosyldiacylglycerol (DGDG), sulfoquinovosyldiacylglycerol (SQDG) and phosphatidylglycerol (PG), which are the building blocks for thylakoids and play specific roles in photosynthesis. During phosphate deprivation DGDG and SQDG accumulate and replace phospholipids to save phosphate for other processes. In addition to glycerolipids, chloroplasts contain a large variety of isoprenoid lipids derived from the methyl-erythritol phosphate pathway. Prenylquinones (tocopherol, plastoquinone, phylloquinone) are one class of chloroplast isoprenoid lipids and serve as antioxidants or electron carriers of photosystems II and I, respectively. β-Carotene (an isoprenoid-derived carotenoid), tocopherol, and phylloquinone (provitamin A, vitamin E and vitamin K1, respectively) are essential components of the human diet. Chlorophyll, the main photosynthetic pigment, carries an isoprenoid-derived phytyl chain, which is released during chlorophyll breakdown. Phytol is employed for the synthesis of fatty acid phytyl esters and tocopherol. In contrast to plants, chloroplasts of eukaryotic algae contain further lipids and employ additional routes for lipid synthesis. Research on chloroplast lipids has seen major progress in the recent past due to the development of highly sensitive and accurate mass spectrometers for the analysis of extremely low abundant lipids or lipids only accumulating during stress conditions.

Introduction

Chloroplasts of plants and algae are highly specialized organelles performing oxygenic photosynthesis. Two main groups of lipids, glycerolipids and isoprenoid lipids, are characteristic for chloroplasts, and directly or indirectly involved in photosynthesis. The glycerolipids constitute the building blocks for chloroplast membranes (inner and outer envelope, thylakoids). Glycerolipids contain two fatty acids linked to the *sn*1 and *sn*2 positions of glycerol, and a polar head group linked to *sn*3. The specialization on oxygenic photosynthesis requires a glycerolipid composition highly conserved in thylakoids of plants, algae and cyanobacteria (Block *et al.*, 1983). Cyanobacteria are, according to the endosymbiont theory, the ancestors of chloroplasts. Surprisingly, a simple set of four lipids is sufficient for establishing the photosynthetic membranes, i.e. the two galactolipids monogalactosyldiacylglycerol (MGDG), digalactosyldiacylglycerol (DGDG), the sulfolipid sulfoquinovosyldiacylglycerol (SQDG), and the phospholipid phosphatidylglycerol (PG) (Fig. 1.1) (Li-Beisson *et al.*, 2013). SQDG and PG are anionic lipids at physiological pH. Plant chloroplasts further contain the phospholipid phosphatidylcholine (PC) in their outer envelope. In thylakoids isolated from spinach chloroplasts, MGDG constitutes about 52%, DGDG 26%, SQDG 6.5% and PG 9.5% of total membrane lipids (Block *et al.*, 1983). The anomeric structures of the glycolipid head groups are also highly conserved. The galactose residue in MGDG is always bound

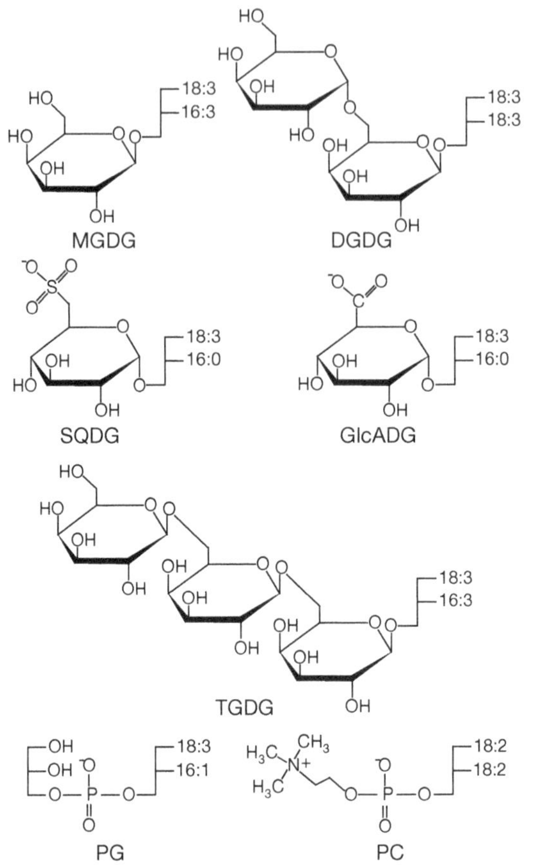

Figure 1.1 Structures of chloroplast glycerolipids. Monogalactosyldiacylglycerol (MGDG), digalactosyldiacylglycerol (DGDG), sulfoquinovosyldiacylglycerol (SQDG) and phosphatidylglycerol (PG) are bulk membrane lipids, while glucuronosyldiacylglycerol (GlcADG) and trigalactosyldiacylglycerol (TGDG) are produced under different stress conditions. Phosphatidylcholine (PC) is only present in the outer chloroplast envelope.

in β-configuration to the glycerol (Fig. 1.1), while DGDG harbours a terminal α-galactose moiety (1→6)-linked to the inner β-galactose. The sulfoquinovose (6-deoxy-6-sulfoglucose) residue of SQDG is derived from α-glucose with a sulfonate group linked to the C-6 carbon of the sugar. The different glycerolipids are further characterized by their unique fatty acid compositions (Li-Beisson *et al.*, 2013). Galactolipids contain high amounts of polyunsaturated fatty acids. The main molecular species of MGDG in plants are 18:3/18:3 (*sn*1/*sn*2) and 18:3/16:3-MGDG (Fig. 1.1). The main DGDG molecular species is 18:3/18:3. SQDG is always more saturated than galactolipids, with 18:3/16:0 as one main molecular species. A special feature of chloroplast PG is the presence of trans Δ3 hexadecenoic acid ($16{:}1^{\Delta 3\mathrm{trans}}$) linked to *sn*2. The fatty acid composition of green algae such as *Chlamydomonas* is similar to higher plants, but often with high amounts of 16:4 linked to *sn*2 of MGDG (Giroud *et al.*, 1988).

Isoprenoid lipids play crucial roles in photosynthetic electron transfer of photosystems I and II (PSI, PSII) and in the light-harvesting complex II (LHCII). Some isoprenoid lipids are stored in plastoglobules, lipid subcompartments found in chloroplasts, especially under stress conditions or in senescent leaves (Tevini and Steinmüller, 1985). Chloroplasts harbour a plethora of different isoprenoid lipids. Photosynthetic pigments such as chlorophyll and carotenoids are derived from isoprenoid metabolism and provide the means for capturing photons for the conversion into chemical energy and for carbon fixation. Carotenoids, in particular xanthophylls, are involved in dissipating excess light energy. Chlorophyll and carotenoids confer the green coloration of leaves or the yellow/red coloration of fruits, petals, and senescent leaves, respectively. Prenylquinones, such as tocopherol or phylloquinone and plastoquinone, protect the photosynthetic membranes against oxidative stress or are components of the photosynthetic electron-transfer chain, respectively. β-Carotene (provitamin A), tocopherol (vitamin E) and phylloquinone (vitamin K1) are lipid-soluble vitamins essential for human and animal nutrition (Grusak and DellaPenna, 1999). The biosynthetic pathways for isoprenoid lipids are tightly connected. Moreover, links between plastidial glycerolipid and isoprenoid pathways exist. Fatty acid phytyl esters are synthesized in senescent leaves or in leaves under stress, from phytol and fatty acids derived from chlorophyll and glycerolipid degradation, respectively (Lippold *et al.*, 2012). Carotenoid esters were found in yellow tomato petals and in microalgae (Ariizumi *et al.*, 2014; Chen *et al.*, 2015). The large variety of glycerolipid molecular species and of isoprenoid lipids in chloroplasts presents a challenge for the comprehensive analysis of these compounds, in particular as the lipids differ in their physical and chemical properties. Glycerolipids can be analysed by thin-layer chromatography (TLC), gas chromatography (GC), gas chromatography–mass spectrometry (GC-MS) and liquid chromatography–tandem mass spectrometry (LC-MS/MS). Many isoprenoid lipids contain conjugated double bonds as chromophores and can therefore be analysed by spectrophotometry or high-pressure liquid chromatography (HPLC) with ultraviolet (UV) or fluorescence detectors.

Glycerolipids in chloroplasts

Glycerolipid synthesis and lipid trafficking

The prokaryotic pathway of glycerolipid synthesis

Fatty acids in plants and algae are synthesized *de novo* in the stroma of chloroplasts by a type II fatty acid synthase (FASII) (Li-Beisson *et al.*, 2013). The growing acyl chains are attached to acyl carrier protein (ACP). Chain elongation reactions are catalysed by β-ketoacyl-ACP synthases (KAS) III and KASI, resulting in 16:0-ACP production, while the elongation from 16:0-ACP to 18:0-ACP is KASII-dependent (Wu *et al.*, 1994). 18:0-ACP is desaturated to 18:1-ACP by a soluble acyl-ACP Δ9-desaturase (SAD, FAB2). The further relative proportions and incorporation into lipids of the main products 16:0-ACP and 18:1-ACP depends on the activities of KASII (synonymous: FAB1), two plastidial acyltransferases, glycerol-3-phosphate acyltransferase (GPAT) and lysophosphatidic acid acyltransferase (LPAAT), and two acyl-ACP thioesterases, FATA and FATB (Fig. 1.2), with substrate specificities for 18:1-ACP, and long-chain acyl-ACPs, respectively. The thioesterases FATA and FATB are specific for 18:1-ACP and long-chain acyl-ACPs, respectively, and release free fatty acids which are exported to the ER for incorporation into extraplastidial glycerolipids ('eukaryotic lipids') (Bonaventure *et al.*, 2004).

Fatty acids that are not released from acyl-ACP by thioesterases are incorporated into plastidial ('prokaryotic') glycerolipids by GPAT and LPAAT. Glycerolipid biosynthesis in chloroplasts starts with the acylation of glycerol-3-phosphate (G3P) at the *sn*1 position by GPAT using acyl-ACP, forming lysophosphatidic acid (LPA) (Fig. 1.2). The plastidial GPAT, also designated ATS1 or ACT1, is a soluble enzyme in the stroma and is specific for 18:1-ACP (Frentzen *et al.*, 1983; Joyard and Douce, 1977). In the *ats1* (synonym: *act1*) mutant, the deficiency in plastidial GPAT results in the loss of plastidial lipids, which is compensated by an increased synthesis of ER-derived lipids (Kunst *et al.*, 1988; Xu *et al.*, 2006). The second acylation at the *sn*2 position of LPA is carried out by LPAAT to form PA, which is an important intermediate in glycerolipid metabolism and furthermore in signal transduction. The plastidial LPAAT (designated ATS2) is a membrane-bound enzyme localized to the chloroplast envelope. ATS2 is highly specific for 16:0-ACP (Andrews and Mudd, 1985; Frentzen *et al.*, 1983). Disruption of ATS2 in the *ats2* mutant leads to embryo lethality, demonstrating that plastidial LPAAT is essential for embryo development (Yu *et al.*, 2004). PA produced in the chloroplasts is employed for the synthesis of plastidial PG, and after dephosphorylation to diacylglycerol (DAG) by PA phosphatases (PAP), for the synthesis of MGDG, DGDG and SQDG (Joyard and Douce,

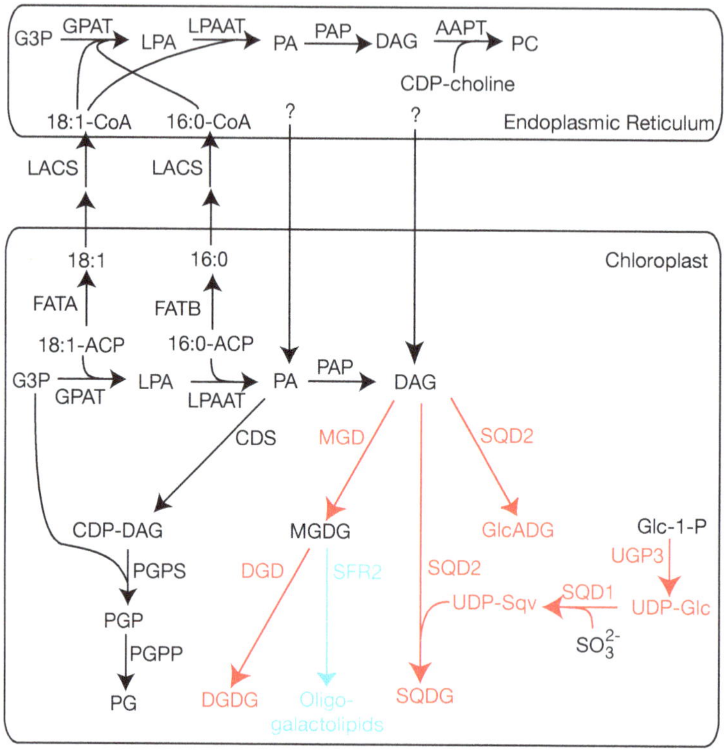

Figure 1.2 Glycerolipid biosynthesis in chloroplasts. Four major classes of glycerolipids are synthesized in chloroplasts from prokaryotic/chloroplast-derived or eukaryotic/ER-derived precursors. During freezing stress, SFR2 activity is stimulated resulting in the production of oligogalactolipids (highlighted in blue). Under phosphate deprivation, DGDG, SQDG and GlcADG are produced to replace phospholipids and to save phosphate for other metabolic pathways (in red). Glycerolipids and glycerolipid precursors: G3P, glycerol-3-phosphate; LPA, lysophosphatidic acid; PC, phosphatidylcholine; PA, phosphatidic acid; DAG, diacylglycerol; CDP-DAG, CDP-diacylglycerol; PGP, phosphatidylglycerol phosphate; PG, phosphatidylglycerol; MGDG, monogalactosyldiacylglycerol; DGDG, digalactosyldiacylglycerol; SQDG, sulfoquinovosyldiacylglycerol; UDP-Glc, UDP-glucose; UDP-Sqv, UDP-sulfoquinovose, Glc-1-P, glucose-1-phosphate; GlcADG, glycuronosyldiacylglycerol. Enzymes of glycerolipid synthesis in chloroplasts: AAPT, aminoalcoholphospho-transferase; FATA, FATB, acyl-ACP thioesterase A or B; GPAT, glycerol-3-phosphate acyltransferase; LPAAT, lysophosphatidic acid acyltransferase; PAP, phosphatidic acid phosphatase; LACS, long-chain acyl-CoA synthetase; CDS, CDP-diacylglycerol synthase; PGPS, phosphatidylglycerol phosphate synthase; PGPP, phosphatidylglycerol phosphate phosphatase; MGD, monogalactosyldiacylglycerol synthase; DGD, digalactosyldiacylglycerol synthase; SFR2, SENSITIVE TO FREEZING 2/galactolipid:galactolipid galactosyltransferase; UPG3, UDP-Glc pyrophosphorylase 3; SQD1, UDP-sulfoquinovose synthase 1; SQD2, sulfoquinovosyl transferase.

1977). Thus, plastidial PA and all glycerolipids derived from it harbour a C_{16} fatty acid at the *sn*2 position (prokaryotic lipids) (C_{18}/C_{16}) (Boudiére *et al.*, 2014; Ohlrogge and Browse, 1995).

The eukaryotic pathway of glycerolipid synthesis

Fatty acids released from acyl-ACP by FATA or FATB are exported from the chloroplast. Recently, a novel inner envelope membrane protein, FAX1 (fatty acid export 1), was identified and suggested to be involved in the export of fatty acids from the chloroplast (Li *et al.*, 2015a). Fatty acids need to be converted into their coenzyme A (CoA) esters for further metabolism at the ER. *Arabidopsis* contains nine members of the long-chain acyl-CoA synthase (LACS) family, but only LACS9 is localized to the chloroplast (Fig. 1.2) (Schnurr *et al.*, 2002). Recent results, however, demonstrated that LACS9 and LACS4, which show overlapping specificities, might not be involved in the export, but rather in the import of fatty acids into the plastid by activating free fatty acids for subsequent incorporation into PC via an LPAAT reaction (Jessen *et al.*, 2015). The ATP-binding cassette (ABC) transporter ABCA9 was suggested to mediate the uptake of acyl groups into the ER. Thus, *Arabidopsis abca9* mutant plants show strongly reduced seed triacylglycerol (TAG) content (Kim *et al.*, 2013). Contact sites between the chloroplast outer envelope and the ER (PLAM, plastid-associated membranes) were identified by fluorescence microscopy with micromanipulation and suggested to facilitate acyl and lipid exchange (Andersson *et al.*, 2007). An alternative way for acyl lipid transport was suggested for acyl-CoA-binding proteins that possibly transport acyl-CoAs to the ER (Xiao and Chye, 2009). At the ER, acyl groups from acyl-CoAs are incorporated into LPA, PA and PC by acyltransferases (ER-localized GPAT, LPAAT) and by acyl editing (Li-Beisson *et al.*, 2013). PC has a central function in glycerolipid metabolism, because it serves as substrate for fatty acid desaturation and it represents the intermediate for the synthesis of the other glycerolipids.

PA produced at the ER contains a C18 fatty acid at the *sn*2 position, because the ER-localized LPAAT is specific for 18:1-CoA. Therefore, ER-derived glycerolipids usually carry a C_{18} acyl group at the *sn*2 position (eukaryotic lipids) (C18/C18 or 16:0/C18). ER-derived lipid precursors can be re-imported into the chloroplast where they are employed for the synthesis of plastidial glycerolipids (MGDG, DGDG, SQDG) with eukaryotic structure. The location of the two pathways in the different compartments requires close coordination to supply lipid precursors for chloroplast lipid assembly and thylakoid expansion. Plants which employ the two pathways for glycerolipid synthesis are called '16:3 plants', because MGDG in these plants is enriched with 16:3 fatty acids, while '18:3 plants' use mainly or exclusively the eukaryotic pathway for galactolipid synthesis and therefore lack 16:3 (Li-Beisson *et al.*, 2013). Analysis of fatty acid composition from 468 plant species revealed that different proportions of glycerolipids in the various plant species are produced from the plastidial and ER pathway (Mongrand *et al.*, 1998).

The major pathways for fatty acid synthesis and glycerolipid assembly are conserved between plants and green algae including *Chlamydomonas*, as derived from the presence of orthologous genes in the genomes (Li-Beisson *et al.*, 2015). Nevertheless, major differences exist between lipid metabolism of plants and green algae. As *Chlamydomonas* lacks PC, MGDG was proposed to play a key role in fatty acid export from the chloroplast. A specific lipase, PGD1 (plastid galactoglycerolipid degradation 1), specifically releases acyl groups with low desaturation degree from MGDG. These fatty acids (mainly 18:1) are activated to acyl-CoA and used for the production of extraplastidial glycerolipids, especially TAG (Li *et al.*, 2012). In addition, *Chlamydomonas* uses only the prokaryotic pathway for lipid synthesis. Therefore, the import of lipid precursors into the plastid (i.e. the eukaryotic pathway) is not relevant for the production of chloroplast lipids. In contrast to plants, *Chlamydomonas* harbours the betaine lipid diacylglyceroltrimethylhomoserine (DGTS) that is believed to functionally replace PC in cytosolic membranes (Moellering *et al.*, 2010a). Furthermore, *Chlamydomonas* contains the unusual fatty acid $16{:}4^{\Delta4,7,10,13}$, which is present only in the *sn*2 position of MGDG. The Δ4 double bond is introduced by a Δ4 desaturase specific for $16{:}3^{\Delta7,10,13}$ bound to MGDG (Zäuner *et al.*, 2012). The introduction of this double bond has a regulatory role for the synthesis and content of MGDG.

Lipid transfer from the inner envelope to the thylakoids

The inner and outer envelopes of chloroplasts are the sites of galactolipid biosynthesis. In contrast to the outer envelope, which contains PC and only low amounts of MGDG, the lipid compositions of the inner envelope and the thylakoid membranes are highly similar (Block *et al.*, 1983). For this reason, it is likely that thylakoids are derived from the inner envelope. Three hypotheses were discussed for lipid transport from the inner envelope to the thylakoids (Hurlock *et al.*, 2014). The first hypothesis implies that the inner envelope forms stromal invaginations followed by fusions of these invaginations leading to the establishment of thylakoid lamellae, which finally are separated from the inner envelope. This pathway is assumed to be active during the early stages in chloroplast development. Secondly, stromal vesicles were suggested to be involved in thylakoid establishment. Vesicles can bud off the inner envelope and fuse with thylakoids, possibly by involvement of VIPP1 (vesicle-inducing protein in plastids 1) (Kroll *et al.*, 2001). Thirdly, the existence of a soluble lipid transport system was discussed, but all these hypotheses need further verification (Hurlock *et al.*, 2014).

Lipid transport from the ER to the chloroplast: the TGD transport complex

Because chloroplasts lack the capacity for PC biosynthesis, which exclusively occurs at the ER, a transport mechanism must exist for the transfer of PC or a related glycerolipid from the ER to the chloroplast outer membrane that contains a considerable amount of PC (Block *et al.*, 1983). Additional evidence for lipid transport from the ER to the chloroplast stems from the fact that chloroplasts utilize eukaryotic DAG backbones for galactolipid biosynthesis (see above). Lipid transfer from the ER to the chloroplast might be facilitated by membrane contact sites (Andersson *et al.*, 2007) (see above), lipid transport proteins or diffusion of individual lipid molecules. Radioactive labelling experiments suggested that lyso-PC might be the ER-derived precursor for the synthesis of eukaryotic chloroplast lipids (Mongrand *et al.*, 2000). Lyso-PC is water-soluble and thus might diffuse from the ER to the chloroplast. This hypothesis was recently corroborated by a study on *Arabidopsis lacs* mutants deficient in long-chain acyl-CoA synthase activity (Jessen *et al.*, 2015).

The ER-derived eukaryotic lipid precursors need to be transferred through the outer to the inner envelope, where many of the enzymes for chloroplast lipid synthesis are localized. The trigalactosyldiacylglycerol protein (TGD) transport complex is involved in the transfer of eukaryotic lipid precursors from the ER to the chloroplast. The *tgd* mutant plants were originally isolated by screening for suppressor mutants of the DGDG-deficient *dgd1* mutant of *Arabidopsis*. The *tgd* mutants accumulate unusual oligogalactolipids, in particular trigalactosyldiacylglycerol (TGDG) (Fig. 1.1), the reason for this remaining enigmatic (Xu *et al.*, 2003). The *tgd* mutants also contain increased amounts of PA and TAG originating from the eukaryotic pathway. The synthesis of ER-derived (eukaryotic) chloroplast lipids is disrupted in the *tgd* mutants (Awai *et al.*, 2006; Lu *et al.*, 2007; Xu *et al.*, 2005). Four TGD proteins were identified in *Arabidopsis*. TGD1, TGD2 and TGD3 form a multisubunit ABC transporter (ATP-binding-cassette transporter) spanning the chloroplast inner envelope, while TGD4 is a homodimer localized to the outer envelope (Roston *et al.*, 2011, 2012; Wang *et al.*, 2013). TGD1 is predicted to be the permease component of the transport complex (Xu *et al.*, 2005). The *TGD2* gene encodes a PA-binding protein anchored to the inner envelope with an N-terminal membrane spanning domain and facing the intermembrane space with its C-terminal domain. TGD3 is a small ATPase localized at the stromal site of the inner envelope (Lu *et al.*, 2007). TGD proteins are part of a large complex containing two copies of TGD1, 8–10 copies of TGD2 and two copies of TGD3, with a total size larger than 500 kDa (Roston *et al.*, 2012). The finding that TGD2 and TGD4 bind PA, and that PA accumulates in *tgd* mutant plants, led to the suggestion that PA is the molecule transported from the ER to the chloroplast, but this needs further confirmation.

Galactolipids

Galactolipid synthesis

Galactolipid synthesis starts with the formation of MGDG by transfer of galactose from UDP-galactose (UDP-Gal) onto DAG (Fig. 1.2) (Dubots *et al.*, 2010). Angiosperms contain two types (A and B) of nuclear-encoded MGDG synthases, while gymnosperms lack type A enzymes (Awai *et al.*, 2001; Yuzawa *et al.*, 2012). MGD1 from *Arabidopsis* (type A) is characterized by a cleavable N-terminal chloroplast targeting sequence, which is removed after insertion into the inner chloroplast envelope. MGD1 produces the bulk of MGDG in green tissues and during all developmental stages. It is required for chloroplast development and the formation of the photosynthetic apparatus (Jarvis *et al.*, 2000; Kobayashi *et al.*, 2007). MGD1 activity depends on the presence of PG and PA (Dubots *et al.*, 2010). MGDG synthesis is also cooperatively regulated by light, by cytokinin and by the redox status. Reduction of the cysteine thiol groups leads to activation of MGD1. The importance of MGD1 for plant development was demonstrated by the characterization of *Arabidopsis* mutant lines. MGD1 activity is strongly decreased in the *mgd1-1* mutant, and completely abolished in *mgd1-2* (Jarvis *et al*., 2000; Kobayashi *et al*., 2007). The decrease of MGDG content by 42% in *mgd1-1* leads to leaf chlorosis and decreased photosynthetic activity. The development of *mdg1-2* mutant seeds is strongly affected, because most seeds are arrested in the early embryo stage. Some *mgd1-2* seeds germinate and produce albino plants incapable of photoautotrophic growth. MGD2 and MGD3 from *Arabidopsis* belong to the type B MGDG synthases containing a shorter non-canonical and non-cleavable transit sequence for insertion into the outer chloroplast envelope (Awai *et al*., 2001). MGD2 and MGD3 are specialized enzymes and only expressed in inflorescences and roots, or at certain developmental stages or conditions, like phosphate deprivation. Deletion mutants of MGD2 and MGD3 do not show lipid alterations under normal conditions. However, the loss of type B enzymes (especially of MGD3) affects the lipid response during phosphate limitation, because *mgd2* or *mgd3* mutants accumulate less DGDG (Kobayashi *et al*., 2009).

DGDG is synthesized by transfer of galactose from UDP-Gal onto MGDG (Fig. 1.2). *Arabidopsis* contains two nuclear encoded DGDG synthases, DGD1 and DGD2 (Dörmann *et al*., 1999; Kelly and Dörmann, 2002; Kelly *et al*., 2003). DGD1 exhibits the main activity and is responsible for the bulk synthesis of DGDG. The N-terminal sequence of DGD1 contains the information for targeting to the chloroplast outer envelope, but is not processed or cleaved. The C-terminal part harbours the glycosyltransferase domain with the DGDG synthase activity. The mutation of DGD1 in the *dgd1* mutant leads to a 90% reduction of DGDG accompanied by stunted growth and impaired photosynthetic activity. The remaining amount of DGDG in *dgd1* is synthesized by DGD2. DGD2 differs from DGD1 by the presence of a short N-terminal cleavable chloroplast transit peptide (Kelly *et al*., 2003) targeting the protein to the outer envelope. The deletion of DGD2 in the *dgd2* mutant has no consequences for growth, lipid composition and photosynthesis under normal conditions (Kelly *et al*., 2003). Under phosphate deprivation, when DGD2 expression is increased in the wild type, *dgd2* plants accumulate less DGDG. The complete loss of DGDG in the double mutant *dgd1 dgd2* leads to a very strong reduction in growth and photosynthetic activity (Kelly *et al*., 2003).

Independent from the DGD1 and DGD2 pathways, a third DGDG-producing activity was observed, designated galactolipid:galactolipid galactosyltransferase and localized to the outer chloroplast envelope (Fig. 1.2) (Heemskerk *et al*., 1988). This enzyme belongs to the glycosylhydrolase family producing galactolipids with two (DGDG) three (TGDG) or more galactoses (oligogalactolipids) in a processive manner, independent of UDP-Gal. The reaction starts with MGDG as the primary acceptor and another MGDG molecule as galactose donor (Moellering *et al*., 2010b). The sugars are linked in β-anomeric configuration and therefore different from the configuration of the terminal galactose in DGDG introduced by DGD1 or DGD2 (Xu *et al*., 2003) (Fig. 1.1). Interestingly, the galactolipid:galactolipid galactosyltransferase is encoded by the gene *SFR2* (sensitive to freezing 2) and is activated in the cold, conferring freezing tolerance to *Arabidopsis* (Moellering *et al*., 2010b; Thorlby *et al*., 2004). SFR2 does not contribute to galactolipid biosynthesis under normal conditions. The removal of a certain proportion of MGDG from the membranes and the formation of oligogalactolipids and TAG under freezing conditions is thought to prevent membrane damage and aggregation (Moellering and Benning, 2011).

After wounding of leaves, a further galactolipid accumulates which carries a third fatty acid esterified to the C-6 carbon of galactose in MGDG (6-acyl-MGDG) (Heinz, 1967; Vu *et al*., 2014). The additional acyl group of 6-acyl-MGDG is derived from another MGDG molecule by a disproportioning reaction. However, the identity of the responsible acyltransferase and the role of 6-acyl-MGDG remain enigmatic.

Galactolipid functions in photosynthesis

The two galactolipids MGDG and DGDG provide building blocks for thylakoid assembly and establish the matrix for embedding the photosynthetic complexes. MGDG and DGDG are involved in the maintenance of membrane fluidity of the thylakoid bilayers. Photosynthetic membranes are densely packed with proteins that constitute about 70% of the membrane. The remainder of 30% is established by lipids, which form a shell around the proteins thereby separating them and allowing lateral diffusion of proteins (Kirchhoff *et al.*, 2002; Páli *et al.*, 2003). The characteristic properties of the bilayer are also determined by the physicochemical characteristics of the individual lipids. MGDG with its small head group is a non-bilayer forming lipid, which forms so-called reverse micelles or Hex_{II} structures in aqueous environments, whereas DGDG is bilayer forming (Webb and Green, 1991). A certain ratio of non-bilayer to bilayer forming lipids is crucial for the functional integrity of the membrane (Israelachvili *et al.*, 1980). In addition, DGDG is important for membrane stacking of thylakoid grana, by forming hydrogen bonds between polar head groups of adjacent bilayers (Demé *et al.*, 2014). Besides of bulk lipid bilayer functions, chloroplast lipids have specific functions with regard to the interactions with photosynthetic protein complexes (Boudiére *et al.*, 2014; Dörmann and Hölzl, 2009). For example, X-ray analysis of crystallized PSI and PSII complexes from plants and cyanobacteria revealed the occurrence of all four chloroplast lipids at specific binding sites as possible co-factors (Fyfe *et al.*, 2005). MGDG and SQDG were also found in the crystal structures of cytochrome b_6f complexes from plants and *Chlamydomonas*, and are possibly involved in maintaining the dimeric structure (Boudiére *et al.*, 2014). Chloroplast lipids (PG, DGDG) further contribute to the assembly and stabilization of LHCII, and might play a role in non-photochemical quenching of excess light, with DGDG mediating the dissipation of excitation energy between two neighbouring LHCII dimers (Boudiére *et al.*, 2014; Dörmann and Hölzl, 2009). The specific function of DGDG in photosynthesis as well as its bulk lipid bilayer function were also demonstrated *in vivo* by expression of the glucosylgalactosyldiacylglycerol (GlcGalDG) synthase from *Chloroflexus* in the DGDG-free *Arabidopsis dgd1 dgd2* mutant (Hölzl *et al.*, 2009). The accumulation of GlcGalDG complemented the severe growth retardation of *dgd1 dgd2*, but the defect in photosynthesis especially under high light conditions was only partially complemented. Therefore, a major proportion of DGDG serves as bulk bilayer lipid, but the unique DGDG head group is specifically involved in interactions with photosynthetic protein complexes, a role that cannot be conveyed by GlcGalDG. Galactolipid accumulation is strictly regulated and coordinated with the synthesis of chlorophyll and photosynthetic proteins (Kobayashi *et al.*, 2013). Defects in galactolipid metabolism lead to the down-regulation of chlorophyll synthesis, while impaired accumulation of chlorophyll leads to reduced MGD1 and DGD1 expression.

The role of galactolipids during phosphate deprivation

Natural habitats are often characterized by low availability of phosphate, which is a limiting factor for growth. Plants, algae and some bacteria are able to replace a certain proportion of their phospholipids with glycolipids to reduce their phosphate requirement and therefore to save phosphate for other essential processes. The most important surrogate lipid for phospholipids in plants is DGDG, which accumulates in the membranes, accompanied by the reduction of the amounts of phospholipids (Härtel *et al.*, 2000). Angiosperms have evolved an extra pathway for the synthesis of additional amounts of DGDG, which is activated under phosphate deprivation. This pathway involves MGD2/MGD3 and DGD2 and results in the accumulation of DGDG in extraplastidial membranes (plasma membrane, tonoplast, mitochondrial membranes) (Andersson *et al.*, 2005; Awai *et al.*, 2001; Härtel *et al.*, 2000; Jouhet *et al.*, 2004; Kelly and Dörmann, 2002). In parallel, the activity of the MGD1/DGD1 pathway is also up-regulated resulting in DGDG accumulation in thylakoids (Kelly *et al.*, 2003). MGDG does not accumulate during phosphate starvation, because the additional amount of MGDG is consumed for DGDG synthesis. The additional DGDG formed by DGD1 and DGD2 is derived from the eukaryotic pathway with 18:3/18:3 and 16:0/18:3 as characteristic species, respectively (Härtel *et al.*, 2000).

During phosphate deprivation, DGDG accumulates in extraplastidial membranes, but the mechanism of transport is unknown. It obviously discriminates between DGDG on one hand, and MGDG and SQDG on the other hand, which are not exported. However, the bacterial glycolipid GlcGalDG, produced by expression of the GlcGalDG synthase from *Chloroflexus* in *Arabidopsis*, was detected in extraplastidial membranes after phosphate deprivation, demonstrating that the components involved in DGDG export do not discriminate between DGDG and GlcGalDG (Hölzl *et al.*, 2009).

Galactolipids and adaptation to high and low temperatures

Chloroplast lipids play a role in temperature adaptation by changing their degree of fatty acid desaturation or by adjusting the flux through the prokaryotic and eukaryotic pathways of glycerolipid synthesis. Adaptation to low temperatures can be achieved by increasing the desaturation degree of acyl groups in galactolipids (Wallis and Browse, 2002). *Arabidopsis* mutants with defects in fatty acid desaturation show different growth defects at low temperatures ranging from leaf chlorosis to plant death (Wallis and Browse, 2002). An interesting adaptation mechanism to different temperatures includes the alteration of flux through the eukaryotic and prokaryotic pathways (Li *et al.*, 2015b; Pearcy, 1978). *Arabidopsis* responds to low temperature by up-regulation of the prokaryotic pathway, primarily for MGDG synthesis, while the eukaryotic pathway is up-regulated under high temperatures. A similar and stronger effect was observed in the 16:3 plant *Atriplex lentiformis*, because the prokaryotic pathway is fully suppressed under high temperatures, leading to a change in the lipid profile from a 16:3 to a 18:3 plant. *Arabidopsis* and *Atriplex* increase the proportion of the non-bilayer lipid MGDG accompanied with the increase in the amounts of 16:3, to adjust to low temperatures. The eukaryotic pathway is activated for the production of more phospholipids and DGDG (16:0/C_{18} or C_{18}/C_{18}) with lower desaturation degree, leading to bilayer stabilizing effects at high temperatures.

In addition, DGDG is involved in establishing thermotolerance of *Arabidopsis*, as the DGDG to MGDG ratio increases at elevated temperatures. Consequently, mutations in the *DGD1* gene result in thermosensitivity of *dgd1* mutant alleles (Chen *et al.*, 2006). Furthermore, low temperatures affect growth and photosynthesis of WT plants, while the *dgd1* mutant plants grow similarly as compared to normal temperatures. Therefore, the strong difference in growth between WT and *dgd1* observed at normal temperature becomes less pronounced at lower temperatures (Hendrickson *et al.*, 2006). Therefore, reduced amounts of DGDG as found in *dgd1* have less severe consequences on photosynthesis and growth in the cold as compared to normal temperatures.

Sulfolipid and glucuronosyldiacylglycerol

Biosynthesis of sulfolipid and glucuronosyldiacylglycerol

SQDG is an anionic glycolipid in the thylakoid membranes. The pathway for SQDG synthesis includes three enzymes, UDP-glucose (UDP-Glc) pyrophosphorylase (UGP3), UDP-sulfoquinovose (UDP-Sqv) synthase (SQD1) and sulfoquinovosyl transferase (SQD2) (Fig. 1.2). The first step is the synthesis of UDP-Glc from glucose-1-phosphate and UTP by UGP3 (Okazaki *et al.*, 2009). Subsequently, UDP-Glc and sulfite are converted into UDP-Sqv by SQD1 (Essigmann *et al.*, 1998). Finally, SQD2 transfers sulfoquinovose from UDP-Sqv onto DAG (Yu *et al.*, 2002). All three enzymes are nuclear encoded with N-terminal transit peptides for chloroplast import. After processing, UGP3 and SQD1 are localized as soluble proteins to the stroma, while SQD2 is inserted into the inner envelope facing the stroma side (Essigmann *et al.*, 1998). Expression of all three enzymes is up-regulated during phosphate deprivation to increase the amount of SQDG (Essigmann *et al.*, 1998; Okazaki *et al.*, 2009). SQD2 harbours a second function, because it is involved in the synthesis of the anionic glycolipid glucuronosyldiacylglycerol (GlcADG) (Fig. 1.1), which accumulates in different plants under phosphate deprivation (Okazaki *et al.*, 2015; Okazaki *et al.*, 2013b). GlcADG contains a glucuronic acid moiety bound in α-anomeric linkage to diacylglycerol (Fig. 1.1). The sugar donor for GlcADG presumably is UDP-glucuronic acid and is synthesized by an unknown UDP-Glc dehydrogenase (Fig. 1.2). The deletion of UGP3, SQD1 and SQD2 in the *Arabidopsis* mutants *ugp3*, *sqd1* and *sqd2* respectively leads to the complete loss of SQDG but has no strong consequences for growth or photosynthesis under normal conditions, because the loss of SQDG is complemented by PG (Okazaki *et al.*, 2009, 2013b; Yu *et al.*, 2002) (see below).

In *Chlamydomonas*, the pathway for SQDG synthesis is similar to plants, because *Chlamydomonas* harbours SQD1 and SQD2 orthologues with high sequence similarity (Riekhof *et al.*, 2003, 2005). SQD1 and SQD2 are also nuclear encoded, and SQD1 contains an N-terminal transit peptide for transport to the chloroplast (Riekhof *et al.*, 2003; Sato *et al.*, 2003). *Chlamydomonas* also contains a gene orthologous to UGP3 which was designated LPB1 (Chang *et al.*, 2005; Sato *et al.*, 2013).

Under sulfur-limiting condition, SQDG is degraded in *Chlamydomonas*, providing a source of sulfur for protein synthesis (Sugimoto *et al.*, 2010). However, in *Arabidopsis* SQDG is not involved in the metabolic response to sulfur deprivation. *Chlamydomonas* contains a further unusual sulfolipid with a fatty acid linked to the C6 carbon of the SQDG head group. The function of this acyl-SQDG is unknown, and the enzyme catalysing the acylation reaction and the acyl donor have not been identified (Riekhof *et al.*, 2005). There are no reports on the presence of GlcADG in *Chlamydomonas*.

Phosphatidylglycerol

Biosynthesis of phosphatidylglycerol

In plants, PG is synthesized in different organelles including chloroplasts, the ER and mitochondria (Andrews and Mudd, 1985; Griebau and Frentzen, 1994; Tanoue *et al.*, 2014). PA is the precursor for PG biosynthesis in the chloroplasts. Cytidine diphosphate diacylglycerol synthase (CDP-DAG synthase, CDS) catalyses the condensation of PA and CTP accompanied with the elimination of diphosphate, giving rise to CDP-DAG (Fig. 1.2). Plastidial CDS activity localizes to the inner envelope (Andrews and Mudd, 1985). *Arabidopsis* contains two closely related plastidial CDS isoforms, CDS4 and CDS5. The double mutant *cds4 cds5* shows a strongly decreased PG content accompanied with the deficiency in thylakoid membrane biogenesis, and loss of photoautotrophic growth (Haselier *et al.*, 2010). PG in *cds4 cds5* is devoid of $16{:}1^{\Delta 3trans}$. Instead, C_{18} fatty acids were detected at the *sn*2 position of PG. On the other hand, C_{16} acyl groups accumulated in the *sn*2 position of MGDG and DGDG, indicating that prokaryotic DAG was channelled into MGDG and DGDG synthesis in *cds4 cds5* (Zhou *et al.*, 2013).

CDP-DAG is employed for PG biosynthesis via phosphatidylglycerol phosphate (PGP) synthase (PGPS) and phosphatidylglycerol phosphate phosphatase (PGPP). In *Arabidopsis*, PGP synthase is encoded by two closely related genes, *PGP1* and *PGP2*. The two plant enzymes share high sequence similarity with bacterial PGP synthases, rather than with yeast and mammalian enzymes. Furthermore, enzymatic characteristics of the plant PGP synthases are similar with those of the bacterial enzymes. Transient expression in tobacco protoplasts showed that PGP2 is located to the ER, while PGP1 is produced as a preprotein targeted to both chloroplasts and mitochondria (Babiychuk *et al.*, 2003; Tanoue *et al.*, 2014). Analysis of different *Arabidopsis pgp1* mutant alleles showed that PGP1 is indispensable for plastidial PG biosynthesis. The *pgp1-1* mutant carrying a point mutation in the conserved CDP-alcohol phosphotransferase motif shows 80% reduction in plastidial PGP synthase activity and a 30% reduction of total PG (Xu *et al.*, 2002). In other *pgp1* mutant alleles carrying T-DNA insertions, the PG content is decreased to 12% of WT levels (Hagio *et al.*, 2002) with the remainder of PG originating from PGP2 at the ER.

The final step for PG biosynthesis is the dephosphorylation of PGP. A putative PGP phosphatase was identified in cyanobacteria. The corresponding *pgpB* mutant showed a reduction of PG content and decreased photoautotrophic growth (Wu *et al.*, 2006). PGP phosphatase activity was observed in chloroplast of plants, but the corresponding gene is still missing.

The role of PG in photosynthesis

Trans-Δ3-hexadecenoic acid ($16{:}1^{\Delta 3trans}$) is a specific fatty acid accumulating only at the *sn*2 position of PG of light-grown plants. The biosynthesis of this $16{:}1^{\Delta 3trans}$ containing PG is modulated by temperature and light as shown for winter rye (*Secale cereale* L.) (Grey *et al.*, 2005). The *Arabidopsis* mutant *fad4* (fatty acid desaturase 4) is deficient in $16{:}1^{\Delta 3trans}$ PG (Browse *et al.*, 1985). The corresponding *FAD4* gene encodes a desaturase introducing the trans Δ3 double bond into 16:0 at the *sn*2 position of PG (Gao *et al.*, 2009). $16{:}1^{\Delta 3trans}$ PG was previously considered to play a specific role in photosynthesis. However, absence of $16{:}1^{\Delta 3trans}$ in PG in the *fad4* mutant had only minor consequences for growth and photosynthesis (Browse *et al.*, 1985).

In plants, PG is the only phospholipid in thylakoids and in the inner envelope of chloroplasts while the outer envelope contains two phospholipids, PG and PC. Previously the role of PG was studied after phospholipase digestion of PG in isolated chloroplasts. Approximately 75% of PG and 60% of PC in pea chloroplasts were hydrolysed after phospholipase A2 treatment. As a result, PSII electron transport was inhibited, while PSI activity was not affected (Chen *et al.*, 2006). Similar results were obtained when isolated pea thylakoids were incubated with phospholipase C because the elimination of 40–50% of phospholipids (PC and PG) resulted in a decrease of photosynthetic electron transport, due to the inactivation of PSII reaction centres (Droppa *et al.*, 1995). Since PC is mainly distributed to the outer envelope, the effects on PSII must be derived from the reduction in PG content in the thylakoids. The content of PG in PSII dimers is 4-times of that in monomers (Kruse *et al.*, 2000). After removing most of PG by phospholipase A2 treatment, the PSII core dimers dissociate into monomers. *In vitro* assays also showed that $16{:}1^{\Delta 3trans}$ PG, but not other lipids, induced dimerization of isolated PSII monomers. Therefore, PG is essential for maintaining the stability of PSII core dimers, which are the functional form in thylakoids (Kruse *et al.*, 2000).

Analysis of PG-deficient mutants was employed to evaluate the role of PG in photosynthesis. Different *Arabidopsis pgp1* mutant alleles show disruptions of the structure of thylakoid membranes, and the development of chloroplasts is severely affected. Moreover, the photosynthetic activity of chloroplasts is defective, resulting in the loss of the ability of photoautotrophic growth (Babiychuk *et al.*, 2003; Hagio *et al.*, 2002; Xu *et al.*, 2002).

PG serves as the precursor for the biosynthesis of cardiolipin, which exclusively occurs in inner membranes of mitochondria. As PGP1 is dually targeted to both chloroplasts and mitochondria, the lack of PGP1 activity in *pgp1* could affect chloroplast and mitochondrial functions. Indeed, most of mitochondrial PGP synthase activity is lost in the *pgp1* mutant, but the amount of cardiolipin is not affected (Babiychuk *et al.*, 2003).

Analysis of *pgp1 pgp2* double mutants in which PG biosynthesis is totally abolished indicates that PG biosynthesis is essential for embryo development and for the establishment of membranes in chloroplasts and mitochondria (Tanoue *et al.*, 2014). Further *in vivo* analyses showed that the PG content is related to the degree of LCHII oligomerization. The LHCII oligomer to monomer ratio in thylakoids of rye cultivars is highest when $16{:}1^{\Delta 3trans}$ content in PG is maximal (Grey *et al.*, 2005). Mutants of *Chlamydomonas* devoid of $16{:}1^{\Delta 3trans}$ PG are almost incapable to form trimeric LHCII complexes or to develop thylakoid appressions. In different *Chlamydomonas* mutants, the amount of $16{:}1^{\Delta 3trans}$ PG is correlated with the increase in PSII activity, the increase in LHCII trimerization and stacking of photosynthetic membranes (El Maanni *et al.*, 1998).

Mutual replacement of anionic glycerolipids in chloroplasts

In contrast to normal growth conditions, SQDG is not dispensable for growth under phosphate deprivation, because the sulfolipid deficient *Arabidopsis sqd2* mutant shows reduced growth with increased levels of PG, in contrast to WT, where SQDG increases accompanied with a decrease in PG (Yu *et al.*, 2002). A more severe phenotype is observed for the double mutant *sqd2 pgp1-1* (Xu *et al.*, 2002; Yu and Benning, 2003). Leaves of *sqd2 pgp-1* plants lack SQDG and contain only 70% PG, with negative effects on growth and photosynthesis even under phosphate replete conditions. The synthetic electron transfer through PSII is affected, and thus *sqd2 pgp1-1* plants are incapable of photoautotrophic growth (Yu and Benning, 2003). Therefore, a certain level of anionic lipids (PG, SQDG) in the chloroplasts is required for optimal growth and photosynthesis (Frentzen, 2004). SQDG can be substituted by PG, as both SQDG and PG are anionic lipids, providing a negatively charged head group at the lipid–water interface, important for the assembly and function of the photosynthetic apparatus. This reciprocal relationship between PG and SQDG is also observed in *Chlamydomonas* (Riekhof *et al.*, 2003). SQDG is the main surrogate for phospholipids in *Chlamydomonas* (Riekhof *et al.*, 2003; Sato *et al.*, 2000).

During phosphate deprivation, the synthesis of SQDG is increased, to replace a certain proportion of PG in the chloroplast (Essigmann *et al.*, 1998; Frentzen, 2004; Riekhof *et al.*, 2003; Yu *et al.*, 2002). Furthermore, GlcADG is newly formed during phosphate deprivation (Okazaki *et al.*, 2015; Okazaki *et al.*, 2013b). Interestingly, growth of *sqd2* is more affected than that of *sqd1* or *ugp3* under phosphate deprivation. This difference is attributed to the additional loss of GlcADG in *sqd2*, which is still present in *sqd1* and *ugp3*. Therefore, GlcADG fulfils an important function under phosphate deprivation (Okazaki *et al.*, 2013b).

Degradation of phospholipids under phosphate deprivation

Under phosphate deprivation, phospholipids are degraded to release phosphate for other essential processes, and phospholipids are replaced by non-phosphorous lipids. Two pathways for phospholipid degradation were described for plants: (i) two-step degradation by phospholipase D (PLD) and phosphatidic acid phosphatase (PAP); (ii) one-step degradation by phospholipase C (PLC). *In vitro* assays with isolated chloroplasts and radiolabelled PC demonstrated that cytosolic PLD activity is required for PC degradation for MGDG synthesis (Andersson *et al.*, 2004). Six types of PLD enzymes were described in *Arabidopsis* based on sequences similarities, requirements of calcium and substrate specificities (Wang, 2005). PLDζ1 selectively hydrolyses PC, while PLDζ2 is involved in PC and PE degradation (Cruz-Ramírez *et al.*, 2006). PLDζ1 and PLDζ2 expression is induced, and the reduced capacity of phospholipid hydrolysis in the *pldζ2* mutant affects the capability to accumulate galactolipids under phosphate limiting conditions (Cruz-Ramírez *et al.*, 2006; Li *et al.*, 2006).

PAP catalyses PA dephosphorylation, releasing DAG for glycerolipid biosynthesis and signal transduction. Two PAP enzymes exist in plants. PAP1 is a soluble and Mg^{2+}-dependent phosphatase, while PAP2 is membrane-bound and Mg^{2+}-independent (Carman and Han, 2006). Furthermore, lipid phosphate phosphatases (LPP) and PA hydrolases (PAH) are involved in dephosphorylation of PA. LPP exhibits DAG-pyrophosphate phosphatase and PA phosphatase activities. Four membrane-bound LPPs (LPP1 to LPP4) with sequence similarities to mammalian enzymes and five homologues of cyanobacterial LPPs were identified in *Arabidopsis*. All three plastidial LPP isoforms (LPPγ, LPPε1, LPPε2) show PAP activities. Disruption of LPPγ is lethal, while *lppε1 lppε2* double mutant plants show no changes in lipid composition or growth (Nakamura *et al.*, 2007). *Arabidopsis* contains two ER-localized PAHs. PAH1 and PAH2 act redundantly to repress of phospholipid synthesis at the ER. Loss of PAH activity in the *pah1 pah2* double mutant leads to a 2-fold increase in phospholipid contents in leaves (Eastmond *et al.*, 2010). Furthermore, PAH1 and PAH2 are key enzymes of lipid remodelling during phosphate deprivation (Nakamura *et al.*, 2009).

Hydrolysis of phospholipids by PLC provides an alternative route for phospholipid degradation under phosphate limiting condition. Six putative PC-PLCs (non-specific phospholipase C, NPC1 to NPC6) were identified in *Arabidopsis*, based on amino acid sequence similarity to bacterial PC-PLCs (Nakamura *et al.*, 2005). NPC4 and NPC5 are up-regulated by phosphate starvation. NPC4 is enriched in the plasma membrane, and disruption of NPC4 in the *npc4* mutant severely compromises the increase in PC-PLC activity during phosphate limitation (Nakamura *et al.*, 2005). In contrast, NPC5 is a soluble cytosol-localized enzyme. The accumulation of DGDG under phosphate limitation is reduced in the *npc5* mutant, indicating that NPC5 is involved in the lipid response under -P conditions (Gaude *et al.*, 2008).

Analysis of chloroplast glycerolipids

Because of their high abundance (chloroplast lipids make up about 80% of lipids in a green leaf), chloroplast lipids can be directly isolated from whole leaves with chloroform/methanol, instead from isolated chloroplasts. Separation by thin-layer chromatography (TLC) provides the means to obtain an overview over the distribution and relative amounts of the chloroplast lipids. Solvent systems have been optimized for the separation of polar lipids (phospholipids, galactolipids; e.g. acetone/toluene/water, 91:30:8, v/v/v) or neutral lipids (e.g. fatty acids, TAG, DAG, carotenoids, fatty acid phytyl esters, tocopherol; e.g. hexane/diethylether/acetic acid, 85:15:1, v/v/v). With the exception of chlorophyll and carotenoids, lipids need to be stained on TLC plates for visualization. α-Naphthol specifically stains glycolipids while iodine vapour can be used to stain all lipids, with high affinity for unsaturated and aromatic compounds. Furthermore, lipids can be stained with sulfuric acid/heating. Lipids separated by TLC can be purified from the silica for subsequent analysis by gas chromatography (GC), NMR or mass spectrometry (MS).

More detailed information on the glycerolipid composition can be obtained by fatty acid analysis using GC. Both total lipid extracts and lipids purified by TLC can be employed for fatty acids analysis. Fatty acids bound to glycerol are converted into their methyl esters by acidic transmethylation (Browse *et al.*, 1986). In many cases, GC coupled to a flame ionization detector provides a good linear signal response over a large range of concentrations. Fatty acid analysis by GC can reveal valuable information on changes in chloroplast metabolism. For example, 16:3 is only found in prokaryotic MGDG, while 16:0 is mostly found in ER-derived glycerolipids. In addition, the content of prokaryotic PG can be estimated by the abundance of $16{:}1^{\Delta 3\mathrm{trans}}$. More structural information on fatty acids, e.g. concerning the existence of methyl branches, hydroxylation or positions of double bonds, for example in algal lipids, can be obtained by GC coupled to mass spectrometry (GC-MS). Here, fatty acid methyl esters are separated by GC and are ionized and fragmented by electron impact ionization and the masses separated in a quadrupole or time-of-flight analyser.

Methods for tandem mass spectrometry (MS/MS) of plant glycerolipids were developed in the last 15 years, including liquid chromatography (LC) for lipid separation, or direct infusion of the sample without prior separation (Okazaki *et al.*, 2013a; Welti *et al.*, 2002). Using MS/MS-based systems, membrane glycerolipids (e.g. MGDG, DGDG, PC, etc.) can be measured in total lipid extracts without prior purification. In contrast to fatty acid analysis by GC, intact lipid molecules rather than their acyl groups are directly measured by mass spectrometry, greatly advancing our understanding of glycerolipid structure, biosynthesis and function. This method provides detailed information on the intact molecule (molecular ion) and on its characteristic fragments after collision-induced dissociation. Using MS/MS detection for chloroplast glycerolipids, information on absolute amounts of MGDG, DGDG, SQDG, PG and PC and their molecular species composition can be obtained.

Isoprenoid lipids in chloroplasts

Tocochromanols (vitamin E) and plastoquinol

The group of tocochromanols encompasses the four forms of tocopherols and tocotrienols, depicted as α, β, γ and δ, depending on the number and position of methyl groups on the chromanol ring, and plastochromanol-8 (PC-8) which is derived from plastochinol-9 (PQ-9) (Fig. 1.3). Tocopherol biosynthesis is localized to the plastidial envelopes and to plastoglobules, while tocopherols are stored in the thylakoids and plastoglobules (Soll *et al.*, 1985; Vidi *et al.*, 2006). The biosynthesis of tocochromanols

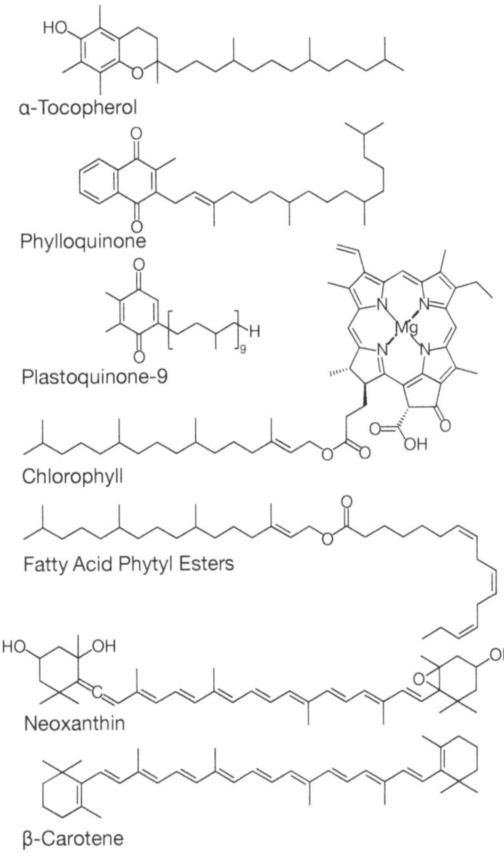

Figure 1.3 Structures of isoprenoid lipids in chloroplasts. Tocopherol and phylloquinone are prenyl quinol lipids harbouring phytyl side chains, while plastoquinol-9 contains a solanesyl chain. Chlorophyll and fatty acid phytyl esters contain phytol in ester linkage (here shown: 16:3-phytol). Neoxanthin and β-carotene are presented as examples for the xanthophylls and carotenes, respectively.

(vitamin E) and plastoquinol depends on precursors derived from the shikimate pathway (Fig. 1.4A). Tyrosine is converted into p-hydroxypyruvate by tyrosine aminotransferase (Riewe *et al.*, 2012). Alternatively, p-hydroxyphenylpyruvate can be directly produced from prephenate (de la Torre *et al.*, 2014). Homogentisate is synthesized from p-hydroxyphenylpyruvate by p-hydroxyphenylpyruvate dioxygenase (HPPD) (Norris *et al.*, 1995). In the next step, one of three different prenyl-moieties (phytyl, geranylgeranyl, solanesyl) can be transferred to homogentisate. For tocopherol and tocotrienol synthesis, homogentisate is prenylated by homogentisate phytyltransferase (HPT, VTE2) or homogentisate geranylgeranyltransferase (HGGT), using phytyl-diphosphate or geranylgeranyl-diphosphate as substrates, respectively (Cahoon *et al.*, 2003; Savidge *et al.*, 2002). Prenylation of homogentisate by HPT or HGGT produces 2-methyl-6-phytyl-benzoquinol or 2-methyl-6-geranylgeranyl-benzoquinol. The methyltransferase VTE3 transfers an additional methyl group to the chromanol ring to synthesize 2,3-dimethyl-6-phytyl-benzoquinol or 2,3-dimethyl-6-geranylgeranyl-benzoquinol. Finally, tocopherol cyclase VTE1 forms the second ring of the chromanol headgroup (Porfirova *et al.*, 2002), and a further methyl group can be attached to the chromanol ring by VTE4. For the synthesis of PQ-9, homogentisate is condensed with solanesyl-PP, a C_{45} isoprenoid, producing 2-methyl-6-solanesyl-1,4-benzoquinol (Sadre *et al.*, 2006). Methylation by VTE3 results in the synthesis of PQ-9, the essential electron carrier in PSII. Finally, cyclization of PQ-9 by VTE1 results in plastochromanol-8 (PC-8) synthesis (Mène-Saffrané and DellaPenna, 2010; Zbierzak *et al.*, 2010). PC-8 represents an alternative antioxidant with overlapping functions with the tocopherols.

Different functions were assigned to tocopherols in seeds and leaves. Tocopherols protect the photosynthetic apparatus against oxidative stress under high light conditions (Havaux *et al.*, 2005). Reactive oxygen species (ROS) can attack polyunsaturated fatty acids in membrane lipids and TAG. The presence of tocopherol is required to protect fatty acids against lipid peroxidation in seed storage lipids and to establish seed longevity (Sattler *et al.*, 2004). For deactivation of ROS, a hydrogen atom is transferred from the chromanol ring to free radicals (e.g. peroxyl radicals), thereby forming a tocopheroxyl radical. Subsequent reduction generates α-tocopheroylquinol (Kobayashi and DellaPenna, 2008). After dehydration of the side chain, α-tocopherol is regenerated by ring closure catalysed by VTE1 (Fig. 1.4A).

Tocopherols play an important role in regulating photoassimilate export from leaf cells independent from its antioxidant activity. *Arabidopsis vte2* mutant plants show strongly reduced growth, accumulation of callose at plasmodesmata, and sugar accumulation in leaf cells when grown at low temperature (Maeda *et al.*, 2006). Interestingly, this low-temperature phenotype was associated with a reduced amount of 18:3 and an increase in 18:2, in particular in PC. The low-temperature growth retardation could be suppressed by introducing fatty acid desaturase mutations, in particular *fad2* or *fad6*, into the *vte2* mutant plant (Maeda *et al.*, 2008). In addition, this phenotype of the *vte2* mutation could also be suppressed by introducing *tgd* mutations, suggesting that lipid transfer from the ER to the chloroplast is involved in this process (Song *et al.*, 2010). The link between tocopherol deficiency, desaturation, lipid transport and the low temperature phenotype of *vte2* is not fully understood.

In animals, tocopherol serves as an antioxidant and is required for reproduction. The potency of tocochromanols for human nutrition is quantitatively determined as 'vitamin E activity'. Vitamin E activity is measured using animal test systems, for example rats, chicken and guinea pigs (Leth and Søndergaard, 1983). α-Tocopherol is the form with the highest vitamin E activity. One international unit of vitamin E activity (IU) is defined as the activity of 1 mg of synthetic all-racemic α-tocopherol acetate. The activity of α-tocopherol isolated from plants (RRR-α-tocopherol) is 1.5-fold higher than that of synthetic origin.

Phylloquinone (vitamin K1)

The pathway of phylloquinone synthesis has been established in cyanobacteria and, more recently, in *Arabidopsis* (Fig. 1.4.B). First, chorismate derived from the shikimate pathway is converted into isochorismate by isochorismate synthase (ICS1/ICS2, orthologues of cyanobacterial MenF) in the chloroplast. Isochorismate is also the precursor for salicylate synthesis (Wildermuth *et al.*, 2001).

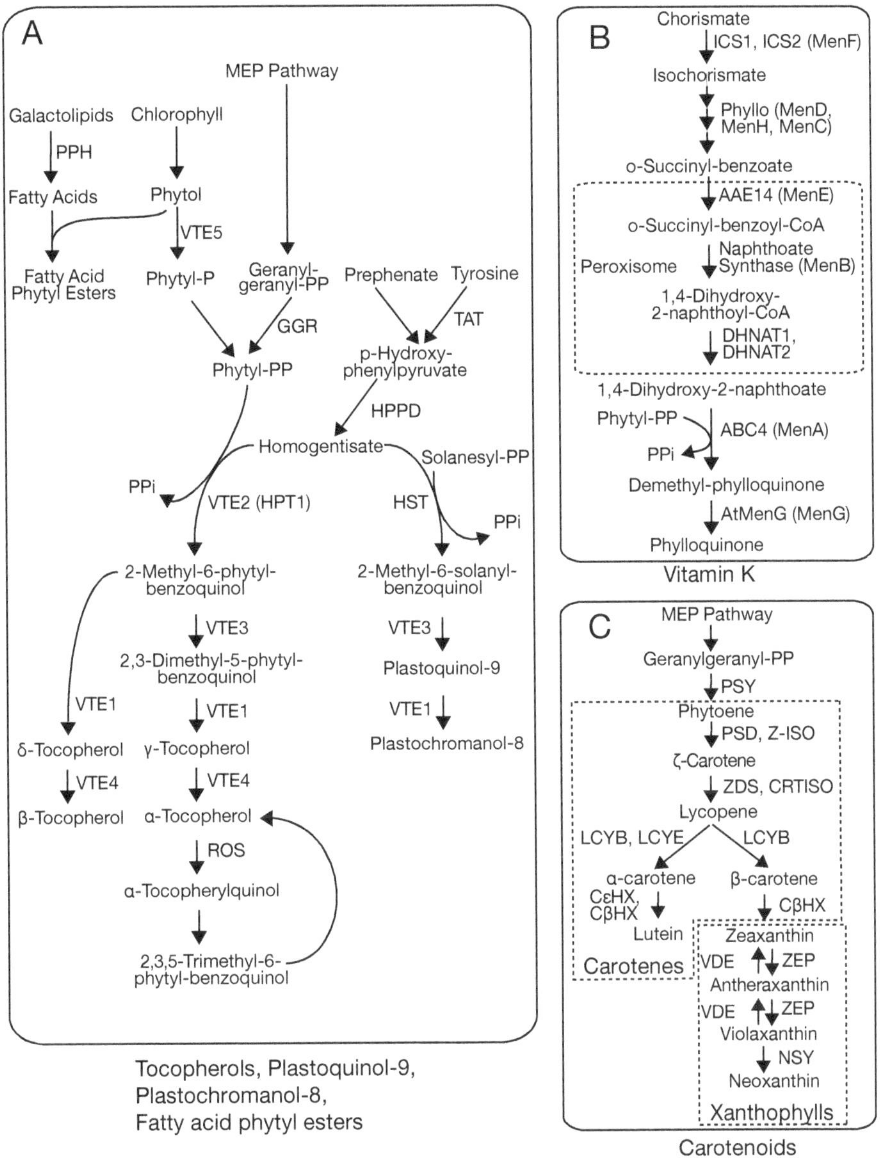

Figure 1.4 Biosynthesis of isoprenoid lipids in chloroplasts. (A) Phytyl-PP derived from the chloroplast methylerythritol-4-phosphate (MEP) pathway and homogentisate derived from the shikimate pathway are employed for tocopherol synthesis. Plastoquinol-9 is derived from solanesyl-PP and homogentisate. Free phytol from chlorophyll degradation can be converted into fatty acid phytyl esters, or after phosphorylation, used for tocopherol synthesis. (B) Phylloquinone synthesis is based on chorismate derived from the shikimate pathway. Three enzymatic steps are presumably localized to peroxisomes, while the other enzymes of phylloquinone synthesis are located to chloroplasts. (C) Carotenoids (carotenes, xanthophylls) are produced from phytoene, derived from head-to-head condensation of two molecules of geranylgeranyl-PP. PPH, pheophytin pheophorbide hydrolase; VTE5, phytol kinase; GGR, geranylgeranyl reductase; TAT, tyrosine aminotransferase; HPPD, p-hydroxyphenylpyruvate dioxygenase; VTE2, homogentisate phytyltransferase; HST, homogentisate solanesyltransferase; VTE3, VTE4, methyltransferases involved in tocopherol synthesis; VTE1, tocopherol cyclase; ROS, reactive oxygen species. ICS1, ICS2, isochorismate synthase 1, 2; Phyllo, multidomain protein harbouring MenD, MenH, MenC activities; AAE14, acyl-activating enzyme 14; DHNAT1, DHNAT2, 1,4-dihydroxy-2-naphthoyl-CoA thioesterase 1, 2; ABC4, 1,4-dihydroxy-2-naphthoate phytyltransferase; AtMenG, demethylphylloquinone methyltransferase, PSY, phytoene synthase; PSD, phytoene desaturase; Z-ISO, 15-*cis*-ζ-carotene isomerase; ZDS, ζ-carotene desaturase; CRTISO, carotenoid isomerase; LCYB, LCYE, lycopene β or ε cyclase; CβHX; CεHX, β- or ε-carotene hydroxylase; ZEP, zeaxanthin epoxidase; VDE, violaxanthin de-epoxidase; NSY, neoxanthin synthase.

Next, isochorismate is converted to o-succinyl benzoate in three steps by the Phyllo protein which harbours three activities according to the cyanobacterial MenD, MenH and MenC enzymes (Gross *et al.*, 2006). *O*-Succinyl benzoate is converted into its CoA ester by acyl activating enzyme 14 (AAE14, MenE) (Kim *et al.*, 2008). The then the naphthoate ring is formed by naphthoate synthase (NS, MenB), and the CoA ester cleaved by a thioesterase (1,4-dihydroxy-2-naphthoyl-CoA thioesterase 1 and 2 – DHNAT1, DHNAT2) (Widhalm *et al.*, 2012). The phytyl chain is attached by a phytyltransferase (ABC4, MenA) (Shimada *et al.*, 2005), and finally, demethylphylloquinone is converted into phylloquinone by methylation (AtMenG, MenG). For a long time, it was assumed that all enzymes of phylloquinone synthesis are derived from the cyanobacterial progenitor and localize to the chloroplasts in plants. However, proteomic studies and localization experiments using green and yellow fluorescent proteins (GFP, YFP) fusions revealed that AAE14 (MenE), naphthoate synthase (NS, MenB) and the two thioesterases (DHNAT1, DHNAT2) show peroxisomal localization (Babujee *et al.*, 2010; Hooks *et al.*, 2012; Widhalm *et al.*, 2012). Furthermore, DHNAT1 and DHNAT2 are of lactobacillales origin (Widhalm *et al.*, 2012). This raises the question of the mechanism for interorganellar transport of phylloquinone precursors.

In plants, phylloquinone serves as the electron carrier in PSI. Therefore, lack of phylloquinone synthesis has severe consequences for photosynthesis and plant growth, as shown for the *ics1 ics2*, *phyllo*, *aae14* and *abc4* mutants (Garcion *et al.*, 2008; Gross *et al.*, 2006; Kim *et al.*, 2008; Shimada *et al.*, 2005). Phylloquinone (vitamin K1) is essential for the human diet, because, it is required as a co-factor for the γ-carboxylation of glutamate residues in certain proteins involved in blood coagulation and bone formation.

Carotenoids

Carotenoids are light-absorbing pigments in photosynthetic organisms, including plants, algae and cyanobacteria, but also in some non-photosynthetic bacteria and fungi. Their colours range from yellow to red and are derived from the presence of conjugated double bonds. Carotenoids are responsible for the coloration of many petals, fruits and senescent leaves and can be organized into carotenes (hydrocarbons) and xanthophylls (containing oxygen). Carotenes are hardly soluble in water, because they are rather lipophilic. The oxidized xanthophylls are less hydrophobic.

Carotenoids are derived from geranylgeranyl-diphosphate synthesized by the plastidial methylerythritol-4-phosphate (MEP) pathway (Fig. 1.4C). Two molecules of geranylgeranyl-diphosphate are condensed by phytoene synthase (PSY) to produce phytoene (C_{40}), the common precursor for all carotenoids. Lycopene is synthesized from phytoene via several desaturation and isomerization steps (phytoene desaturase, PDS; 15-*cis*-ζ-carotene isomerase, Z-ISO; ζ-carotene desaturase, ZDS; carotenoid isomerase, CRTISO). Subsequently, β-carotene or α-carotene are produced by lycopene ε cyclase (LCYE) and lycopene β cyclase (LCYB). α-Carotene contains one β-ionone and one ε-ionone ring, while β-carotene contains two β-ionone rings. Lutein is synthesized from α-carotene by hydroxylation of the β-ionone and ε-ionone rings (β-carotene hydroxylase, CβHx; ε-carotene hydroxylase, CεHx). β-carotene is employed for the synthesis of xanthophylls. Firstly, the β-ionone rings are hydroxylated by β-carotene hydroxylase (CβHx), and then, zeaxanthin epoxidase (ABA1, ZEP, NPQ2) introduces an epoxy group into the β-ionone rings. The products are antheraxanthin and violaxanthin. Finally, neoxanthin is produced via isomerization of violaxanthin by neoxanthin synthase (NSY).

Carotenoids are present in the thylakoid membrane and in plastoglobules of chloroplasts and they are also found in chromoplasts of fruits and petals. Plants contain a large variety of carotenoids. The most abundant carotenoids are the carotenes lutein and β-carotene, and the xanthophylls neoxanthin and violaxanthin. Carotenoids, which absorb light at wavelengths between 400 and 500 nm, serve as accessory light-harvesting pigments in addition to chlorophyll, or are involved in dissipation of excess light energy. Zeaxanthin, antheraxanthin and violaxanthin play a specific role during non-photochemical quenching (NPQ), because they participate in the xanthophyll cycle catalysed by zeaxanthin epoxidase (ZEP, NPQ2) and violaxanthin de-epoxidase (VDE, NPQ1) (Niyogi *et al.*, 1998). In humans, β-carotene serves as provitamin A and is therefore an essential dietary component. β-Carotene is cleaved into two molecules of retinol which are crucial for the visual effect in the eye and to prevent, amongst others, xerophthalmia and infections.

Xanthophylls carrying hydroxyl groups can be esterified to fatty acids. In senescent leaves, xanthophyll esters are enriched in plastoglobules (Tevini and Steinmüller, 1985). In tomato, two enzymes were identified (PYP1, PYP2) that catalyse the acylation of xanthophylls in petals (Li *et al.*, 2015a). PYP1 and PYP2 are homologous to the phytyl ester synthases PES1 and PES2 from *Arabidopsis* (Lippold *et al.*, 2012). Xanthophyll esters were also identified in the unicellular microalgae *Haematococcus pluvialis* (Chen *et al.*, 2015).

Changes in phytyl lipid composition during stress

Chlorophyll is subject to continuous turnover, and chlorophyll degradation increases during leaf senescence or chlorotic stress (Csupor, 1971). During fruit ripening (e.g. tomato), chlorophyll is also degraded (Hörtensteiner and Kräutler, 2011). The initial step for chlorophyll degradation is the conversion of chlorophyll *b* to chlorophyll *a* by chlorophyll reductase followed by hydrolysis of the phytol ester bond and by removal of the magnesium cation (Hörtensteiner and Kräutler, 2011). In chlorotic leaves, the thylakoid membranes are disintegrated and large plastoglobules and starch granules appear (Gaude *et al.*, 2007). Plastoglobules contain TAG and different isoprenoid lipids, including tocopherol, fatty acid phytyl esters, plastoquinone and plastochromanol (Tevini and Steinmüller, 1985; Vidi *et al.*, 2006). The breakdown products of galactolipids and chlorophyll, i.e. fatty acids and phytol, are further metabolized to avoid their accumulation as they are detrimental to membranes (Löbbecke and Cevc, 1995). Therefore, phytol accumulation is prevented by channelling phytol into the synthesis of tocopherol and fatty acid phytyl esters (Ischebeck *et al.*, 2006).

Phytol is directed towards tocopherol biosynthesis by conversion into phytyl-P and phytyl-PP. VTE5 encodes a phytol kinase involved in the phosphorylation of phytol to phytyl-P (Valentin *et al.*, 2006). The reaction product of this pathway, phytyl-PP, is used for tocopherol synthesis, similarly as phytyl-PP produced by reduction of geranylgeranyl-PP derived from the MEP pathway (Fig. 1.4A).

Another route for phytol metabolism is the esterification to fatty acids, yielding fatty acid phytyl esters. Fatty acid phytyl esters were first described in senescent leaves of *Acer platanoides* (Csupor, 1971). In *Arabidopsis*, fatty acid phytyl esters have a distinct fatty acid composition, dominated by 16:3 and medium-chain fatty acids (10:0, 12:0 and 14:0) (Gaude *et al.*, 2007; Lippold *et al.*, 2012). The fatty acid phytyl ester synthases from *Arabidopsis*, PES1 and PES2, were recently identified (Lippold *et al.*, 2012) (Fig. 1.4A).

In animals, the enzymes and intermediates involved in phytol degradation have been characterized (Wanders *et al.*, 2011). Animals take up chlorophyll via their diet. After hydrolysis, phytol is converted to phytenal and phytenic acid, which is activated to phytenoyl-CoA. Phytenoyl-CoA is transported to the peroxisomes and reduced to phytanoyl-CoA by trans-2-enoyl-CoA reductase. During peroxisomal α-oxidation, pristanic acid is produced and broken down by β-oxidation. Plants contain genes homologous to the animal genes involved in α- and β-oxidation. Therefore, it is likely that a similar pathway for phytol degradation operates in plants. In line with this scenario, the *Arabidopsis etfqo* (electron-transfer flavoprotein:ubiquinone oxidoreductase) mutant accumulates phytanoyl-CoA (Ishizaki *et al.*, 2005).

Analysis of chloroplast isoprenoid lipids

For prenylquinones and photosynthetic pigments, sample preparation and storage is critical, because the lipids have to be kept cold and in the dark as they are sensitive to photodegradation. Prenylquinones are measured by liquid chromatography (LC) and subsequent analysis by fluorescence detection (tocopherol, tocotrienol, plastochromanol, phylloquinone) or UV/visible light detection (plastoquinone) (Balz *et al.*, 1992; Lohmann *et al.*, 2006; Zbierzak *et al.*, 2010). A comprehensive analysis of photosynthetic pigments (chlorophyll, carotenoids) can be conducted using HPLC with UV/visible light detection (Thayer and Björkman, 1990).

Conclusion

Photosynthetic membranes in chloroplasts contain four major glycerolipids and a complex set of isoprenoid lipids. The biosynthetic pathways of most lipid components in chloroplasts were elucidated in the recent years. Comprehensive biochemical studies were instrumental to understand the role of glycerolipids and isoprenoids as building blocks for thylakoid membrane assembly or as integral components of the photosynthetic protein complexes. The characterization of *Arabidopsis* mutants has contributed to the elucidation of the biosynthetic routes and of the involvement of lipid transport systems. Chloroplast lipids are major contributors to maintain the integrity of the photosynthetic membrane during stress, in particular during phosphate deprivation, low and high temperature, and during senescence.

Future trends

Research on plant lipids has seen major progress in the past years, with *Arabidopsis* as the most important model system. Algae are a highly diverse group of photosynthetic organisms, and this diversity is also reflected in the presence of lipid molecules and metabolic pathways. Therefore, algae become the focus of interest. Based on the endosymbiont theory, algae are characterized by primary or complex plastids with up to four envelope membranes, and thus with different pathways for chloroplast membrane lipid synthesis (Petroutsos *et al.*, 2014). The simple definition of prokaryotic and eukaryotic pathways for lipid synthesis cannot be applied to many algae. Furthermore, a major goal of current lipid research is the increase of oil content in crop plants by breeding or genetic engineering for food or biodiesel production. In addition to increasing the oil content in seeds, attempts are undertaken to increase TAG production in leaves of forage crops. Again, algae become the focus of interest for the biotechnological production of oil. Many algae, such as *Chlamydomonas* or *Nannochloropsis*, accumulate oil, but only under abiotic stress like nitrogen deprivation, which affects growth and photosynthesis. Therefore, current research focuses on the mechanisms of uncoupling oil accumulation and nutrient deficiency. Moreover, algae offer a rich source of unusual or very long polyunsaturated ω3 fatty acids (Lang *et al.*, 2011). ω3 Fatty acids play an important role in the human nutrition as 'essential fatty acids'. Fish oil represents a rich source of these fatty acids, however, resources are threatened by overfishing. Therefore, investigations on the biosynthetic pathways in algae and the transfer of these pathways into crop plants for production of unusual ('fish oil') fatty acids are of high interest (Abbadi *et al.*, 2004).

Additional progress on analytical instrumentation, in particular the development of highly sensitive mass spectrometers, provides the means for the detection of low abundant lipids in chloroplasts in minute sample amounts. In addition, sensitive MS/MS methods are required for lipid analysis in small volumes of isolated membrane fractions (e.g. plasma membrane, mitochondria or peroxisomes), as was already demonstrated in yeast (Schneiter *et al.*, 1999). A similar approach can be used to analyse the distribution of lipid molecular species to different chloroplast membranes (thylakoids, plastoglobules and envelopes). Organelle mass spectrometry provides the possibility to select a single organelle, e.g. a lipid droplet, by means of a nanomanipulator, and to subject this isolated organelle to mass spectrometry (Horn *et al.*, 2011). This method could be applied for the analysis of a single chloroplast. Furthermore, techniques are developed to track metabolic fluxes in chloroplasts, via feeding with ^{13}C-substrates and subsequent analysis by mass spectrometry. Recent developments in mass spectrometry, e.g. via multiple rounds of mass spectrometry/fragmentation (MS^n experiments) facilitate the structural elucidation without the need for NMR analysis. These techniques are extremely useful for broadening our current view on chloroplast lipid metabolism, as was recently demonstrated by the isolation of GlcADG as a novel chloroplast lipid from *Arabidopsis* (Okazaki *et al.*, 2013b).

Acknowledgements

Funding was in part provided by the Deutsche Forschungsgemeinschaft (SFB645) and by University of Bonn.

References

Abbadi, A., Domergue, F., Bauer, J., Napier, J.A., Welti, R., Zähringer, U., Cirpus, P., and Heinz, E. (2004). Biosynthesis of very-long-chain polyunsaturated Fatty acids in transgenic oilseeds: constraints on their accumulation. Plant Cell *16*, 2734–2748.

Andersson, M.X., Goksor, M., and Sandelius, A.S. (2007). Optical manipulation reveals strong attracting forces at membrane contact sites between endoplasmic reticulum and chloroplasts. J. Biol. Chem. *282*, 1170–1174.

Andersson, M.X., Kjellberg, J.M., and Sandelius, A.S. (2004). The involvement of cytosolic lipases in converting phosphatidyl choline to substrate for galactolipid synthesis in the chloroplast envelope. Biochim. Biophys. Acta *1684*, 46–53.

Andersson, M.X., Larsson, K.E., Tjellström, H., Liljenberg, C., and Sandelius, A.S. (2005). Phosphate-limited oat. The plasma membrane and the tonoplast as major targets for phospholipid-to-glycolipid replacement and stimulation of phospholipases in the plasma membrane. J. Biol. Chem. *280*, 27578–27586.

Andrews, J., and Mudd, J.B. (1985). Phosphatidylglycerol synthesis in pea chloroplasts. Pathways and localization. Plant Physiol. *79*, 259–265.

Ariizumi, T., Kishimoto, S., Kakami, R., Maoka, T., Hirakawa, H., Suzuki, Y., Ozeki, Y., Shirasawa, K., Bernillon, S., Okabe, Y., *et al.* (2014). Identification of the carotenoid modifying gene *PALE YELLOW PETAL 1* as an essential factor in xanthophyll esterification and yellow flower pigmentation in tomato (*Solanum lycopersicum*). Plant J. *79*, 453–465.

Awai, K., Maréchal, E., Block, M.A., Brun, D., Masuda, T., Shimada, H., Takamiya, K., Ohta, H., and Joyard, J. (2001). Two types of MGDG synthase genes, found widely in both 16:3 and 18:3 plants, differentially mediate galactolipid syntheses in photosynthetic and nonphotosynthetic tissues in *Arabidopsis thaliana*. Proc. Natl. Acad. Sci. U.S.A. *98*, 10960–10965.

Awai, K., Xu, C., Tamot, B., and Benning, C. (2006). A phosphatidic acid-binding protein of the chloroplast inner envelope membrane involved in lipid trafficking. Proc. Natl. Acad. Sci. U.S.A. *103*, 10817–10822.

Babiychuk, E., Müller, F., Eubel, H., Braun, H.-P., Frentzen, M., and Kushnir, S. (2003). *Arabidopsis* phosphatidylglycerophosphate synthase 1 is essential for chloroplast differentiation, but is dispensable for mitochondrial function. Plant J. *33*, 899–909.

Babujee, L., Wurtz, V., Ma, C., Lueder, F., Soni, P., van Dorsselaer, A., and Reumann, S. (2010). The proteome map of spinach leaf peroxisomes indicates partial compartmentalization of phylloquinone (vitamin K1) biosynthesis in plant peroxisomes. J. Exp. Bot. *61*, 1441–1453.

Balz, M., Schulte, E., and Thier, H.-P. (1992). Trennung von Tocopherolen und Tocotrienolen duch die HPLC. Fat Sci. Technol. *94*, 209–213.

Block, M.A., Dorne, A.J., Joyard, J., and Douce, R. (1983). Preparation and characterization of membane fractions enriched in outer and inner envelope membranes from spinach chloroplasts. II. Biochemical characterization. J. Biol. Chem. *258*, 13281–13286.

Bonaventure, G., Bao, X., Ohlrogge, J., and Pollard, M. (2004). Metabolic responses to the reduction in palmitate caused by disruption of the *FATB* gene in *Arabidopsis*. Plant Physiol. *135*, 1269–1279.

Boudiére, L., Michaud, M., Petroutsos, D., Rébeillé, F., Falconet, D., Bastien, O., Roy, S., Finazzi, G., Rolland, N., Jouhet, J., *et al.* (2014). Glycerolipids in photosynthesis: composition, synthesis and trafficking. Biochim. Biophys. Acta *1837*, 470–480.

Browse, J., McCourt, P., and Somerville, C.R. (1985). A mutant of *Arbidopsis* lacking a chloroplast-specific lipid. Science *227*, 763–765.

Browse, J., McCourt, P.J., and Somerville, C.R. (1986). Fatty acid composition of leaf lipids determined after combined digestion and fatty acid methyl ester formation from fresh tissue. Anal. Biochem. *152*, 141–145.

Cahoon, E.B., Hall, S.E., Ripp, K.G., Ganzke, T.S., Hitz, W.D., and Coughlan, S.J. (2003). Metabolic redesign of vitamin E biosynthesis in plants for tocotrienol production and increased antioxidant content. Nat. Biotechnol. *21*, 1082–1087.

Carman, G.M., and Han, G.S. (2006). Roles of phosphatidate phosphatase enzymes in lipid metabolism. Trends Biochem. Sci. *31*, 694–699.

Chang, C.W., Moseley, J.L., Wykoff, D., and Grossman, D. (2005). The LPB1 gene is important for acclimation of *Chlamydomonas reinhardtii* to phosphorus and sulfur deprivation. Plant Physiol. *138*, 319–329.

Chen, G., Wang, B., Han, D., Sommerfeld, M., Lu, Y., Chen, F., and Hu, Q. (2015). Molecular mechanisms of the coordination between astaxanthin and fatty acid biosynthesis in *Haematococcus pluvialis* (Chlorophyceae). Plant J. *81*, 95–107.

Chen, J., Burke, J.J., Xin, Z., Xu, C., and Velten, J. (2006). Characterization of the *Arabidopsis* thermosensitive mutant *atts02* reveals an important role for galactolipids in thermotolerance. Plant Cell Environm. *29*, 1437–1448.

Cruz-Ramírez, A., Oropeza-Aburto, A., Razo-Hernández, F., Ramírez-Chávez, E., and Herrera-Estrella, L. (2006). Phospholipase DZ2 plays an important role in extraplastidic galactolipid biosynthesis and phosphate recycling in *Arabidopsis* roots. Proc. Natl. Acad. Sci. U.S.A. *103*, 6765–6770.

Csupor, L. (1971). Das Phytol in vergilbten Blättern. Planta Med. *19*, 37–41.

de la Torre, F., El-Azaz, J., Ávila, C., and Cánovas, F.M. (2014). Deciphering the role of aspartate and prephenate aminotransferase activities in plastid nitrogen metabolism. Plant Physiol. *164*, 92–104.

Demé, B., Cataye, C., Block, M.A., Maréchal, E., and Jouhet, J. (2014). Contribution of galactoglycerolipids to the 3-dimensional architecture of thylakoids. FASEB J. *28*, 3373–3383.

Dörmann, P., Balbo, I., and Benning, C. (1999). *Arabidopsis* galactolipid biosynthesis and lipid trafficking mediated by DGD1. Science *284*, 2181–2184.

Dörmann, P., and Hölzl, G. (2009). The role of glycolipids in photosynthesis. In Lipids in Photosynthesis, H. Wada, and N. Murata, eds. (Dordrecht: Springer), pp. 265–282.

Droppa, M., Horváth, G., Hideg, É., and Farkas, T. (1995). The role of phospholipids in regulating photosynthetic electron transport activities: Treatment of thylakoids with phospholipase C. Photosynth. Res. *46*, 287–293.

Dubots, E., Audry, M., Yamaryo, Y., Bastien, O., Ohta, H., Breton, C., Maréchal, E., and Block, M.A. (2010). Activation of the chloroplast monogalactosyldiacylglycerol synthase MGD1 by phosphatidic acid and phosphatidylglycerol. J. Biol. Chem. *285*, 6003–6011.

Eastmond, P., Quettier, A.-L., Kroon, J., Craddock, C., Adams, N., and Slabas, A. (2010). Phosphatidic Acid Phosphohydrolase 1 and 2 regulate phospholipid synthesis at the endoplasmic reticulum in *Arabidopsis*. Plant Cell *22*, 2796–2811.

El Maanni, A., Dubertret, G., Delrieu, M.-J., Roche, O., and Trémolières, A. (1998). Mutants of *Chlamydomonas reinhardtii* affected in phosphatidylglycerol metabolism and thylakoid biogenesis. Plant Physiol. Biochem. *36*, 609–619.

Essigmann, B., Güler, S., Narang, R.A., Linke, D., and Benning, C. (1998). Phosphate availability affects the thylakoid lipid composition and the expression of *SQD1*, a gene required for sulfolipid biosynthesis in *Arabidopsis thaliana*. Proc. Natl. Acad. Sci. U.S.A. *95*, 1950–1955.

Frentzen, M. (2004). Phosphatidylglycerol and sulfoquinovosyldiacylglycerol: anionic membrane lipids and phosphate regulation. Curr. Opin. Plant Biol. *7*, 270–276.

Frentzen, M., Heinz, E., McKeon, T.A., and Stumpf, P.K. (1983). Specificities and selectivities of glycerol-3-phosphate acyltransferase and monoacylglycerol-3-phosphate acyltransferase from pea and spinach chloroplasts. Eur. J. Biochem. *129*, 629–636.

Fyfe, P.K., Hughes, A.V., Heathcote, P., and Jones, M.R. (2005). Proteins, chlorophylls and lipids: X-ray analysis of a three-way relationship. Trends Plant Sci. *10*, 275–282.

Gao, J., Ajjawi, I., Manoli, A., Sawin, A., Xu, C., Froehlich, J.E., Last, R.L., and Benning, C. (2009). fatty acid desaturase 4 of *Arabidopsis* encodes a protein distinct from characterized fatty acid desaturases. Plant J. *60*, 832–839.

Garcion, C., Lohmann, A., Lamodière, E., Catinot, J., Buchala, A., Doermann, P., and Métraux, J.-P. (2008). Characterization and giological function of the *Isochorismate Synthase 2* gene of *Arabidopsis*. Plant Physiol. *147*, 1279–1287.

Gaude, N., Bréhélin, C., Tischendorf, G., Kessler, F., and Dörmann, P. (2007). Nitrogen deficiency in *Arabidopsis* affects galactolipid composition and gene expression and results in accumulation of fatty acid phytyl esters. Plant J. *49*, 729–739.

Gaude, N., Nakamura, Y., Scheible, W.-R., Ohta, H., and Dörmann, P. (2008). Phospholipase C5 (NPC5) is involved in galactolipid accumulation during phosphate limitation in leaves of *Arabidopsis*. Plant J. *56*, 28–39.

Giroud, C., Gerber, A., and Eichenberger, W. (1988). Lipids of *Chlamydomonas reinhardtii*. Analysis of molecular species and intracellular site(s) of biosynthesis. Plant Cell Physiol. *29*, 587–595.

Gray, G.R., Ivanov, A.G., Król, M., Williams, J.P., Kahn, M.U., Myscich, E.G., and Huner, N.P.A. (2005). Temperature and light modulate the *trans*-Δ^3-Hexadecenoic acid content of phosphatidylglycerol: light-harvesting complex II organization and non-photochemical quenching. Plant Cell Physiol. *46*, 1272–1282.

Griebau, R., and Frentzen, M. (1994). Biosynthesis of phosphatidylglycerol in isolated mitochondria of etiolated mung bean (*Vigna radiata* L.) seedlings. Plant Physiol. *105*, 1269–1274.

Gross, J., Cho, W.K., Lezhneva, L., Falk, J., Krupinska, K., Shinozaki, K., Seki, M., Herrmann, R.G., and Meurer, J. (2006). A plant locus essential for phylloquinone (vitamin K1) biosynthesis originated from a fusion of four eubacterial genes. J. Biol. Chem. *281*, 17189–17196.

Grusak, M.A., and DellaPenna, D. (1999). Improving the nutrient composition of plants to enhance human nutrition and health. Annu. Rev. Plant Physiol. Plant Mol. Biol. *50*, 133–161.

Hagio, M., Sakurai, I., Sato, S., Kato, T., Tabata, S., and Wada, H. (2002). Phosphatidylglycerol is essential for the development of thylakoid membranes in *Arabidopsis thaliana*. Plant Cell Physiol. *43*, 1456–1464.

Härtel, H., Dörmann, P., and Benning, C. (2000). DGD1-independent biosynthesis of extraplastidic galactolipids after phosphate deprivation in *Arabidopsis*. Proc. Natl. Acad. Sci. U.S.A. *97*, 10649–10654.

Haselier, A., Akbari, H., Weth, A., Baumgartner, W., and Frentzen, M. (2010). Two closely related genes of *Arabidopsis* encode plastidial cytidinediphosphate diacylglycerol synthases essential for photoautotrophic growth. Plant Physiol. *153*, 1372–1384.

Havaux, M., Eymery, F., Porfirova, S., Rey, P., and Dörmann, P. (2005). Vitamin E protects against photoinhibition and photooxidative stress in *Arabidopsis thaliana*. Plant Cell *17*, 3451–3469.

Heemskerk, J.W.M., Wintermans, J.F., G.M., Joyard, J., Block, M.A., Dorne, A.J., and Douce, R. (1988). Localization of galactolipid:galactolipid galactosyltransferase and acyltransferase in outer envelope membrane of spinach chloroplasts. Biochim. Biophys. Acta *877*, 281–289.

Heinz, E. (1967). Acylgalaktosyldiglycerid aus Blatthomogenaten. Biochim. Biophys. Acta *144*, 321–332.

Hendrickson, L., Vlčková, A., Selstam, E., Huner, N., Öquist, G., and Hurry, V. (2006). Cold acclimation of the *Arabidopsis dgd1* mutant results in recovery from photosystem I-limited photosynthesis. FEBS Lett. *580*, 4959–4968.

Hölzl, G., Witt, S., Gaude, N., Melzer, M., Schöttler, M.A., and Dörmann, P. (2009). The role of diglycosyl lipids in photosynthesis and membrane lipid homeostasis in *Arabidopsis*. Plant Physiol. *150*, 1147–1159.

Hooks, K.B., Turner, J.E., Graham, I.A., Runions, J., and Hooks, M.A. (2012). GFP-tagging of *Arabidopsis* acyl-activating enzymes raises the issue of peroxisome-chloroplast import competition versus dual localization. J. Plant Physiol. *169*, 1631–1638.

Horn, P.J., Ledbetter, N.R., James, C.N., Hoffman, W.D., Case, C.R., Verbeck, G.F., and Chapman, K.D. (2011). Visualization of lipid droplet composition by direct organelle mass spectrometry. J. Biol. Chem. *286*, 3298–3306.

Hörtensteiner, S., and Kräutler, B. (2011). Chlorophyll breakdown in higher plants. Biochim. Biophys. Acta *1807*, 977–988.

Hurlock, A.K., Roston, R.L., Wang, K., and Benning, C. (2014). Lipid trafficking in plant cells. Traffic *15*, 915–932.

Ischebeck, T., Zbierzak, A.M., Kanwischer, M., and Dormann, P. (2006). A salvage pathway for phytol metabolism in *Arabidopsis*. J. Biol. Chem. *281*, 2470–2477.

Ishizaki, K., Larson, T.R., Schauer, N., Fernie, A.R., Graham, I.A., and Leaver, C.J. (2005). The critical role of *Arabidopsis* electron-transfer flavoprotein:ubiquinone oxidoreductase during dark-induced starvation. Plant Cell *17*, 2587–2600.

Israelachvili, J.N., Marcelja, S., and Horn, R.G. (1980). Physical principles of membrane organization. Quart. Rev. Biophys. *13*, 121–200.

Jarvis, P., Dörmann, P., Peto, C.A., Lutes, J., Benning, C., and Chory, J. (2000). Galactolipid deficiency and abnormal chloroplast development in the *Arabidopsis* MGD synthase 1 mutant. Proc. Natl. Acad. Sci. U.S.A. *97*, 8175–8179.

Jessen, D., Roth, C., Wiermer, M., and Fulda, M. (2015). Two activities of long-chain acyl-coenzyme A synthetase are involved in lipid trafficking between the endoplasmic reticulum and the plastid in *Arabidopsis*. Plant Physiol. *167*, 351–366.

Jouhet, J., Maréchal, E., Baldan, B., Bligny, R., Joyard, J., and Block, M.A. (2004). Phosphate deprivation induces transfer of DGDG galactolipid from chloroplast to mitochondria. J. Cell Biol. *167*, 863–874.

Joyard, J., and Douce, R. (1977). Site of synthesis of phosphatidic acid and diacylglycerol in spinach chloroplasts. Biochim. Biophys. Acta *486*, 273–285.

Kelly, A.A., and Dörmann, P. (2002). DGD2, an *Arabidopsis* gene encoding a UDP-galactose-dependent digalactosyldiacylglycerol synthase is expressed during growth under phosphate-limiting conditions. J. Biol. Chem. *277*, 1166–1173.

Kelly, A.A., Froehlich, J.E., and Dörmann, P. (2003). Disruption of the two digalactosyldiacylglycerol synthase genes DGD1 and DGD2 in *Arabidopsis* reveals the existence of an additional enzyme of galactolipid synthesis. Plant Cell *15*, 2694–2706.

Kim, H.U., Oostende, C.v., Basset, G.J.C., and Browse, J. (2008). The *AAE14* gene encodes the *Arabidopsis* o-succinylbenzoyl-CoA ligase that is essential for phylloquinone synthesis and photosystem-I function. Plant J. *54*, 272–283.

Kim, S., Yamaoka, Y., Ono, H., Kim, H., Shim, D., Maeshima, M., Martinoia, E., Cahoon, E.B., Nishida, I., and Lee, Y. (2013). AtABCA9 transporter supplies fatty acids for lipid synthesis to the endoplasmic reticulum. Proc. Natl. Acad. Sci. U.S.A. *110*, 773–778.

Kirchhoff, H., Mukherjee, U., and Galla, H.J. (2002). Molecular architecture of the thylakoid membrane: lipid diffusion space for plastoquinone. Biochemistry *41*, 4872–4882.

Kobayashi, K., Awai, K., Nakamura, M., Nagatani, A., Masuda, T., and Ohta, H. (2009). Type-B monogalactosyldiacylglycerol synthases are involved in phosphate starvation-induced lipid remodeling, and are crucial for low-phosphate adaptation. Plant J. *57*, 322–331.

Kobayashi, K., Kondo, M., Fukuda, H., Nishimura, M., and Ohta, H. (2007). Galactolipid synthesis in chloroplast inner envelope is essential for proper thylakoid biogenesis, photosynthesis, and embryogenesis. Proc. Natl. Acad. Sci. U.S.A. *104*, 17216–17221.

Kobayashi, K., Narise, T., Sonoike, K., Hashimoto, H., Sato, N., Kondo, M., Nishimura, M., Sato, M., Toyooka, K., Sugimoto, K., *et al.* (2013). Role of galactolipid biosynthesis in coordinated development of photosynthetic complexes and thylakoid membranes during chloroplast biogenesis in *Arabidopsis*. Plant J. 73, 250–261.

Kobayashi, N., and DellaPenna, D. (2008). Tocopherol metabolism, oxidation and recycling under high light stress in *Arabidopsis*. Plant J. *55*, 607–618.

Kroll, D., Meierhoff, K., Bechtold, N., Kinoshita, M., Westphal, S., Vothknecht, U.C., Soll, J., and Westhoff, P. (2001). VIPP1, a nuclear gene of *Arabidopsis thaliana* essential for thylakoid membrane formation. Proc. Natl. Acad. Sci. U.S.A. *98*, 4238–4242.

Kruse, O., Hankamer, B., Konczak, C., Gerle, C., Morris, E., Radunz, A., Schmid, G.H., and Barber, J. (2000). Phosphatidylglycerol is involved in the dimerization of photosystem II. J. Biol. Chem. *275*, 6509–6514.

Kunst, L., Browse, J., and Somerville, C. (1988). Altered regulation of lipid biosynthesis in a mutant of *Arabidopsis* deficient in chloroplast glycerol-3-phosphate acyltransferase activity. Proc. Natl. Acad. Sci. U.S.A. *85*, 4143–4147.

Lang, I., Hodac, L., Friedl, T., and Feussner, I. (2011). Fatty acid profiles and their distribution patterns in microalgae: a comprehensive analysis of more than 2000 strains from the SAG culture collection. BMC Plant Biol. *11*, 124.

Leth, T., and Søndergaard, H. (1983). Biological activity of all-rac-a-tocopherol and RRR-α-tocopherol determined by three different rat bioassays. International Journal for Vitamin and Nutrition Research 53, 297–311.

Li-Beisson, Y., Beisson, F., and Riekhof, W. (2015). Metabolism of acyl-lipids in *Chlamydomonas reinhardtii*. Plant, J., 1–19.

Li-Beisson, Y., Shorrosh, B., Beisson, F., Andersson, M.X., Arondel, V., Bates, P.D., Baud, S., Bird, D., DeBono, A., Durrett, T.P., *et al.* (2013). Acyl-Lipid Metabolism (The American Society of Plant Biologists).

Li, M., Welti, R., and Wang, X. (2006). Quantitative profiling of *Arabidopsis* polar glycerolipids in response to phosphorus starvation. Roles of phospholipases Dζ1 and Dζ2 in phosphatidylcholine hydrolysis and digalactosyldiacylglycerol accumulation in phosphorus-starved plants. Plant Physiol. *142*, 750–761.

Li, N., Gügel, I.L., Giavalisco, P., Zeisler, V., Schreiber, L., Soll, J., and Philippar, K. (2015a). FAX*1*, a novel membrane protein mediating plastid fatty acid export. PLoS Biol. *13*, e1002053.

Li, Q., Zheng, Q., Shen, W., Cram, D., Fowler, D.B., Wei, Y., and Zou, J. (2015b). Understanding the biochemical basis of temperature-induced lipid pathway adjustments in plants. Plant Cell *27*, 86–103.

Li, X., Moellering, E.R., Liu, B., Johnny, C., Fedewa, M., Sears, B.B., Kuo, M.H., and Benning, C. (2012). A galactoglycerolipid lipase is required for triacylglycerol accumulation and survival following nitrogen deprivation in *Chlamydomonas reinhardtii*. Plant Cell *24*, 4670–4686.

Lippold, F., vom Dorp, K., Abraham, M., Hölzl, G., Wewer, V., Lindberg Yilmaz, J., Lager, I., Montandon, C., Besagni, C., Kessler, F., *et al.* (2012). Fatty acid phytyl ester synthesis in chloroplasts of *Arabidopsis*. Plant Cell *24*, 2001–2014.

Löbbecke, L., and Cevc, G. (1995). Effects of short-chain alcohols on the phase behavior and interdigitation of phosphatidylcholine bilayer membranes. Biochim. Biophys. Acta *1237*, 59–69.

Lohmann, A., Schöttler, M.A., Brehelin, C., Kessler, F., Bock, R., Cahoon, E.B., and Dörmann, P. (2006). Deficiency in phylloquinone (vitamin K1) methylation affects prenyl quinone distribution, photosystem I abundance, and anthocyanin accumulation in the *Arabidopsis* AtmenG mutant. J. Biol. Chem. *281*, 40461–40472.

Lu, B., Xu, C., Awai, K., Jones, A., and Benning, C. (2007). A small ATPase protein of *Arabidopsis*, TGD3, involved in chloroplast lipid import. J. Biol. Chem. *282*, 35945–35953.

Maeda, H., Sage, T.L., Isaac, G., Welti, R., and Dellapenna, D. (2008). Tocopherols modulate extraplastidic polyunsaturated fatty acid metabolism in *Arabidopsis* at low temperature. Plant Cell *20*, 452–470.

Maeda, H., Song, W., Sage, T.L., and DellaPenna, D. (2006). Tocopherols play a crucial role in low-temperature adaptation and phloem loading in *Arabidopsis*. Plant Cell *18*, 2710–2732.

Mène-Saffrané, L., and DellaPenna, D. (2010). Biosynthesis, regulation and functions of tocochromanols in plants. Plant Physiol. Biochem. *48*, 301–309.

Moellering, E.R., and Benning, C. (2011). Galactoglycerolipid metabolism under stress: a time for remodeling. Trends Plant Sci. *16*, 98–107.

Moellering, E.R., Miller, R., and Benning, C. (2010a). Molecular genetics of lipid metabolism in the model green alga *Chlamydomonas reinhardtii*. In Lipids in photosynthesis, H. Wada, and N. Murata, eds. (Dordrecht: Springer), pp. 139–155.

Moellering, E.R., Muthan, B., and Benning, C. (2010b). Freezing tolerance in plants requires lipid remodeling at the outer chloroplast membrane. Science *330*, 226–228.

Mongrand, S., Bessoule, J.-J., Cabantous, F., and Cassagne, C. (1998). The C16:3/C18:3 fatty acid balance in photosynthetic tissues from 468 plant species. Phytochemistry *49*, 1049–1064.

Mongrand, S., Cassagne, C., and Bessoule, J.J. (2000). Import of lyso-phosphatidylcholine into chloroplasts likely at the origin of eukaryotic plastidial lipids. Plant Physiol. *122*, 845–852.

Nakamura, Y., Awai, K., Masuda, T., Yoshioka, Y., Takamiya, K., and Ohta, H. (2005). A novel phosphatidylcholine-hydrolyzing phospholipase C induced by phosphate starvation in *Arabidopsis*. J. Biol. Chem. *280*, 7469–7476.

Nakamura, Y., Koizumi, R., Shui, G., Shimojima, M., Wenk, M.R., Ito, T., and Ohta, H. (2009). *Arabidopsis* lipins mediate eukaryotic pathway of lipid metabolism and cope critically with phosphate starvation. Proceedings of the National Academy of Sciences *106*, 20978–20983.

Nakamura, Y., Tsuchiya, M., and Ohta, H. (2007). Plastidic phosphatidic acid phosphatases identified in a distinct subfamily of lipid phosphate phosphatases with prokaryotic origin. J. Biol. Chem. *282*, 29013–29021.

Niyogi, K.K., Grossman, A.R., and Björkman, O. (1998). *Arabidopsis* mutants define a central role for the xanthophyll cycle in the regulation of photosynthetic energy conversion. Plant Cell *10*, 1121–1134.

Norris, S.R., Barrette, T.R., and Dellapenna, D. (1995). Genetic dissection of carotenoid synthesis in *Arabidopsis* defines plastoquinone as an essential component of phytoene desaturation. Plant Cell 7, 2139–2148.

Ohlrogge, J., and Browse, J. (1995). Lipid biosynthesis. Plant Cell 7, 957–970.

Okazaki, Y., Kamide, Y., Hirai, M., and Saito, K. (2013a). Plant lipidomics based on hydrophilic interaction chromatography coupled to ion trap time-of-flight mass spectrometry. Metabolomics 9, 121–131.

Okazaki, Y., Nishizawa, T., Takano, K., Ohnishi, M., Mimura, T., and Saito, K. (2015). Induced accumulation of glucuronosyldiacylglycerol in tomato and soybean under phosphorus deprivation. Physiol. Plant. *155*, 33–42.

Okazaki, Y., Otsuki, H., Narisawa, T., Kobayashi, M., Sawai, S., Kamide, Y., Kusano, M., Aoki, T., Hirai, M.Y., and Saito, K. (2013b). A new class of plant lipid is essential for protection against phosphorus depletion. Nat. Commun. *4*, 1510.

Okazaki, Y., Shimojima, M., Sawada, Y., Toyooka, K., Narisawa, T., Mochida, K., Tanaka, H., Matsuda, F., Hirai, A., Hirai, M.Y., *et al.* (2009). A chloroplastic UDP-glucose pyrophosphorylase from *Arabidopsis* is the committed enzyme for the first step of sulfolipid biosynthesis. Plant Cell *21*, 892–909.

Páli, T., Garab, G., Horváth, L.I., and Kóta, Z. (2003). Functional significance of the lipid–protein interface in photosynthetic membranes. Cell. Mol. Life Sci. *60*, 1591–1606.

Pearcy, R.W. (1978). Effect of growth temperature on the fatty acid composition of the leaf lipids in *Atriplex lentiformis* (Torr.) wats. Plant Physiol. *61*, 484–486.

Petroutsos, D., Amiar, S., Abida, H., Dolch, L.J., Bastien, O., Rébeillé, F., Jouhet, J., Falconet, D., Block, M.A., McFadden, G.I., *et al.* (2014). Evolution of galactoglycerolipid biosynthetic pathways – from cyanobacteria to primary plastids and from primary to secondary plastids. Prog. Lipid Res. *54*, 68–85.

Porfirova, S., Bergmüller, E., Tropf, S., Lemke, R., and Dörmann, P. (2002). Isolation of an *Arabidopsis* mutant lacking vitamin E and identification of a cyclase essential for all tocopherol biosynthesis. Proc. Natl. Acad. Sci. U.S.A. *99*, 12495–12500.

Riekhof, W.R., Ruckle, M.E., Lydic, T.A., Sears, B.B., and Benning, C. (2003). The sulfolipids 2′-O-acyl-sulfoquinovosyldiacylglycerol and sulfoquinovosyldiacylglycerol are absent from a *Chlamydomonas reinhardtii* mutant deleted in *SQD1*. Plant Physiol. *133*, 864–874.

Riekhof, W.R., Sears, B.B., and Benning, C. (2005). Annotation of genes involved in glycerolipid biosynthesis in *Chlamydomonas reinhardtii*: Discovery of the betaine lipid synthase BTA1Cr. Eukaryotic Cell *4*, 242–252.

Riewe, D., Koohi, M., Lisec, J., Pfeiffer, M., Lippmann, R., Schmeichel, J., Willmitzer, L., and Altmann, T. (2012). A tyrosine aminotransferase involved in tocopherol synthesis in *Arabidopsis*. Plant J. *71*, 850–859.

Roston, R., Gao, J., Xu, C., and Benning, C. (2011). *Arabidopsis* chloroplast lipid transport protein TGD2 disrupts membranes and is part of a large complex. Plant J. *66*, 759–769.

Roston, R.L., Gao, J., Murcha, M.W., Whelan, J., and Benning, C. (2012). TGD*1*, -2 and -3 proteins involved in lipid trafficking form ATP-binding cassette (ABC) transporter with multiple substrate-binding proteins. J. Biol. Chem. *287*, 21406–21415.

Sadre, R., Gruber, J., and Frentzen, M. (2006). Characterization of homogentisate prenyltransferases involved in plastoquinone-9 and tocochromanol biosynthesis. FEBS Lett. *580*, 5357–5362.

Sato, A., Sugimoto, K., Tsuzuki, M., and Sato, N. (2013). UDP-glucose pyrophosphorylase responsible for sulfolipid synthesis in a green alga *Chlamydomonas reinhardtii*. In Photosynthesis Research for Food, Fuel and the Future, T. Kuang, C. Lu, and L. Zhang, eds. (Berlin: Springer), pp. 616–619.

Sato, N., Hagio, M., Wada, H., and Tsuzuki, M. (2000). Environmental effects on acidic lipids of thylakoid membranes. Biochem. Soc. Trans. *28*, 912–914.

Sato, N., Sugimoto, K., Meguro, A., and Tsuzuki, M. (2003). Identification of a gene for UDP-sulfoquinovose synthase of a green alga, *Chlamydomonas reinhardtii*, and its phylogeny. DNA Res. *10*, 229–237.

Sattler, S.E., Gilliland, L.U., Magallanes-Lundback, M., Pollard, M., and DellaPenna, D. (2004). Vitamin E is essential for seed longevity and for preventing lipid peroxidation during germination. Plant Cell *16*, 1419–1432.

Savidge, B., Weiss, J.D., Wong, Y.-, H.H., Lassner, M.W., Mitsky, T.A., Shewmaker, C.K., Post-Beittenmiller, D., and Valentin, H.E. (2002). Isolation and characterization of homogentisate phytyltransferase genes from *Synechocystis* sp. PCC 6803 and *Arabidopsis*. Plant Physiol. *129*, 321–332.

Schneiter, R., Brügger, B., Sandhoff, R., Zellnig, G., Leber, A., Lampl, M., Athenstaedt, K., Hrastnik, C., Eder, S., Daum, G., *et al.* (1999). Electrospray ionization tandem mass spectrometry (ESI-MS/MS) analysis of the lipid molecular species composition of yeast subcellular membranes reveals acyl chain-based sorting/remodeling of distinct molecular species en route to the plasma membrane. J. Cell Biol. *146*, 741–754.

Schnurr, J.A., Shockey, J.M., de Boer, G.J., and Browse, J.A. (2002). Fatty acid export from the chloroplast. Molecular characterization of a major plastidial acyl-coenzyme A synthetase from *Arabidopsis*. Plant Physiol. *129*, 1700–1709.

Shimada, H., Ohno, R., Shibata, M., Ikegami, I., Onai, K., Ohto, M.-A., and Takamiya, K.-I. (2005). Inactivation and deficiency of core proteins of photosystems I and II caused by genetical phylloquinone and plastoquinone deficiency but retained lamellar structure in a T-DNA mutant of *Arabidopsis*. Plant J. *41*, 627–637.

Soll, J., Schultz, G., Joyard, J., Douce, R., and Block, M.A. (1985). Localization and synthesis of prenylquinones in isolated outer and inner envelope membranes from spinach chloroplasts. Arch. Biochem. Biophys. *238*, 290–299.

Song, W., Maeda, H., and DellaPenna, D. (2010). Mutations of the ER to plastid lipid transporters *TGD1, 2, 3 and 4* and the ER oleate desaturase *FAD2* suppress the low temperature-induced phenotype of *Arabidopsis* tocopherol-deficient mutant *vte2*. Plant J. *62*, 1004–1018.

Sugimoto, K., Tsuzuki, M., and Sato, N. (2010). Regulation of synthesis and degradation of a sulfolipid under sulfur-starved conditions and its physiological significance in *Chlamydomonas reinhardtii*. New Phytol. *185*, 676–686.

Tanoue, R., Kobayashi, M., Katayama, K., Nagata, N., and Wada, H. (2014). Phosphatidylglycerol biosynthesis is required for the development of embryos and normal membrane structures of chloroplasts and mitochondria in *Arabidopsis*. FEBS Lett. *588*, 1680–1685.

Tevini, M., and Steinmüller, D. (1985). Composition and function of plastoglobuli. II. Lipid composition of leaves and plastoglobuli during beech leaf senescence. Planta *163*, 91–96.

Thayer, S., and Björkman, O. (1990). Leaf xanthophyll content and composition in sun and shade determined by HPLC. Photosynth. Res. *23*, 331–343.

Thorlby, G., Fourrier, N., and Warren, C. (2004). The *Sensitive To Freezing 2* gene, required for freezing tolerance in *Arabidopsis thaliana*, encodes a β-glucosidase. Plant Cell *16*, 2192–2203.

Valentin, H.E., Lincoln, K., Moshiri, F., Jensen, P.K., Qi, Q., Venkatesh, T.V., Karunanandaa, B., Baszis, S.R., Norris, S.R., Savidge, B., *et al.* (2006). The *Arabidopsis vitamin E pathway gene5-1* mutant reveals a critical role for phytol kinase in seed tocopherol biosynthesis. Plant Cell *18*, 212–224.

Vidi, P.A., Kanwischer, M., Baginsky, S., Austin, J.R., Csucs, G., Dörmann, P., Kessler, F., and Bréhélin, C. (2006). Tocopherol cyclase (VTE1) localization and vitamin E accumulation in chloroplast plastoglobule lipoprotein particles. J. Biol. Chem. *281*, 11225–11234.

Vu, H.S., Roth, M.R., Tamura, P., Samarakoon, T., Shiva, S., Honey, S., Lowe, K., Schmelz, E.A., Williams, T.D., and Welti, R. (2014). Head-group acylation of monogalactosyldiacylglycerol is a common stress response, and the acyl-galactose acyl composition varies with the plant species and applied stress. Physiol. Plant. *150*, 517–528.

Wallis, J.G., and Browse, J. (2002). Mutants of *Arabidopsis* reveal many roles for membrane lipids. Prog. Lipid Res. *41*, 254–278.

Wanders, R.J.A., Komen, J., and Ferdinandusse, S. (2011). Phytanic acid metabolism in health and disease. Biochim. Biophys. Acta *1811*, 498–507.

Wang, X. (2005). Regulatory functions of phospholipase D and phosphatidic acid in plant growth, development, and stress responses. Plant Physiol. *139*, 566–573.

Wang, Z., Anderson, N.S., and Benning, C. (2013). The phosphatidic acid binding site of the *Arabidopsis* trigalactosyldiacylglycerol 4 (TGD4) protein required for lipid import into chloroplasts. J. Biol. Chem. *288*, 4763–4771.

Webb, M.S., and Green, B.R. (1991). Biochemical and biophysical properties of thylakoid acyl lipids. Biochim. Biophys. Acta *1060*, 133–158.

Welti, R., Li, W., Li, M., Sang, Y., Biesiada, H., Zhou, H.-E., Rajashekar, C., Williams, T., and Wang, X. (2002). Profiling membrane lipids in plant stress responses. Role of phospholipase Da in freezing-induced lipid changes in *Arabidopsis*. J. Biol. Chem. 277, 31994–32002.

Widhalm, J.R., Ducluzeau, A.-L., Buller, N.E., Elowsky, C.G., Olsen, L.J., and Basset, G.J.C. (2012). Phylloquinone (vitamin K1) biosynthesis in plants: two peroxisomal thioesterases of lactobacillales origin hydrolyze 1,4-dihydroxy-2-naphthoyl-CoA. Plant J. *71*, 205–215.

Wildermuth, M., Dewdney, J., Wu, G., and Ausubel, F. (2001). Isochorismate synthase is required to synthesize salicylic acid for plant defence. Nature *417*, 562–565.

Wu, F., Yang, Z., and Kuang, T. (2006). Impaired photosynthesis in phosphatidylglycerol-deficient mutant of cyanobacterium *Anabaena* sp. PCC7120 with a disrupted gene encoding a putative phosphatidylglycerophosphatase. Plant Physiol. *141*, 1274–1283.

Wu, J., James, D.W., Jr., Dooner, H.K., and Browse, J. (1994). A mutant of *Arabidopsis* deficient in the eongation of palmitic acid. Plant Physiol. *106*, 143–150.

Xiao, S., and Chye, M.L. (2009). An *Arabidopsis* family of six acyl-CoA-binding proteins has three cytosolic members. Plant Physiol. Biochem. *47*, 479–484.

Xu, C., Fan, J., Froehlich, J.E., Awai, K., and Benning, C. (2005). Mutation of the TGD1 chloroplast envelope protein affects phosphatidate metabolism in *Arabidopsis*. Plant Cell *17*, 3094–3110.

Xu, C., Fan, J., Riekhof, W., Froehlich, J.E., and Benning, C. (2003). A permease-like protein involved in ER to thylakoid lipid transfer in *Arabidopsis*. EMBO J. *22*, 2370–2379.

Xu, C., Härtel, H., Wada, H., Hagio, M., Yu, B., Eakin, C., and Benning, C. (2002). The *pgp1* mutant locus of *Arabidopsis* encodes a phosphatidylglycerolphosphate synthase with impaired activity. Plant Physiol. *129*, 594–604.

Xu, C., Yu, B., Cornish, A.J., Froehlich, J.E., and Benning, C. (2006). Phosphatidylglycerol biosynthesis in chloroplasts of *Arabidopsis* mutants deficient in acyl-ACP glycerol-3- phosphate acyltransferase. Plant J. *47*, 296–309.

Yu, B., and Benning, C. (2003). Anionic lipids are required for chloroplast structure and function in *Arabidopsis*. Plant J. *36*, 762–770.

Yu, B., Wakao, S., Fan, J., and Benning, C. (2004). Loss of plastidic lysophosphatidic acid acyltransferase causes embryo-lethality in *Arabidopsis*. Plant Cell Physiol. *45*, 503–510.

Yu, B., Xu, C., and Benning, C. (2002). *Arabidopsis* disrupted in *SQD2* encoding sulfolipid synthase is impaired in phosphate-limited growth. Proc. Natl. Acad. Sci. U.S.A. *99*, 5732–5737.

Yuzawa, Y., Nishihara, H., Haraguchi, T., Masuda, S., Shimojima, M., Shimoyama, A., Yuasa, H., Okada, N., and Ohta, H. (2012). Phylogeny of galactolipid synthase homologs together with their enzymatic analyses revealed a possible origin and divergence time for photosynthetic membrane biogenesis. DNA Res. *19*, 91–102.

Zäuner, S., Jochum, W., Bigorowski, T., and Benning, C. (2012). A Cytochrome b_5-containing plastid-located fatty acid desaturase from *Chlamydomonas reinhardtii*. Eukaryotic Cell *11*, 856–863.

Zbierzak, A.M., Kanwischer, M., Wille, C., Vidi, P.A., Giavalisco, P., Lohmann, A., Briesen, I., Porfirova, S., Brehelin, C., Kessler, F., *et al.* (2010). Intersection of the tocopherol and plastoquinol metabolic pathways at the plastoglobule. Biochem. J. *425*, 389–399.

Zhou, Y., Weth, A., Peisker, H., Baumgartner, W., Dörmann, P., and Frentzen, M. (2013). Extraplastidial cytidinediphosphate diacylglycerol synthases of *Arabidopsis thaliana*. Plant J. *75*, 867–879.

Assembly and Degradation of Pigment-binding Proteins

2

Tania Tibiletti, Miguel A. Hernández-Prieto, Trudi A. Semeniuk and Christiane Funk

Abstract

In organisms performing oxygenic photosynthesis, the light absorbed by the photosystem II reaction centre is used to power the extraction of electrons from water, and initiate a series of oxidation-reduction reactions that power the synthesis of energy-rich molecules (ATP and NAD(P)H). These energy-rich molecules are transported to other cellular compartments to power all the metabolic reactions required to maintain growth. To ensure a constant supply of energy and avoid deleterious reactions caused by the absorption of excess sunlight, photosynthetic organisms have evolved different mechanisms to reduce or increase the number of pigment-binding molecules present in their thylakoids. This chapter offers a current overview of the processes underlying assembly, degradation, and repair of the pigment-binding complexes present in higher plants, green algae and cyanobacteria. We focus specifically on the chlorophyll-binding complexes in these organisms.

Introduction

Photosynthesis provides raw materials that result in a constant supply of food, shelter, cloth, fibre, energy, etc., for the increasing human population (Janssen *et al.*, 2014). However, population growth, at its current rate, has been predicted to result in future shortages of food and other plant-derived materials. Initiatives, such as reforestation of disturbed areas or substitution of photosynthetically more efficient plant species, could ameliorate these effects, also slowing down climate change through enhanced CO_2 fixation. Such increased photosynthetic efficiency would also facilitate a transition from fossil to bio-fuels of plant origin, which will further slow climate change (Janssen *et al.*, 2014; Long *et al.*, 2015).

Harvesting of (sun) light is the first step in the photosynthetic process. In the light-dependent photosynthetic reaction, antennae complexes rapidly transfer absorbed light energy to the reaction centres of photosystems I (PSI) and II (PSII), where it is used for charge separation, initiating the transport of electrons across a specialized membrane system – the thylakoids. Four major photosynthetic complexes can be found in the thylakoid membrane: PSI, PSII, cytochrome b_6f, and ATP synthase. Both photosystems organize into two moieties: a core complex that contains the reaction centre (RC) where the photochemical reaction takes place; and a peripheral antenna system that increases and regulates their light harvesting capacity, and participates in their protection from high-intensity light damage (photo-damage). PSI and PSII core complexes were highly conserved during evolution, such that photosynthesis operates in a similar manner in different species, while various pigments and antenna systems have evolved (Hohmann-Marriott and Blankenship, 2011). Prokaryotic cyanobacteria (except for prochlorophytes) and eukaryotic red algae contain a soluble complex called the phycobilisome, as their major outer antennae; the complex is formed by linear tetrapyrrole molecules, or phycobilins, covalently bound to a protein matrix (Bailey and Grossman, 2008). Typically, phycobilisomes are associated with the external part of the thylakoid membrane, absorbing light in the range between 550 and 660 nm (i.e. in the green gap, not absorbed by chlorophylls). In green algae and higher plants, the most abundant antenna is the chlorophyll *a*/*b*-binding light-harvesting complex (referred to as LHC) that is nuclear encoded by a multigene family, with a range of copy numbers (Neilson and Durnford, 2010). Exceptionally, Cryptophyte algae contain a chlorophyll *a/c*-binding membrane-intrinsic antenna homologous to the ones of green algae and higher plants, in addition to phycobiliproteins attached to the inner (luminal) side of the thylakoid membrane (Broughton *et al.*, 2006). In this chapter, we focus on the structure, assembly and degradation of chlorophyll-binding (light-harvesting) proteins in the green lineage (cyanobacteria, green algae and higher plants), rather than non-chlorophyll binding antennae complexes.

The thylakoids of cyanobacteria and chloroplast

The distribution of cyanobacteria thylakoids inside the cell differs from that of chloroplasts. In contrast to chloroplasts thylakoids that are dedicated to photosynthesis-related reactions, the cyanobacteria thylakoid membrane contains components of both the respiratory and the photosynthetic electron transport chain. Furthermore, in cyanobacteria, the thylakoid membrane is generally distributed parallel to the cellular membrane; while in chloroplasts, the thylakoids form different structural areas with highly stacked areas (granas) connected by un-stacked thylakoids or stroma lamellae. In chloroplasts, the grade of stacking of the thylakoid membranes correlates with the distribution of photosynthetic complexes. PSII is localized mainly in the grana thylakoid membranes, while PSI and the ATP synthase are found in the stroma lamellae structures with the mobile fraction of the LHCII pool shifting between them depending of the light conditions. The cytochrome b_6f is distributed evenly between these two structures (Dekker and Boekema, 2005; Nelson and Yocum, 2006).

Chlorophyll *a*-binding core complexes, PSI and PSII

Although the crystallographic structure of PSI has been determined for both cyanobacteria and higher plants, only the PSII structure for cyanobacteria has been resolved at high resolution (Zouni *et al.*, 2001; Ferreira *et al.*, 2004; Loll *et al.*, 2005; Guskov *et al.*, 2009; Umena *et al.*, 2011).

Photosystem I core complex

The core complex of PSI is composed of 12 subunits in cyanobacteria and 19 subunits in higher plants (Fig. 2.1). The X-ray structure of PSI for the cyanobacterium *Synechococcus elongatus* identified the location of 96 chlorophyll *a*, 22 β-carotenes, four lipids, three iron–sulfur clusters and two phylloquinones per monomeric core complex (Jordan *et al.*, 2001; Fig. 2.1A); while the crystal structure of PSI extracted from the pea plant *Pisum sativum* resolved the location of 168 chlorophyll molecules (Amunts *et al.*, 2007; Fig. 2.1B). The pigment organization in the PSI–core complex is very similar for plants and cyanobacteria. The additional 33 to 28 chlorophyll found in plants are either associated with plant-specific subunits arranged in a belt surrounding the core complex connecting the PSI core with the LHCI ('gap' chlorophylls), or between the Lhca monomers within LHCI ('linker' chlorophylls).

Recently, two new X-ray structures of PSI from *Pisum sativum* have been published at 2.8 Å resolution containing a lower number of chlorophyll molecules (152 or 155 chlorophylls, respectively) (Qin *et al.*, 2015; Mazor *et al.*, 2015). These differences have been attributed to a decreased number of 'gap chlorophylls', which are thought to have a profound effect on the light transfer from LHCI to PSI core. The central part of the core complex is occupied by two subunits, PsaA and PsaB (Fig.2.1, red and yellow respectively), forming a heterodimer comprising 22 transmembrane helices. This heterodimer hosts most of the cofactors in PSI and functions as RC and core-antenna (Barber *et al.*, 2000). The PsaA and PsaB N- and C-terminal regions can be seen as two separate functional units. The six N-terminal transmembrane helices function as antennae, binding most of the chlorophyll *a* molecules in PSI. While the five transmembrane helices in the C-terminal region host the cofactors involved in the initial charge separation needed to extract electrons from the mobile plastocyanin: P700 (Primary electron donor absorbing at 700 nm), A_0 (chlorophyll *a*), A_1 (phylloquinone), and the iron–sulfur (Fe_4S_4) cluster (F_x). The terminal electron acceptors (F_A and F_B, two iron–sulfur clusters) are bound to the PsaC subunit on the cytoplasmic/stromal side. In addition to PsaC, other small Psa subunits function in different processes, such as plastocyanin docking (PsaF, PsaN),

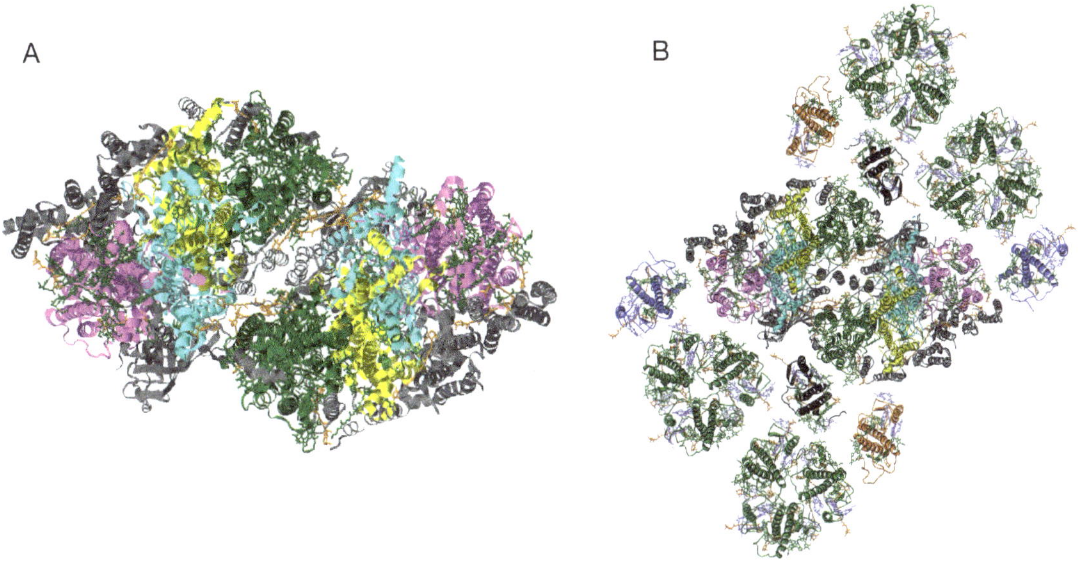

Figure 2.1 Crystal structures of photosystem I from cyanobacteria and plants. (A) Subunit and pigment organization of the PSI complex from *Synechococcus elongates*, viewed from the membrane plan. The major subunits are: PsaA in red, PsaB in yellow, and small subunits in grey. The image was created with the software Jmol and the PDB file 1JB0 (Jordan *et al*., 2001). (B) Subunits and pigment organization of the PSI complex from *Pea*, viewed from the stromal side of the membrane. The major subunits are: PsaA in red, PsaB in yellow, LHCa in blue, and small subunits in grey. The image was created with the software Jmol and the PDB file 2WSF (Amunts *et al*., 2007).

ferredoxin docking (PsaC, PsaD), formation of the docking site for the LHCII (PsaH), correct assembly and stabilization of the system, and chlorophyll-binding (Nelson and Ben-Shem, 2005; Jensen *et al*., 2007). In PSI, several low molecular mass proteins directly bind chlorophyll molecules and carotenoids (Jensen *et al*., 2007). In green algae and higher plants, PSI functions as a monomer, while in cyanobacteria PSI cores occur as trimers (Bibby *et al*., 2001; Boekema *et al*., 2001). In cyanobacteria, PsaL has been shown to be important for such trimerization (Jensen *et al*., 2007). Trimeric PSI is thought to be more effectively protected from PSII energy spillover compared with monomeric PSI, a problem that plants do not encounter due to the spatial separation of their two photosystems in the thylakoid membrane (Mullineaux, 2005). Although there are homologies between the PSI core complexes of higher plants and cyanobacteria, the subunits PsaX and PsaM are only present in cyanobacteria (Jordan *et al*., 2001), while the PsaG, PsaH, PsaN and PsaR subunits of higher plants are not present in photosynthetic prokaryotes (Amunts and Nelson, 2009).

Photosystem II core complex

PSII, which normally exists as a dimer, is a complex composed of 25–30 subunits (reviewed in Pagliano *et al*., 2013). Most of them have a molecular mass under 10 kDa, and some are species specific (Shi and Schroder, 2004; Shi *et al*., 2012). Its smallest biochemical unit able to evolve oxygen is known as the PSII core complex. It hosts 35 chlorophyll *a*, 2 phaeophytins, 11 β-carotenes and two plastoquinones, as well as the [4Mn–1Ca] cluster (Umena *et al*., 2011). The chlorophyll *a*-binding subunits comprise the RC proteins D1 (PsbA) and D2 (PsbD), as well as the core antenna proteins CP43 (PsbC) and CP47 (PsbB) (Fig. 2.2). D1 and D2 (Fig. 2.2, cyan and yellow respectively) are each composed of five transmembrane helices organized in a handshake motif, resembling the C-terminal region of PsaA and PsaB. The D1–D2 heterodimer binds six chlorophyll *a*, two pheophytins, two β-carotenes and two phylloquinones. These proteins also stabilize the metallic catalytic core, the [4Mn–1Ca] cluster, of the oxygen-evolving complex (OEC). Adjacent to the D1/D2 heterodimer, there are two low molecular weight proteins essential for PSII

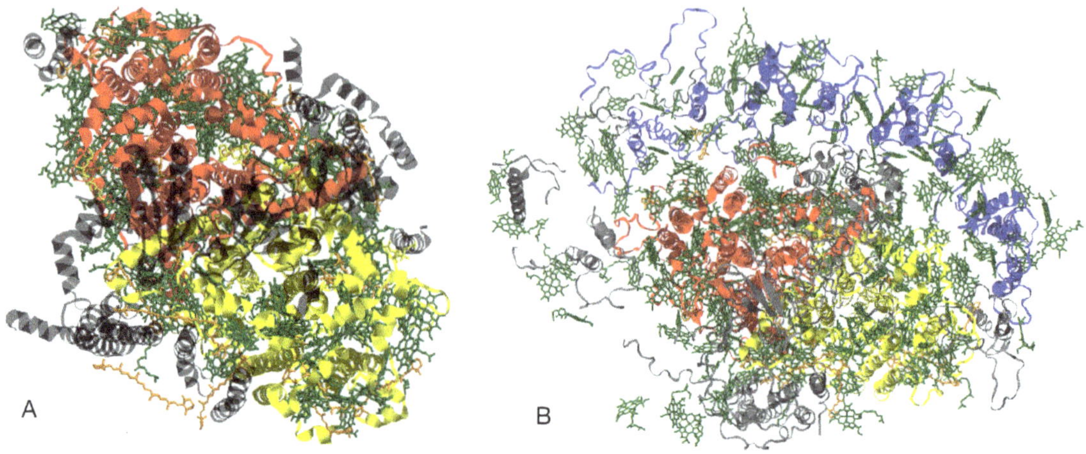

Figure 2.2 Crystal structures of photosystem II from cyanobacteria and plants. (A) Subunit and pigment organization of the dimeric PSII complex isolated from *Thermosynechoccus vulcanus*, and viewed from the cytoplasmic site of the membrane. The major subunits are: D1 in cyan, D2 in yellow, CP43 in pink, CP47 in green, and small subunits in grey. Pigments are shown in green, chlorophyll *a*; and orange, carotenoids. The image was created with the software Jmol and the PDB file 3ARC (Umena *et al.*, 2011). (B) Subunit and pigment organization of the PSII supercomplex in plants, viewed from the stromal site of the membrane. The major subunits of the core complex are coloured as in (A). The antenna system is coloured as follow: CP29 in black, CP24 in orange, CP26 in blue, and the S and the M trimers in green. Pigments are coloured: green, chlorophyll *a*; violet, chlorophyll *b*; and orange, carotenoids. Phytol tails of chlorophyll molecules are omitted for clarity. The image was modified after Caffarri *et al.* (2014), using the software WebLab ViewerPro.

activity (PsbE and PsbF) that are the subunits α and β of the cyt b_{559} (Swiatek *et al.*, 2003), also forming part of the PSII RC. The RC is surrounded by the inner antenna, formed by the two subunits PsbB (CP47) and PsbC (CP43), composed of six transmembrane helices, binding 16 and 13 chlorophyll *a*, respectively (Fig. 2.2, CP43 in pink and CP47 in green). These chlorophyll are arranged to maximize the transfer of the harvested energy to the RC (Umena *et al.*, 2011). The final members of the PSII core complex are the subunits on the luminal side, forming the OEC. In green algae and higher plants, the OEC is composed of four subunits, PsbO, PsbP, PsbQ and PsbR, while, in cyanobacteria, PsbR is replaced by PsbU and PsbV (Roose *et al.*, 2007). Several intrinsic low-molecular-weight proteins also are associated with the PSII core complex in cyanobacteria, but none has been shown to bind chlorophyll *a* (Shi and Schroder, 2004). Only PsbH and PsbI are known to form hydrogen bonds with the chlorophyll *a* molecules of CP47 and CP43, respectively (Muh *et al.*, 2008).

The outer antenna complexes – the LHC family

The shift from an aquatic to terrestrial environment during the evolution of photosynthetic eukaryotes was accompanied by substitution of the soluble antenna system (the phycobilisome) with membrane-integral light-harvesting complexes (LHC; Hohmann-Marriott and Blankenship, 2011). One of the main reasons for such replacement is the more efficient photo-protective mechanisms inherent to LHCs (detailed later). An efficient non-photochemical quenching is of crucial importance in the areal environment, where a higher risk of singlet oxygen ($^1O_2$) production exists (Neilson and Durnford, 2010). The high homology between sequences of the nuclear-encoded LHC proteins indicates a common evolutionary origin (Jansson, 2006). LHC proteins comprise three transmembrane helices that coordinate chlorophyll *a* and chlorophyll *b* (or chlorophyll *c*) and carotenoids in different amounts (Fig. 2.3). A common feature of the extended family of LHC proteins is the highly conserved chlorophyll *a*/*b*-binding (CAB) domain, containing the amino acid residues involved in pigment binding (see below). This domain is not only

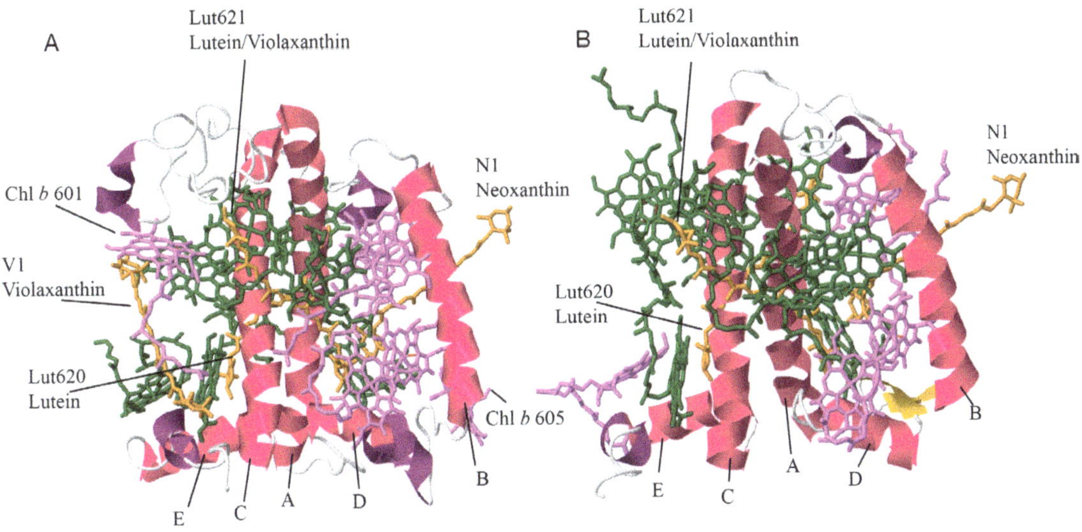

Figure 2.3 Light-harvesting complex II monomer. (A) Molecular model of LHCII shown with ligands. The model was generated using PMol from crystal structures of LHCII in the PDB file 1RWT (Liu *et al.*, 2004). (B) Molecular model of Lhcb4, shown with ligands. The model was generated using PMol from crystal structures of Lhcb4 in the PDB file 3PL9 (Pan *et al.*, 2011). Coloured regions are: pink, polypeptides; green, chlorophyll *a*; violet, chlorophyll *b*; orange, carotenoids; and blue, lipids.

identified in the chlorophyll *b*-binding green lineage, but also in LHCs in algae, after secondary or tertiary endosymbiosis, such as diatoms, dinoflagellates, cryptophytes and haptophytes; although chlorophyll *c* is bound instead of chlorophyll *b* (Hohmann-Marriott and Blankenship, 2011). In higher plants, six genes (*Lhca1–6*) encode the distal antenna of PSI (LHCI), and six (*Lhcb1–6*) the distal antenna of PSII (LHCII). In the green alga *Chlamydomonas reinhardtii* (hereafter *C. reinhardtii*), nine genes encode the antenna of PSI (*Lhca1–9*), while 11 genes (*Lhcbm1–6, Lhcbm8–11, Lhcb4–5*) encode the subunits of LHCII. In the moss *Physcomitrella patens* (hereafter *P. patens*), four genes encode the PSI antenna (*Lhca1–3, Lhcb5*), while 13 *Lhcbm* genes have been identified in its genome, together with *Lhcb3, Lhcb4, Lhcb5* and *Lhcb9* (Alboresi *et al.*, 2008, 2011). Each LHC can have different isoforms, e.g. Lhcb1 and Lhcb2 are encoded by several nuclear genes in plants (Jansson, 1999; Caffarri *et al.*, 2004; Alboresi *et al.*, 2008).

LHCI

The LHCI complexes in plants are organized into two adjacent heterodimers, Lhca1–Lhca4 and Lhca2–Lhca3, arranged in a half-moon-shaped belt attached to one side of the PSI core complex (Neilson and Durnford, 2010). Both Lhca5 and Lhca6 present high homology with Lhca1–4, and are found in sub-stoichiometric amounts in the PSI complex (Klimmek *et al.*, 2006). The organization of the PSI super-complex is conserved in the moss *P. patens* (Busch *et al.*, 2013), while monomers of the LHCI antenna in the alga *C. reinhardtii* can interact at alternative sites, resulting in larger and more heterogeneous PSI supercomplexes (Neilson and Durnford, 2010). The recently published high-resolution structures of PSI show the transmembrane structure of LHCI to be similar to LHCII with three major transmembrane helices (A, B, C) and an amphipathic helix D at the luminal side. However, there are notable differences in the loop regions as well as in the N-terminal region (Qin *et al.*, 2015).

LHCII

The peripheral antenna complex of PSII in green algae and higher plants, LHCII, coordinate chlorophyll *a*, chlorophyll *b* and xanthophylls (Jansson, 1999). Two types of Lhc antenna proteins associated with PSII can be distinguished based on their relative abundance, as major LHCII antenna complex or minor LHCII antennae (Dekker and Boekema, 2005). In *Arabidopsis thaliana* (hereafter *A. thaliana*), the major LHCII is a heterotrimer of

Lhcb1, Lhcb2 and Lhcb3 (Peter and Thornber, 1991; Caffarri *et al.*, 2004), often referred as the most common membrane protein complex on earth (Fig. 2.2B, in green). It is located distal to the PSII-core complex, connected via the monomeric minor antennae CP24, CP26, and CP29 (Lhcb6 in orange, Lhcb5 in blue and Lhcb4 in black; Fig. 2.2B).

High-resolution X-ray structures of LHCII (Liu *et al.*, 2004; Standfuss *et al.*, 2005), and more recently CP29 (Pan *et al.*, 2011), have been obtained (Fig. 2.3). The monomeric complex shows three transmembrane helices (A, B, C) connected by both stroma and lumen-exposed loops, as well as two amphipathic helices along the luminal surface of the membrane (D, E). Based on this structure, the LHCII monomer binds 14 chlorophyll (8 chlorophyll *a* and 6 chlorophyll *b*), and four xanthophylls (one neoxanthin, two luteins and one violaxanthin). The CP29 monomer only seems to bind 13 chlorophyll (nine chlorophyll *a* and four chlorophyll *b*) and three xanthophylls (two luteins and one neoxanthin) (Pan *et al.*, 2011).

Two characteristic LHC motif/CAB domains can be found in a LHCII monomer localized within the 32 (28 in Lhcb4) amino acid helices A (–ExxHxxR–) and C (–ExxNxxR–). The three dimensional structure of the monomer facilitates the interaction of these two helices, with a chlorophyll molecule establishing a salt bridge between the glutamate (E) of one helix and the arginine (R) of the other, further stabilizing the monomer conformation. The four carotenoids form different domains within the monomer, with the two luteins (Lut620 and Lut621, as denoted in Liu *et al.*, 2004) bound in parallel to the A and B transmembrane helices, neoxanthin bound in proximity to helix C, and violaxanthin on the periphery (Fig. 2.3).

The organization of the LHCII antenna varies under different environmental and stress conditions. Thus, the amounts of various antenna subunits change to maintain efficient photosynthesis; in low light, plants build a larger antenna, while in high-intensity light, its size decrease (Anderson and Andersson, 1988; Pfannschmidt, 2003). The PSII core with its peripheral antenna forms super-complexes, which differ in the number of LHCII bound to it (Caffarri *et al.*, 2009). Depending on the minor antenna proteins acting as linkers, LHCII can bind to the PSII core strongly (S), moderately (M), or loosely (L). The monomeric linker associated with the S-form is CP26; while for the M-form, they are CP29 and CP24. A dimeric PSII core with 2S- and 2M-type LHCII is the most abundant super-complex in plant photosynthetic membranes (Boekema *et al.*, 1999; Kouril *et al.*, 2012). In contrast, the antenna composition of the PSI super-complex does not vary with environmental changes, although its structure supports weak interactions between the antenna complexes and the core (Ballottari *et al.*, 2004).

In low light conditions, a process called state transition counterbalances the absorbed energy reaching PSI or PSII (Bonaventura and Myers, 1969; Allen and Staehelin, 1992; Wollman, 2001). When PSII is preferentially excited, the Lhcb2 and Lhcb1 monomers are phosphorylated by a kinase (Stn7), which is activated upon reduction of the plastoquinol pool. The phosphorylated L-bound LHCII trimer diffuses from PSII (in the grana) to PSI (in the stroma lamellae) (state 2). When a change in light conditions results in the preferable excitation of PSI (far-red or high-intensity light) the plastoquinol pool is oxidized and thus Stn7 becomes inactive. In this new conditions LHCII is dephosphorylated by the constitutively active LHCII phosphatase (TAP38/PPH1) and it moves back to PSII (state 1) (Allen and Staehelin, 1992). The recent development of a protocol for the purification of PSI-LHCII has facilitated the detailed biophysical characterization of both these states (Galka *et al.*, 2012). Results showed that LHCII attached to PSI transfers energy more efficiently than when it is attached to PSII (Galka *et al.*, 2012).

In vitro LHC reconstitution

Although it is believed that all proteins of the LHC family are structurally similar differences in their biochemical and spectroscopic properties have been observed. For example, all Lhcs bind chlorophyll *a*, chlorophyll *b*, lutein and violaxantin, but to different degrees (Dainese and Bassi, 1991). Based on early experiments, the chlorophyll/protein ratio of Lhcb6, Lhcb5, Lhcb4 or LHCII was determined to be 5, 9, 8 or 13, respectively. While there is a good correspondence between the X-ray structure of the LHCII and biochemical analyses, more recently, the minor Lhcb was shown to contain more chlorophyll/protein, using milder isolation procedures (Ballottari *et al.*, 2012). This suggests that some of the pigments are more loosely bound than others, and are easily removed by detergents during the purification process. Light-harvesting pigments have also been found on the periphery of complexes, not directly coordinated with protein residues, but bound to lipids or water molecules (Muh *et al.*, 2008; Pan *et al.*, 2011; Umena *et al.*, 2011). *In vitro* pigment reconstitution experiments of proteins with mutations on the chlorophyll-binding ligands were remarkably successful at determining the occupancy of each binding site. No difference was observed between the reconstituted complexes and those found in the thylakoid membranes, using absorbance or circular dichroism (Giuffra *et al.*, 1997). Such experiments established that pigments were necessary to induce native folding of the protein (Paulsen *et al.*, 1993; Remelli *et al.*, 1999) and that chlorophyll *a* was necessary to obtain stable folded complexes (Croce *et al.*, 1999; Kleima *et al.*, 1999); no antenna complex was able to fold in the absence of chlorophyll *a*, even in presence of excess chlorophyll *b*, while chlorophyll *b* stabilized the pigment–protein complex. Furthermore, it was observed that the chlorophyll-binding sites located in the centre (helix A and B) of all investigated LHCs bind chlorophyll *a* (chlorophyll 610, 612, 602 and 603), while the domains around helix C accommodate chlorophyll *b* (Bassi *et al.*, 1999; Remelli *et al.*, 1999; Liu *et al.*, 2004; Standfuss *et al.*, 2005; Pan *et al.*, 2011). Some binding sites can be occupied by either chlorophyll *a* or chlorophyll *b* (Giuffra *et al.*, 1997; Bassi *et al.*, 1999; Remelli *et al.*, 1999), however, this mixed occupancy is probably induced by competition between chlorophyll *a* and chlorophyll *b in vitro*. Stopped flow experiments showed that Lhcb1 is folded in two temporal steps: the first one occurs on the order of 10 s, while the other is slower, occurring over minutes (Reinsberg *et al.*, 2000). The binding of chlorophyll *a* occurred during the fast phase, followed by the slower integration of chlorophyll *b* (Reinsberg *et al.*, 2000). Chlorophyll *b* forms stronger coordination bonds with water than chlorophyll *a*, which slows down its incorporation into the protein complex. In the cyanobacterium *Synechocystis* sp. PCC 6803 (hereafter *Synechocystis* 6803), which naturally does not contains chlorophyll *b*, the chlorophyllide *a* oxygenase (CAO) gene, that catalyses the transformation of chlorophyllide *a* to *b*, was expressed to demonstrate that there is no specificity for most binding sites (Satoh *et al.*, 2001; Xu *et al.*, 2001); incorporation of chlorophyll *b* instead of chlorophyll *a* into a binding site was shown to be dependent only on the availability of chlorophyll *b*. In fact, the activity and degradation of CAO is regulated by accumulating chlorophyll *b* (Yamasato *et al.*, 2005; Sakuraba *et al.*, 2007).

Carotenoids play a very basic role in the assembly and stabilization of LHC proteins. The number of carotenoids bound per LHC complex varies between two and four (Ruban *et al.*, 1999; Morosinotto *et al.*, 2003). The Lut620 site, with a very high binding affinity for lutein, plays a major role in folding of the pigment–protein complex; only Lhcb4 mutants with one lutein attached to Lut620 were stable enough to be purified (Pesaresi *et al.*, 1997). In contrast, the Lut621 site is highly variable in its carotenoid binding preference; violaxanthin or lutein can be found here. Typically, violaxanthin is more common in the minor antenna complexes compared with LHCII, consistent with their prominent role in the xanthophyll cycle (Morosinotto *et al.*, 2002; Betterle *et al.*, 2010). This cycle is a photoprotective mechanism, in which violaxanthin is converted to zeaxanthin by the high light-activated violaxanthin depoxidase (Havaux and Niyogi, 1999). The Lut621 site was also shown to be important for protein stability (Morosinotto *et al.*, 2002). Neoxanthin (Caffarri *et al.*, 2007) and violaxanthin (Caffarri *et al.*, 2001) binding sites are not present in all LHCs, which suggests they are not crucial in the folding and stability of the monomers (Croce *et al.*, 1999; Caffarri *et al.*, 2001). Lhcb1 could be refolded *in vitro* with one or both sites empty (Formaggio *et al.*, 2001).

Besides lacking the violaxanthin domain, CP29 also is deficient in chl601 and chl605 found in LHCII (Fig. 2.3B; Liu *et al.*, 2011; Pan *et al.*, 2011). Reconstitution experiments of CP29 showed that it binds only eight chlorophyll molecules (six chlorophyll *a*, two chlorophyll *b*). *In vitro* pigment composition of CP29 was quite sensitive to the chlorophyll *a*/*b* ratio used in the pigment mixture (Giuffra *et al.*, 1997; Pagano *et al.*, 1998). The four central sites were coordinated by chlorophyll *a* (Pascal *et al.*, 2002), while the other four sites had mixed occupancy (Bassi *et al.*, 1999). CP24 is a very special minor antenna complex, found only in land plants, suggesting a more recent evolution (Ballottari *et al.*, 2012). It was reconstituted *in vitro* with 10–11 chlorophylls, with a chlorophyll *a*/*b* ratio of 1.0, the lowest ratio found in any LHCb. CP24 contains only two luteins, on the sites Lut620 and Lut621 (Passarini *et al.*, 2009); it lacks neoxanthin, and can be reconstituted without lutein. CP26 reconstitution showed that it binds 6 chlorophyll *a*, 3 chlorophyll *b*, and can bind lutein, violaxanthin or zeaxanthin (Croce *et al.*, 2002). More recent data suggest it binds 13–14 chlorophylls. It has been proposed that CP26 contains a violaxanthin site similar to LHCII (Ruban and Horton, 1999), as well as sites for Lut620, Lut621, and neoxanthin; however, CP26 was co-purified with only three carotenoids (Dall'Osto *et al.*, 2005).

LHCI complexes also have been reconstituted *in vitro* and characterized (Croce *et al.*, 2002; Castelletti *et al.*, 2003). Both dimers contain a chlorophyll *a*/*b* ratio of 3.7 and a chlorophyll/car ratio of 4.6–4.8 (Wientjes *et al.*, 2011). Twenty-eight chlorophyll molecules were found to bind to the Lhca1/Lhca3 dimer and 29 chlorophyll molecules to the Lhca2/Lhca4 dimer (Qin *et al.*, 2015). Lhca2 and Lhca4 have higher affinity for chlorophyll *b* than Lhca1 and Lhca3 in agreement with reconstitution experiments (Croce *et al.*, 2002; Castelletti *et al.*, 2003). A striking feature of Lhca is the presence of red chlorophyll forms that are important for energy trapping and energy transfer (Morosinotto *et al.*, 2003). Lhca antenna complexes contain less chlorophyll *b* than Lhcb (Croce *et al.*, 2002), and instead of neoxanthin that is found in Lhcb, Lhca binds β-carotene.

In vivo assembly of LHC

The high instability of chlorophyll-binding proteins when depleted of chlorophyll (Horn and Paulsen, 2002), together with the danger of ROS generation due to the accumulation of unquenched chlorophyll molecules (Krieger-Liszkay *et al.*, 2008) requires a high coordination between both pathways to avoid futile cycles and undesired reactions. In *C. reinhardtii*, accumulation of LHC does not occur *in vivo* in the absence of chlorophyll or xanthophylls, although its mRNA is present (Herrin *et al.*, 1992). The nuclear-encoded apoproteins of LHCII are synthetized on cytosolic ribosomes as hydrophilic precursors and imported into the chloroplast via the translocons TOC (translocon on the outer chloroplast membrane) and TIC (translocon on the inner chloroplast membrane) within the outer and inner chloroplast envelope, respectively (Andres *et al.*, 2010; Kovacs-Bogdan *et al.*, 2010). Once within the stroma, the transit peptide is removed by the stromal processing peptidase (SPP) (Schunemann, 2004) and the protein targets the thylakoid membranes through the chloroplast signal recognition particle (cpSRP) (Schunemann, 2004). This cpSRP is a heterodimer composed of a conserved 54 kDa protein (cpSRP54) and a 43 kDa subunit (cpSRP43) unique to chloroplasts (Nguyen *et al.*, 2011). A model proposes that the precursor LHC traverses the chloroplast envelope, becoming a 'soluble complex' in the presence of cpSRP. In addition to cpSRP, the receptor cpFtsY helps target the LHC precursor to the translocon, composed of the Alb3 (Albino3) translocase – an integral multi-spanning thylakoid membrane protein (Kogata *et al.*, 1999; Moore *et al.*, 2000, 2003). Alb3 catalyses the GTP-dependent co-translational targeting of the protein into the thylakoids, which will form the mature complex after binding the pigments (Zhang *et al.*, 2009; Nguyen *et al.*, 2011). The *tla3* mutant in *C. reinhardtii* was mapped as a homologous gene of cpSRP43; it shows strong reduction of Lhcb (Kirst *et al.*, 2012). An Alb3.1 mutant of *C. reinhardtii* contains strongly reduced amounts of Lhca, Lhcb, PSII, and chlorophyll (Bellafiore *et al.*, 2002). While the *A. thaliana* double mutant, depleted of cpSRP54 and cpSRP43, appears pale yellow, with a drastically reduced amount of chlorophyll and carotenoids (Hutin *et al.*, 2002). Interestingly, in this double mutant, LHC is unevenly distributed, lacking Lhcb3 and Lhca1; while Lhcb1 and Lhcb6 are partially retained, and Lhcb4 is enriched (Hutin *et al.*, 2002). Notably, some of the light-harvesting like (LIL) proteins [i.e. the early light induced proteins (ELIPs)] and the PsbS (four transmembrane helices), which show high homology to LHC proteins but have no function in light harvesting, do not require cpSRP for their insertion (Kim *et al.*, 1999). Their amounts also were found to be diminished in the double mutant cpSRP54/cpSRP43 (Hutin *et al.*, 2002) indicating that more than one targeting mechanism of LHC and LIL exists in the chloroplast, although this still need to be experimentally confirmed.

Hoober and co-workers (Hoober *et al.*, 2007) proposed another scenario for LHC assembly, which occurs in the chloroplast envelope. When helix A of Lhcb exits the TIC translocon on the stromal site of the chloroplast envelope, its transit sequence is removed; further transport of the LHC precursor through the envelope is arrested by binding chlorophyll *a* to the conserved motif located in the helix A (Eggink and Hoober, 2000; Hoober and Eggink, 2001). Since chlorophyll *b* has been shown to be important in stabilizing the LHC in higher plants (Thornber and Highkin, 1974; Reinbothe *et al.*, 2006), it has been proposed that chlorophyll *b* could bind to the site chl601 at the N-terminus of the protein. This would provide an additional anchor for the protein within the chloroplast envelope (Hoober *et al.*, 2007). An additional chlorophyll *b* could bind to chl605 at the luminal site of helix B. However, CP29 does not possess the chl601 or the chl605 binding sites (Fig. 2.3), although it is reported to need chlorophyll *b* for stable accumulation in the chloroplast (Thornber and Highkin, 1974). Alternatively, during protein import and before the formation of the cross-helix Glu-Arg ion pairs, chlorophyll *b* may be bound by chl602 to enter the chloroplast envelope. Subsequent folding of the protein would create an environment more favourable for chlorophyll *a*, which would then displace chlorophyll *b* (Hoober and Argyroudi-Akoyunoglou, 2004). Even in CP26 chl601 or chl605 are missing; in contrast to CP29, CP26 accumulates to WT level in the absence of chlorophyll *b*. Chl601 and chl605, therefore, cannot be crucial for retaining chlorophyll-binding proteins in the chloroplast, and chlorophyll *b* might not be that important in stabilizing LHCs during their import into the chloroplast. In fact, it was shown that chlorophyll *b* depletion in a CAO mutant did not impair import of the LHC precursor, its processing or insertion into the thylakoid membrane (Nick *et al.*, 2013). In *C. reinhardtii*, the lack of chlorophyll *b* does not impair assembly of any LHC (Polle *et al.*, 2000). In conclusion, an exhaustive picture of LHC import mechanism can still not be drawn.

While the insertion of chlorophyll into the protein scaffold has been the focus of detailed studies, not much information is available on the insertion of carotenoids *in vivo*. The *A. thaliana* mutant npq2lut2, which is unable to synthetize lutein and violaxanthin, has a decreased PSII antenna size because of the destabilized LHCII (Havaux *et al.*, 2004). Several mutations in xanthophyll biosynthesis, leading to decreased xanthophyll/carotenoids ratios, showed a strong decrease in the ratio of LHCII to PSII, but no change in LHCI to PSI stoichiometry (Fiore *et al.*, 2012). This effect could be partially explained by the presence of β-carotene in LHCI, which does not occur in the antenna of PSII.

Assembly of photosystem II core complex

The assembly of PSII is a stepwise process, where subcomplexes are initially formed and later fused to form the active complex. In *Synechocystis* 6803, this process has been well studied, since mutants unable to grow autotrophically can be grown under mixo- or heterotrophic conditions. This stepwise mechanism of PSII assembly seems to be largely conserved in the chloroplasts of green algae and plants (Fig. 2.4). Numerous accessory proteins are involved in this assembly binding only transiently to the growing complex, and cannot be detected in the X-ray structure of mature PSII (Nixon *et al.*, 2010; Komenda *et al.*, 2012; Nickelsen and Rengstl, 2013). Some of them are conserved in both cyanobacteria and chloroplasts, confirming their importance in this process. Cytochrome (Cyt) b_{559} functions as a nucleation factor in PSII assembly, to which D2, PsbE and PsbF attach, forming the D2/cyt b_{559} sub-complex (Komenda *et al.*, 2004). The D1 precursor, together with PsbI, attach later, forming a PSII-like RC (Dobakova *et al.*, 2007). During the formation of the PSII-like RC, the D1 precursor is processed by proteases, yielding mature D1 (reviewed in Kmiec *et al.*, 2014). In cyanobacteria, the CP47-PsbH/PsbL/PsbT subcomplex was found to bind to the PSII-like RC in the absence of CP43, forming the so-called RC47 complex. In chloroplasts, PsbH joins this complex after CP47 (Rokka *et al.*, 2005; Boehm *et al.*, 2012b). The subsequent attachment of the sub-complexes CP43-PsbZ/PsbK/Psb30 in cyanobacteria, or CP43/PsbK in *C. reinhardtii*, forms monomeric PSII, inducing assembly of the OEC, which consists of PsbO, PsbP, PsbU, PsbQ and PsbV in cyanobacteria or PsbO, PsbP and PsbQ in plants (Bricker *et al.*, 2012). Dimerization occurs as the last step, with the involvement of several low-mass subunits (PsbI, PsbM and PsbW). At the same time, the antenna system is attached, i.e. the peripheral phycobilisomes in cyanobacteria, or the

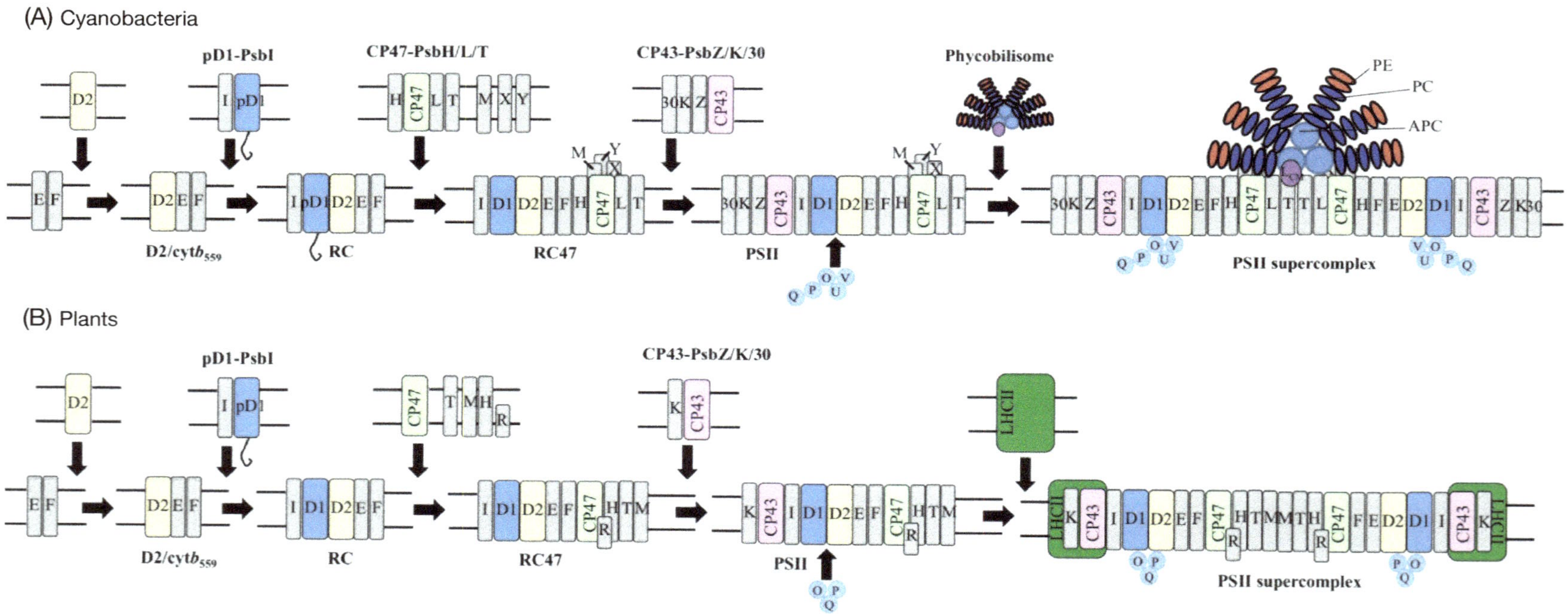

Figure 2.4 Model for the assembly of photosystem II in cyanobacteria (A) and higher plants (B). The D2 subunit forms a pre-complex with PsbE, PsbF, and subunits of cyt b_{559}, called the D2/cyt b_{559} complex. While the pD1-PsbI pre-complex is formed by the D1 precursor and PsbI. The D2/cyt b_{559} complex binds to the pD1-PsbI pre-complex, forming the reaction centre (RC) complex. PsbH, PsbL and PsbT bind to CP47 and then attach to the RC complex, forming RC47 in cyanobacteria. PsbH joins CP47 later in plants. Monomeric PSII is formed by a CP43-complex, containing CP43 and other small subunits. The final step prior the PSII dimerization is the assembly of the Oxygen Evolving Complex (OEC), followed by the attachment of the phycobilisome antenna in cyanobacteria and LHCs in plants. The major subunits are: D1 in cyan, D2 in yellow, CP43 in pink, CP47 in light green, and small subunits in grey. LHCII are in dark green. The letters or numbers at the ends of their respective Psb protein names (e.g. I for PsbI) designate PSII subunits, except for the chlorophyll-binding subunits that are designated by their common names (D1, D2, CP47, and CP43). Abbreviations: cyt b_{559}, cytochrome b_{559}; LHCII, light-harvesting complex II; pD1, precursor form of D1; RC47, reaction–centre complex lacking CP43.

transmembrane LHC in green chloroplasts (Kouril *et al.*, 2012).

Cyanobacterial and plastid-encoded membrane proteins, such as the apoproteins of the core complexes, are inserted co-translationally into the membrane (Zhang *et al.*, 1999). For example, D1 protein was found attached to polysomes in cyanobacteria (Tyystjarvi *et al.*, 2001). During translation, nascent D1 protein interacts with components of cpSRP54 and the secretory TM protein transport systems for chloroplast proteins (Nilsson *et al.*, 1999; Zhang *et al.*, 1999). The members of the machinery that inserts membrane proteins into the membrane have been identified as SRP, FtsY, SecYEG, YidD/Alb3; they share similar components with the LHC assembly machinery (see above). When the first segment of a membrane protein is translated, the ribosome–SRP complex is targeted to the membrane-associated receptor FtsY. This complex is transferred to the translocation apparatus; the SecYEG translocon forms the essential channel to insert the protein through the lipid bilayer. In association with a translocon, SecYEG membrane proteins are inserted into the membrane and properly folded (Sobotka, 2014). YidC is involved in partitioning the membrane segment and is able to work in a Sec-independent manner (Welte *et al.*, 2012). It seems that an important role for the assembly of chlorophyll-binding proteins into the membranes is accomplished by the YidC/Alb3 insertase. Direct interaction between YidC/Alb3 and the proteins D1, D2, CP43, and PsaA was shown in *A. thaliana* (Pasch *et al.*, 2005); these proteins also were shown to be involved in post-translational targeting of LHCs into the membranes (Moore *et al.*, 2003). Recently, a complex containing chlorophyll synthase (ChlG), one-helical LIL proteins (SCPs, see below), the Ycf39 protein, the YidC/Alb3 insertase, and pigments have been co-isolated from *Synechocystis* 6803 membranes. YidC/Alb3, therefore, may assist chlorophyll insertion into the apoprotein (Chidgey *et al.*, 2014). Pigment availability is an important factor limiting the accumulation of PSII core complexes. PSII function and assembly relies on carotenoids that mainly influence the synthesis and integration of CP47 and CP43, without affecting PsaA, PsaB, D1 or D2 (Sozer *et al.*, 2010). Previous work has shown that *Synechocystis* 6803 and *C. reinhardtii* depleted of the enzyme protochlorophyllide oxidoreductase (POR) that mediates the conversion of protochlorophyllide *a* to chlorophyllide *a*, are unable to accumulate PSII in darkness (Malnoe *et al.*, 1988; He and Vermaas, 1998). However, cyanobacterial RC-like complexes are able to perform light-induced charge separation, indicating that pigments are bound to these D1/D2 complexes at early stages of assembly (Keren *et al.*, 2005). Also, non-assembled complexes of *Synechocystis* 6803 were isolated using His-tagged CP47 and CP43, and shown to bind molecules of β-carotene and chlorophyll (Boehm *et al.*, 2011). Therefore, pigment binding may stabilize these sub-complexes.

Psb28 has been proposed to link chlorophyll synthesis and insertion into the protein scaffold; since a *Synechocystis* 6803 mutant deficient in Psbs28 was shown to contain reduced amounts of CP47, PsaA and PsaB (Dobáková *et al.*, 2009). POR may also be involved in the chlorophyll insertion in PSII, via a factor called Pitt. Pitt may stabilize POR and help it to position POR in the correct orientation for chlorophyll insertion into the apoprotein (Schottkowski *et al.*, 2009).

Chlorophyll storage proteins – the LIL family

An active exchange of chlorophyll molecules between different chlorophyll-binding proteins has been proposed for both plants and cyanobacteria (Tzinas *et al.*, 1987; Kada *et al.*, 2003). It has been calculated that the average lifetime of a chlorophyll molecule is twice as long as the protein subunits in a photosystem, indicating that chlorophyll has to be recycled and transferred between different protein subunits (Vavilin *et al.*, 2007; Yao *et al.*, 2007, 2012). CP47 was proposed to function as a hub to store and deliver chlorophyll molecules to the pigment-binding proteins (Sobotka *et al.*, 2008a; Dobáková *et al.*, 2009; Boehm *et al.*, 2011). LIL proteins are highly homologous to the LHC proteins, and have been proposed to function similarly. LIL proteins also are thought to be involved in photo-protection (Jansson, 2006), rather than in harvesting light energy. The large family of LIL proteins in plants consists of proteins predicted to have four (PsbS; Funk *et al.*, 1995a,b, 2001), three (ELIPs; Adamska *et al.*, 2001), two (stress enhanced proteins, SEPs; Heddad and Adamska, 2002), or one membrane spanning helix [one-helix proteins

(OHPs), high-light-induced proteins (HLIPs) or small CAB-like proteins (SCPs)] (Funk and Vermaas, 1999; Heddad and Adamska, 2000; Jansson *et al.*, 2000). The LIL proteins are expressed under different stress conditions in an opposite manner to the LHC proteins; i.e. under high-intensity light conditions, when the expression of the LHC proteins is repressed, the LIL proteins are up-regulated. This suggests they function in high-intensity light protection in the broadest sense; they may provide direct protection (as shown for PsbS; Niyogi *et al.*, 2005), affect pigment stability and/or act as pigment carriers (Xu *et al.*, 2002, 2004; Rossini *et al.*, 2006). They may even have novel functions, yet not determined. It is likely they all bind pigments (chlorophylls and carotenoids; Funk, 2001), although indisputable evidence for this is lacking. One-helical LHC-like proteins have been detected in all organisms performing oxygenic photosynthesis, while multi-helix LIL proteins are absent in cyanobacteria (Funk and Vermaas, 1999; Bhaya *et al.*, 2002). The single transmembrane domain of the cyanobacterial SCPs shows significant sequence similarity to the first and third membrane-spanning region of proteins belonging to the LHC family and may be the evolutionary ancestor of the LHC proteins (Green *et al.*, 1991). It has been proposed that LIL proteins evolved after a series of duplications, fusions, and deletion of an ancestral *SCP*-like gene (Green and Pichersky, 1994), and only later gained their light-harvesting function (Jansson, 2006). Another model based on more recent phylogenetic analyses suggests that the two-helix SEPs are the best candidates for the ancestor of the multi-helix LHC/LIL proteins (Engelken *et al.*, 2010).

The importance of the HLIPs/SCPs to adapt and survive to stress conditions is evident in the much reduced genome of the high-light ecotype *Prochlorococcus marinus* strain Med4, which contains 24 *scp*-genes (Bhaya *et al.*, 2002). SCPs also have been identified in cyanophages of marine cyanobacteria (Bailey *et al.*, 2004; Lindell *et al.*, 2004). In *Synechocystis* 6803, the SCPs were shown to play an important role in chlorophyll recycling. In *Synechocystis* 6803, five SCPs were identified (ScpA–E). ScpB-E (HliA–D) are proteins of around 6 kDa, with a single transmembrane helix containing the CAB domain; they have been shown to bind chlorophyll and carotenoids (Funk and Vermaas, 1999; Storm *et al.*, 2008; Knoppova *et al.*, 2014). ScpA is actually the C-terminal domain of the ferrochelatase (FC, HemH), the enzyme inserting Fe^{2+} into the porphyrin ring during haem biosynthesis (Woodson *et al.*, 2011). Interestingly, two different FCs exist in plants, one enzyme is located in the mitochondria, while the other is found in plastids. Only the plastid-imported paralogue contains the C-terminal extension with the CAB domain. This CAB domain regulates the activity and oligomerization of FC (Sobotka *et al.*, 2008b; Storm *et al.*, 2013).

SCPs temporal association with PSII have been proven by various studies (Promnares *et al.*, 2006; Yao *et al.*, 2007; Kufryk *et al.*, 2008; Shi *et al.*, 2012). ScpD and ScpC were found to be located closely to CP47 and PsbH (Promnares *et al.*, 2006; Yao *et al.*, 2007); they have been proposed to stabilize PSII assembly/repair recycling of chlorophyll, therefore, avoiding oxidative damage (Hernandez-Prieto *et al.*, 2011; Sinha *et al.*, 2012; Yao *et al.*, 2012; Tibiletti *et al.*, 2016). ScpB and ScpE have been proposed to regulate chlorophyll biosynthesis at the 5-aminolaevulinic acid (ALA) step (Xu *et al.*, 2002; Xu *et al.*, 2004), and recently they have been found to be associated with unassembled D1 (Knoppova *et al.*, 2014). Finally, SCPs have been proposed to function in light dissipation (Havaux *et al.*, 2003).

Two groups of OHPs can be distinguished in plants: ones with prokaryotic origin, such as the OHP1/HLIP/SCP-type present in cyanophages, cyanobacteria, and plants; and the OHP2-type, restricted to eukaryotes (Andersson *et al.*, 2003). While some OHPs are located in the plastid genome, others are nuclear encoded (Neilson and Durnford, 2010). In the cryptophyte *Guillardia theta*, which occurs after two endosymbiotic events, one of its two OHPs was found to be encoded in the reduced genome from a red algal nucleus (termed the nucleomorph; Douglas *et al.*, 2001), while the other is encoded by the chloroplast genome (Douglas *et al.*, 2001; Neilson and Durnford, 2010). Based on their expression pattern, they do not seem to be involved in photo-protective mechanisms (Funk *et al.*, 2011). In *A. thaliana*, two genes codify the one-helix proteins, *Ohp1* and *Ohp2*; *Ohp2* is expressed in low light, although both its transcription and translation were enhanced in high light-exposed plants. OHP2 was found to be associated with PSI in high light conditions (Andersson *et al.*, 2003).

Our knowledge of SEPs proteins is sparse. In

A. thaliana, six two-helical proteins, named Sep1, Sep2, Sep3-1 (Lil3:1), Sep3:2 (Lil3:2), Sep4 and Sep5, have been detected. Transcripts of Sep1 and Sep2 were detected in plants grown at low light, but increased on exposure to high light (Heddad and Adamska, 2000). Lil3 transcript remained unchanged when plants were transferred to high light (Jansson, 1999). Lil3 was found to bind chlorophyll *a*, protochlorophyll *a*, and carotenoids in barley seedlings during de-etiolation (Reisinger *et al.*, 2008). Both isoforms of Lil3 in *A. thaliana* were found to stabilize geranylgeranyl reductase (GGR), the enzyme synthesizing phytol-pyrophosphate, an intermediate of the chlorophyll and tocopherol biosynthesis (Tanaka *et al.*, 2010). Recently, the Lil3 isoforms were found associated with the LHCII antenna; it was proposed that they could provide a link between chlorophyll biosynthesis and the chlorophyll-binding protein assembly, similar to SCPs (Lohscheider *et al.*, 2015).

Historically, the ELIPs were the first LIL family members to be discovered in etiolated pea and barley seedlings exposed to high light (Meyer and Kloppstech, 1984; Green *et al.*, 1991). ELIPs have three transmembrane helices, and like the LHC proteins, the first and the third helices are homologous to each other. ELIPs are present in Viriplantae, but they are absent in cyanobacteria, as well as from members of the red chloroplast evolution branch (Neilson and Durnford, 2010). They accumulate transiently during different stress conditions that would cause photo-inhibition (Heddad and Adamska, 2000, 2002). In *A. thaliana*, two *ELIP* genes (Elip1 and Elip2), with 81% sequence similarity, are differentially expressed during high light, greening, and senescence (Montane and Kloppstech, 2000; Heddad *et al.*, 2006). ELIPs have been proposed to prevent photooxidative damage during stress exposure by binding free chlorophyll and/or by dissipating excess energy to protect PSII (Hutin *et al.*, 2003). However, an *A. thaliana elip1/elip2* double mutant was not sensitive towards high light, suggesting the existence of other compensatory processes (Casazza *et al.*, 2005; Rossini *et al.*, 2006). Measurements carried out on an Elip2 overexpressor mutant suggest that ELIPs decrease photo-inhibition by interfering with the chlorophyll biosynthesis pathway (Tzvetkova-Chevolleau *et al.*, 2007). ELIPs were shown to associate with different LHCII subpopulations. In pea plants, they could be isolated with chlorophyll *a* and bound lutein (Adamska *et al.*, 1999; Heddad *et al.*, 2006); however, evidence for *in vivo* pigment binding is still lacking.

The PsbS protein, spanning the thylakoid membrane with four helices, originally was proposed to function as a chlorophyll storage protein. It has been found to bind pigments (Funk *et al.*, 1994, 1995b), but is also stable in the absence of pigments (Adamska, 1995; Funk *et al.*, 1995a). Today, the role of PsbS in non-photochemical quenching, a photo-protective mechanism to dissipate excitation energy, is well established (Muller *et al.*, 2001). Although its biochemical role is unclear, lumen acidification triggers the protonation of two lumen-exposed glutamate residues on PsbS that induce conformational changes in the protein that might open quenching sites on the antenna of PSII super-complex in the grana (Bonente *et al.*, 2008a; Betterle *et al.*, 2009). Not only higher plants, but even moss and green algae, have at least one *PsbS* gene. Interestingly, in the unicellular green alga *C. reinhardtii*, the PsbS protein has never been found to be expressed (Bonente *et al.*, 2008b). Instead, in *C. reinhardtii* and the diatom *Pheodactylum tricornutum*, another LIL protein LHCRS (or Li818), absent in vascular plants, was found to be necessary for non-photochemical quenching (Peers *et al.*, 2009; Bailleul *et al.*, 2010). LHCRS is not present in higher plants (Neilson and Durnford, 2010). It comprises three transmembrane helices, and is able to bind pigments (Bonente *et al.*, 2011). LHCRS proteins seem to be more ancient (found in green algae) than PsbS of higher plants, and mosses are able to perform both the PSBS-dependent and the LHCSR-dependent non-photochemical quenching processes independently of each other (Alboresi *et al.*, 2010; Gerotto *et al.*, 2011, 2012).

Spatial organization of PSII assembly

Recent progress has been made on the localization of PSII biogenesis centres in cyanobacteria and plants (reviewed in Rast *et al.*, 2015). In cyanobacteria, PSII assembly was detected in special biogenesis centres connecting the thylakoid membrane and the periplasmic membrane (Stengel *et al.*, 2012). In plants, PSII assembly seems to be located in the stroma-exposed thylakoid regions. Immunological studies identified the cyanobacterial PSII subunits D1, D2 and cyt b_{559} in the periplasmic membrane, while CP47 and CP43 were localized only in the thylakoid membrane (Zak *et al.*, 2001). A membrane sub-fraction, called PratA-defined membrane or thylakoid centre (Stengel *et al.*, 2012), was identified containing pD1, PratA (an early assembly factor of PSII), POR, together with Pitt and chlorophyllide *a*. This PratA-defined membrane seems to be important, not only for chlorophyll synthesis, but also for the integration of pigment into apoprotein (Schottkowski *et al.*, 2009; Armbruster *et al.*, 2010). The importance of spatial differentiation between PSII assembly and PSII activity was highlighted by the finding of special assembly centres in the cyanobacterium *Gloeobacter violaceus*, which does not contain any physical separation between the periplasmic and thylakoid membranes (Rexroth *et al.*, 2011).

In *C. reinhardtii* special translation zones have been shown to be the site of PSII biogenesis (Uniacke and Zerges, 2007). Two models have been proposed to explain assembly of chlorophyll-binding protein in plants; one suggests the thylakoid membranes as the site of assembly (Keegstra and Cline, 1999), while the other model proposes assembly of the photosynthetic units in the inner envelope membrane surrounding the chloroplast (Hoober *et al.*, 2007). The envelope is the major source of lipids, and clearly the site where the latter steps of chlorophyll synthesis occur (see below). Immuno-labelling experiments in *C. reinhardtii* cells identified LHC precursors along the chloroplast inner envelope (White *et al.*, 1996). Transportation of pigment-binding proteins or pigment-storage proteins from the inner envelope to the thylakoid membrane occurs via vesicles. The protein VIPP1 was postulated to mediate such vesicle formation (Kroll *et al.*, 2001). However, recent data indicate that VIPP inserts lipid molecules in the newly assembled complex (Kroll *et al.*, 2001).

Cyanobacteria, algae and plants synthesize chlorophyll and haem via a common branched pathway (Tanaka and Tanaka, 2011). Not only chlorophyll, but also intermediates of the chlorophyll biosynthetic pathway can potentially cause cell damage. Therefore, the existence of chlorophyll biosynthesis centres have been postulated (Shlyk, 1971), containing all enzymes involved in the chlorophyll biosynthetic pathway. The first step in the process is the synthesis of ALA from glutamate, which involves three enzymes: glutamyl-tRNA-reductase (GluTR), glutamate-1-semialdeyde-aminotransferase (GSAT), and glutamyl-tRNA-synthase (GluTS). GluTR and GSAT were co-immunoprecipitated, but also GluTS may belong to this complex (Nogaj and Beale, 2005). Protoporphyrinogen oxidase (PPOX), the last common enzyme between the haem and the chlorophyll branch, was identified with FC (Masoumi *et al.*, 2008). Mg-chelatase and Mg-protoporphyrin methyltransferase (Mg-PMT) were found to interact with each other (Shepherd *et al.*, 2005), as were POR and the Mg-protoporphyrin IX methylester cyclase (Mg-PVC, Kauss *et al.*, 2012). Finally, POR and chlorophyll-synthase, as well as chlorophyll-synthase and geranylgeranyl reductase (GGR) were proposed to interact via substrate channelling (Domanskii *et al.*, 2003; Rudiger *et al.*, 2005). However, some enzymes involved in the synthesis of ALA were found to be soluble, while others are membrane bound (Manohara and Tripathy, 2000). Transient interactions of the stromal proteins and the intrinsic ones also has been observed. GUN4, for example, is involved in anchoring the stromal Mg-chelatase into the thylakoid membrane (Sobotka *et al.*, 2008a). In the cyanobacterium *Thrichodesmium eritreum*, GluRS was found to be attached to the membrane via an evolutionary conserved protein (Olmedo-Verd *et al.*, 2011). Thus, it is a feasible hypothesis that protein synthesis and translation could assemble in the chlorophyll-biosynthesis centres to synchronize pigment and protein production.

Assembly of photosystem I core complex

Compared with the detailed knowledge of PSII assembly, PSI assembly is not well understood, although considerable progress has been made in the recent years (reviewed by Schottler *et al.*, 2011; Yang *et al.*, 2015) (Fig. 2.5). The first step in PSI biogenesis is the co-translational membrane insertion of PsaB into the thylakoids, which is essential for the translation and integration of PsaA. Subsequently, the two subunits form the large heterodimer (Wollman *et al.*, 1999; Wostrikoff *et al.*, 2004). This early step is assisted by the chaperones Alb3, VIPP1, and PPD1, protecting the nascent protein from unspecific binding to other membrane proteins during folding (Zhang *et al.*, 2012). The phylloquinone and 4Fe–4S (Fx) cluster are integrated into this

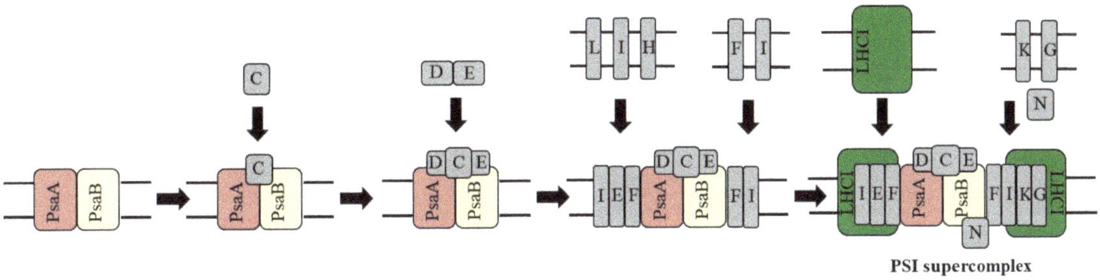

Figure 2.5 Model for the assembly of photosystem I in higher plants. The first step involves the formation of the reaction centre heterodimer by PsaA and PsaB. Later, PsaC, PsaD and PsaE assemble on the stromal side. Between these small subunits, PsaH, PsaI, PsaL and PsaJ bind and also other small subunits, forming the complex. PsaA is shown in red, PsaB in yellow, LHCI in green, and small subunits in grey. Subunits are designated by the letters after their respective Psb protein names, except for PsaA and PsaB.

heterodimer (Shen *et al.*, 2002a,b). Mediated by Ycf3 and Y3IP1, PsaC assembles with the PsaA/PsaB heterodimer, induced by conformational changes, then PsaD and PsaE attach at the stromal side. Subsequently, PsaF is inserted, with the help of Ycf4 (Ozawa *et al.*, 2009). Later, other small subunits, such as PsaH, PsaI, PsaL and PsaJ in plants, bind to this intermediate complex PsaL and PsaJ (Ozawa *et al.*, 2010). The binding of LHCI and the remaining small subunits (PsaN, PsaK, PsaG, PsaO, PsaP) seems to be a slower process, although their assembly pathway is still unknown; Alb3 may assist their integration into the complex (Chidgey *et al.*, 2014).

Retrograde signalling

The expression of genes encoding chlorophyll-binding proteins and enzymes involved in chlorophyll biosynthesis has to be tightly coordinated within a cell to avoid photooxidative damage. Enzymes of the chlorophyll biosynthesis as well as the LHC encoding genes are located in the nucleus. Therefore, enzymes of chlorophyll biosynthesis and LHC precursors have to be post-translationally imported into the chloroplast where they are functionally active; this means that a strict communication pathway between the chloroplast and the nucleus is required. What is known as retrograde signalling is the feedback control that reports the functional state of the organelle to the nucleus (Kropat *et al.*, 2000; Strand, 2004; Fernandez and Strand, 2008). Chlorophyll precursors have been shown to be important in such chloroplast-to-nucleus signalling. Mg-protoporphyrin IX (Mg-PPIX) was proposed to be directly involved in this signalling, since all disturbances in the chlorophyll biosynthesis pathway lead to accumulation of Mg-PPIX, provoking a down-regulation of the LHC genes in wild type *A. thaliana* (Strand *et al.*, 2003). For instance, transcription of the nuclear-encoded genes Hsp70a/b/c was induced after feeding *C. reinhardtii* with Mg-PPIX (Kropat *et al.*, 2000). Moreover, several *A. thaliana* lines with mutations in the tetrapyrrole biosynthesis were shown to be *gun* mutants; *gun2* and *gun3* encode the haem oxygenase (HO) and phytochromobilin synthase, respectively; while *gun4* is an activator of the Mg-chelatase (Larkin *et al.*, 2003, Fig. 2.6). GUN4 is proposed to regulate the chlorophyll biosynthesis flux by binding Mg-PPIX and may also export Mg-PPIX from the chloroplast to the cytosol, where it is bound to signalling proteins or could act as a transcription factor (Strand, 2004). However, experimental proof of either hypothesis is still lacking. An orthologue to *gun4* was found in *Synechocystis* 6803, its disruption led to decreased cellular chlorophyll (Wilde *et al.*, 2004). The *A. thaliana gun5* mutant carries a single amino acid mutation in the ChlH subunit of the Mg-chelatase, leading to a deficiency of Mg-PPIX. Recently the importance of Mg-PPIX as a signal molecule was questioned in plants, since the correlation between Mg-PPIX and a decrease of expression of Lhcb genes was not replicable (Mochizuki *et al.*, 2008; Moulin *et al.*, 2008). It is still not clear why the mutation in Mg-PVC cyclase did not give a *gun* phenotype (Strand, 2004). Recently, the isolation of a *gun6* mutant was reported; this mutation was identified as being in the upstream regulatory region of the

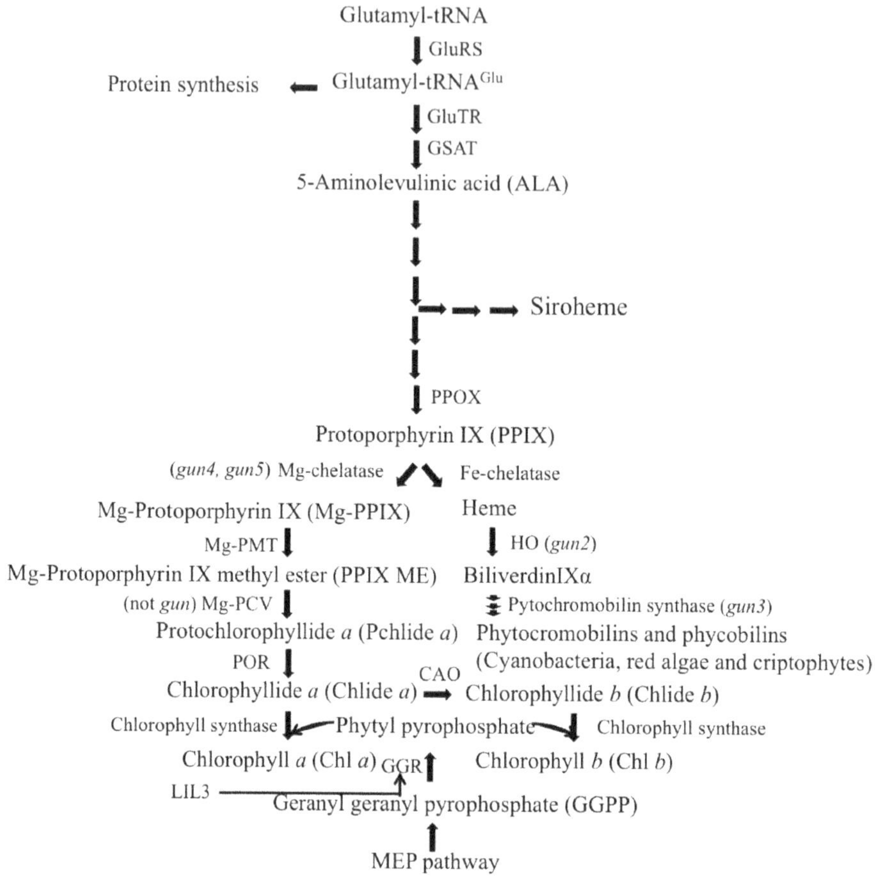

Figure 2.6 The biosynthetic pathway of tetrapyrroles in photosynthetic organisms. Glutamyl-tRNA is converted to aminolaevulinic acid (ALA) by a 3-step process, involving the three enzymes, GluRS, GSAT and GluTR; this is an important control point on this pathway. After several modifications, ALA is converted to protoporphyrin IX (PPIX), the common substrate used by either Fe-chelatase or Mg-chelatase. Fe-chelatase inserts Fe^{2+} into the porphyrin ring of PPIX within the haem pathway, while Mg-chelatase inserts Mg^{2+} within the chlorophyll pathway. Mg-PPIX formed by Fe-chelatase is converted to Pchlide (protochlorophyllide) via two enzymatic reactions. PPIX ME is the intermediate of these two reactions. Pchlide is converted to Chlide *a* (chlorophyllide *a*) by POR. In organisms containing chlorophyll *b*, CAO converts Chlide *a* to Chlide *b*. Both are phytylated by GGR (geranylgeranyl reductase), using phytyl pyrophosphate of the non-mevalonate pathway (MEP pathway). POR, NADPH:protochlorophyllide oxidoreductase; CAO, chlorophyllide *a* oxygenase.

gene encoding FC (Woodson *et al.*, 2011). The mutation causes the *gun6* mutant to overexpress FC, causing an accumulation of haem, which may act as a retrograde signal molecule, carrying out positive regulation of the nuclear-encoded genes. Haem is a good candidate for a signal molecule, as it is known to be exported from the plastid and it is not photodynamic like Mg-PPIX (Thomas and Weinstein, 1990). A role for haem as a positive regulator of nuclear-encoded genes involved in photosynthesis may also provide an explanation for the behaviour of other *gun* mutants (Woodson *et al.*, 2011).

To avoid photooxidative damage, chlorophyll has to be located in proximity to a quenching molecule like a carotenoid. Therefore, the carotenoid biosynthetic pathway has to be linked to chlorophyll and chlorophyll-binding protein biosynthesis. Mutants with carotenoid deficiency accumulate less *lhc* transcripts (Mayfield and Taylor, 1984). In mature chloroplasts, chlorophyll seems to regulate the activity of phytoene synthase (PSY), catalysing the condensation of geranylgeranyl diphosphate to phytoene (Fraser *et al.*, 2000). The next step of the carotenoid biosynthesis is the desaturation of phytoene to zeta-carotene performed by phytoene desaturase (PDS). This enzyme requires plastoquinones and an active alternative terminal oxidase (AOX) to generate electrons. A link between AOX and PDS has been observed in the variegated *immutans A. thaliana* mutants (Wetzel *et al.*, 1994). Thus, PDS seems to be a key regulatory component in the coordination of carotenoid biosynthesis, nuclear signalling, and chloroplast development.

Singlet oxygen and hydrogen peroxide can activate different pathways of chloroplast-to-nucleus signalling (Meskauskiene *et al.*, 2001; op den Camp *et al.*, 2003). Hydrogen peroxide is a relatively stable product, able to traverse membranes; therefore, it was also suggested as a signalling compound (Kropat *et al.*, 2000). Singlet oxygen has a shorter lifetime and may react with cellular compounds directly at the place of its formation. The *flu* mutant in *A. thaliana* accumulates large amounts of free protochlorophyllide in darkness (Meskauskiene *et al.*, 2001). Transition of the mutant from dark to light led to an immediate release of singlet oxygen that caused an arrest of plant growth. Other studies showed that this effect was not directly dependent on the presence of singlet oxygen, but was activated by a genetically determined cell programme mediated by the gene *EXECUTER1* (*EX1*) (Wagner *et al.*, 2004). Inactivation of *EX1* resulted in a reduced sensitivity to singlet oxygen. Although *EX1* may transduce signals from the chloroplast to nuclear gene expression, it requires a linker within the envelope membrane, which is still not identified.

Optimal photosynthesis requires continuous acclimation to changing environmental conditions. Therefore, plants and algae adjust their PSII/PSI stoichiometry and change their antennae size. The redox state of the different components of the photosynthetic apparatus has been identified as a regulating parameter of these mechanisms. Therefore, the redox signal must be involved in chloroplast-to-nucleus communication. The impact of the redox state on the nuclear gene expression, in particular expression of the LHCs encoding genes, has been studied extensively (reviewed in Fey *et al.*, 2005). Several lines of evidence suggest that the phosphorylation cascade transduces the redox state (Durnford and Falkowski, 1997). The Stn7 kinase is involved in transducing the signal via phosphorylation to further signalling components (Bonardi *et al.*, 2005). Although candidates for sensing and transduction have been identified in the chloroplast, little is known about how the signal is transferred through the chloroplast envelope to the cytosol. Further experimental data are required to understand this process.

An interesting hypothesis about the correlation between plastid gene expression and Lhcb regulation has been developed by Brautigam *et al.* (2008), based on the observation that chlorophyll *b* made by CAO is necessary for the import of Lhcb (Reinbothe *et al.*, 2006). When the *trnE* gene (that encodes the glutamyl-tRNA) is not transcribed, chlorophyll biosynthesis is interrupted. A lack of chlorophyllide will cause the Lhcb precursor to remain attached to the translocon, and expression of *Lhc* genes in the nucleus will be repressed via a feedback mechanism. However, the localization of CAO has not been clarified, nor has the effectors between the chloroplast and nucleus; therefore many questions remain concerning this hypothesis.

Degradation of chlorophyll-binding proteins

Cells have evolved extensive molecular chaperones and proteases to control protein quality and prevent protein damage. Many factors trigger degradation of proteins, including changes in environmental conditions, genetic mutations, and limitations in the availability of cofactors. Cleavage by peptidases, which activates proteins, has been shown to be equally important in progressive degradation. Such 'housekeeping' by enzymes maintains stoichiometric amounts of protein subunits or controls metabolic pathways. In plants, light serves as an environmental signal that regulates physiological and developmental processes, and therefore chlorophyll-binding proteins are important substrates for regulation.

Genome information has revealed various plant proteases that are likely localized in the plant chloroplast (Garcia-Lorenzo *et al.*, 2006). Although proteases play crucial roles in many processes in plant cells, very little information is available on their substrate specificity or the physiological roles of various proteases in photosynthetic organisms. Despite identification of many genes with sequences similar to known proteases, protease activity has only been demonstrated for a few of these genes. Conversely, proteolytic processes are frequently observed in plant biology, but have rarely been ascribed to specific proteases. Even though biochemical evidence for most of proteases and peptidases is limited, specific enzymes have been identified in all major compartments of the chloroplast. Given the evolutionary origin of chloroplasts, it is not surprising that all plastid proteases identified to date are orthologues of bacterial proteases (reviewed in Adam *et al.*, 2001). However, plants contain a significantly higher number of these genes than bacteria, or mitochondria, suggesting that a diversification occurred during the evolution of photosynthetic organisms (Garcia-Lorenzo *et al.*, 2006). In this respect, cyanobacteria are interesting, as a link between plants and the non-photosynthetic prokaryotes, they exhibit a diversity of proteases similar to the range found in chloroplasts. About 3% of all genes of *A. thaliana* were found to code for putative proteases, while the single-celled cyanobacterium *Synechocystis* 6803 has more than 90 proteases. In the chloroplast, there are at least 11 different protease families represented (e.g. Kmiec *et al.*, 2014; Nishimura and van Wijk, 2015; Yoshioka-Nishimura and Yamamoto, 2014; Adam, 2015). Here, we will only summarize information available on the degradation of chlorophyll-binding proteins, rather than include the degradation of enzymes involved in pigment-synthesis (e.g. Nishimura and van Wijk, 2015).

Like all proteins, the chlorophyll-binding membrane proteins also undergo constant turnover: damaged or misfolded proteins are removed and the protein set is adjusted to changing conditions. However, the lipid bilayer prevents access to these proteins. Therefore, specialized membrane-bound proteases play a decisive role in membrane protein regulation. Surprisingly, of all the proteases present in the plant chloroplast, or in the cyanobacterial cell, only two proteases appear to be involved in the degradation of chlorophyll-binding proteins.

FtsH proteases

FtsH (filamentation temperature-sensitive) proteases are ATP-dependent metalloproteases found in eubacteria, mitochondria and chloroplasts (for recent reviews see Wagner *et al.*, 2012; Yoshioka-Nishimura and Yamamoto, 2014). The *FtsH* gene was first identified in *E. coli*, as an essential protein, involved in degrading various short-lived proteins (Santos and De Almeida, 1975). Later, eukaryotic FtsH homologues were detected in the mitochondria of yeast (Weber *et al.*, 1996), suggesting that this family of proteases might act as chaperones during assembly of membrane proteins (Rep *et al.*, 1996). Today, FtsH proteases have been found in eubacteria, mitochondria, and chloroplasts. FtsH proteases are membrane-inserted, attached by one or two transmembrane helices. They expose their main cytosolic region, consisting of AAA-ATPase and Zn^{2+}-metalloprotease domains, into the cytoplasm or stroma (Garcia-Lorenzo *et al.*, 2008). FtsH can work either as an exo- or endoprotease (Okuno *et al.*, 2006); therefore, it can easily degrade unnecessary membrane proteins. In the genome of *A. thaliana*, 17 *FtsH* genes have been identified, encoding 12 active FtsH proteases and five FtsH homologues, with mutations in their active site. Nine of these active FtsH target chloroplast, four target the mitochondria, and one, FtsH11, targets both organelles (see Wagner *et al.*, 2012). Within the chloroplast, the FtsH proteases FtsH1, 2, 5, 6 and 8 are localized in the thylakoid membrane, with their ATPase and proteolytic domains exposed to the stroma. In contrast, proteomic studies have identified FtsH7, 9, 11, 12 in the chloroplast envelope (Ferro *et al.*, 2010). The importance of FtsH in general maintenance of the functional photosynthetic membranes was demonstrated by mutations leading to the phenotype yellow *variegated 1* (*var1*) and *var2*, which impaired the genes coding for *FtsH5* and *FtsH2*, respectively (Chen *et al.*, 2000; Takechi *et al.*, 2000). The VAR1/VAR2 group was found to contain FtsH 1, 2, 5 and 8 (e.g. Adam *et al.*, 2006; Kato and Sakamoto, 2010; Liu *et al.*, 2010). The four genes *FtsH1*, *FtsH2*, *FtsH5* and *FtsH8* have higher expression than other *FtsH* genes with similar expression patterns (Yu *et al.*, 2004). In fact, under normal growth conditions, only expression of these four *FtsH* genes was observed. Transcription was previously described as being strictly light dependent (Lindahl *et al.*, 1996). For example, the *FtsH8* gene is highly induced by high light (Sinvany-Villalobo *et al.*, 2004); while *FtsH1* and *FtsH5* are reported to be down-regulated in darkness (Lin and Wu, 2004). However, on a protein level, none of the thylakoid-located FtsH proteases were up-regulated in high light (Zaltsman *et al.*, 2005a). In addition, three homologues to the VAR1/VAR2 sub-family were detected in a proteomic analysis of the etioplast inner membranes of dark-grown wheat leaves (Blomqvist *et al.*, 2006), suggesting that FtsH might be a protease that rapidly degrades protochlorophyllide-oxido-reductase upon illumination, the key enzyme in chlorophyll biosynthesis. Later, it was shown that not only light, but also heat affects the plastidic FtsH proteases (Garcia-Lorenzo *et al.*, 2008). Active FtsH proteases are known to form functional homo- or heterohexamers. FtsH 1 and 5 (type A), as well as FtsH 2 and 8 (type B), form homologous pairs with a very high degree of sequence identity. In heterohexamers, type A and type B subunits were found to possess a ratio of 1:2 (Sakamoto *et al.*, 2003; Zaltsman *et al.*, 2005b). FtsH5 was shown to be 4–5 times more abundant than FtsH1 in type A, whereas in type B subunits, the FtsH2 amount was 2–3 times higher than FtsH8 (Moldavski *et al.*, 2012).

In *Synechocystis* 6803, four isomers of FtsH have been identified. Inactivation of two of these isomers (FtsH1/Slr1390 and FtsH3/Slr1604) is lethal, while the other two (FtsH2/Slr0228 and FtsH4/Sll1463) are dispensable (Nixon *et al.*, 2005). After membrane fractionation, FtsH2 and FtsH4 were detected in the thylakoid membrane (Komenda *et al.*, 2006; Pisareva *et al.*, 2007), while FtsH1 and FtsH3 were found in the plasma membrane (Pisareva *et al.*, 2007). Furthermore, FtsH2 was co-purified with FtsH3 (Barker *et al.*, 2008), and FtsH1/FtsH3 and FtsH4 complexes (Boehm *et al.*, 2012a) were identified.

Interestingly, further multiplication of plant FtsHs has occurred since chloroplasts diverged from their prokaryotic progenitors, such that duplicate genes have evolved from the different cyanobacterial genes. In the green algae *C. reinhardtii*, seven active and two inactive *FtsH* genes were identified. In monocot rice, at least nine active and one inactive *FtsH* genes were identified, while in the *Populus* tree, 18 active and six inactive ones were found (Garcia-Lorenzo *et al.*, 2008).

Deg proteases

Deg proteases, formerly DegP (degradation of periplasmic proteins) or HtrA (high temperature requirement A), are ATP-independent serine proteases (Rawlings *et al.*, 2012). Their structure consists of an N-terminal proteolytic domain containing the His–Asp–Ser catalytic triad, followed by one or two C-terminal PDZ domains, mediating a wide range of protein–protein interactions. Deg proteases were first described in *E. coli* (Lipinska *et al.*, 1989; Strauch *et al.*, 1989), but were later detected in other bacteria, archaea and eukaryota. They are involved in various cellular pathways related to stress tolerance or protein folding during stress (Clausen *et al.*, 2002). *Synechocystis* 6803 contains three Deg proteases (see Kieselbach and Funk, 2003); while in *A. thaliana*, there are 16 Deg protease members located in various organelles (e.g. the nucleus, the chloroplast, the mitochondria and the peroxisome) (Schuhmann and Adamska, 2012). At least four of these *A. thaliana* Deg proteases are located in the plastid (Deg2 on the stromal side of the thylakoid membrane, and Deg1, 5, 8 in the thylakoid lumen) (Tanz *et al.*, 2014). The best characterized Deg proteases are located in the thylakoid lumen: Deg1 was shown to support the assembly of Photosystem II (PSII), while Deg5 and Deg8 form hetero-complexes and have synergistic functions (Sun *et al.*, 2007).

Deg proteases form stable homotrimeric complexes via hydrophobic interactions between the protease domains. PDZ domains and/or the protruding LA loop of the protease domain facilitate further oligomerization of these basic trimer units into bigger complexes, such as hexamers, 12-mers or 24-mers. These cage-like oligomeric structures are thought to trap substrate within their inner cavity, until protein procession is complete. Formation of high-order oligomers can be triggered by the presence of substrate or a change in pH (Huesgen *et al.*, 2011; Kley *et al.*, 2011; Sun *et al.*, 2012), which normally involves activation of the protease.

D1 degradation

When photosynthetic organisms absorb more light than can be used for chemical conversion, this excess energy can cause damage to the cell. The PSII reaction centre polypeptide D1 is particularly susceptible to photodamage (Barber and Andersson, 1992). Notably, any photodamaged D1 protein is rapidly replaced by a *de novo* synthesized copy, while all other 30 PSII subunits stay intact. The continuous photo-inactivation and repair of PSII takes place under all light conditions, although it proceeds at a much higher rate under high light. Two mechanisms of photoinhibition have been proposed: acceptor-side photoinhibition, as a consequence of the generation of singlet oxygen after $Phe^-/P680^+$ charge recombination; and donor-side photoinhibition, involving generation of superoxide anion radicals on the donor side of PSII (reviewed in Andersson and Aro, 2001). Repair processes include the degradation of damaged D1 polypeptides, *de novo* synthesis of D1 on chloroplast ribosomes, processing of newly synthesized D1, association of D1 with chlorophyll and its reaction centre partners, and assembly of the heterodimeric complex with other PSII polypeptides. Recent biochemical and molecular biological studies identified several of the repair factors involved in the degradation of damaged D1 subunits (Nixon *et al.*, 2010). *In vitro* experiments identified different proteases responsible for the degradation of photodamaged D1, as a consequence of donor-side photoinhibition or acceptor-side photoinhibition events (see Vass and Cser, 2009); however, it is still unclear which proteases perform the primary cleavage (Nixon *et al.*, 2010). Despite uncertainty about the precise mechanisms involved in D1 degradation, it is known that the FtsH and Deg protease families are involved in the process (reviewed in Huesgen *et al.*, 2009). Although, there is controversy over the exact roles of Deg and FtsH in the proteolytic process of the D1 protein. For example, an *A. thaliana* FtsH2 deletion mutant was unable to repair damaged D1 polypeptide (Bailey *et al.*, 2001; Bailey and Grossman, 2008). Interestingly, FtsH2 contains a unique luminal loop domain of 81 amino acids that is exclusively found in FtsH proteins from oxygenic phototrophs and may be crucial to the function of oxygenic photosynthetic organisms (Bailey *et al.*, 2001; Garcia-Lorenzo *et al.*, 2008). The same domain was also observed in FtsH6 and FtsH8 of *A. thaliana*, as well as FtsH2 of *Synechocystis* 6803, but is missing in *A. thaliana* FtsH1 or FtsH5. Since FtsH proteases are poor at unfolding, the first proteolytic cleavage was assumed to be performed by a Deg protease in plants (Kato and Sakamoto, 2010), before the VAR1/VAR2 FtsH hetero-complex further digests D1. The stromal Deg2 protease also was proposed to perform primary cleavage (Haussuhl *et al.*, 2001), although later it was shown that an *A. thaliana* mutant lacking Deg2 was able to degrade D1 (Huesgen *et al.*, 2006). Instead, the luminal Deg1 (Kapri-Pardes *et al.*, 2007) or the Deg5/Deg7 pair could introduce single cuts in the lumen-exposed loops of D1, given that a *deg5/deg8* double mutant grew poorly when exposed to high light and was shown to be impaired in D1 degradation (Sun *et al.*, 2007). More recently, studies on *A. thaliana* Deg/FtsH double mutants proposed a reverse order of these proteases (Adam *et al.*, 2011; Kato *et al.*, 2012). Alternatively, the order of cleavage might be dependent on the damage site of D1 (Yoshioka-Nishimura and Yamamoto, 2014).

Current evidence suggests that damaged D1 protein is removed by FtsH alone in *Synechocystis* 6803 (Fig. 2.7). Contrary to higher plants, D1 is degraded at the N-terminal in cyanobacteria (Yamamoto *et al.*, 2001). The amount of D1 degradation, which occurs rapidly, was reduced dramatically in the absence of FtsH2 (Silva *et al.*, 2003). Furthermore, full-length D1 was stabilized in the *ftsh2* deletion mutant, and there was no evidence for the accumulation of D1 fragments (Silva *et al.*, 2003). FtsH protease was found to accumulate during high light treatment (Singh *et al.*, 2008) and FtsH2, 3 and 4 have been co-isolated with PSII in light-stressed cells (Kashino *et al.*, 2002; Yao *et al.*, 2007). Studies of mutants depleted of all three Deg proteases

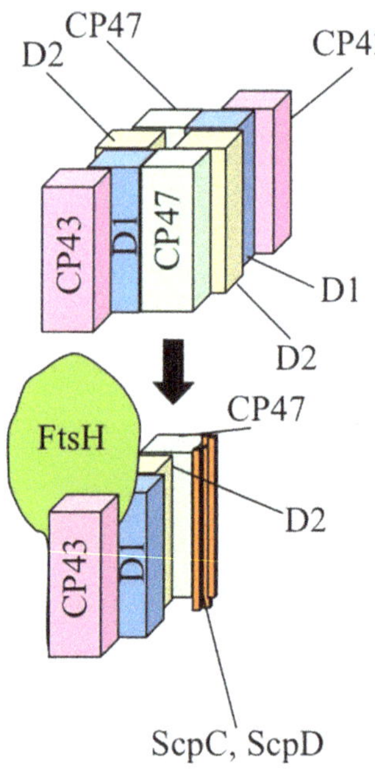

Figure 2.7 Model of D1 degradation in the cyanobacterium *Synechocystis* sp. PCC 6803. Current evidence suggests that damaged D1 protein is removed by FtsH2 in *Synechocystis* 6803. FtsH2, FtsH3 and FtsH4, have been co-isolated with His-tagged ScpD, ScpC and PSII complexes exposed to high light stress; ScpC and ScpD may therefore be involved in chlorophyll storage during PSII repair.

showed that Deg proteases are not crucial for the repair of cyanobacterial D1 (Barker *et al.*, 2006), and proteomic studies comparing wild type cells with single or triple Deg mutants did not identify any chlorophyll-binding protein as hypothetical substrate (Tam *et al.*, 2015; Miranda *et al.*, 2013). However, Deg proteases may play an indirect role in optimizing repair at higher temperatures in cyanobacteria.

In any case, damaged D1 has to be transported out of the grana towards the margins where the repair machinery is presumed to be located. Recently, it was shown that in the stromal regions of the thylakoid membrane, FtsH proteins do not form hexameric complexes, but monomers and homo- or heteromeric dimers (Yoshioka *et al.*, 2010); hence, hexameric complexes were identified only in the grana regions. In plants, general structural changes of the thylakoid membrane were observed after stress exposure, including shrinkage of the grana and unstacking (Khatoon *et al.*, 2009; Kirchhoff *et al.*, 2013). It is assumed that FtsHs action takes place either near the grana margins or in the thylakoid membranes when they are less compressed, e.g. during high light or heat stress.

In cyanobacteria, D1, D2, cyt b_{559} and PsbO have been detected in the plasma membrane, after two-phase partition followed by immune staining (Zak *et al.*, 2001), while CP47 and CP43 were detected exclusively in the thylakoid membrane fraction. Confocal microscopy studies on cells exposed to different light intensities even locate FtsH1 in the cytoplasmic membrane, while FtsH2 and an FtsH1 subpopulation were detected in distinct regions of the thylakoid membrane. FtsH3 and FtsH4 were observed at the distal edges of the thylakoid membrane under low light, shifting towards the cell periphery after high light exposure (Sacharz *et al.*, 2015). While FtsH3 and FtsH4 are co-localized with assembly centres (Stengel *et al.*, 2012), FtsH2 might be located in distinct repair zones (Sacharz *et al.*, 2015). A FtsH2-deletion mutant was shown to be sensitive towards high light, while its photodamaged D1 protein was found to accumulate within the thylakoid (Silva *et al.*, 2003; Komenda *et al.*, 2006); FtsH2 appears to have no function in degrading unassembled D1 (Komenda *et al.*, 2010). Finally, FtsH2, but also FtsH3 and FtsH4, have been co-isolated with His-tagged ScpD and PSII complexes exposed to high light stress (Yao *et al.*, 2007). In this context, ScpC and ScpD may be involved in chlorophyll storage during PSII repair (Yao *et al.*, 2007; Hernandez-Prieto *et al.*, 2011), while ScpE is important for the assembly of chlorophyll-binding proteins (Yao *et al.*, 2012).

LHC degradation

Despite the very high content of LHC proteins in the plant cell and their importance for the adaptation and viability of plants, very little is known about LHC proteins degradation and regulation (reviewed in Garcia-Lorenzo *et al.*, 2005; Adam, 2013; Dall'Osto *et al.*, 2015). Under normal growth conditions, turnover of the LHC proteins is very low. However, acclimative decline of LHCII in plants grown under high-intensity light was shown to occur not only on a transcriptional level, but also by degradation of pre-accumulated LHCII apoproteins (Yang *et al.*, 1998; Jansson *et al.*, 2000). Monomerization and/or exposure to certain amino acid sequences within LHCII may trigger degradation of LHC proteins (Yang *et al.*, 1998; Mick *et al.*, 2004). Alternatively, the degradation of chlorophyll (Park *et al.*, 2007), specifically of chlorophyll *b* (Horie *et al.*, 2009) may induce proteolysis. Several different proteases have been identified that might be involved in the proteolysis of LHCII apoproteins under various environmental conditions (reviewed in Garcia-Lorenzo *et al.*, 2005; Dall'Osto *et al.*, 2015). More recently, a metalloprotease was identified to be responsible for Lhcb3 degradation in barley as well as *A. thaliana* (Zelisko and Jackowski, 2004; Zelisko *et al.*, 2005). FtsH6 was identified to be able to degrade Lhcb3 *in vitro* (Zelisko *et al.*, 2005); however, an *ftsH6* knock-out mutant did not display any phenotype when grown in the chamber or in the field (Wagner *et al.*, 2011). Furthermore, the VAR1/VAR2 hetero-complex was shown to be involved in the stress-dependent degradation of Lhcb1–3 in isolated PSII-membrane particles (Lucinski and Jackowski, 2013). In contrast, the minor antenna proteins were shown to be degraded by Deg proteases. For example, Deg1 of *A. thaliana* or maize degraded CP26 and CP29 in a light-dependent manner (Zienkiewicz *et al.*, 2012, 2013), while Deg2 was shown to degrade CP24, when exposed to high temperatures or high-intensity light (Lucinski *et al.*, 2011).

Concluding remarks and future trends

During the last decades, considerable understanding of assembly and disassembly of chlorophyll-binding proteins has been achieved. The function of numerous accessory proteins has been identified (reviewed in Nixon *et al.*, 2010). X-ray structures of LHCII, CP29, and cyanobacterial PSII at high resolution have provided new insights into the structural and functional organization of these complexes, and super-molecular organization in the thylakoids (Pan *et al.*, 2011; Umena *et al.*, 2011). However, the X-ray structure of the PSII super-complex in plants, which differs from the PSII core of cyanobacteria, is yet to be determined (Caffarri *et al.*, 2009). Much information was obtained on the function of LILs, which seem to be involved in different aspects of photo-protection. Some of them are directly involved in insertion/regulation of pigments into the polypeptides, others are involved in light dissipation without tightly binding pigments. However, the molecular function and localization of some LILs remains elusive, as do the roles of PsbS, and ELIPs in plants. To understand at molecular level the process of light dissipation is important to enhance the use of sunlight to increase crop productivity (Horton, 2000). An increased capacity of light dissipation could increase crop productivity of 30% (Zhu *et al.*, 2004). Several advances in the cpSRP pathway were gained recently; specifically, mutations in this pathway may increase algal biomass production substantially (Kirst and Melis, 2014).

Several aspects of the LHCs role in targeting and folding *in vivo* are not well understood; for example, the role of chlorophyll *b* in this processes is yet to be determined. Furthermore, although retrograde signalling from chloroplast to nucleus is a widely accepted mechanism, the signal molecules are mostly unknown.

Evolution has shaped the photosynthetic apparatus creating a plethora of chlorophyll-binding proteins to adapt to different environmental cues. Light harvesting in general and chlorophyll-binding proteins in particular represent an important resource to increase the efficiency of light-to-biomass conversion in photosynthetic microorganisms, algae and plants.

References

Adam, Z. (2013). Emerging roles for diverse intramembrane proteases in plant biology. Biochim. Biophys. Acta *1828*, 2933–2936.

Adam, Z. (2015). Plastid intramembrane proteolysis. Biochim. Biophys. Acta *1847*, 910–914.

Adam, Z., Adamska, I., Nakabayashi, K., Ostersetzer, O., Haussuhl, K., Manuell, A., Zheng, B., Vallon, O., Rodermel, S.R., Shinozaki, K., and Clarke, A.K. (2001). Chloroplast and mitochondrial proteases in *Arabidopsis*. A proposed nomenclature. Plant Physiol. *125*, 1912–1918.

Adam, Z., Frottin, F., Espagne, C., Meinnel, T., and Giglione, C. (2011). Interplay between N-terminal methionine excision and FtsH protease is essential for normal chloroplast development and function in *Arabidopsis*. Plant Cell *23*, 3745–3760.

Adam, Z., Rudella, A., and van Wijk, K.J. (2006). Recent advances in the study of Clp, FtsH and other proteases located in chloroplasts. Curr. Opin. Plant Biol. *9*, 234–240.

Adamska, I. (1995). Regulation of early light-inducible protein gene expression by blue and red light in etiolated seedlings involves nuclear and plastid factors. Plant Physiol. *107*, 1167–1175.

Adamska, I., Kruse, E., and Kloppstech, K. (2001). Stable insertion of the early light-induced proteins into etioplast membranes requires chlorophyll a. J. Biol. Chem. *276*, 8582–8587.

Adamska, I., Roobol-Boza, M., Lindahl, M., and Andersson, B. (1999). Isolation of pigment-binding early light-inducible proteins from pea. Eur. J. Biochem. *260*, 453–460.

Alboresi, A., Caffarri, S., Nogue, F., Bassi, R., and Morosinotto, T. (2008). In silico and biochemical analysis of *Physcomitrella patens* photosynthetic antenna: identification of subunits, which evolved upon land adaptation. PLoS One 3, e2033.

Alboresi, A., Dall'osto, L., Aprile, A., Carillo, P., Roncaglia, E., Cattivelli, L., and Bassi, R. (2011). Reactive oxygen species and transcript analysis upon excess light treatment in wild-type *Arabidopsis thaliana* vs a photosensitive mutant lacking zeaxanthin and lutein. BMC Plant Biol. *11*, 62.

Alboresi, A., Gerotto, C., Giacometti, G.M., Bassi, R., and Morosinotto, T. (2010). *Physcomitrella patens* mutants affected on heat dissipation clarify the evolution of photoprotection mechanisms upon land colonization. Proc. Natl. Acad. Sci. U.S.A. *107*, 11128–11133.

Allen, K.D., and Staehelin, L.A. (1992). Biochemical Characterization of Photosystem II Antenna Polypeptides in Grana and Stroma Membranes of Spinach. Plant Physiol. *100*, 1517–1526.

Amunts, A., Drory, O., and Nelson, N. (2007). The structure of a plant photosystem I supercomplex at 3.4A resolution. Nature *447*, 58–63.

Amunts, A., and Nelson, N. (2009). Plant photosystem I design in the light of evolution. Structure *17*, 637–650.

Anderson, J.M., and Andersson, B. (1988). The dynamic photosynthetic membrane and regulation of solar energy conversion. Trends Biochem Sci. *13*, 351–355.

Andersson, B., and Aro, E.-M. (2001). Photodamage and D1 protein turnover in photosystem II. In Regulation of

Photosynthesis, E.M. Aro, B., and Andersson, eds. Vol 11. (Dordrecht: Springer), pp. 377–393.

Andersson, U., Heddad, M., and Adamska, I. (2003). Light stress-induced one-helix protein of the chlorophyll a/b-binding family associated with photosystem I. Plant Physiol. *132*, 811–820.

Andres, C., Agne, B., and Kessler, F. (2010). The TOC complex: preprotein gateway to the chloroplast. Biochim. Biophys. Acta *1803*, 715–723.

Armbruster, U., Zuhlke, J., Rengstl, B., Kreller, R., Makarenko, E., Ruhle, T., Schunemann, D., Jahns, P., Weisshaar, B., Nickelsen, J., and Leister, D. (2010). The *Arabidopsis* thylakoid protein PAM68 is required for efficient D1 biogenesis and photosystem II assembly. Plant Cell *22*, 3439–3460.

Bailey, S., Clokie, M.R., Millard, A., and Mann, N.H. (2004). Cyanophage infection and photoinhibition in marine cyanobacteria. Res. Microbiol. *155*, 720–725.

Bailey, S., and Grossman, A. (2008). Photoprotection in cyanobacteria: regulation of light harvesting. Photochem. Photobiol. *84*, 1410–1420.

Bailey, S., Walters, R.G., Jansson, S., and Horton, P. (2001). Acclimation of *Arabidopsis thaliana* to the light environment: the existence of separate low light and high light responses. Planta *213*, 794–801.

Bailleul, B., Rogato, A., de Martino, A., Coesel, S., Cardol, P., Bowler, C., Falciatore, A., and Finazzi, G. (2010). An atypical member of the light-harvesting complex stress-related protein family modulates diatom responses to light. Proc. Natl. Acad. Sci. U.S.A. *107*, 18214–18219.

Ballottari, M., Girardon, J., Dall'osto, L., and Bassi, R. (2012). Evolution and functional properties of photosystem II light harvesting complexes in eukaryotes. Biochim. Biophys. Acta *1817*, 143–157.

Ballottari, M., Govoni, C., Caffarri, S., and Morosinotto, T. (2004). Stoichiometry of LHCI antenna polypeptides and characterization of gap and linker pigments in higher plants Photosystem I. Eur. J. Biochem. *271*, 4659–4665.

Barber, J., and Andersson, B. (1992). Too much of a good thing: light can be bad for photosynthesis. Trends Biochem. Sci. *17*, 61–66.

Barber, J., Morris, E., and Buchel, C. (2000). Revealing the structure of the photosystem II chlorophyll binding proteins, CP43 and CP47. Biochim. Biophys. Acta *1459*, 239–247.

Barker, M., Boehm, M., Nixon, P.J., and Nield, J. (2008). Structural analysis of an FtsH2/FtsH3 complex isolated from *Synechocystis* sp. PCC 6803. In Photosynthesis. Energy from the Sun, J.F. Allen, E. Gannt, J. Golbeck, and B. Osmond, eds. (Dordrecht: Springer), pp. 738–740.

Barker, M., de Vries, R., Nield, J., Komenda, J., and Nixon, P.J. (2006). The deg proteases protect *Synechocystis* sp. PCC 6803 during heat and light stresses but are not essential for removal of damaged D1 protein during the photosystem two repair cycle. J. Biol. Chem. *281*, 30347–30355.

Bassi, R., Croce, R., Cugini, D., and Sandona, D. (1999). Mutational analysis of a higher plant antenna protein provides identification of chromophores bound into multiple sites. Proc. Natl. Acad. Sci. U.S.A. *96*, 10056–10061.

Bellafiore, S., Ferris, P., Naver, H., Gohre, V., and Rochaix, J.D. (2002). Loss of Albino3 leads to the specific depletion of the light-harvesting system. Plant Cell *14*, 2303–2314.

Betterle, N., Ballottari, M., Hienerwadel, R., Dall'Osto, L., and Bassi, R. (2010). Dynamics of zeaxanthin binding to the photosystem II monomeric antenna protein Lhcb6 (CP24) and modulation of its photoprotection properties. Arch. Biochem. Biophys. *504*, 67–77.

Betterle, N., Ballottari, M., Zorzan, S., de Bianchi, S., Cazzaniga, S., Dall'osto, L., Morosinotto, T., and Bassi, R. (2009). Light-induced dissociation of an antenna hetero-oligomer is needed for non-photochemical quenching induction. J. Biol. Chem. *284*, 15255–15266.

Bhaya, D., Dufresne, A., Vaulot, D., and Grossman, A. (2002). Analysis of the hli gene family in marine and freshwater cyanobacteria. FEMS Microbiol. Lett. *215*, 209–219.

Bibby, T.S., Nield, J., and Barber, J. (2001). Iron deficiency induces the formation of an antenna ring around trimeric photosystem I in cyanobacteria. Nature *412*, 743–745.

Blomqvist, L.A., Ryberg, M., and Sundqvist, C. (2006). Proteomic analysis of the etioplast inner membranes of wheat (Triticum aestivum) by two-dimensional electrophoresis and mass spectrometry. Physiol. Plant., 368–381.

Boehm, M., Romero, E., Reisinger, V., Yu, J., Komenda, J., Eichacker, L.A., Dekker, J.P., and Nixon, P.J. (2011). Investigating the early stages of photosystem II assembly in *Synechocystis* sp. PCC 6803: isolation of CP47 and CP43 complexes. J. Biol. Chem. *286*, 14812–14819.

Boehm, M., Yu, J., Krynicka, V., Barker, M., Tichy, M., Komenda, J., Nixon, P.J., and Nield, J. (2012a). Subunit organization of a *synechocystis* hetero-oligomeric thylakoid FtsH complex involved in photosystem II repair. Plant Cell *24*, 3669–3683.

Boehm, M., Yu, J., Reisinger, V., Beckova, M., Eichacker, L.A., Schlodder, E., Komenda, J., and Nixon, P.J. (2012b). Subunit composition of CP43-less photosystem II complexes of *Synechocystis* sp. PCC 6803: implications for the assembly and repair of photosystem II. Philos. Trans. R. Soc. Lond. B. Biol. Sci. *367*, 3444–3454.

Boekema, E.J., Jensen, P.E., Schlodder, E., van Breemen, J.F., van Roon, H., Scheller, H.V., and Dekker, J.P. (2001). Green plant photosystem I binds light-harvesting complex I on one side of the complex. Biochemistry *40*, 1029–1036.

Boekema, E.J., van Roon, H., Calkoen, F., Bassi, R., and Dekker, J.P. (1999). Multiple types of association of photosystem II and its light-harvesting antenna in partially solubilized photosystem II membranes. Biochemistry *38*, 2233–2239.

Bonardi, V., Pesaresi, P., Becker, T., Schleiff, E., Wagner, R., Pfannschmidt, T., Jahns, P., and Leister, D. (2005). Photosystem II core phosphorylation and photosynthetic acclimation require two different protein kinases. Nature *437*, 1179–1182.

Bonaventura, C., and Myers, J. (1969). Fluorescence and oxygen evolution from *Chlorella pyrenoidosa*. Biochim. Biophys. Acta *189*, 366–383.

Bonente, G., Ballottari, M., Truong, T.B., Morosinotto, T., Ahn, T.K., Fleming, G.R., Niyogi, K.K., and Bassi, R. (2011). Analysis of LhcSR3, a protein essential for feedback de-excitation in the green alga *Chlamydomonas reinhardtii*. PLoS Biol. *9*, e1000577.

Bonente, G., Howes, B.D., Caffarri, S., Smulevich, G., and Bassi, R. (2008a). Interactions between the photosystem II subunit PsbS and xanthophylls studied *in vivo* and *in vitro*. J. Biol. Chem. *283*, 8434–8445.

Bonente, G., Passarini, F., Cazzaniga, S., Mancone, C., Buia, M.C., Tripodi, M., Bassi, R., and Caffarri, S. (2008b). The occurrence of the psbS gene product in *Chlamydomonas reinhardtii* and in other photosynthetic organisms and its correlation with energy quenching. Photochem. Photobiol. *84*, 1359–1370.

Brautigam, A., Hoffmann-Benning, S., and Weber, A.P. (2008). Comparative proteomics of chloroplast envelopes from C3 and C4 plants reveals specific adaptations of the plastid envelope to C4 photosynthesis and candidate proteins required for maintaining C4 metabolite fluxes. Plant. Physiol. *148*, 568–579.

Bricker, T.M., Roose, J.L., Fagerlund, R.D., Frankel, L.K., and Eaton-Rye, J.J. (2012). The extrinsic proteins of Photosystem II. Biochim. Biophys. Acta *1817*, 121–142.

Broughton, M.J., Howe, C.J., and Hiller, R.G. (2006). Distinctive organization of genes for light-harvesting proteins in the cryptophyte alga Rhodomonas. Gene *369*, 72–79.

Busch, A., Petersen, J., Webber-Birungi, M.T., Powikrowska, M., Lassen, L.M., Naumann-Busch, B., Nielsen, A.Z., Ye, J., Boekema, E.J., Jensen, O.N., *et al.* (2013). Composition and structure of photosystem I in the moss *Physcomitrella patens*. J. Exp. Bot. *64*, 2689–2699.

Caffarri, S., Croce, R., Breton, J., and Bassi, R. (2001). The major antenna complex of photosystem II has a xanthophyll binding site not involved in light harvesting. J. Biol. Chem. *276*, 35924–35933.

Caffarri, S., Croce, R., Cattivelli, L., and Bassi, R. (2004). A look within LHCII: differential analysis of the Lhcb1-3 complexes building the major trimeric antenna complex of higher-plant photosynthesis. Biochemistry *43*, 9467–9476.

Caffarri, S., Kouril, R., Kereiche, S., Boekema, E.J., and Croce, R. (2009). Functional architecture of higher plant photosystem II supercomplexes. EMBO J. *28*, 3052–3063.

Caffarri, S., Passarini, F., Bassi, R., and Croce, R. (2007). A specific binding site for neoxanthin in the monomeric antenna proteins CP26 and CP29 of Photosystem II. FEBS Lett. *581*, 4704–4710.

Caffarri, S., Tibiletti, T., Jennings, R.C., and Santabarbara, S. (2014). A comparison between plant photosystem I and photosystem II architecture and functioning. Curr. Prot. Peptide Sci. *15*, 296–331.

Casazza, A.P., Rossini, S., Rosso, M.G., and Soave, C. (2005). Mutational and expression analysis of ELIP1 and ELIP2 in *Arabidopsis thaliana*. Plant Mol. Biol. *58*, 41–51.

Castelletti, S., Morosinotto, T., Robert, B., Caffarri, S., Bassi, R., and Croce, R. (2003). Recombinant Lhca2 and Lhca3 subunits of the photosystem I antenna system. Biochemistry *42*, 4226–4234.

Chen, M., Choi, Y., Voytas, D.F., and Rodermel, S. (2000). Mutations in the *Arabidopsis* VAR2 locus cause leaf variegation due to the loss of a chloroplast FtsH protease. Plant J. *22*, 303–313.

Chidgey, J.W., Linhartova, M., Komenda, J., Jackson, P.J., Dickman, M.J., Canniffe, D.P., Konik, P., Pilny, J., Hunter, C.N., and Sobotka, R. (2014). A cyanobacterial chlorophyll synthase–HliD complex associates with the Ycf39 protein and the YidC/Alb3 insertase. Plant Cell *26*, 1267–1279.

Clausen, T., Southan, C., and Ehrmann, M. (2002). The HtrA family of proteases: implications for protein composition and cell fate. Mol. Cell. *10*, 443–455.

Croce, R., Morosinotto, T., Castelletti, S., Breton, J., and Bassi, R. (2002). The Lhca antenna complexes of higher plants photosystem I. Biochim. Biophys. Acta *1556*, 29–40.

Croce, R., Remelli, R., Varotto, C., Breton, J., and Bassi, R. (1999). The neoxanthin binding site of the major light harvesting complex (LHCII) from higher plants. FEBS Lett. *456*, 1–6.

Dainese, P., and Bassi, R. (1991). Subunit stoichiometry of the chloroplast photosystem II antenna system and aggregation state of the component chlorophyll a/b binding proteins. J. Biol. Chem. *266*, 8136–8142.

Dall'Osto, L., Bressan, M., and Bassi, R. (2015). Biogenesis of light harvesting proteins. Biochim. Biophys. Acta. *1847*, 861–871.

Dall'Osto, L., Caffarri, S., and Bassi, R. (2005). A mechanism of nonphotochemical energy dissipation, independent from PsbS, revealed by a conformational change in the antenna protein CP26. Plant Cell *17*, 1217–1232.

Dekker, J.P., and Boekema, E.J. (2005). Supramolecular organization of thylakoid membrane proteins in green plants. Biochim. Biophys. Acta *1706*, 12–39.

Dobáková, M., Sobotka, R., Tichý, M., and Komenda, J. (2009). Psb28 Protein Is Involved in the Biogenesis of the Photosystem II Inner Antenna CP47 (PsbB) in the Cyanobacterium *Synechocystis* sp. PCC 6803. Plant Physiol. *149*, 1076–1086.

Dobakova, M., Tichy, M., and Komenda, J. (2007). Role of the PsbI protein in photosystem II assembly and repair in the cyanobacterium *Synechocystis* sp. PCC 6803. Plant Physiol. *145*, 1681–1691.

Domanskii, V., Rassadina, V., Gus-Mayer, S., Wanner, G., Schoch, S., and Rudiger, W. (2003). Characterization of two phases of chlorophyll formation during greening of etiolated barley leaves. Planta *216*, 475–483.

Douglas, S., Zauner, S., Fraunholz, M., Beaton, M., Penny, S., Deng, L.T., Wu, X., Reith, M., Cavalier-Smith, T., and Maier, U.G. (2001). The highly reduced genome of an enslaved algal nucleus. Nature *410*, 1091–1096.

Durnford, D., and Falkowski, P. (1997). Chloroplast redox regulation of nuclear gene transcription during photoacclimation. Photosynth. Res. *53*, 229–241.

Eggink, L.L., and Hoober, J.K. (2000). Chlorophyll binding to peptide maquettes containing a retention motif. J. Biol. Chem. *275*, 9087–9090.

Engelken, J., Brinkmann, H., and Adamska, I. (2010). Taxonomic distribution and origins of the extended LHC (light-harvesting complex) antenna protein superfamily. BMC Evol. Biol. *10*, 233.

Fernandez, A.P., and Strand, A. (2008). Retrograde signaling and plant stress: plastid signals initiate cellular stress responses. Curr. Opin. Plant Biol. *11*, 509–513.

Ferreira, K.N., Iverson, T.M., Maghlaoui, K., Barber, J., and Iwata, S. (2004). Architecture of the photosynthetic oxygen-evolving center. Science *303*, 1831–1838.

Ferro, M., Brugière, S., Salvi, D., Seigneurin-Berny, D., Court, M., Moyet, L., Ramus, C., Miras, S., Mellal, M., Le

Gall, S., *et al.* (2010). AT_CHLORO, a comprehensive chloroplast proteome database with subplastidial localization and curated information on envelope proteins. Mol. Cell. Proteom. *9*, 1063–1084.

Fey, V., Wagner, R., Brautigam, K., and Pfannschmidt, T. (2005). Photosynthetic redox control of nuclear gene expression. J. Exp. Bot. *56*, 1491–1498.

Fiore, A., Dall'Osto, L., Cazzaniga, S., Diretto, G., Giuliano, G., and Bassi, R. (2012). A quadruple mutant of *Arabidopsis* reveals a beta-carotene hydroxylation activity for LUT1/CYP97C1 and a regulatory role of xanthophylls on determination of the PSI/PSII ratio. BMC Plant Biol. *12*, 50.

Formaggio, E., Cinque, G., and Bassi, R. (2001). Functional architecture of the major light-harvesting complex from higher plants. J. Mol. Biol. *314*, 1157–1166.

Fraser, P.D., Schuch, W., and Bramley, P.M. (2000). Phytoene synthase from tomato (Lycopersicon esculentum) chloroplasts – partial purification and biochemical properties. Planta *211*, 361–369.

Funk, C. (2001). The PsbS protein: A Cab-protein with a function of its own. In Regulation of Photosynthesis, E.M. Aro and B. Andersson, eds., Vol. 11 (Dordrecht: Springer), pp. 453–467.

Funk, C., Adamska, I., Green, B.R., Andersson, B., and Renger, G. (1995a). The nuclear-encoded chlorophyll-binding photosystem II-S protein is stable in the absence of pigments. J. Biol. Chem. *270*, 30141–30147.

Funk, C., Alami, M., Tibiletti, T., and Green, B.R. (2011). High light stress and the one-helix LHC-like proteins of the cryptophyte *Guillardia theta*. Biochim. Biophys. Acta *1807*, 841–846.

Funk, C., Schröder, W.P., Green, B.R., Renger, G., and Andersson, B. (1994). The intrinsic 22 kDa protein is a chlorophyll-binding subunit of photosystem II. FEBS Lett. *342*, 261–266.

Funk, C., Schröder, W.P., Napiwotzki, A., Tjus, S.E., Renger, G., and Andersson, B. (1995b). The PSII-S protein of higher plants: a new type of pigment-binding protein. Biochemistry *34*, 11133–11141.

Funk, C., and Vermaas, W. (1999). A cyanobacterial gene family coding for single-helix proteins resembling part of the light-harvesting proteins from higher plants. Biochemistry *38*, 9397–9404.

Funk, C., Wiklund, R., Schröder, W.P., and Jansson, C. (2001). D1' centers are less efficient than normal photosystem II centers. FEBS Lett. *505*, 113–117.

Galka, P., Santabarbara, S., Khuong, T.T., Degand, H., Morsomme, P., Jennings, R.C., Boekema, E.J., and Caffarri, S. (2012). Functional analyses of the plant photosystem I-light-harvesting complex II supercomplex reveal that light-harvesting complex II loosely bound to photosystem II is a very efficient antenna for photosystem I in state II. Plant Cell *24*, 2963–2978.

Garcia-Lorenzo, M., Pruzinska, A., and Funk, C. (2008). ATP-dependent proteases in the plant chloroplast. In ATP-dependent Proteases, E. Kutejova, ed. (Kerala: Research Signpost), pp. 145–176.

Garcia-Lorenzo, M., Sjödin, A., Jansson, S., and Funk, C. (2006). Protease gene families in Populus and *Arabidopsis*. BMC Plant Biol. *6*, 30.

Garcia-Lorenzo, M., Zelisko, A., Jackowski, G., and Funk, C. (2005). Degradation of the main Photosystem II light-harvesting complex. Photochem. Photobiol. Sci *4*, 1065–1071.

Gerotto, C., Alboresi, A., Giacometti, G.M., Bassi, R., and Morosinotto, T. (2011). Role of PSBS and LHCSR in *Physcomitrella patens* acclimation to high light and low temperature. Plant Cell Environ. *34*, 922–932.

Gerotto, C., Alboresi, A., Giacometti, G.M., Bassi, R., and Morosinotto, T. (2012). Coexistence of plant and algal energy dissipation mechanisms in the moss *Physcomitrella patens*. New Phytol. *196*, 763–773.

Giuffra, E., Zucchelli, G., Sandona, D., Croce, R., Cugini, D., Garlaschi, F.M., Bassi, R., and Jennings, R.C. (1997). Analysis of some optical properties of a native and reconstituted photosystem II antenna complex, CP29: pigment binding sites can be occupied by chlorophyll a or chlorophyll b and determine spectral forms. Biochemistry *36*, 12984–12993.

Green, B.R., Pichersky, E., and Kloppstech, K. (1991). Chlorophyll a/b-binding proteins: an extended family. Trends Biochem. Sci. *16*, 181–186.

Green, R.R., and Pichersky, E. (1994). Hypothesis for the evolution of three-helix Chlorophyll a/b and Chlorophyll a/c light-harvesting antenna proteins from two-helix and four-helix ancestors. Photosynth. Res. *39*, 149–162.

Guskov, A., Kern, J., Gabdulkhakov, A., Broser, M., Zouni, A., and Saenger, W. (2009). Cyanobacterial photosystem II at 2.9-A resolution and the role of quinones, lipids, channels and chloride. Nat. Struct. Mol. Biol. *16*, 334–342.

Haussuhl, K., Andersson, B., and Adamska, I. (2001). A chloroplast DegP2 protease performs the primary cleavage of the photodamaged D1 protein in plant photosystem II. EMBO J. *20*, 713–722.

Havaux, M., Dall'Osto, L., Cuine, S., Giuliano, G., and Bassi, R. (2004). The effect of zeaxanthin as the only xanthophyll on the structure and function of the photosynthetic apparatus in *Arabidopsis thaliana*. J. Biol. Chem. *279*, 13878–13888.

Havaux, M., Guedeney, G., He, Q., and Grossman, A.R. (2003). Elimination of high-light-inducible polypeptides related to eukaryotic chlorophyll a/b-binding proteins results in aberrant photoacclimation in *Synechocystis* PCC6803. Biochim. Biophys. Acta *1557*, 21–33.

Havaux, M., and Niyogi, K.K. (1999). The violaxanthin cycle protects plants from photooxidative damage by more than one mechanism. Proc. Natl. Acad. Sci. U.S.A. *96*, 8762–8767.

He, Q., and Vermaas, W. (1998). Chlorophyll a availability affects psbA translation and D1 precursor processing *in vivo* in *Synechocystis* sp. PCC 6803. Proc. Natl. Acad. Sci. U.S.A. *95*, 5830–5835.

Heddad, M., and Adamska, I. (2000). Light stress-regulated two-helix proteins in *Arabidopsis thaliana* related to the chlorophyll a/b-binding gene family. Proc. Natl. Acad. Sci. U.S.A. *97*, 3741–3746.

Heddad, M., and Adamska, I. (2002). The evolution of light stress proteins in photosynthetic organisms. Comp. Funct. Genom. 3, 504–510.

Heddad, M., Noren, H., Reiser, V., Dunaeva, M., Andersson, B., and Adamska, I. (2006). Differential expression and localization of early light-induced proteins in *Arabidopsis*. Plant Physiol. *142*, 75–87.

Hernandez-Prieto, M.A., Tibiletti, T., Abasova, L., Kirilovsky, D., Vass, I., and Funk, C. (2011). The small CAB-like proteins of the cyanobacterium *Synechocystis* sp. PCC 6803: their involvement in chlorophyll biogenesis for Photosystem II. Biochim. Biophys. Acta *1807*, 1143–1151.

Herrin, D.L., Battey, J.F., Greer, K., and Schmidt, G.W. (1992). Regulation of chlorophyll apoprotein expression and accumulation. Requirements for carotenoids and chlorophyll. J. Biol. Chem. *267*, 8260–8269.

Hohmann-Marriott, M.F., and Blankenship, R.E. (2011). Evolution of photosynthesis. Annu. Rev. Plant Biol. *62*, 515–548.

Hoober, J.K., and Argyroudi-Akoyunoglou, J. (2004). Assembly of light-harvesting complexes of photosystem II and the role of chlorophyll b. in chlorophyll *a* fluorescence, G. Papageorgiou and Govindjee, eds., Vol. 19 (Dordrecht: Springer), pp. 679–712.

Hoober, J.K., and Eggink, L.L. (2001). A potential role of chlorophylls b and c in assembly of light-harvesting complexes. FEBS Lett. *489*, 1–3.

Hoober, J.K., Eggink, L.L., and Chen, M. (2007). Chlorophylls, ligands and assembly of light-harvesting complexes in chloroplasts. Photosynth. Res. *94*, 387–400.

Horie, Y., Ito, H., Kusaba, M., Tanaka, R., and Tanaka, A. (2009). Participation of chlorophyll b reductase in the initial step of the degradation of light-harvesting chlorophyll a/b-protein complexes in *Arabidopsis*. J. Biol. Chem. *284*, 17449–17456.

Horn, R., and Paulsen, H. (2002). Folding *in vitro* of light-harvesting chlorophyll a/b protein is coupled with pigment binding. J. Mol. Biol. *318*, 547–556.

Horton, P. (2000). Prospects for crop improvement through the genetic manipulation of photosynthesis: morphological and biochemical aspects of light capture. J. Exp. Bot. *51*, 475–485.

Huesgen, P.F., Miranda, H., Lam, X., Perthold, M., Schuhmann, H., Adamska, I., and Funk, C. (2011). Recombinant Deg/HtrA proteases from *Synechocystis* sp. PCC 6803 differ in substrate specificity, biochemical characteristics and mechanism. Biochem. J. *435*, 733–742.

Huesgen, P.F., Schuhmann, H., and Adamska, I. (2006). Photodamaged D1 protein is degraded in *Arabidopsis* mutants lacking the Deg2 protease. FEBS Lett. *580*, 6929–6932.

Huesgen, P.F., Schuhmann, H., and Adamska, I. (2009). Deg/HtrA proteases as components of a network for photosystem II quality control in chloroplasts and cyanobacteria. Res. Microbiol. *160*, 726–732.

Hutin, C., Havaux, M., Carde, J.P., Kloppstech, K., Meiherhoff, K., Hoffman, N., and Nussaume, L. (2002). Double mutation cpSRP43 /cpSRP54 – is necessary to abolish the cpSRP pathway required for thylakoid targeting of the light-harvesting chlorophyll proteins. Plant J. *29*, 531–543.

Hutin, C., Nussaume, L., Moise, N., Moya, I., Kloppstech, K., and Havaux, M. (2003). Early light-induced proteins protect *Arabidopsis* from photooxidative stress. Proc. Natl. Acad. Sci. U.S.A. *100*, 4921–4926.

Janssen, P.J.D., Lambreva, M.D., Plumeré, N., Bartolucci, C., Antonacci, A., Buonasera, K., Frese, R.N., Scognamiglio, V., and Rea, G. (2014). Photosynthesis at the forefront of a sustainable life. Front. Chem. *2*, 36.

Jansson, S. (1999). A guide to the Lhc genes and their relatives in *Arabidopsis*. Trends Plant Sci. *4*, 236–240.

Jansson, S. (2006). A Protein Family Saga: From Photoprotection to Light-Harvesting (and Back?). In Photoprotection, Photoinhibition, Gene Regulation, and Environment, B. Demmig-Adams, W. Adams III and A. Mattoo, eds., Vol 21. (Dordrecht, The Netherlands: Springer), pp. 145–153.

Jansson, S., Andersson, J., Kim, S.J., and Jackowski, G. (2000). An *Arabidopsis thaliana* protein homologous to cyanobacterial high-light-inducible proteins. Plant Mol. Biol. *42*, 345–351.

Jensen, P.E., Bassi, R., Boekema, E.J., Dekker, J.P., Jansson, S., Leister, D., Robinson, C., and Scheller, H.V. (2007). Structure, function and regulation of plant photosystem I. Biochim. Biophys. Acta *1767*, 335–352.

Jordan, P., Fromme, P., Witt, H.T., Klukas, O., Saenger, W., and Krausz, N. (2001). Three-dimensional structure of cyanobacterial photosystem I at 2.5Å resolution. Nature *411*, 909–917.

Kada, S., Koike, H., Satoh, K., Hase, T., and Fujita, Y. (2003). Arrest of chlorophyll synthesis and differential decrease of Photosystems I and II in a cyanobacterial mutant lacking light-independent protochlorophyllide reductase. Plant Mol. Biol. *51*, 225–235.

Kapri-Pardes, E., Naveh, L., and Adam, Z. (2007). The thylakoid lumen protease Deg1 is involved in the repair of photosystem II from photoinhibition in *Arabidopsis*. Plant Cell *19*, 1039–1047.

Kashino, Y., Lauber, W.M., Carroll, J.A., Wang, Q., Whitmarsh, J., Satoh, K., and Pakrasi, H.B. (2002). Proteomic analysis of a highly active Photosystem II preparation from the cyanobacterium *Synechocystis* sp. PCC 6803 reveals the presence of novel polypeptides. Biochemistry *41*, 8004–8012.

Kato, Y., and Sakamoto, W. (2010). New insights into the types and function of proteases in plastids. Int. Rev. Cell Mol. Biol. *280*, 185–218.

Kato, Y., Sun, X., Zhang, L., and Sakamoto, W. (2012). Cooperative D1 Degradation in the Photosystem II Repair Mediated by Chloroplastic Proteases in *Arabidopsis*. Plant Physiol. *159*, 1428–1439.

Kauss, D., Bischof, S., Steiner, S., Apel, K., and Meskauskiene, R. (2012). FLU, a negative feedback regulator of tetrapyrrole biosynthesis, is physically linked to the final steps of the Mg(++)-branch of this pathway. FEBS Lett. *586*, 211–216.

Keegstra, K., and Cline, K. (1999). Protein import and routing systems of chloroplasts. Plant Cell *11*, 557–570.

Keren, N., Ohkawa, H., Welsh, E.A., Liberton, M., and Pakrasi, H.B. (2005). Psb29, a conserved 22-kD protein, functions in the biogenesis of Photosystem II complexes in *Synechocystis* and *Arabidopsis*. Plant Cell *17*, 2768–2781.

Khatoon, M., Inagawa, K., Pospisil, P., Yamashita, A., Yoshioka, M., Lundin, B., Horie, J., Morita, N., Jajoo, A., Yamamoto, Y., and Yamamoto, Y. (2009). Quality control of photosystem II: Thylakoid unstacking is necessary to avoid further damage to the D1 protein and to facilitate D1 degradation under light stress in spinach thylakoids. J. Biol. Chem. *284*, 25343–25352.

Kieselbach, T., and Funk, C. (2003). The family of Deg/HtrA proteases: from Escherichia coli to *Arabidopsis*. Physiol. Plant. *119*, 337–346.

Kim, S.J., Jansson, S., Hoffman, N.E., Robinson, C., and Mant, A. (1999). Distinct 'assisted' and 'spontaneous' mechanisms for the insertion of polytopic chlorophyll-binding proteins into the thylakoid membrane. J. Biol. Chem. *274*, 4715–4721.

Kirchhoff, H., Sharpe, R.M., Herbstova, M., Yarbrough, R., and Edwards, G.E. (2013). Differential mobility of pigment–protein complexes in granal and agranal thylakoid membranes of C(3) and C(4) plants. Plant Physiol. *161*, 497–507.

Kirst, H., Garcia-Cerdan, J.G., Zurbriggen, A., and Melis, A. (2012). Assembly of the light-harvesting chlorophyll antenna in the green alga *Chlamydomonas reinhardtii* requires expression of the TLA2-CpFTSY gene. Plant Physiol. *158*, 930–945.

Kirst, H., and Melis, A. (2014). The chloroplast signal recognition particle (CpSRP) pathway as a tool to minimize chlorophyll antenna size and maximize photosynthetic productivity. Biotechnol. Adv. *32*, 66–72.

Kleima, F.J., Hobe, S., Calkoen, F., Urbanus, M.L., Peterman, E.J., van Grondelle, R., Paulsen, H., and van Amerongen, H. (1999). Decreasing the chlorophyll a/b ratio in reconstituted LHCII: structural and functional consequences. Biochemistry *38*, 6587–6596.

Kley, J., Schmidt, B., Boyanov, B., Stolt-Bergner, P.C., Kirk, R., Ehrmann, M., Knopf, R.R., Naveh, L., Adam, Z., and Clausen, T. (2011). Structural adaptation of the plant protease Deg1 to repair photosystem II during light exposure. Nat. Struct. Mol. Biol. *18*, 728–731.

Klimmek, F., Sjodin, A., Noutsos, C., Leister, D., and Jansson, S. (2006). Abundantly and rarely expressed Lhc protein genes exhibit distinct regulation patterns in plants. Plant Physiol. *140*, 793–804.

Kmiec, B., Teixeira, P.F., and Glaser, E. (2014). Shredding the signal: targeting peptide degradation in mitochondria and chloroplasts. Trends Plant Sci. *19*, 771–778.

Knoppova, J., Sobotka, R., Tichy, M., Yu, J., Konik, P., Halada, P., Nixon, P.J., and Komenda, J. (2014). Discovery of a chlorophyll binding protein complex involved in the early steps of photosystem II assembly in Synechocystis. Plant Cell *26*, 1200–1212.

Kogata, N., Nishio, K., Hirohashi, T., Kikuchi, S., and Nakai, M. (1999). Involvement of a chloroplast homologue of the signal recognition particle receptor protein, FtsY, in protein targeting to thylakoids. FEBS Lett. *447*, 329–333.

Komenda, J., Barker, M., Kuviková, S., de Vries, R., Mullineaux, C.W., Tichý, M., and Nixon, P.J. (2006). The FtsH protease slr0228 Is Important for quality control of photosystem II in the thylakoid membrane of *Synechocystis* sp. PCC 6803. J. Biol. Chem. *281*, 1145–1151.

Komenda, J., Knoppova, J., Krynicka, V., Nixon, P.J., and Tichy, M. (2010). Role of FtsH2 in the repair of Photosystem II in mutants of the cyanobacterium *Synechocystis* PCC 6803 with impaired assembly or stability of the CaMn(4) cluster. Biochim. Biophys. Acta *1797*, 566–575.

Komenda, J., Reisinger, V., Muller, B.C., Dobakova, M., Granvogl, B., and Eichacker, L.A. (2004). Accumulation of the D2 protein is a key regulatory step for assembly of the photosystem II reaction center complex in *Synechocystis* PCC 6803. J. Biol. Chem. *279*, 48620–48629.

Komenda, J., Sobotka, R., and Nixon, P.J. (2012). Assembling and maintaining the photosystem II complex in chloroplasts and cyanobacteria. Curr. Opin. Plant Biol. *15*, 245–251.

Kouril, R., Dekker, J.P., and Boekema, E.J. (2012). Supramolecular organization of photosystem II in green plants. Biochim. Biophys. Acta *1817*, 2–12.

Kovacs-Bogdan, E., Soll, J., and Bolter, B. (2010). Protein import into chloroplasts: the Tic complex and its regulation. Biochim. Biophys. Acta *1803*, 740–747.

Krieger-Liszkay, A., Fufezan, C., and Trebst, A. (2008). Singlet oxygen production in photosystem II and related protection mechanism. Photosynth. Res. *98*, 551–564.

Kroll, D., Meierhoff, K., Bechtold, N., Kinoshita, M., Westphal, S., Vothknecht, U.C., Soll, J., and Westhoff, P. (2001). VIPP1, a nuclear gene of *Arabidopsis thaliana* essential for thylakoid membrane formation. Proc. Natl. Acad. Sci. U.S.A. *98*, 4238–4242.

Kropat, J., Oster, U., Rudiger, W., and Beck, C.F. (2000). Chloroplast signalling in the light induction of nuclear HSP70 genes requires the accumulation of chlorophyll precursors and their accessibility to cytoplasm/nucleus. Plant J. *24*, 523–531.

Kufryk, G., Hernandez-Prieto, M., Kieselbach, T., Miranda, H., Vermaas, W., and Funk, C. (2008). Association of small CAB-like proteins (SCPs) of *Synechocystis* sp. PCC 6803 with Photosystem II. Photosynth. Res. *95*, 135–145.

Larkin, R.M., Alonso, J.M., Ecker, J.R., and Chory, J. (2003). GUN4, a regulator of chlorophyll synthesis and intracellular signaling. Science *299*, 902–906.

Lin, J.F., and Wu, S.H. (2004). Molecular events in senescing *Arabidopsis* leaves. Plant J. *39*, 612–628.

Lindahl, M., Tabak, S., Cseke, L., Pichersky, E., Andersson, B., and Adam, Z. (1996). Identification, characterization, and molecular cloning of a homologue of the bacterial FtsH protease in chloroplasts of higher plants. J. Biol. Chem. *271*, 29329–29334.

Lindell, D., Sullivan, M.B., Johnson, Z.I., Tolonen, A.C., Rohwer, F., and Chisholm, S.W. (2004). Transfer of photosynthesis genes to and from Prochlorococcus viruses. Proc. Natl. Acad. Sci. U.S.A. *101*, 11013–11018.

Lipinska, B., Fayet, O., Baird, L., and Georgopoulos, C. (1989). Identification, characterization, and mapping of the Escherichia coli htrA gene, whose product is essential for bacterial growth only at elevated temperatures. J. Bacteriol. *171*, 1574–1584.

Liu, X., Yu, F., and Rodermel, S. (2010). An *Arabidopsis* pentatricopeptide repeat protein, Suppressor Of Variegation 7, is required for FtsH-mediated chloroplast biogenesis. Plant Physiol. *154*, 1588–1601.

Liu, Z., Klatt, C.G., Wood, J.M., Rusch, D.B., Ludwig, M., Wittekindt, N., Tomsho, L.P., Schuster, S.C., Ward, D.M., and Bryant, D.A. (2011). Metatranscriptomic analyses of chlorophototrophs of a hot-spring microbial mat. ISME J. *5*, 1279–1290.

Liu, Z., Yan, H., Wang, K., Kuang, T., Zhang, J., Gui, L., An, X., and Chang, W. (2004). Crystal structure of spinach major light-harvesting complex at 2.72 A resolution. Nature *428*, 287–292.

Lohscheider, J.N., Rojas-Stütz, M.C., Rothbart, M., Andersson, U., Funck, D., Mendgen, K., Grimm, B., and Adamska, I. (2015). Altered levels of LIL3 isoforms in *Arabidopsis* lead to disturbed pigment–protein assembly and chlorophyll synthesis, chlorotic phenotype and impaired photosynthetic performance. Plant Cell Environ. *38*, 2115–2127.

Loll, B., Kern, J., Saenger, W., Zouni, A., and Biesiadka, J. (2005). Towards complete cofactor arrangement in the 3.0 A resolution structure of photosystem II. Nature *438*, 1040–1044.

Long, S.P., Marshall-Colon, A., and Zhu, X.-G. (2015). Meeting the global food demand of the future by engineering crop photosynthesis and yield potential. Cell *161*, 56–66.

Lucinski, R., and Jackowski, G. (2013). AtFtsH heterocomplex-mediated degradation of apoproteins of the major light harvesting complex of photosystem II (LHCII) in response to stresses. J. Plant Physiol. *170*, 1082–1089.

Lucinski, R., Misztal, L., Samardakiewicz, S., and Jackowski, G. (2011). The thylakoid protease Deg2 is involved in stress-related degradation of the photosystem II light-harvesting protein Lhcb6 in *Arabidopsis thaliana*. New Phytol. *192*, 74–86.

Malnoe, P., Mayfield, S.P., and Rochaix, J.D. (1988). Comparative analysis of the biogenesis of photosystem II in the wild-type and Y-1 mutant of *Chlamydomonas reinhardtii*. J. Cell Biol. *106*, 609–616.

Manohara, M.S., and Tripathy, B.C. (2000). Regulation of protoporphyrin IX biosynthesis by intraplastidic compartmentalization and adenosine triphosphate. Planta *212*, 52–59.

Masoumi, A., Heinemann, I.U., Rohde, M., Koch, M., Jahn, M., and Jahn, D. (2008). Complex formation between protoporphyrinogen IX oxidase and ferrochelatase during haem biosynthesis in *Thermosynechococcus elongatus*. Microbiology *154*, 3707–3714.

Mayfield, S.P., and Taylor, W.C. (1984). Carotenoid-deficient maize seedlings fail to accumulate light-harvesting chlorophyll a/b binding protein (LHCP) mRNA. Eur. J. Biochem. *144*, 79–84.

Mazor, Y., Borovikova, A., and Nelson, N. (2015). The structure of plant photosystem I super-complex at 2.8 Å resolution. eLife *4*, e07433.

Meskauskiene, R., Nater, M., Goslings, D., Kessler, F., op den Camp, R., and Apel, K. (2001). FLU: a negative regulator of chlorophyll biosynthesis in *Arabidopsis thaliana*. Proc. Natl. Acad. Sci. U.S.A. *98*, 12826–12831.

Meyer, G., and Kloppstech, K. (1984). A rapidly light-induced chloroplast protein with a high turnover coded for by pea nuclear DNA. Eur. J. Biochem. *138*, 201–207.

Mick, V., Eggert, K., Heinemann, B., Geister, S., and Paulsen, H. (2004). Single amino acids in the luminal loop domain influence the stability of the major light-harvesting chlorophyll a/b complex. Biochemistry *43*, 5467–5473.

Miranda, H., Cheregi, O., Netotea, S., Hvidsten, T.R., Moritz, T., and Funk, C. (2013). Co-expression analysis, proteomic and metabolomic study on the impact of a Deg/HtrA protease triple mutant in *Synechocystis* sp. PCC 6803 exposed to temperature and high light stress. J. Proteom. *78*, 294–311.

Mochizuki, N., Tanaka, R., Tanaka, A., Masuda, T., and Nagatani, A. (2008). The steady-state level of Mg-protoporphyrin IX is not a determinant of plastid-to-nucleus signaling in *Arabidopsis*. Proc. Natl. Acad. Sci. U.S.A. *105*, 15184–15189.

Moldavski, O., Levin-Kravets, O., Ziv, T., Adam, Z., and Prag, G. (2012). The hetero-hexameric nature of a chloroplast AAA+ FtsH protease contributes to its thermodynamic stability. PLoS One *7*, e36008.

Montane, M.H., and Kloppstech, K. (2000). The family of light-harvesting-related proteins (LHCs, ELIPs, HLIPs): was the harvesting of light their primary function? Gene *258*, 1–8.

Moore, M., Goforth, R.L., Mori, H., and Henry, R. (2003). Functional interaction of chloroplast SRP/FtsY with the ALB3 translocase in thylakoids: substrate not required. J. Cell Biol. *162*, 1245–1254.

Moore, M., Harrison, M.S., Peterson, E.C., and Henry, R. (2000). Chloroplast Oxa1p homolog albino3 is required for post-translational integration of the light harvesting chlorophyll-binding protein into thylakoid membranes. J. Biol. Chem. *275*, 1529–1532.

Morosinotto, T., Baronio, R., and Bassi, R. (2002). Dynamics of chromophore binding to Lhc proteins *in vivo* and *in vitro* during operation of the xanthophyll cycle. J. Biol. Chem. *277*, 36913–36920.

Morosinotto, T., Breton, J., Bassi, R., and Croce, R. (2003). The nature of a chlorophyll ligand in Lhca proteins determines the far-red fluorescence emission typical of photosystem I. J. Biol. Chem. *278*, 49223–49229.

Moulin, M., McCormac, A.C., Terry, M.J., and Smith, A.G. (2008). Tetrapyrrole profiling in *Arabidopsis* seedlings reveals that retrograde plastid nuclear signaling is not due to Mg-protoporphyrin IX accumulation. Proc. Natl. Acad. Sci. U.S.A. *105*, 15178–15183.

Muh, F., Renger, T., and Zouni, A. (2008). Crystal structure of cyanobacterial photosystem II at 3.0 A resolution: a closer look at the antenna system and the small membrane-intrinsic subunits. Plant Physiol. Biochem. *46*, 238–264.

Muller, P., Li, X.P., and Niyogi, K.K. (2001). Non-photochemical quenching. A response to excess light energy. Plant Physiol. *125*, 1558–1566.

Mullineaux, C.W. (2005). Function and evolution of grana. Trends Plant. Sci. *10*, 521–525.

Neilson, J.A., and Durnford, D.G. (2010). Evolutionary distribution of light-harvesting complex-like proteins in photosynthetic eukaryotes. Genome *53*, 68–78.

Nelson, N., and Ben-Shem, A. (2005). The structure of photosystem I and evolution of photosynthesis. Bioessays *27*, 914–922.

Nelson, N., and Yocum, C.F. (2006). Structure and function of photosystems I and II. Annu. Rev. Plant Biol. *57*, 521–565.

Nguyen, T.X., Chandrasekar, S., Neher, S., Walter, P., and Shan, S.O. (2011). Concerted complex assembly and GTPase activation in the chloroplast signal recognition particle. Biochemistry *50*, 7208–7217.

Nick, S., Meurer, J., Soll, J., and Ankele, E. (2013). Nucleus-encoded light-harvesting chlorophyll a/b proteins are imported normally into chlorophyll b-free chloroplasts of *Arabidopsis*. Mol. Plant *6*, 860–871.

Nickelsen, J., and Rengstl, B. (2013). Photosystem II assembly: from cyanobacteria to plants. Annu. Rev. Plant Biol. *64*, 609–635.

Nilsson, R., Brunner, J., Hoffman, N.E., and van Wijk, K.J. (1999). Interactions of ribosome nascent chain complexes of the chloroplast-encoded D1 thylakoid membrane protein with cpSRP54. EMBO J. *18*, 733–742.

Nishimura, K., and van Wijk, K.J. (2015). Organization, function and substrates of the essential Clp protease system in plastids. Biochim. Biophys. Acta *1847*, 915–930.

Nixon, P.J., Barker, M., Boehm, M., de Vries, R., and Komenda, J. (2005). FtsH-mediated repair of the photosystem II complex in response to light stress. J. Exp. Bot. *56*, 357–363.

Nixon, P.J., Michoux, F., Yu, J., Boehm, M., and Komenda, J. (2010). Recent advances in understanding the assembly and repair of photosystem II. Annals.Bot. *106*, 1–16.

Niyogi, K.K., Li, X.P., Rosenberg, V., and Jung, H.S. (2005). Is PsbS the site of non-photochemical quenching in photosynthesis? J. Exp. Bot. *56*, 375–382.

Nogaj, L.A., and Beale, S.I. (2005). Physical and kinetic interactions between glutamyl-tRNA reductase and glutamate-1-semialdehyde aminotransferase of *Chlamydomonas reinhardtii*. J. Biol. Chem. *280*, 24301–24307.

Okuno, T., Yamanaka, K., and Ogura, T. (2006). Flavodoxin, a new fluorescent substrate for monitoring proteolytic activity of FtsH lacking a robust unfolding activity. J. Struct. Biol. *156*, 115–119.

Olmedo-Verd, E., Santamaria-Gomez, J., Ochoa de Alda, J.A., Ribas de Pouplana, L., and Luque, I. (2011). Membrane anchoring of aminoacyl-tRNA synthetases by convergent acquisition of a novel protein domain. J. Biol. Chem. *286*, 41057–41068.

op den Camp, R.G., Przybyla, D., Ochsenbein, C., Laloi, C., Kim, C., Danon, A., Wagner, D., Hideg, E., Gobel, C., Feussner, I., Nater, M., and Apel, K. (2003). Rapid induction of distinct stress responses after the release of singlet oxygen in *Arabidopsis*. Plant Cell *15*, 2320–2332.

Ozawa, S., Nield, J., Terao, A., Stauber, E.J., Hippler, M., Koike, H., Rochaix, J.D., and Takahashi, Y. (2009). Biochemical and structural studies of the large Ycf4-photosystem I assembly complex of the green alga *Chlamydomonas reinhardtii*. Plant Cell *21*, 2424–2442.

Ozawa, S., Onishi, T., and Takahashi, Y. (2010). Identification and characterization of an assembly intermediate subcomplex of photosystem I in the green alga *Chlamydomonas reinhardtii*. J. Biol. Chem. *285*, 20072–20079.

Pagano, A., Cinque, G., and Bassi, R. (1998). *In vitro* reconstitution of the recombinant photosystem II light-harvesting complex CP24 and its spectroscopic characterization. J. Biol. Chem. *273*, 17154–17165.

Pagliano, C., Saracco, G., and Barber, J. (2013). Structural, functional and auxiliary proteins of photosystem II. Photosynth. Res. *116*, 167–188.

Pan, X., Li, M., Wan, T., Wang, L., Jia, C., Hou, Z., Zhao, X., Zhang, J., and Chang, W. (2011). Structural insights into energy regulation of light-harvesting complex CP29 from spinach. Nat. Struct. Mol. Biol. *18*, 309–315.

Park, S.Y., Yu, J.W., Park, J.S., Li, J., Yoo, S.C., Lee, N.Y., Lee, S.K., Jeong, S.W., Seo, H.S., Koh, H.J., Jeon, J.S., Park, Y.I., and Paek, N.C. (2007). The senescence-induced staygreen protein regulates chlorophyll degradation. Plant Cell *19*, 1649–1664.

Pascal, A., Caffarri, S., Croce, R., Sandona, D., Bassi, R., and Robert, B. (2002). A structural investigation of the central chlorophyll a binding sites in the minor photosystem II antenna protein, Lhcb4. Biochemistry *41*, 2305–2310.

Pasch, J.C., Nickelsen, J., and Schunemann, D. (2005). The yeast split-ubiquitin system to study chloroplast membrane protein interactions. Appl. Microbiol. Biotechnol. *69*, 440–447.

Passarini, F., Wientjes, E., Hienerwadel, R., and Croce, R. (2009). Molecular basis of light harvesting and photoprotection in CP24: unique features of the most recent antenna complex. J. Biol. Chem. *284*, 29536–29546.

Paulsen, H., Finkenzeller, B., and Kuhlein, N. (1993). Pigments induce folding of light-harvesting chlorophyll a/b-binding protein. Eur. J. Biochem. *215*, 809–816.

Peers, G., Truong, T.B., Ostendorf, E., Busch, A., Elrad, D., Grossman, A.R., Hippler, M., and Niyogi, K.K. (2009). An ancient light-harvesting protein is critical for the regulation of algal photosynthesis. Nature *462*, 518–521.

Pesaresi, P., Sandona, D., Giuffra, E., and Bassi, R. (1997). A single point mutation (E166Q) prevents dicyclohexylcarbodiimide binding to the photosystem II subunit CP29. FEBS Lett. *402*, 151–156.

Peter, G.F., and Thornber, J.P. (1991). Biochemical composition and organization of higher plant photosystem II light-harvesting pigment-proteins. J. Biol. Chem. *266*, 16745–16754.

Pfannschmidt, T. (2003). Chloroplast redox signals: how photosynthesis controls its own genes. Trends Plant. Sci. *8*, 33–41.

Pisareva, T., Shumskaya, M., Maddalo, G., Ilag, L., and Norling, B. (2007). Proteomics of *Synechocystis* sp. PCC 6803. Identification of novel integral plasma membrane proteins. FEBS J. *274*, 791–804.

Polle, J.E., Benemann, J.R., Tanaka, A., and Melis, A. (2000). Photosynthetic apparatus organization and function in the wild type and a chlorophyll b-less mutant of *Chlamydomonas reinhardtii*. Dependence on carbon source. Planta *211*, 335–344.

Promnares, K., Komenda, J., Bumba, L., Nebesarova, J., Vacha, F., and Tichy, M. (2006). Cyanobacterial small chlorophyll-binding protein ScpD (HliB) is located on the periphery of photosystem II in the vicinity of PsbH and CP47 subunits. J. Biol. Chem. *281*, 32705–32713.

Qin, X., Suga, M., Kuang, T., and Shen, J.R. (2005). Structural basis for energy transfer pathways in the plant PSI-LHCI supercomplex. Science *348*, 989–995.

Rast, A., Heinz, S., and Nickelsen, J. (2015). Biogenesis of thylakoid membranes. Biochim. Biophys. Acta *1847*, 821–830.

Rawlings, N.D., Barrett, A.J., and Bateman, A. (2012). MEROPS: the database of proteolytic enzymes, their

substrates and inhibitors. Nucleic. Acids Res. *40*, D343–350.
Reinbothe, C., Bartsch, S., Eggink, L.L., Hoober, J.K., Brusslan, J., Andrade-Paz, R., Monnet, J., and Reinbothe, S. (2006). A role for chlorophyllide a oxygenase in the regulated import and stabilization of light-harvesting chlorophyll a/b proteins. Proc. Natl. Acad. Sci. U.S.A. *103*, 4777–4782.
Reinsberg, D., Booth, P.J., Jegerschold, C., Khoo, B.J., and Paulsen, H. (2000). Folding, assembly, and stability of the major light-harvesting complex of higher plants, LHCII, in the presence of native lipids. Biochemistry *39*, 14305–14313.
Reisinger, V., Ploscher, M., and Eichacker, L.A. (2008). Lil3 assembles as chlorophyll-binding protein complex during deetiolation. FEBS Lett. *582*, 1547–1551.
Remelli, R., Varotto, C., Sandona, D., Croce, R., and Bassi, R. (1999). Chlorophyll binding to monomeric light-harvesting complex. A mutation analysis of chromophore-binding residues. J. Biol. Chem. *274*, 33510–33521.
Rep, M., van Dijl, J.M., Suda, K., Schatz, G., Grivell, L.A., and Suzuki, C.K. (1996). Promotion of mitochondrial membrane complex assembly by a proteolytically inactive yeast Lon. Science *274*, 103–106.
Rexroth, S., Mullineaux, C.W., Ellinger, D., Sendtko, E., Rogner, M., and Koenig, F. (2011). The plasma membrane of the cyanobacterium Gloeobacter violaceus contains segregated bioenergetic domains. Plant Cell *23*, 2379–2390.
Rokka, A., Suorsa, M., Saleem, A., Battchikova, N., and Aro, E.M. (2005). Synthesis and assembly of thylakoid protein complexes: multiple assembly steps of photosystem II. Biochem. J. *388*, 159–168.
Roose, J.L., Kashino, Y., and Pakrasi, H.B. (2007). The PsbQ protein defines cyanobacterial Photosystem II complexes with highest activity and stability. Proc..Natl. Acad. Sci. *104*, 2548–2553.
Rossini, S., Casazza, A.P., Engelmann, E.C.M., Havaux, M., Jennings, R.C., and Soave, C. (2006). Suppression of both ELIP1 and ELIP2 in *Arabidopsis* does not affect tolerance to photoinhibition and photooxidative stress. Plant Physiol. *141*, 1264–1273.
Ruban, A.V., and Horton, P. (1999). The xanthophyll cycle modulates the kinetics of nonphotochemical energy dissipation in isolated light-harvesting complexes, intact chloroplasts, and leaves of spinach. Plant Physiol. *119*, 531–542.
Ruban, A.V., Lee, P.J., Wentworth, M., Young, A.J., and Horton, P. (1999). Determination of the stoichiometry and strength of binding of xanthophylls to the photosystem II light harvesting complexes. J. Biol. Chem. *274*, 10458–10465.
Rudiger, W., Bohm, S., Helfrich, M., Schulz, S., and Schoch, S. (2005). Enzymes of the last steps of chlorophyll biosynthesis: modification of the substrate structure helps to understand the topology of the active centers. Biochemistry *44*, 10864–10872.
Sacharz, J., Bryan, S.J., Yu, J., Burroughs, N.J., Spence, E.M., Nixon, P.J., and Mullineaux, C.W. (2015). Sub-cellular location of FtsH proteases in the cyanobacterium *Synechocystis* sp. PCC 6803 suggests localised PSII repair zones in the thylakoid membranes. Mol. Microbiol. *96*, 448–462.
Sakamoto, W., Zaltsman, A., Adam, Z., and Takahashi, Y. (2003). Coordinated regulation and complex formation of yellow variegated1 and yellow variegated 2, chloroplastic FtsH metalloproteases involved in the repair cycle of photosystem II in *Arabidopsis* thylakoid membranes. Plant Cell *15*, 2843–2855.
Sakuraba, Y., Yamasato, A., Tanaka, R., and Tanaka, A. (2007). Functional analysis of N-terminal domains of *Arabidopsis* chlorophyllide a oxygenase. Plant Physiol. Biochem. *45*, 740–749.
Santos, D., and De Almeida, D.F. (1975). Isolation and characterization of a new temperature-sensitive cell division mutant of Escherichia coli K-12. J. Bacteriol. *124*, 1502–1507.
Satoh, S., Ikeuchi, M., Mimuro, M., and Tanaka, A. (2001). Chlorophyll b expressed in Cyanobacteria functions as a light-harvesting antenna in photosystem I through flexibility of the proteins. J. Biol. Chem. *276*, 4293–4297.
Schottkowski, M., Gkalympoudis, S., Tzekova, N., Stelljes, C., Schunemann, D., Ankele, E., and Nickelsen, J. (2009). Interaction of the periplasmic PratA factor and the PsbA (D1) protein during biogenesis of photosystem II in *Synechocystis* sp. PCC 6803. J. Biol. Chem. *284*, 1813–1819.
Schottler, M.A., Albus, C.A., and Bock, R. (2011). Photosystem I: its biogenesis and function in higher plants. J. Plant Physiol. *168*, 1452–1461.
Schuhmann, H., and Adamska, I. (2012). Deg proteases and their role in protein quality control and processing in different subcellular compartments of the plant cell. Physiol. Plant *145*, 224–234.
Schunemann, D. (2004). Structure and function of the chloroplast signal recognition particle. Curr. Genet. *44*, 295–304.
Shen, G., Antonkine, M.L., van der Est, A., Vassiliev, I.R., Brettel, K., Bittl, R., Zech, S.G., Zhao, J., Stehlik, D., Bryant, D.A., and Golbeck, J.H. (2002a). Assembly of photosystem I. II. Rubredoxin is required for the *in vivo* assembly of F(X) in Synechococcus sp. PCC 7002 as shown by optical and EPR spectroscopy. J. Biol. Chem. *277*, 20355–20366.
Shen, G., Zhao, J., Reimer, S.K., Antonkine, M.L., Cai, Q., Weiland, S.M., Golbeck, J.H., and Bryant, D.A. (2002b). Assembly of photosystem I. I. Inactivation of the rubA gene encoding a membrane-associated rubredoxin in the cyanobacterium Synechococcus sp. PCC 7002 causes a loss of photosystem I activity. J. Biol. Chem. *277*, 20343–20354.
Shepherd, M., McLean, S., and Hunter, C.N. (2005). Kinetic basis for linking the first two enzymes of chlorophyll biosynthesis. FEBS J. *272*, 4532–4539.
Shi, L.-X., Hall, M., Funk, C., and Schröder, W.P. (2012). Photosystem II, a growing complex: Updates on newly discovered components and low molecular mass proteins. Biochim. Biophys. Acta *1817*, 13–25.
Shi, L.X., and Schroder, W.P. (2004). The low molecular mass subunits of the photosynthetic supracomplex, photosystem II. Biochim. Biophys. Acta *1608*, 75–96.
Shlyk, A.A. (1971). Biosynthesis of chlorophyll b. Ann. Rev. Plant Physiol. *22*, 169–184.

Silva, P., Thompson, E., Bailey, S., Kruse, O., Mullineaux, C.W., Robinson, C., Mann, N.H., and Nixon, P.J. (2003). FtsH is involved in the early stages of repair of photosystem II in *Synechocystis* sp PCC 6803. Plant Cell *15*, 2152–2164.

Singh, M., Satoh, K., Yamamoto, Y., Kanervo, E., and Aro, E.M. (2008). *In vivo* quality control of photosystem II in cyanobacteria *Synechocystis* sp. PCC 6803: D1 protein degradation and repair under the influence of light, heat and darkness. Indian J. Biochem. Biophys. *45*, 237–243.

Sinha, R.K., Komenda, J., Knoppova, J., Sedlarova, M., and Pospisil, P. (2012). Small CAB-like proteins prevent formation of singlet oxygen in the damaged photosystem II complex of the cyanobacterium *Synechocystis* sp. PCC 6803. Plant Cell Environ. *35*, 806–818.

Sinvany-Villalobo, G., Davydov, O., Ben-Ari, G., Zaltsman, A., Raskind, A., and Adam, Z. (2004). Expression in multigene families. Analysis of chloroplast and mitochondrial proteases. Plant Physiol. *135*, 1336–1345.

Sobotka, R. (2014). Making proteins green; biosynthesis of chlorophyll-binding proteins in cyanobacteria. Photosynth. Res. *119*, 223–232.

Sobotka, R., Duhring, U., Komenda, J., Peter, E., Gardian, Z., Tichy, M., Grimm, B., and Wilde, A. (2008a). Importance of the cyanobacterial Gun4 protein for chlorophyll metabolism and assembly of photosynthetic complexes. J. Biol. Chem. *283*, 25794–25802.

Sobotka, R., McLean, S., Zuberova, M., Hunter, C.N., and Tichy, M. (2008b). The C-terminal extension of ferrochelatase is critical for enzyme activity and for functioning of the tetrapyrrole pathway in *Synechocystis* strain PCC 6803. J. Bacteriol. *190*, 2086–2095.

Sozer, O., Komenda, J., Ughy, B., Domonkos, I., Laczko-Dobos, H., Malec, P., Gombos, Z., and Kis, M. (2010). Involvement of carotenoids in the synthesis and assembly of protein subunits of photosynthetic reaction centers of *Synechocystis* sp. PCC 6803. Plant Cell Physiol. *51*, 823–835.

Standfuss, J., Terwisscha van Scheltinga, A.C., Lamborghini, M., and Kuhlbrandt, W. (2005). Mechanisms of photoprotection and nonphotochemical quenching in pea light-harvesting complex at 2.5 A resolution. EMBO J. *24*, 919–928.

Stengel, A., Gugel, I.L., Hilger, D., Rengstl, B., Jung, H., and Nickelsen, J. (2012). Initial steps of photosystem II *de novo* assembly and preloading with manganese take place in biogenesis centers in *Synechocystis*. Plant Cell *24*, 660–675.

Storm, P., Hernandez-Prieto, M.A., Eggink, L.L., Hoober, J.K., and Funk, C. (2008). The small CAB-like proteins of *Synechocystis* sp. PCC 6803 bind chlorophyll. *In vitro* pigment reconstitution studies on one-helix light-harvesting-like proteins. Photosynth. Res. *98*, 479–488.

Storm, P., Tibiletti, T., Hall, M., and Funk, C. (2013). Refolding and enzyme kinetic studies on the ferrochelatase of the cyanobacterium *Synechocystis* sp. PCC 6803. PLoS One *8*, e55569.

Strand, A. (2004). Plastid-to-nucleus signalling. Curr. Opin. Plant Biol. *7*, 621–625.

Strand, A., Asami, T., Alonso, J., Ecker, J.R., and Chory, J. (2003). Chloroplast to nucleus communication triggered by accumulation of Mg-protoporphyrinIX. Nature *421*, 79–83.

Strauch, K.L., Johnson, K., and Beckwith, J. (1989). Characterization of degP, a gene required for proteolysis in the cell envelope and essential for growth of *Escherichia coli* at high temperature. J. Bacteriol. *171*, 2689–2696.

Sun, H., Sako, K., Suzuki, Y., Maekawa, S., Yasuda, S., Chiba, Y., Sato, T., and Yamaguchi, J. (2012). Sugar-inducible RPT2a, a subunit of 26S proteasome, participates in sugar response in *Arabidopsis*. Plant Biotech. *29*, 279–284.

Sun, X., Peng, L., Guo, J., Chi, W., Ma, J., Lu, C., and Zhang, L. (2007). Formation of DEG5 and DEG8 complexes and their involvement in the degradation of photodamaged photosystem II reaction center D1 protein in *Arabidopsis*. Plant Cell *19*, 1347–1361.

Swiatek, M., Regel, R.E., Meurer, J., Wanner, G., Pakrasi, H.B., Ohad, I., and Herrmann, R.G. (2003). Effects of selective inactivation of individual genes for low-molecular-mass subunits on the assembly of photosystem II, as revealed by chloroplast transformation: the psbEFLJoperon in *Nicotiana tabacum*. Mol. Genet. Genom. *268*, 699–710.

Takechi, K., Sodmergen, Murata, M., Motoyoshi, F., and Sakamoto, W. (2000). The yellow variegated (VAR2) locus encodes a homologue of FtsH, an ATP-dependent protease in *Arabidopsis*. Plant Cell Physiol. *41*, 1334–1346.

Tam, L.X., Aigner, H., Timmerman, E., Gevaert, K., and Funk, C. (2015). Proteomic approaches to identify substrates of the three Deg/HtrA proteases of the cyanobacterium *Synechocystis* sp. PCC 6803. Biochem. J. *468*, 373–384.

Tanaka, R., Rothbart, M., Oka, S., Takabayashi, A., Takahashi, K., Shibata, M., Myouga, F., Motohashi, R., Shinozaki, K., Grimm, B., and Tanaka, A. (2010). LIL3, a light-harvesting-like protein, plays an essential role in chlorophyll and tocopherol biosynthesis. Proc. Natl. Acad. Sci.U.S.A. *107*, 16721–16725.

Tanaka, R., and Tanaka, A. (2011). Chlorophyll cycle regulates the construction and destruction of the light-harvesting complexes. Biochim. Biophys. Acta *1807*, 968–976.

Tanz, S.K., Castleden, I., Hooper, C.M., Small, I., and Millar, A.H. (2014). Using the SUBcellular database for *Arabidopsis* proteins to localize the Deg protease family. Front. Plant Sci. *5*, 396.

Thomas, J., and Weinstein, J.D. (1990). Measurement of heme efflux and heme content in isolated developing chloroplasts. Plant Physiol. *94*, 1414–1423.

Thornber, J.P., and Highkin, H.R. (1974). Composition of the photosynthetic apparatus of normal barley leaves and a mutant lacking chlorophyll b. Eur. J. Biochem. *41*, 109–116.

Tibiletti, T., Hernandez-Prieto, M., Matthijs, H., Niyogi, K.K., and Funk, C. (2016). Deletion of the gene family of small chlorophyll-binding proteins (ScpABCDE) offsets C/N homeostasis in *Synechocystis* PCC 6803, Biochim. Biophys. Acta *1857*, 396–407.

Tyystjarvi, T., Herranen, M., and Aro, E.M. (2001). Regulation of translation elongation in cyanobacteria: membrane targeting of the ribosome nascent-chain complexes controls the synthesis of D1 protein. Mol. Microbiol. *40*, 476–484.

Tzinas, G., Argyroudi-Akoyunoglou, J.H., and Akoyunoglou, G. (1987). The effect of the dark interval in intermittent light on thylakoid development: photosynthetic unit formation and light harvesting protein accumulation. Photosynth. Res. *14*, 241–258.

Tzvetkova-Chevolleau, T., Hutin, C., Noel, L.D., Goforth, R., Carde, J.P., Caffarri, S., Sinning, I., Groves, M., Teulon, J.M., Hoffman, N.E., *et al.* (2007). Canonical signal recognition particle components can be bypassed for posttranslational protein targeting in chloroplasts. Plant Cell *19*, 1635–1648.

Umena, Y., Kawakami, K., Shen, J.R., and Kamiya, N. (2011). Crystal structure of oxygen-evolving photosystem II at a resolution of 1.9 A. Nature *473*, 55–60.

Uniacke, J., and Zerges, W. (2007). Photosystem II assembly and repair are differentially localized in *Chlamydomonas*. Plant Cell *19*, 3640–3654.

Vass, I., and Cser, K. (2009). Janus-faced charge recombinations in photosystem II photoinhibition. Trends Plant Sci. *14*, 200–205.

Vavilin, D., Yao, D., and Vermaas, W. (2007). Small Cab-like proteins retard degradation of photosystem II-associated chlorophyll in *Synechocystis* sp. PCC 6803: kinetic analysis of pigment labeling with ^{15}N and ^{13}C. J. Biol. Chem. *282*, 37660–37668.

Wagner, D., Przybyla, D., Op den Camp, R., Kim, C., Landgraf, F., Lee, K.P., Wursch, M., Laloi, C., Nater, M., Hideg, E., and Apel, K. (2004). The genetic basis of singlet oxygen-induced stress responses of *Arabidopsis thaliana*. Science *306*, 1183–1185.

Wagner, R., Aigner, H., and Funk, C. (2012). FtsH proteases located in the plant chloroplast. Physiol. Plant *145*, 203–214.

Wagner, R., Aigner, H., Pruzinska, A., Jankanpaa, H.J., Jansson, S., and Funk, C. (2011). Fitness analyses of *Arabidopsis thaliana* mutants depleted of FtsH metalloproteases and characterization of three FtsH6 deletion mutants exposed to high light stress, senescence and chilling. New Phytol. *191*, 449–458.

Weber, E.R., Hanekamp, T., and Thorsness, P.E. (1996). Biochemical and functional analysis of the YME1 gene product, an ATP and zinc-dependent mitochondrial protease from *S. cerevisiae*. Mol. Biol. Cell *7*, 307–317.

Welte, T., Kudva, R., Kuhn, P., Sturm, L., Braig, D., Muller, M., Warscheid, B., Drepper, F., and Koch, H.G. (2012). Promiscuous targeting of polytopic membrane proteins to SecYEG or YidC by the *Escherichia coli* signal recognition particle. Mol. Biol. Cell *23*, 464–479.

Wetzel, C.M., Jiang, C.Z., Meehan, L.J., Voytas, D.F., and Rodermel, S.R. (1994). Nuclear–organelle interactions: the immutans variegation mutant of *Arabidopsis* is plastid autonomous and impaired in carotenoid biosynthesis. Plant J. *6*, 161–175.

White, R.A., Wolfe, G.R., Komine, Y., and Hoober, J.K. (1996). Localization of light-harvesting complex apoproteins in the chloroplast and cytoplasm during greening of *Chlamydomonas reinhardtii* at 38 degrees C. Photosynth. Res. *47*, 267–280.

Wientjes, E., van Stokkum, I.H., van Amerongen, H., and Croce, R. (2011). The role of the individual Lhcas in photosystem I excitation energy trapping. Biophys. J. *101*, 745–754.

Wilde, A., Mikolajczyk, S., Alawady, A., Lokstein, H., and Grimm, B. (2004). The gun4 gene is essential for cyanobacterial porphyrin metabolism. FEBS Lett. *571*, 119–123.

Wollman, F.A. (2001). State transitions reveal the dynamics and flexibility of the photosynthetic apparatus. EMBO J. *20*, 3623–3630.

Wollman, F.A., Minai, L., and Nechushtai, R. (1999). The biogenesis and assembly of photosynthetic proteins in thylakoid membranes1. Biochim. Biophys. Acta *1411*, 21–85.

Woodson, J.D., Perez-Ruiz, J.M., and Chory, J. (2011). Heme synthesis by plastid ferrochelatase I regulates nuclear gene expression in plants. Curr. Biol. *21*, 897–903.

Wostrikoff, K., Girard-Bascou, J., Wollman, F.A., and Choquet, Y. (2004). Biogenesis of PSI involves a cascade of translational autoregulation in the chloroplast of *Chlamydomonas*. EMBO J. *23*, 2696–2705.

Xu, H., Vavilin, D., Funk, C., and Vermaas, W. (2002). Small Cab-like proteins regulating tetrapyrrole biosynthesis in the cyanobacterium *Synechocystis* sp. PCC 6803. Plant Mol. Biol. *49*, 149–160.

Xu, H., Vavilin, D., Funk, C., and Vermaas, W. (2004). Multiple deletions of small Cab-like proteins in the cyanobacterium *Synechocystis* sp. PCC 6803: consequences for pigment biosynthesis and accumulation. J. Biol. Chem. *279*, 27971–27979.

Xu, H., Vavilin, D., and Vermaas, W. (2001). Chlorophyll b can serve as the major pigment in functional photosystem II complexes of cyanobacteria. Proc. Natl. Acad. Sci. U.S.A. *98*, 14168–14173.

Yamamoto, Y., Inagaki, N., and Satoh, K. (2001). Overexpression and characterization of carboxyl-terminal processing protease for precursor D1 protein: regulation of enzyme–substrate interaction by molecular environments. J. Biol. Chem. *276*, 7518–7525.

Yamasato, A., Nagata, N., Tanaka, R., and Tanaka, A. (2005). The N-terminal domain of chlorophyllide a oxygenase confers protein instability in response to chlorophyll B accumulation in *Arabidopsis*. Plant Cell *17*, 1585–1597.

Yang, D.H., Webster, J., Adam, Z., Lindahl, M., and Andersson, B. (1998). Induction of acclimative proteolysis of the light-harvesting chlorophyll a/b protein of photosystem II in response to elevated light intensities. Plant Physiol. *118*, 827–834.

Yang, H., Liu, J., Wen, X., and Lu, C. (2015). Molecular mechanism of photosystem I assembly in oxygenic organisms. Biochim. Biophys. Acta *1847*, 838–848.

Yao, D., Kieselbach, T., Komenda, J., Promnares, K., Prieto, M.A., Tichy, M., Vermaas, W., and Funk, C. (2007). Localization of the small CAB-like proteins in photosystem II. J. Biol. Chem. *282*, 267–276.

Yao, D.C., Brune, D.C., Vavilin, D., and Vermaas, W.F. (2012). Photosystem II component lifetimes in the cyanobacterium *Synechocystis* sp. strain PCC 6803: small Cab-like proteins stabilize biosynthesis intermediates and affect early steps in chlorophyll synthesis. J. Biol. Chem. *287*, 682–692.

Yoshioka-Nishimura, M., and Yamamoto, Y. (2014). Quality control of Photosystem II: the molecular basis for the action of FtsH protease and the dynamics of the thylakoid membranes. J. Photochem. Photobiol. B *137*, 100–106.

Yoshioka, M., Nakayama, Y., Yoshida, M., Ohashi, K., Morita, N., Kobayashi, H., and Yamamoto, Y. (2010).

Quality control of photosystem II: FtsH hexamers are localized near photosystem II at grana for the swift repair of damage. J. Biol. Chem. *285*, 41972–41981.

Yu, F., Park, S., and Rodermel, S.R. (2004). The *Arabidopsis* FtsH metalloprotease gene family: interchangeability of subunits in chloroplast oligomeric complexes. Plant J. *37*, 864–876.

Zak, E., Norling, B., Maitra, R., Huang, F., Andersson, B., and Pakrasi, H.B. (2001). The initial steps of biogenesis of cyanobacterial photosystems occur in plasma membranes. Proc. Natl. Acad. Sci. U.S.A. *98*, 13443–13448.

Zaltsman, A., Feder, A., and Adam, Z. (2005a). Developmental and light effects on the accumulation of FtsH protease in *Arabidopsis* chloroplasts – implications for thylakoid formation and photosystem II maintenance. Plant J. *42*, 609–617.

Zaltsman, A., Ori, N., and Adam, Z. (2005b). Two types of FtsH protease subunits are required for chloroplast biogenesis and Photosystem II repair in *Arabidopsis*. Plant Cell *17*, 2782–2790.

Zelisko, A., Garcia-Lorenzo, M., Jackowski, G., Jansson, S., and Funk, C. (2005). AtFtsH6 is involved in the degradation of the light-harvesting complex II during high-light acclimation and senescence. Proc. Natl. Acad. Sci. U.S.A. *102*, 13699–13704.

Zelisko, A., and Jackowski, G. (2004). Senescence-dependent degradation of Lhcb3 is mediated by a thylakoid membrane-bound protease. J. Plant Physiol. *161*, 1157–1170.

Zhang, L., Paakkarinen, V., van Wijk, K.J., and Aro, E.M. (1999). Co-translational assembly of the D1 protein into photosystem II. J. Biol. Chem. *274*, 16062–16067.

Zhang, Y., Liu, C., and Yang, C. (2012). Analysis of heat-induced disassembly process of three different monomeric forms of the major light-harvesting chlorophyll a/b complex of photosystem II. Photosynth. Res. *111*, 103–111.

Zhang, Y.J., Tian, H.F., and Wen, J.F. (2009). The evolution of YidC/Oxa/Alb3 family in the three domains of life: a phylogenomic analysis. BMC Evol. Biol. *9*, 137.

Zhu, X.G., Ort, D.R., Whitmarsh, J., and Long, S.P. (2004). The slow reversibility of photosystem II thermal energy dissipation on transfer from high to low light may cause large losses in carbon gain by crop canopies: a theoretical analysis. J. Exp. Bot. *55*, 1167–1175.

Zienkiewicz, M., Ferenc, A., Wasilewska, W., and Romanowska, E. (2012). High light stimulates Deg1-dependent cleavage of the minor LHCII antenna proteins CP26 and CP29 and the PsbS protein in *Arabidopsis thaliana*. Planta *235*, 279–288.

Zienkiewicz, M., Kokoszka, N., Baclawska, I., Drozak, A., and Romanowska, E. (2013). Light intensity and quality stimulated Deg1-dependent cleavage of PSII components in the chloroplasts of maize. Plant Physiol. Biochem. *67*, 126–136.

Zouni, A., Witt, H.T., Kern, J., Fromme, P., Krauss, N., Saenger, W., and Orth, P. (2001). Crystal structure of photosystem II from Synechococcus elongatus at 3.8 A resolution. Nature *409*, 739–743.

Structural and Functional Dynamics of the Thylakoid Membrane System

3

Sujith Puthiyaveetil, Helmut Kirchhoff and Ricarda Höhner

Abstract

The thylakoid membrane harbours the photosynthetic machinery and is the site of light reactions in plants, algae, and cyanobacteria. In plants and green algae it stacks up to form the remarkable granal structures, the origin of which is likely to be a compromise between efficient light harvesting and diffusional freedom of electron carriers. Diffusional limitation in the highly crowded thylakoid membrane may also keep the grana diameter in check. The semi-crystalline array formation of photosystems and compositional changes in thylakoid lipids are strategies to overcome diffusional limitations under a variety of abiotic stresses. Under light and nutrient stress, the thylakoid membrane is profoundly reconfigured in ultrastructure, organization and composition. This striking plasticity in the structure and function of the thylakoid membrane tunes photosynthetic energy conversion in changing environmental and developmental conditions. A complete understanding of thylakoid membrane flexibility may illuminate and inform reengineering of photosynthesis and the design of artificial photosynthetic membranes.

Introduction

Chloroplasts are double membrane-bound cytoplasmic organelles and the sites of photosynthesis in plants and algae. They originated from free-living cyanobacteria, which established a symbiotic relationship with a eukaryotic host cell around 1.2 billion years ago. In both cyanobacteria and chloroplasts, photosynthesis takes place on specialized membranes known as the thylakoids, which are essentially flattened sacs. Thylakoid membranes are contained within chloroplasts in plants and algae, but in cyanobacteria they are the invaginations of the plasma membrane. Although our understanding of thylakoid structure and function is nearing atomic resolution, the thylakoid biogenesis is still shrouded in mystery. Thylakoid biogenesis should involve the concerted assembly and incorporation of multi-subunit protein complexes, lipids and a variety of inorganic and organic cofactors. There is an increasing realization that the synthesis and assembly of protein complexes takes place in distinct membrane 'zones' or 'centres' (Nickelsen and Zerges, 2013; Rast *et al.*, 2015). In cyanobacteria 'thylakoid centres' are rod-like membrane structures whose ends are tethered to the plasma membrane (Van De Meene *et al.*, 2006). These rod-like structures consist of assemblies of homo-oligomeric vesicle-inducing protein in plastids 1 (VIPP1) proteins (Fuhrmann *et al.*, 2009; Nordhues *et al.*, 2012). VIPP1 participates in the assembly of protein complexes by facilitating lipid insertion into them (Nordhues *et al.*, 2012). The important role of VIPP1 in thylakoid biogenesis is underlined by the fact that it is absent in the non-thylakoid-forming cyanobacterium *Gloeobacter violaceus* (Rippka *et al.*, 1974; Nakamura *et al.*, 2003). VIPP1 is a nuclear-encoded chloroplast protein in plants and algae. In *Chlamydomonas reinhardtii* it assembles into dot- or rod-like structures in an area comprising the so-called translation zones (T zones), where the translation of some major photosynthetic protein complexes takes place (Nordhues *et al.*, 2012). It is not clear whether such dedicated thylakoid membrane zones or biogenesis centres occur in plant chloroplasts.

There are four major electron transport complexes embedded in the thylakoid membrane. They are photosystem II (PS II), photosystem I

(PS I), cytochrome b_6f complex (cyt b_6f) and ATP synthase (ATPase). PS II is the water: PQ oxidoreductase of oxygenic photosynthesis. Its monomeric core is 350 kDa in size, a massive structure with nearly 30 protein subunits (Suga *et al.*, 2014; Umena *et al.*, 2011). The reaction centre (RC) is formed by the heterodimeric D1-D2 proteins. CP43 and CP47 proteins constitute an internal antenna ring around the RC. Other subunits are mostly minor that sit in and around the RC, with roles in stabilization, dimerization, and biogenesis of PS II. Some of the minor subunits form the luminal housing – the oxygen evolving complex (OEC) – that harbours the Mn^4Ca cluster catalysing the water oxidation. The beta barrel protein PsbO is a major component of the OEC. PS II core also contains the redox cofactors involved in electron transport and the special pair of chlorophylls, which initiate electron transfer reactions. The trimeric light-harvesting complex II (LHC II) is the light harvesting antenna of PS II in plants and green algae but under certain conditions it can also serve PS I. Each LHC II monomer binds eight chlorophyll *a*, six chlorophyll *b*, two luteins, one neoxanthin and one additional xanthophyll, to the tune of a chlorophyll concentration of 0.3 M (Liu *et al.*, 2004). LHC II tethers to the core through the monomeric antennae CP29, CP24 and CP26. The core itself is dimeric, to which LHC II may bind strongly (through CP26), moderately (through CP29 and CP24), or loosely, forming the S, M, and L trimers of the LHC II (Boekema *et al.*, 1999). The membrane-extrinsic phycobilisome substitutes LHC II in cyanobacteria and red algae. Certain non-green algae contain PS II antennae that are quite different from LHC II and phycobilisomes (Hohmann-Marriott and Blankenship, 2011).

PS I is the PC:ferredoxin oxidoreductase, which generates the most negative redox potential in nature (Nelson and Ben-Shem, 2004). It is monomeric in plant and green algae, trimeric in cyanobacteria and even tetrameric in a thermophilic cyanobacterium (Li *et al.*, 2014). The PS I RC is formed by the heterodimeric PsaA-PsaB proteins. The PsaC, D and E subunits comprise the stromal ferredoxin docking site and the PsaF and N subunits constitute the luminal plastocyanin (PC) binding site. The PS I antenna is made up of four light-harvesting complex I (LHC I) monomers that form a crescent-shaped belt around the core (Amunts *et al.*, 2007; Qin *et al.*, 2015).

The cyt b_6f complex is the PQH_2:PC oxidoreductase that bridges the two photosystems in the electron transport chain. The cytochrome b_6 subunit binds the two haems that participate in the electron transport. It also binds an additional haem, haem x, whose function remains uncertain (Kurisu *et al.*, 2003; Stroebel *et al.*, 2003). The other major subunits of cyt b_6f are subunit IV, cytochrome *f* and the Rieske iron–sulfur protein. Subunit IV together with cyt b_6 and Rieske protein form the two PQ binding sites, Q_i and Q_o, for the operation of the Q-cycle. The cyt *f* makes up the luminal PC docking site and the Rieske protein transfers an electron from the Q_o-bound PQH_2 to cyt *f* through its mobile head. In addition to these four large subunits, cyt b_6f also contains four small (~ 3 kDa each) single membrane-spanning subunits, PetG, -L, -M and -N, which form a 'picket-fence' at the periphery of the complex (Kurisu *et al.*, 2003). A beta-carotene molecule inserts through this and is likely to aid in the assembly of the whole complex. A chlorophyll *a* molecule, with an assumed structural role, is also found within the cyt b_6f (Hasan and Cramer, 2012).

The ATPase couples the proton gradient to ATP formation through a rotary mechanism (Junge *et al.*, 2009). It comprises a membrane embedded F_o and membrane extrinsic F_1 portions. The rotor is made up of the rod-shaped central gamma subunit and the C-subunit ring, whose ring size varies between species. The major components of the stator are the catalytic A and B subunits, the stalks I and II, and the proton channel forming IV subunit (Abrahams *et al.*, 1994). In addition to these four major electron transport complexes, chloroplasts also contain the NADH dehydrogenase complex, homologous to the complex I of mitochondria. It seems that the plant NADH dehydrogenase has lost the NADH-binding subunit and its NADH dehydrogenase function (Shikanai, 2014). Its functional role is discussed below.

PS II, cyt b_6f and the PS I participate in the linear electron transport from water to $NADP^+$, producing ATP and the reducing power NADPH for the fixation of CO_2 into carbohydrates. The vectorial electron transport reactions of photosynthesis generate a proton gradient, required for ATP synthesis by the ATPase. Electrons could also flow around PS I in cyclic electron transport, producing just the ATP in the process. This cyclic mode also requires the cyt b_6f complex, with the NADH dehydrogenase complex thought to be one arm of this pathway (see Chapter 4) (Shikanai, 2014). For their efficient functioning in light conversion, electron transport complexes form supercomplexes. Supercomplexes of PS II comprise a PS II core with varying numbers of LHC II attached. The most abundant supercomplex is C2S2M2, which has a dimeric core with two S and M trimers attached (Kouril *et al.*, 2012). Plant PS I can form a supercomplex with NADH dehydrogenase, as part of the cyclic electron transport (Shikanai, 2014). In the green alga *C. reinhardtii* the cyclic mode utilizes a supercomplex that incorporates PS I, LHC II, and cyt b_6f (Iwai *et al.*, 2010).

The functional design of the thylakoid membrane reveals it to be anything but static. A remarkable dynamism, reminding of the grand design of photosynthesis, is built into the fabric of the photosynthetic machinery across the phylogenetic spectrum (Anderson *et al.*, 1995). This is to ensure its smooth functioning and preservation in changing environmental and developmental conditions. The dynamic responses of thylakoids include changes in its structure and function on a timescale of seconds to days. Abiotic factors such as light stress, drought and salinity are major drains on photosynthetic efficiency, contributing to 70% of the productivity loss in crop plants (Boyer, 1982). A thorough understanding of the thylakoid acclimatory responses to abiotic factors should reveal important pathways and mechanisms that could be targets for reengineering photosynthesis in crops and biofuel algae. Abiotic stress tolerant pathways could also be transferred from tolerant to susceptible species. The dynamic design principles of the thylakoid membrane may also inspire and inform the fabrication of artificial photosynthetic membranes and systems.

This chapter gives an account of the structural and functional flexibility of the thylakoid membrane system. It begins by explaining the ultrastructural and molecular features of the thylakoids, and moves onto the evolutionary selection pressure for the origin of grana, the problem of molecular crowding and its effect on diffusional processes, the constancy of grana diameter, the occurrence and significance of PS II arrays, the acclimatory changes in thylakoid lipid composition, and light and nutrient stress acclimations. The chapter ends by giving some areas for future pursuit. The treatment of some topics is not exhaustive, and if this is the case, relevant chapters and sources are indicated for a more comprehensive coverage. The main thrust is on mechanisms of plant thylakoids. If related processes, however, are known in algae and cyanobacteria, these are also mentioned.

Architecture of the thylakoid membrane

Plant and some green algal thylakoid membranes stack up into remarkable cylindrical granal structures (Fig. 3.1). The thylakoid stacking produces four distinct domains: the core grana, the grana margins, the end membranes, and the unstacked stroma lamellae. PS II is mainly found in the grana (Albertsson, 2001; Andersson and Anderson,

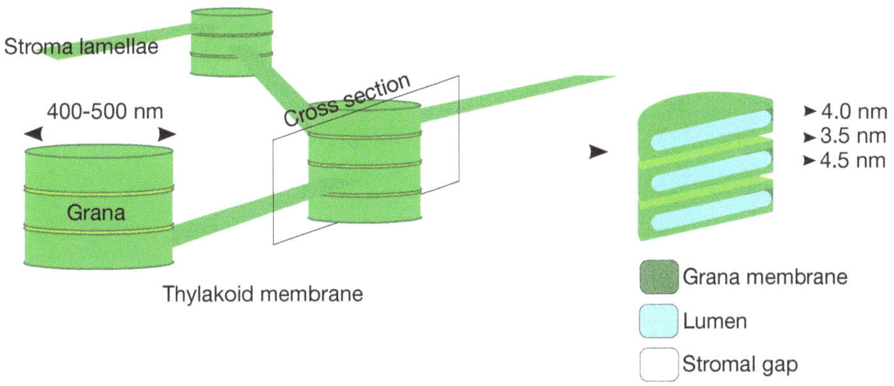

Figure 3.1 Dimensions and organization of the grana thylakoid membrane, about 4 nm thick; modified from (Kirchhoff, 2014). The figure is not to scale.

1980; Dekker and Boekema, 2005; Staehelin, 2003; Staehelin and van der Staay, 1996). The stromal gap in grana is about 3.5 nm (Fig. 3.1), resulting in the exclusion of PS I and the ATPase from the grana, as their protrusions are larger (Abrahams *et al.*, 1994; Amunts *et al.*, 2007). This concentrates them in the stroma lamellae. The separation of PS II and PS I introduces a lateral heterogeneity in the thylakoid membrane of plants and green algae. Cyanobacterial thylakoid membranes, in contrast to plants and green algae, show almost no membrane stacking or lateral heterogeneity (Mullineaux, 2005). The width of the plant thylakoid lumen is 4.5 nm in moderate light (Fig. 3.1), and this is variable under different light conditions. This variability represents an important aspect of the functional flexibility of the thylakoid membrane. The significance of the lumen width will be addressed in later sections 10. In contrast to the photosystems, the location of the cyt b_6f complex is still under debate. It was initially assumed to be evenly distributed across the thylakoid membrane (Allen and Forsberg, 2001; Staehelin and van der Staay, 1996). A recent study using a semi-quantitative proteomics approach, however, shows an enrichment of cyt b_6f complexes in stroma lamellae (Tomizioli *et al.*, 2014). Another recent study, using a PC-functionalized atomic force microscopy (AFM) probe, finds cyt b_6f in nanodomains with PS II in granal regions (Johnson *et al.*, 2014). An immunolocalization of the cyt b_6f using an anti-Rieske antibody, shows an enrichment of cyt b_6f in grana margins instead, contradicting these other reports (Puthiyaveetil, unpublished data). Where exactly the cyt b_6f occurs in the thylakoid membrane thus remains inconclusive, though its precise location is significant for long-range diffusional processes as discussed below (see 'Macromolecular crowding in grana membranes – a challenge for diffusional processes').

Why do thylakoid membranes form grana?

Interestingly, the sophisticated grana membrane structure of plants is not essential for oxygenic photosynthesis, raising the question of the functional advantage of grana stacks. Perhaps as an extreme example in granal stacking, the low light grown *Alocasia macrorrhiza* has as much as 160 thylakoids per granum (Chow *et al.*, 1988a). An earlier suggestion is that grana formation facilitates light harvesting and its regulation via non-photochemical quenching (NPQ), thus protecting PS II from photo-damage (Anderson, 1999; Horton, 1999). The membrane stacking has also been described as a way to physically separate the slow-working PS II from the fast-working PS I, hence avoiding a spillover of excitation energy (Trissl and Wilhelm, 1993). It is further suggested that the lateral heterogeneity, by grana formation, avoids competition between linear and cyclic electron flow (Albertsson, 2001), since the induction of cyclic electron flow requires a strong regulatory input (Bendall and Manasse, 1995). However, most of the cyanobacterial thylakoids do not show lateral heterogeneity nor extensive stacking and have efficient NPQ mechanisms (Bailey *et al.*, 2005; Cadoret *et al.*, 2004). Hence, preventing spillover and optimizing light harvesting are not sufficient reasons for grana evolution (Mullineaux, 2005). Another observation that speaks against the link between grana formation and efficient light harvesting is the demonstration, using artificially destacked grana membranes, that transversal energy transfer between membranes is not significant but the energy transfer is mainly lateral along the membrane plane (Kirchhott et al., 2004a). It has thus been speculated (Mullineaux, 2005) that the evolution of stacked grana membranes might represent a compromise between efficient light harvesting with a relatively large LHC antenna and the diffusional freedom of PQ in the PS II-enriched membrane, since the lateral packing of LHC II around PS II would restrict PQ diffusion.

For many species the fraction of grana membranes, including grana core and grana margins, is 80% of the thylakoid membrane and is remarkably constant (Anderson *et al.*, 1995). The grana diameter in different plant species of higher plants is also remarkably constant, varying only slightly in non-stressed conditions within 400–500 nm (Fig. 3.1) (Anderson *et al.*, 2008; Kaftan *et al.*, 2002; Kirchhoff, 2008a). Thus it seems that the grana diameter is evolutionarily constant, though the significance of this consistency is not known. Grana harbour 80% of the chlorophylls in thylakoid membrane and are densely packed with proteins. The latter has a major effect on diffusional processes that are required for diverse photosynthetic functions, as discussed in detail below.

Macromolecular crowding in grana membranes – a challenge for diffusional processes

LHC II–PS II supercomplexes alone and LHC II–PS II supercomplexes with additional loosely bound LHC II trimers (approximately three per PS II RC) occupy around 50% and 80% of the total grana membrane area, respectively (Kirchhoff, 2008a). EM and AFM studies verify these data (Kirchhoff *et al.*, 2002, 2004b; Tremmel *et al.*, 2003), revealing the protein area of grana to be remarkably constant between 70 and 80%.

For the entire thylakoid membrane protein density is about 69% (Kirchhoff *et al.* 2002). Precise data for the protein density in stroma lamellae is lacking, but is assumed to be one-third lower than that of grana. Dense protein packing in grana membranes is advantageous for light harvesting as sunlight is a dilute energy source on the molecular scale (Blankenship, 2002). High pigment concentration and protein densities are beneficial in terms of efficient light absorption and harvesting, respectively, maintaining the required ultrafast energy transfer and the high quantum yield of PS II (Kirchhoff, 2008; Haferkamp *et al.* 2010). Interestingly, dilution of the protein density from 80% to about 70% increases light harvesting by PS II and the connectivity of PS II reaction centres, indicating that native grana membranes adjust their protein density to about 70% to ensure efficient light harvesting (Haferkamp *et al.*, 2010). A further dilution of the protein density with lipids, however, causes a separation of PS II and LHC II complexes. A decreased excitonic coupling of these complexes led to a lower PS II quantum efficiency, pointing towards the significance of molecular crowding and an optimal protein packing density of 70% (Haferkamp and Kirchhoff, 2008; Haferkamp *et al.*, 2010).

Molecular crowding can be disadvantageous as well, as it hampers lateral diffusion processes (Ellis, 2001). The diffusion space available for molecules is determined by the amount of obstacles in a definite (membrane) area and the size of the diffusion particles (Minton, 2001). In grana membrane bilayers, PS II and LHC II cause crowding, whereas in the lumen, the water splitting complexes of PS II represent obstacles for diffusion (Fig. 3.2). Thus, diffusion of PQ and xanthophylls through the membrane and of PC through the thylakoid lumen are hindered.

Diffusion of PQ through the grana membrane is challenged by macromolecular crowding

PQ diffusion is central for photosynthetic energy conversion as it is the electronic connection between PS II and the cyt b_6f complex. The significance of PQ is not confined to electron transport (Kirchhoff, 2008b), but its redox state also mediates the expression of photosynthetic genes as well as the redistribution of antennae proteins (LHC II) in a process known as state transitions (Allen *et al.*,

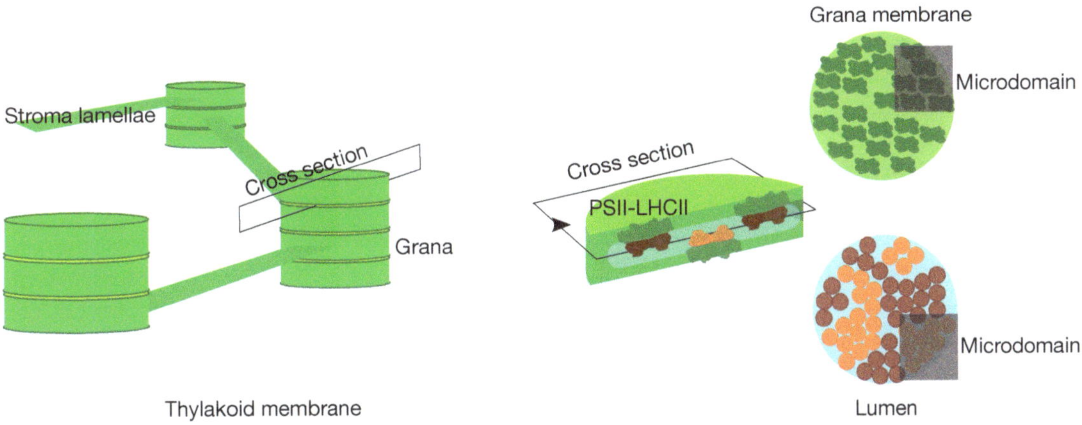

Figure 3.2 Macromolecular crowding in the grana membrane and lumen. Modified from Kirchhoff (2014). Crowding by PS II with its LHC II in the grana membrane is represented by dark green PS II-LHC II complexes occurring as microdomains. Microdomains are also formed in the lumen by the water splitting complex of PS II, depicted in orange for the lower membrane and violet for the upper membrane. Figure is not to scale.

1981; Pfannschmidt *et al.*, 1999). An overreduced PQ pool can also trigger photo-inhibition, the light-induced loss of photosynthetic activity.

A theoretical consideration, assuming randomly distributed proteins in the grana area, predicts restricted PQ diffusion (Tremmel *et al.*, 2003). PQ diffusion through a pure lipid membrane is faster approximately a thousand-fold than through a thylakoid membrane (Kirchhoff *et al.*, 2002). This implies that many thylakoid proteins, considered as 'diffusion obstacles', hinder PQ diffusion and consequently the linear electron transport becomes diffusion-limited. The formation of protein complexes such as the dimeric cyt b_6f, trimeric LHC II, dimeric PS II-LHC II and monomeric PS I-LHC I (Nelson and Ben-Shem, 2004) has been considered to facilitate energy transfer between LHC II complexes and PS II (Haferkamp *et al.*, 2010), and to be a strategy on the molecular scale to decrease the problem of limited diffusion (Kirchhoff, 2008a).

Another hierarchy of organization has been predicted in the microdomain concept, which states that small lipid microdomains occur in close proximity to active PS II (Fig. 3.2) (Lavergne *et al.*, 1992). Within these domains only a few PS II and cyt b_6f complexes share a small PQ pool, allowing fast, short distance electron shuttling between PS II and cyt b_6f complexes and thereby avoiding convoluted diffusion paths (Kirchhoff *et al.*, 2000, 2002).

On a supramolecular level, the chiral arrangements of PS II and LHC II complexes in grana can result in LHC II-containing macrodomains that exclude PS I (Garab and Mustárdy, 2000; Garab *et al.*, 1991). They are considered to not only ensure lateral segregation of the two photosystems but also provide the basis for a controlled, yet, flexible light-harvesting machinery (Anderson and Andersson, 1988; Garab and Mustárdy, 2000). It was shown that LHC II-containing macrodomains have a long-range order. Based on the evidence of long distance migration of energy in lamellar aggregates of LHC II (Barzda *et al.*, 1996), large arrays with long-range order were assumed to represent a structure for long-distance energy migration in membranes (Garab and Mustárdy, 2000). It was later shown by Boekema *et al* that PS II–LHC II arrange in well-ordered crystalline arrays, with PS II–LHC II supercomplexes in one membrane, and the LHC II-only complex in the adjacent part of the opposing membrane (Boekema *et al.*, 2000; Dekker and Boekema, 2005). The authors also speculate that this array formation and structural organization of macrodomains might allow efficient energy transfer from LHC II to PS II (Boekema *et al.*, 2000). The structure and flexibility of macrodomains depends on the lipid content (Simidjiev *et al.*, 1997) as well as light conditions. Light induced structural changes appear as partial destacking of membranes, followed by a detachment of particles or complexes from macrodomains, and diffusion of the free particles such as damaged PS II and LHC II to stroma lamellae (Garab and Mustárdy, 2000). This arrangement thus allows a dynamic response to ever changing light conditions. The formation of protein supercomplexes and their organization into macrodomains could further help to minimize the obstruction of PQ diffusion (Tremmel *et al.*, 2003). This will be discussed in more detail below ('Formation of protein arrays under diverse stress conditions – a mechanism to overcome diffusion limitation?').

The mobility of xanthophylls is restricted by macromolecular crowding

Another process that is diffusion dependent and expected to be restricted in crowded grana membranes is the conversion of violaxanthin to zeaxanthin, a process that is crucial for the proton-gradient-dependent photo-protective quenching of excess excitation energy in PS II (qE). The violaxanthin de-epoxidase, required for the conversion of vioalxanthin to zeaxanthin, is found in the lumen (Jahns *et al.*, 2009). The induction of qE therefore involves the unbinding of violaxanthin from LHC II, its migration to the luminal VDE, and the back migration of zeaxanthin to the LHCII complexes (Jahns *et al.*, 2009). Since this process requires long-distance diffusion, the induction of qE may also be diffusion limited (Macko *et al.*, 2002).

The diffusion of PC can be restricted by the water splitting complex of PS II

Another electron carrier, PC, electrochemically connects the cyt b_6f complex and PS I by its diffusion through the luminal space. In dark-adapted plants the width of the luminal space in thylakoid membranes is 4.5 nm (Fig. 3.1). Considering the size of 3 × 3 × 4 nm of the reduced PC (Guss *et al.*, 1986), it scarcely fits in the luminal space. Further, considering the protrusions of the PS II water splitting complex (Fig. 3.2), the available diffusion space for PC is similarly restricted to microdomains as for PQ. These PC microdomains seem to be dependent on light conditions (Kirchhoff *et al.*, 2011), and disappear at higher light intensities. This aspect is further discussed in 'High light acclimation of the thylakoid membrane', below. Diffusion-dependent processes include not only mobile carriers like PQ and PC, but also protein complexes like LHC II and PS II.

The protein complexes PS II and LHC II need to diffuse through the crowded grana membrane

Compared to PC and PQ, PS II is rather immobile (Kirchhoff *et al.*, 2008). Though PS II resides in stacked grana, its repair takes place in unstacked regions of the membrane. This requires long-range diffusion of PS II. PS II should therefore be mobilized from the granal regions. Phosphorylation of PS II subunits, partial destacking of the granal stacks, and the swelling of thylakoid lumen may all contribute to the mechanism of PS II mobilization (Kirchhoff, 2014). The ultrastructural and molecular logistics of PS II repair are further discussed in 'High light acclimation of the thylakoid membrane', below. Diffusion coefficients of photosynthetic antenna proteins in plants and cyanobacteria are 100 times faster than PS II (Consoli *et al.*, 2005; Kirchhoff *et al.*, 2008). A light quality acclimation requires the diffusion of LHC II between PS II and PS I through the stacked and the unstacked membrane regions. This aspect of antenna migration is further discussed in 'Low light acclimation of the thylakoid membrane', below.

Why is the grana diameter constant?

As mentioned earlier the grana diameter is remarkably constant in several species across the evolutionary scale. Why the grana diameter is constant is an open question. Theoretical considerations show that a grana diameter smaller than 400 nm would lead to high statistical fluctuations of the LHC II:PS II and the PS II:cyt b_6f ratios (Kirchhoff 2008a). This could lead to an inefficient transfer of excitation energy from LHC II to PS II. Further, random distribution of PS II complexes could reduce the efficiency of electron transport (Tremmel *et al.*, 2003) or lead to an overreduction of the PQ Pool that in turn causes, among other things (see section 'Macromolecular crowding in grana membranes'), photo-inhibition. By contrast, a larger grana diameter would lead to longer diffusion times. Hence a grana diameter of 500 nm might reflect a compromise between optimal protein composition and efficient lateral diffusion processes such as that of PQ (Kirchhoff, 2008a). This computer simulation-derived hypothesis was difficult to test, given the lack of a suitable biological model system. This changed recently with the availability of a mutant plant for the Curvature thylakoid1 (CURT1) protein in *Arabidopsis thaliana*.

CURT1 is a highly conserved family of thylakoid proteins found enriched in grana margins (Armbruster *et al.*, 2013). They enable tight membrane curvature at the grana margins. An *A. thaliana* quadruple knockout mutant, in which all four CURT1 proteins (A, B, C and D) are missing, has a grana diameter three times larger, with fewer grana stacks, than the wild-type plants (Armbruster *et al.*, 2013). An overexpressor line of *curt1a*, on the other hand, has a smaller grana diameter with more stacked membranes per granum (Armbruster *et al.*, 2013). *Curt1abc* knockout plants have lower photosynthetic efficiency and a more reduced PQ Pool. Fluorescence induction kinetics measurements and single turnover flash measurements of P700 indicate that due to a larger grana diameter, the diffusion-distance of PQ to the cyt b_6f complex becomes too long to be efficient (Höhner *et al.*, unpublished data). Impaired diffusion therefore increases the reduction level of the PQ pool. The overexpressor line, in contrast, has a lower quantum efficiency of PS II and a reduced biomass. It is tempting to speculate that these defects arise from a non-optimal grana diameter. However, the PQ Pool is slightly more oxidized and the reduction efficiency of the PQ pool in the overexpressor is slightly higher than in the wild type. In conclusion the grana diameter seems to determine the efficiency of PQ mediated electron transport (Höhner *et al.*, unpublished data).

Formation of protein arrays under diverse stress conditions – a mechanism to overcome diffusion limitation?

Semi-crystalline arrays are formed under different abiotic stresses

As plants are sessile, it is challenging for them to adapt to ever-changing environmental conditions. Studies investigating acclimation processes in spinach under low light conditions reveal a down-regulation of PS II and cyt b_6f complexes and an increase in the functional antenna size of up to 60% to allow efficient light harvesting (Anderson, 1986; Kirchhoff *et al.*, 2007). In the same light condition, highly ordered semi-crystalline protein arrays form in the stacked grana membranes (Kirchhoff *et al.*, 2007), comprised exclusively of LHC II-PS II supercomplexes (Daum *et al.*, 2010). Based on computer calculations, diffusion processes are hindered more in membranes with smaller obstacles than in membranes with larger obstacles (Saxton, 1993; Tremmel *et al.*, 2003). Hence, array formation of LHC II-PS II supercomplexes might overcome the potential diffusion problems arising from the increase in smaller LHC II complexes and the decrease in PS II and cyt b_6f complexes. In the same membrane, a fraction of the thylakoid lipids (MGDG) form a non-bilayer phase known as the H_{II} phase (further discussed in 'Thylakoid membrane lipids determine protein assembly and function'). This lipid phase transition could increase the protein density in the thylakoid membrane, contributing to array formation. Additionally, the exclusion of the MGDG from the bilayer as HII phases has been suggested to reduce the lateral membrane pressure and thus permit array formation of PS II (Kirchhoff *et al.*, 2007). It is interesting to note that arrays are more abundant in low-light thylakoids than in moderate light thylakoids, probably to improve the diffusion of PQ and xanthophylls (Kirchhoff *et al.*, 2007; Kouřil *et al.*, 2013). Arrays are, however, also induced by certain buffers (Miller *et al.*, 1976; Tsvetkova *et al.*, 1995) or when exposed to low temperature (4°C) (Garber and Steponkus, 1976; Semenova, 1995). Sznee and co-authors (Sznee *et al.*, 2011) also interpret this as a mechanism to facilitate long-range diffusion in a crowded membrane.

Mutations that affect the abundance of semi-crystalline arrays

The importance of the ability of PS II to form ordered arrays is demonstrated with an antisense plant of *A. thaliana*. This plant is unable to synthesize proteins that form LHC II trimers but harbours PS II supercomplexes with almost the same abundance as the wild-type plant (Ruban *et al.*, 2003). This antisense plant compensates for the absence of LHC II by synthesizing more CP26, and forms arrays with the same frequency and structural features as the wild type (Ruban *et al.*, 2003). Another protein with an important role in array formation is PsbS. PsbS is a LHC II-like thylakoid protein involved in NPQ formation. In its absence the arrays are more abundant, which is consistent with a role of PsbS in controlling the macroorganization of the grana membrane (Kereïche *et al.*, 2010). When PsbS is overexpressed, arrays do not exist at all and the fluidity of the thylakoid membrane increases. This might allow the reorganization of the PS II macrostructure necessary for efficient NPQ (Goral *et al.*, 2012). This is in line with a recent study showing non-existence of arrays and a more random arrangement of PS II supercomplexes in high light-acclimated *A. thaliana* plants (Kouřil *et al.*, 2013). An *A. thaliana* mutant lacking the minor antenna CP24 (de Bianchi *et al.*, 2008) and the barley mutant *zb63* that lacks PS I (Morosinotto *et al.*, 2006) have an increased fraction of grana membranes that form arrays. Counterintuitively, these mutants show a compromised electron transport and an impaired PQ diffusion (de Bianchi *et al.*, 2008; Morosinotto *et al.*, 2006). The reason for this turns out to be a difference in the composition of LHC II–PS II supercomplexes that form the arrays. *A. thaliana* wild type predominantly forms arrays using the C2S2M2 supercomplexes (Dekker and Boekema, 2005), whereas the mutants with increased array formation utilize the C2S2 supercomplexes. This suggests that the array composition and structure determine their functional consequence for photosynthesis.

Significance of semi-crystalline arrays

The significance of the highly ordered semi-crystalline arrays is not fully understood. It is currently assumed that the lipid-filled gap inside the protein arrays provides a 'diffusion channel' for lipophilic molecules such as PQ or xanthophylls. Since protein density is higher in arrays, array formation may reduce macromolecular crowding in disordered regions elsewhere in the membrane. The mobility of PS II in array regions, however, might be restricted, leading to a slowdown of the PS II repair cycle. Depending on the acclimatory needs, abiotic factors might thus select the non-array arrangement to favour PS II repair cycle or the array arrangement for fast electron transport and induction of NPQ. This notion is supported by a recent study on the *A. thaliana fad5* mutant that constitutively forms PS II arrays (Tietz *et al.*, 2015).

Thylakoid membrane lipids determine protein assembly and function

Although lipids occupy only about 20–30% of the grana membrane area, they are of high structural and functional significance (see Chapter 1). Lipids form bilayers thereby separating the two aqueous phases, the cellular interior and the external world. Membranes not only provide a protective barrier for the cell by insulating it from the harsh conditions of the outside world, but also allow a controlled exchange of water, ions and organic molecules necessary for cellular homeostasis (see Chapter 10). Lipid bilayers further allow the embedding of proteins and provide the matrix for supercomplex formation. At the same time lipids prevent non-specific protein aggregations. Moreover, as discussed earlier, lipids also provide the diffusion medium for mobile carriers such as PQ.

Significance of thylakoid lipids for protein stability, function and membrane integrity

Galactolipids are the most abundant lipid class in thylakoid membranes. They are rich in polyunsaturated fatty acyl chains, which significantly contribute to membrane fluidity and flexibility (Nishio *et al.*, 1985). Digalactosyl-diacylglycerol (DGDG), sulfoquinovosyl-diacylglycerol (SQDG) and phosphatidyl-diacylglycerol (PG) are bilayer forming galactolipids that make about 30%, 12% and 12% of the lipid content in thylakoid membranes, respectively (Garab *et al.*, 2000; Simidjiev *et al.*, 2000). PG binds to LHC II and is involved in LHC II trimer formation in *C. reinhardtii* (Dubertret *et al.*, 2002). Though SQDG has a similar fatty acid composition as PG, it does not seem to bind to LHC II. A study in *C. reinhardtii* indicates that SQDG might afford structural protection for PS II and help with its recovery from heat stress (Sato *et al.*, 2003). A DGDG-deficient mutant of *A. thaliana* shows not only stunted growth and pale leaves but also an altered thylakoid membrane ultrastructure, a compromised photosynthetic capacity (Dörmann *et al.*, 1995), and a reduced stability of PS I as the PS I subunits PsaD and PsaE become susceptible to removal by chaotropic agents in the absence of DGDG (Guo *et al.*, 2005).

Importance of MGDG and H_{II} phases

MGDG occupies about 50% of the total lipid area and is the most abundant galactolipid in the thylakoid membrane. It is, however, a non-bilayer lipid. Due to its conical shape, a small head group, and bulky fatty acids, isolated MGDG does not form planar bilayer phases in the presence of water. There is however a long-standing interest in the role of non-bilayer lipids like MGDG, as the relative amount of non-bilayer lipids changes in response to environmental conditions (Harwood, 1998). The variability in non-bilayer lipid content allows acclimation to extreme environmental stresses such as high light and drought (Garab *et al.*, 2000). MGDG has the tendency to form rod-like aggregates packed in hexagonal arrays (Lee, 2000). In the rods the polar head group of MGDG either faces the centre, in which case the structure is called H_{II} phase, or outwards as the H_I phase (reviewed in Lee, 2000). The 'lateral pressure model' suggests that non-bilayer lipids that are forced into a bilayer exist in a state of 'physical frustration' leading to an increase in membrane surface tension (Lee, 2000). This surface tension can influence the protein packing and protein function in the membrane (Gruner, 1989). LHC II complexes, in fact, force MGDG into a bilayer structure (Simidjiev *et al.*, 2000). It is well known that LHC II complexes enable membrane stacking through transversal LHC II-LHC II interaction from adjacent membranes. Magnesium ions are also important for this process, since they mask the negative surface charge of LHC II (Barber, 1982; McDonnel and Staehelin, 1980). Membrane stacking, however, does not occur in the absence of MGDG (Simidjiev *et al.*, 1998), pointing to the crucial role of MGDG in the stacking process. How precisely MGDG contributes to membrane stacking is not understood. The significance of H_{II} phases in the thylakoid membrane is still unclear, decades after it has been observed and described for thylakoid membranes (Larsson, 1989). Jahns *et al* suggest that H_{II} phases in the thylakoid membrane are sites of non-protein-bound violaxanthin in the membrane and the docking site of the VDE. The hydrophobic areas of segregated MGDG phases are in direct contact with the membrane bilayer and its proteins. Such an arrangement could be a strategy to minimize diffusion times of xanthophyll cycle pigments to LHC II and back (see the section above), allowing efficient and fast photo-protection (Jahns *et al.*, 2009).

Light stress acclimation of thylakoids

Although abundant as an energy source, light often becomes limiting for photosynthesis. Only half of the solar radiation that reaches the earth is photosynthetically active radiation, falling within the 400–700 nm wavelength range. Of this in-band wavelength, some of the energy of the short wavelength blue light does not produce any useful photochemistry. This is due to the fact that the blue light excites the chlorophyll to a higher singlet state (S2) and that this results in the loss of the excitation as heat. In essence, before accounting for the losses due to respiration, plants and algae are able to use only about 10% of the energy from sunlight to fix carbon. This meagre efficiency can further be eroded by profound gradients of light quantity and quality that exist in terrestrial and aquatic habitats on a time scale of seconds to seasons. Photosynthetic organisms have evolved a number of strategies to optimize light harvesting and conversion in changing light conditions (Croce and van Amerongen, 2014). They involve changes in the ultrastructure, organization and composition of the photosynthetic thylakoid membrane.

Low light acclimation of the thylakoid membrane

Low light conditions predominate in canopy understories and under cloud cover and shading. Even within a single plant or algal culture, self-shading can produce low light intensities for the lower leaves or low-lying algal cells. At the ultrastructural level, low light condition produces thylakoid membranes with more stacks per grana. As discussed earlier one advantage of the granal stacks may be to enable the vertical connectivity between light harvesting antenna and PS II from adjacent membranes of the stack and thus a larger effective antenna cross-section, without hindering the diffusion of PQ (Mullineaux, 2005). In low light, at the molecular level, PS II supercomplexes tend to form protein arrays and the array formation may be a strategy to evade molecular crowding (Kirchhoff *et al.*, 2007). The lipidic channels within the arrays might facilitate the diffusion of the small lipophilic electron carrier PQ and the photo-protective xanthophyll species in the membrane (Tietz *et al.*, 2015). At low light conditions, the photosynthetic antenna becomes larger due to *de novo* synthesis and assembly of antenna proteins. As harvesting every photon becomes a priority in low light, this antenna size adjustment is a universal feature of low light acclimation in all photosynthetic organisms.

State transitions: an acclimation to changes in light quality

Due to reflection of the short wavelength green and preferential absorption of blue, yellow, orange and red by the upper layers of leaves, the light that reaches the lower leaves becomes enriched in longer wavelength far-red light. Conversely, in aquatic ecosystems longer wavelength light is absorbed as the light penetrates through the water column, and the transmitted light is mainly of shorter wavelength. Low light conditions thus often have pronounced, and dependent on the ecosystem, different gradients of light quality, which poses a particular challenge for the photosynthetic machinery. This challenge comes from the unique pigment environment, pigment composition and function of the two photosystems in oxygenic photosynthesis. With its chlorophyll *b*-rich antenna, PS II prefers shorter wavelength light and is inefficient in energy conversion with wavelengths beyond 680 nm (Myers, 1971). The antennae of PS I, on the other hand, are enriched with chlorophyll *a*, especially the so-called far-red chlorophylls that absorb wavelengths beyond 700 nm (Qin *et al.*, 2015). PS I is therefore the long wavelength photosystem. Under far-red light, PS I thus works faster than PS II. Since the two photosystems are wired in tandem for linear electron transport, they should absorb and convert light energy at equal rates. Light quality gradients reconfigure the light harvesting antenna of photosystems through an acclimatory process known as state transitions (Allen, 2003; Bonaventura and Myers, 1969; Murata, 1969). Under preferential illumination of PS II with a short wavelength light, light 2, a fraction of LHC II that is loosely bound to the PS II core (L-form) is phosphorylated and drifts away from PS II (Fig. 3.3) (Galka *et al.*, 2012). The detached LHC II then associates with PS I as its light harvesting antenna. Since PS I struggles in a PS II illumination, the newly associated LHC II diverts some of the absorbed excitation energy towards PS I at the expense of PS II. This acclimatory state to light 2 is known as state 2. Conversely, if PS I is selectively excited with a long wavelength light, light 1, LHC II becomes dephosphorylated and returns to PS II to assist it with light harvesting in an illumination condition that favours PS I. This acclimatory state to light 1 is state 1 and the transitions between these states are called state transitions.

State transitions unequivocally alters the antenna cross-section of PS II and PS I in a complementary way, and this can be monitored by fluorescence changes at 77K. The physiological relevance of this response, however, has been questioned by some studies, especially the existence of a light 2 illumination condition in nature (Tikkanen and Aro, 2012; Wientjes *et al.*, 2013a). Sun flecks and a sudden transition from dark to light and low light to high light will favour PS II transiently, reduce the PQ pool and promote state 2. These light conditions therefore should be considered as light 2. Electron sink limitation and some altered metabolic pathways, which make the PQ pool over-reduced, could also induce state 2 transition. In principle, any light or metabolic conditions that momentarily reduce the PQ pool should activate the LHC II kinase and promote state 2. Another

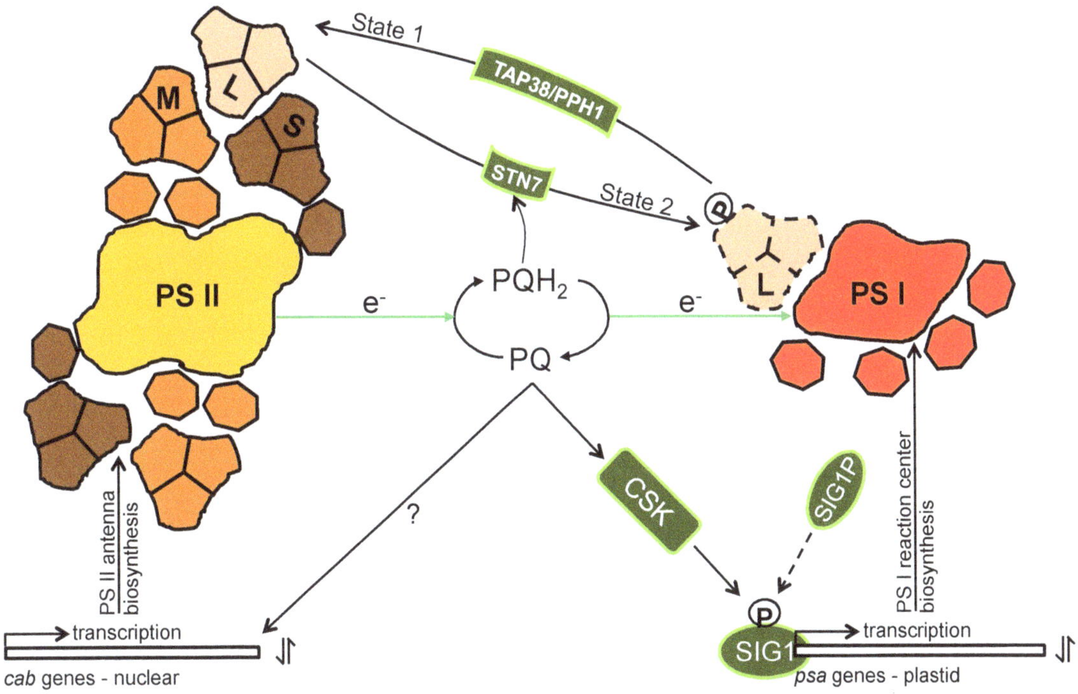

Figure 3.3 A schematic representation of plant state transitions and photosystem stoichiometry adjustment. The approximate contours of PS II, PS I, and their antennae are shown. The mobile LHC II trimer in state transitions is marked as 'L'. The LHC II kinase STN7 drives state 1 to state 2 transition and the LHC II phosphatase TAP38/PPH1 initiates state 2 to state 1 transition. The oxidized PQ regulates the transcription of plastid-encoded PS I reaction centre genes through the CSK-SIG1 signalling system. 'SIG1P' is a putative phospho-SIG1 phosphatase. The PQ pool also regulates, through an unknown pathway, nuclear *cab* genes encoding the PS II antenna.

apparent inconsistency was that plants grown in green houses and growth cabinets tend to have a significant proportion of LHC II associated with PS I, as if they are in state 2 (Wientjes *et al.*, 2013a). This observation has led to the suggestion that the LHC II serves both photosystems as antennae in most of the natural light conditions and there is no clear-cut state 1 or 2 in nature. A simple explanation for the growth-light induced LHC II-PS I association may be the fact that the fluorescent lamps that lit plant growth facilities are conspicuously deficient in the far-red light component and are likely to act as light 2. As an alternative explanation one could imagine a continuum in state transitions, i.e. 'in-between' states, with varying amounts of LHC II getting phosphorylated and associated with PS I in a variety of light conditions. The two extreme ends of this continuum will be complete phosphorylation, state 2, and dephosphorylation of all the LHC II in the mobile fraction, state 1. In such a scenario, prolonged incubation in far-red light should completely dephosphorylate all the LHC II, and the LHC II is then the exclusive antenna of PS II; this is what one actually observes in far-red light.

Plant and green algal state transitions differ in certain respects. In *C. reinhardtii*, some of the minor antenna are also phosphorylated and move along with the trimeric LHC II to PS I in state 2 (Kargul *et al.*, 2005; Takahashi *et al.*, 2006), while in plants there is no evidence for the migration of the minor antenna. In algae, two LHC II trimers seem to associate with PS I in state 2, while in plants there is evidence for only one trimer association (Drop *et al.*, 2014). In plants, some of the S and M LHC II trimers are also phosphorylated in state 2, but it is not clear whether it is part of their state transitions (Wientjes *et al.*, 2013b). In *C. reinhardtii*, as much as 80% of the total LHC II dissociates from PS II in state 2, as opposed to a smaller percentage in plants (Delosme *et al.*, 1996). This large-scale detachment of LHC II from PS II has been taken as evidence for an extensive remodelling of PS I antenna in state 2. Recent algal research, however, shows that only a small proportion of the detached LHC II associates with PS I in state 2 as in plants, and the rest of the dissociated LHC II occurs in a quenched state to protect PS II from photo-damage (Nagy *et al.*, 2014; Ünlü *et al.*, 2014). In algae, the transition to state 2 has been thought to promote cyclic electron transport around PS I to generate ATP (see Chapter 4) (Finazzi *et al.*, 2002). Recent research, however, suggests that the switch to cyclic electron flow and state 2 transition are two independent yet parallel processes that depend on the redox state of the electron transport chain (Takahashi *et al.*, 2013).

The role of PQ as the regulatory signal that controls the reversible phosphorylation of LHC II is well established (Allen *et al.*, 1981). PS II-specific illumination causes transient reduction of the PQ pool, and the reduced PQ (PQH_2) promotes the kinase activity of the LHC II kinase. PS I-specific illumination, in contrast, results in oxidation of the PQ pool, as electrons leave the pool faster than they arrive from PS II. The LHC II kinase is inactive under this condition. The phosphatase activity of the LHC II phosphatase is believed to be slow and be overwhelmed by the kinase activity in state 2 condition. This results in a net phosphorylation of LHC II. While the kinase becomes inactive during state 1 transition, the constant activity of the phosphatase tips the equilibrium towards dephosphorylation of LHC II (Puthiyaveetil *et al.*, 2012). The identities of the LHC II kinase and the phosphatase have come to light. Known as Stt7 in *C. reinhardtii* or as Stn7 in *A. thaliana*, the LHC II kinase is a highly conserved serine/threonine kinase (Bellafiore *et al.*, 2005; Depege *et al.*, 2003). It consists of a transmembrane helix, a luminal stretch of amino acids and a stromally-located kinase domain. The LHC II phosphatase is a PP2C-type phosphatase that is found in both green algae and plants and is known as TAP38 or as PPH1 (Pribil *et al.*, 2010; Shapiguzov *et al.*, 2010). A recent crystal structure of this protein reveals why it is specific in dephosphorylating only the phospho-LHC II (Wei *et al.*, 2015). Thylakoid-associated kinases (TAKs) are a family of three plant protein kinases that exhibit striking similarity to the transforming growth factor receptors of metazoans (Snyders and Kohorn, 1999). TAK1 is chloroplast localized and was initially believed to have a role in state transitions, but its precise function remains uncertain (Snyders and Kohorn, 2001). TAK2 and 3 turn out not to be chloroplast proteins after all (Armbruster *et al.*, 2009).

The mechanism of LHC II kinase regulation

How exactly the PQH_2 activates the Stn7 kinase is unclear. The PQH_2 acts on the kinase from the quinol oxidation site (Q_o pocket) of the cyt b_6f

complex and the kinase is closely associated with subunits of the cyt b_6f, especially the Rieske iron sulfur protein (Lemeille *et al.*, 2009). It has been suggested that the quinol promotes the reduction of a disulfide bridge in the kinase, formed through the conserved luminal cysteine motif, and that this results in a conformational change that ripples across the membrane to the stromally-located catalytic domain to promote kinase activity (Puthiyaveetil, 2011). It is likely that the conformational change on the luminal side results from a quinol-mediated conversion of an intramolecular disulfide bridge in the kinase into an intermolecular one. Transient dimerization of Stn7 monomers is indeed correlated with an increased kinase activity (Wunder *et al.*, 2013a). An alternate model posits that the kinase is activated by conformational changes in the cyt b_6f complex, caused by the to-and-fro movement of the Rieske iron sulfur protein during Q-cycle (Zito *et al.*, 1999). Both of these models await experimental tests.

High light inhibits the plant LHC II kinase through the thioredoxin system, an observation consistent with state transitions as a low light phenomenon (Rintamaki *et al.*, 2000). How precisely thioredoxin inhibits the kinase is not currently understood. The luminal cysteines have been suggested as the target site of thioredoxin (Depege *et al.*, 2003). An additional conserved, stromally-located cysteine motif occurs in the ATP-binding domain of the kinase. This stromal site, instead of the luminal site, has been suggested as to where the thioredoxin acts to prevent the kinase activity, through interference with the binding of ATP (Puthiyaveetil, 2011). Indeed the removal of luminal cysteines does not affect the thioredoxin interaction, supporting the role of the stromal cysteine motif as the real thioredoxin target site (Wunder *et al.*, 2013b). The *C. reinhardtii's* Stt7 curiously lacks the stromal cysteine motif, and it is unclear whether the kinase is inhibited in high light by the thioredoxin system. The high level of LHC II phosphorylation and its large-scale detachment from PS II in state 2 as a photo-protective mechanism are consistent with the lack of inhibition of the Stt7 by thioredoxin in high light (Ünlü *et al.*, 2014). Recently, however, hydrogen peroxide has been suggested to inactivate Stt7 under prolonged high light condition through the oxidation of Stt7's luminal cysteines (Roach *et al.*, 2015). The structure of the *Micromonas* LHC II kinase catalytic domain is now known and should inform the experiments that test different models of kinase regulation (Guo *et al.*, 2013). Neither LHC II nor the LHC II kinase is found in cyanobacteria or red algae, which also undergo state transitions. In these organisms the movement of the large extrinsic antenna, phycobilisome, accounts for state transitions. The PQ redox state, as in green algae and plants, is the all-important regulatory signal that shuttles the phycobilisome between PS II and PS I (Mullineaux and Allen, 1986). The precise post-translational modification that drives the movement of the phycobilisome, however, is unknown.

Photosystem stoichiometry adjustment: tackling light gradients with photosystem numbers

Photosystem stoichiometry adjustment is another remarkable acclimatory response that corrects imbalanced excitation of photosystems in light quality gradients. In this acclimatory response the relative abundance of the two photosystems is adjusted so as to correct the imbalance in their excitation in light conditions that favour one photosystem over the other (Chow *et al.*, 1990). It is a fundamental acclimatory mechanism that is essential for the efficient operation of the linear electron transport – the Z-scheme (Hill and Bendall, 1960). Found in all oxygenic photosynthesizers, photosystem stoichiometry adjustment improves the quantum yield of photosynthesis in light quality environments. Stoichiometry adjustment is initiated not only as a response to light quality changes but also by a differing requirement for NADPH and ATP from the non-cyclic and cyclic electron transport pathways. Both cyanobacteria and chloroplasts achieve photosystem stoichiometry adjustment through regulation of the abundance of PS I and the PS II light harvesting antenna (Fig. 3.3) (Murakami *et al.*, 1997a,b; Pfannschmidt *et al.*, 1999). Whether the amount of PS II reaction centres also changes in some species as part of their stoichiometry adjustment is unclear (Pfannschmidt *et al.*, 1999). In response to light 2 illumination, the amount of PS I increases and that of PS II antenna decreases. While in the opposite light condition, light 1, the synthesis of PS I is repressed and that of PS II antenna is induced. These light quality acclimations shift the PS II/PS I ratio far from unity.

For example, in barley, PS I light shifts the ratio to 2.7 and PS II light to 1.4 (Kim *et al.*, 1993). The complete stoichiometry changes take from days to as much as a week in many plants species. It is not difficult to envisage how an abundance of PS I in light 2 will also promote cyclic electron transport and a concomitant generation of ATP.

For photosystem stoichiometry adjustment, plants modulate the *de novo* synthesis of both PS I and LHC II by transcriptional regulation of their chloroplast and nuclear-encoded genes (Escoubas *et al.*, 1995; Pfannschmidt *et al.*, 1999; Puthiyaveetil and Allen, 2008). As in state transitions, PQ redox state is the signal that initiates the photosystem stoichiometry adjustment (Pfannschmidt *et al.*, 1999). The reduced PQ, in PS II light, activates PS I gene transcription and represses *cab* genes (encoding LHC II). Conversely the oxidized PQ, in PS I light, represses PS I genes and activates *cab* genes. The biogenesis and breakdown of photosystems may require the PQ redox regulation of photosystem gene transcription to act in conjunction with a translational autoregulation known as the control by epistasy of synthesis (CES) (Choquet and Vallon, 2000). The sequential assembly of photosystems and other electron transport complexes require the CES autoregulatory feedback loop in which the translation of key subunits depends on the assembly state of the complex. This means that photosystems can rapidly be made or broken down by controlling the abundance of key subunits, which act as the 'pacemakers' for photosystem biogenesis (Nickelsen and Rengstl, 2013). Long-term decrease in photosystem and antenna abundance may also entail degradation of the already-existing protein complexes. The degradation of LHC II is especially important for PS II light acclimation (Lindahl *et al.*, 1995).

How is the PQ redox state communicated to chloroplast gene transcription?

PS I core reaction centre proteins PsaA and PsaB are invariably chloroplast-encoded and which mechanism connects PQ redox state with *psaAB* gene transcription has been a fascinating question. Two important discoveries seem to provide the answer. One was the discovery of a bacterial-type two-component sensor histidine kinase known as the chloroplast sensor kinase (CSK), which is conserved from cyanobacteria to plants (Puthiyaveetil *et al.*, 2008). The paradigm signalling elements in bacteria are two-component systems, comprising a sensor histidine kinase and a response regulator as signal transduction units (Stock *et al.*, 2000). *A. thaliana* knock-out mutants of csk are unable to repress *psaAB* gene transcription in PS I light (Puthiyaveetil *et al.*, 2008). The second discovery was that the major sigma factor subunit of chloroplast RNA polymerase, SIG1, becomes phosphorylated in PS I light and the phosphorylated SIG1 represses transcription from the *psaAB* gene (Shimizu *et al.*, 2010). The sigma factor helps the RNA polymerase to recognize, melt DNA, and initiate transcription from the bacterial-type promoter (Paget and Helmann, 2003).

A signalling scheme that incorporates CSK and SIG1 has been suggested for the plant and green algal photosystem stoichiometry adjustment (Fig. 3.3) (Puthiyaveetil *et al.*, 2010, 2012, 2013). Its essence is that in PS I light, the oxidized PQ activates CSK, which then phosphorylates SIG1 on one or more serine/threonine residues in its N-terminus. The SIG1 N-terminus contains a latent region 1.1, whose function is to discourage transcription initiation from weaker promoter elements (Vuthoori *et al.*, 2001). The phosphorylation-mediated unmasking of SIG1 region 1.1 in PS I light causes selective repression of transcription from the weaker PS I gene promoter, while leaving transcription from the stronger PS II gene promoter unaffected. This decreases PS I amount and the PS I/PS II stoichiometry. In PS II light, CSK is inactive; and an as-yet-unidentified phosphatase dephosphorylates phospho-SIG1, de-activating the region 1.1 switch and releasing the repression on PS I transcription. This increases PS I amount and the PS I/PS II ratio.

Many aspects of the CSK-SIG1 scheme and the SIG1 gene regulatory phosphoswitch remain to be experimentally validated. The full spectrum of CSK's functional role is also unknown. Whether the CSK homologue in cyanobacteria, the histidine kinase 2 (Hik2), also regulates photosystem stoichiometry remains to be determined. It interacts with Rre1 and RppA, two response regulators with roles in reaction centre and phycobilisome gene transcription (Ibrahim *et al.*, 2016). Hik2 also competitively transfers phosphoryl groups to these response regulators. Many non-green algae retain Rre1 as the Ycf29 response regulator, along with CSK homologues (Puthiyaveetil and Allen, 2009). How this system regulates photosystem stoichiometry in these algae is an open question. The conserved nature of the reaction centre gene regulation by PQ and its signalling circuit in both cyanobacteria and chloroplasts bears heavily on the question of why genes still remain in bioenergetics organelles (Allen, 2015).

Some authors have placed the state transition kinase Stn7 in the regulatory scheme of photosystem stoichiometry adjustment (Dietzel *et al.*, 2008; Pesaresi *et al.*, 2009). It is not clear why and how this is necessary. *stn7* knockout mutants seemingly have an aberrant photosystem gene expression (Bonardi *et al.*, 2005). This may simply be a consequence or a pleiotropic effect of an overreduced PQ pool. Unlike CSK, Stn7 has no homologue in cyanobacteria. Stn7's proposed transcriptional regulatory role has therefore no phylogenetic continuity. It might still be possible that Stn7 and especially its algal homologue Stt7 influence the steady state of *psaAB* transcript accumulation by affecting the transcript stability through phosphorylation of an RNA-binding protein. If this turns out be the case, it qualifies only as a post-transcriptional regulation.

PS I content is down-regulated in certain oceanic diatoms and cyanobacteria due to severe iron deficiency, resulting in an extremely low PS I/PS II ratio (see 'Influence of iron content on the thylakoid membrane and its protein complexes') (Guikema and Sherman, 1983). Whether this regulatory response shares signalling components with the light quality-driven photosystem stoichiometry changes is unknown. It is also not clear how the PQ signal is transduced to the nuclear-encoded *cab* genes in algae and plants as part of the light quality and quantity acclimations. The transcription of *cab* genes in algae responds robustly especially to light quantity changes, with high transcription in low light and low transcription in high light (Escoubas *et al.*, 1995). Experiments with electron transport and kinase inhibitors strongly support the role of the PQ pool as the regulatory signal and protein phosphorylation as the likely signal transduction mechanism. Studies in *C. reinhardtii* implicate calcium as a signalling molecule for a related up-regulation of the photoprotective antenna (Petroutsos *et al.*, 2011). It is unclear whether the transcriptional arm of the *cab* gene expression also occurs in plants, but there is evidence for its post-transcriptional regulation (Flachmann and Kühlbrandt, 1995). Recently, the reduced PQ pool has also been shown to regulate the alternative splicing of nuclear-encoded genes and this response is important for proper plant response to the light environment (Petrillo *et al.*, 2014). The signalling network that transmits the PQ signal to the nuclear splicing machinery is unclear.

Stoichiometry adjustment of non-photosystem complexes

The abundance of electron transport complexes other than photosystems also varies according to light quality and intensity. Their stoichiometry regulation is important for the flux control of electrons through the photosynthetic chain (Schöttler and Tóth, 2014). The stability of the pea *petB* gene transcript, coding for the cytochrome b_6 of the cyt b_6f complex, responds to light quality changes, with a decrease in stability in PS I light and an unchanged stability in PS II light (Alexciev and Tullberg, 1997). Like pea, the amount of *A. thaliana* cyt b_6f per chlorophyll is low in PS I light. PS II light, however, increases the amount of *A. thaliana* cyt b_6f by 50% (Dietzel *et al.*, 2011). The amount of cyt b_6f per PS II can have a major effect on linear electron transport, since the oxidation of PQH_2 is a limiting step in the linear electron transport. When it comes to light quantity, the cyt b_6f per chlorophyll is low in low light and increases with an increase in light intensity (Leong and Anderson, 1984). In response to changes in light intensity, the amount of the ATP synthase per chlorophyll also changes in parallel with the cyt b_6f (Leong and Anderson, 1984). The cyclic electron flow around PS I in chloroplasts involve two independent pathways, one incorporating the NADH dehydrogenase and the other, PGRL1 protein (Shikanai, 2014). Both pathways use the cyt b_6f as a conduit. Varying the amount of NADH dehydrogenase may affect the extent of cyclic electron transport. Like PS I, the NADH dehydrogenase amount may decrease in PS I light and increase in PS II light. It may also increase in high light. Table 3.1 summarizes how light intensity or quality controls the abundance of photosynthetic protein complexes. More study is indeed required to see whether stoichiometry changes of non-photosystem complexes have a recurring

Table 3.1 Light quality and quantity-driven stoichiometry changes in plant photosynthetic complexes, expressed as complexes per chlorophyll. The upwards arrow indicates higher abundance, and the downwards arrow lower abundance. 'UA' stands for unaltered or unresponsive abundance. The dash indicates an unknown response. Only general tendencies are shown; species-specific differences may occur for some complexes.

Protein complex	PS II light	PS I light	Low light	High light
PS II	UA	UA	↓	↑
PS I	↑	↓	UA	UA
LHC II	↓	↑	↑	↓
Cyt b6f	↑	↓	↓	↑
ATPase	-	-	↓	↑
NADH dehydrogenase	-	-	-	-

theme and what physiological significance they might have.

A remarkable far-red light acclimation has been discovered recently in cyanobacteria, which involves extensive reconfiguration of photosystems and phycobilisomes (Gan *et al.*, 2014). In *Leptolyngbya* sp. strain JSC-1, this remodelling includes synthesis of chlorophylls *d* and *f*, replacement of core subunits of PS II and PS I, and modification of the light harvesting antenna phycobilisome core. The redesign of the photosynthetic apparatus helps this cyanobacterial strain to absorb light and thrive in a far-red illumination of 700–750 nm. Unlike the PQ-controlled photosystem stoichiometry adjustment, this light quality acclimation is probably controlled by a knotless red/far-red phytochrome. It is interesting to note that this far-red light acclimation operates independent of, and in addition to, the well-known cyanobacterial light quality response known as the complementary chromatic adaptation (CCA).

High light acclimation of the thylakoid membrane

High light conditions occur under direct sunlight and on top of the canopy. An immediate concern for plants, algae and cyanobacteria growing in high light is the photo-damage to the D1 subunit of PS II, since photo-damage is a function of the photon flux density (Mattoo *et al.*, 1984; Ohad *et al.*, 1984). In plants and algae, a highly efficient repair mechanism known as the PS II repair cycle replaces the damaged D1 with a newly synthesized copy and maintains photosynthetic efficiency in high light. The damage, however, can exceed the rate of repair in the midday sun, resulting in the characteristic midday depression of photosynthesis.

Thylakoid ultrastructural changes in high light

An efficient operation of the PS II repair cycle entails many ultrastructural changes in the thylakoid membrane system. First of all, the number of stacks per grana goes down in the long term (Anderson *et al.*, 1988). This decreases the connectivity between LHC II from adjacent membranes, relieving the excitation pressure on PS II and preventing further damage to D1. Less stacking also means less absorbance and more transmission of light by chloroplasts and leaves. In high light, as a short-term measure, the granal stacks destack partially at the lateral plane, resulting in the shrinkage of grana diameter (Herbstova *et al.*, 2012). The shrinkage of grana, in fact, results from an interconversion of core grana regions to grana margins (Puthiyaveetil *et al.*, 2014). This membrane interconversion is advantageous for the repair since PS II residing in former core granal regions now find themselves in the margins and directly accessible to the repair machinery. A decrease in grana diameter also reduces the diffusion distance of PS II in core grana, since the repair cycle depends on a brisk protein trafficking between appressed and unappressed regions of the membrane (Herbstova *et al.*, 2012). It is not clear which forces unstack the membrane in high-light. Since LHC II trimers act as the glue between adjacent membranes, their migration out of the core grana may simply result in unstacking. An unusual bending of the membranes in the grana periphery and preferential swelling of the lumen in the grana margins accompany the unstacking

(Puthiyaveetil *et al.*, 2014; Yoshioka-Nishimura *et al.*, 2014).

The lumen swelling may involve water influx, but it is not clear why swelling is more prominent in the grana margins. It will be interesting to see whether this has anything to do with the high light-dependent phosphorylation of the membrane-bending, margin-localized CURT proteins. Phosphorylation of CURT may loosen the tight curvature of the membrane, allowing it to bulge. Lumen swelling has also been noted in core granal regions in both moderate and high light (Kirchhoff *et al.*, 2011; Tsabari *et al.*, 2015). The high light, however, seems to further widen the lumen than the moderate light. Lumen swelling facilitates the diffusion of the PC and thereby increases the inter-photosystem electron transport in high light (Kirchhoff *et al.*, 2011). When the high light stress, however, is combined with a CO_2 and O_2 depleted condition, the luminal width decreases (Tsabari *et al.*, 2015). This might restrict the electron flow and protect the photosynthetic machinery when the lack of the electron sink accentuates the effect of the high light stress.

Recent research shows a hitherto unknown level of organization and division of labour in the PS II repair machinery, which probably accounts for the high efficiency of the repair cycle (Puthiyaveetil *et al.*, 2014). The localization of repair cycle components and reactions show a high degree of heterogeneity in the membrane, with the three thylakoid membrane domains – core grana, grana margin, and stroma lamellae – seem to have very distinct roles in the repair cycle. The margin compartment especially is pivotal in the repair in that it harbours the membrane-intrinsic D1 protease FtsH (Puthiyaveetil *et al.*, 2014; Yoshioka *et al.*, 2010). This together with the concerted ultrastructural changes of margin expansion and margin lumen swelling allow controlled access of the damaged D1 to the high light-activated proteases. In an expanded margin lumen the luminal D1 protease DegP may find more room for degradation, given its location in the unstacked regions (Suorsa *et al.*, 2013). The stromal partition gap between adjacent granal membranes (Fig. 3.1) does not widen in high light, as indicated by the electron micrographs and the unchanged FtsH concentration in core grana after high light (Puthiyaveetil *et al.*, 2014). This steric constraint probably occludes FtsH from going towards the core granal regions where it might indiscriminately degrade D1 of active PS II.

Molecular remodelling of the thylakoid membrane in high light

At the molecular level, the high light response involves remodelling of the photosynthetic complexes. In high light, the priority must be for relieving the excitation pressure on PS II. Induction of the NPQ by the photosynthetic antenna is certainly an overriding response (Ruban, 2015). The amount of the PsbS protein, an essential regulator of NPQ in plants, goes up in high light (Li *et al.*, 2000). PsbS is believed to be a luminal pH sensor that couples the proton gradient to NPQ (Li *et al.*, 2004). A recent crystal structure of this protein offers new insights into its photo-protective function (Fan *et al.*, 2015). LhcSR is another high light inducible protein involved in NPQ formation in green algae and diatoms (Peers *et al.*, 2009). Early light-inducible proteins (ELIPs) are a family of widespread chlorophyll-binding proteins that accumulate in high light and other stressful conditions (Meyer and Kloppstech, 1984). The precise mechanism by which ELIPs dissipate light energy is not clear. Orange Carotenoid Protein (OCP) is a cyanobacterial carotenoid binding protein that protects the photosystems in high light by triggering thermal dissipation of the excitation energy at the level of phycobilisomes (Kirilovsky and Kerfeld, 2013).

A cyclic electron flow around PS II has been suggested to mitigate the excitation pressure on PS II and photo-inhibition (Miyake and Yokota, 2001). The components of this pathway remain to be clarified. The better known cyclic flow around PS I is also important for preventing overreduction of the PQ pool and protecting both PS II and PS I from photo-inhibition in high light (Shikanai, 2014). The wate–water cycle – Mehler reaction – is an additional electron pathway that induces NPQ for photo-protection in high light (Asada, 2000). Plastid terminal oxidase (PTOX) is a chloroplast PQH_2 oxidase that shares homology with the mitochondrial alternate oxidase (Wu, 1999). PTOX bypasses both cyt b_6f and PS I in electron transport by directly oxidizing the PQH_2 produced by PS II. It prevents the overreduction of PQ pool and thereby supplies oxidized PQ for carotenoid biosynthesis during chloroplast development. The lack of oxidized PQ, with which to oxidize phytoene, a carotenoid intermediate, stalls carotenoid biosynthesis and causes leaf variegation (Nawrocki *et al.*, 2015). In the absence of photoprotective carotenoids, photo-damage ensues in the photosynthetic apparatus. Since the PQ pool becomes overreduced in high light, its re-oxidation by PTOX restores the redox poise of the thylakoid membrane. This is especially pertinent in developing chloroplasts which, when lacking PTOX, fail to form fully organized thylakoid membranes in high light (Foudree *et al.*, 2012). The rerouting of electrons through PTOX is also crucial for high light-acclimation in marine phytoplankton (Cardol *et al.*, 2008).

The extent of photo-damage directly correlates with the antenna size of PS II (Baroli and Melis, 1998). Reducing the antenna size must therefore be an important strategy in high light. The antenna size decreases in high light either by degradation of the existing antenna or by repression of new antenna synthesis. In high light the Stn8 kinase phosphorylates PS II core, which causes antenna dissociation from the core (Fristedt and Vener, 2011; Puthiyaveetil and Kirchhoff, 2013; Tikkanen *et al.*, 2008). This early step in the PS II repair cycle has an interesting functional parallel with state transitions, which also depends on phosphorylation for antenna dissociation. Given the paralogous nature of the Stn kinases and redundancy in their substrates, it has been suggested that the antenna dissociation by phosphorylation is a common theme that unites state transitions and PS II repair cycle, with both processes trying to minimize the excitation pressure on PS II (Puthiyaveetil and Kirchhoff, 2013). In line with this, the Stn7-mediated phosphorylation of the CP29 monomeric antenna is important for both green algal state transitions and high light-dependent disassembly of plant PS II supercomplexes (Fristedt and Vener, 2011; Kargul *et al.*, 2005; Takahashi *et al.*, 2006). The photoprotective role of *Chlamydomonas* state 2 transition also underscores this functional unity (Ünlü *et al.*, 2014). Cyanobacteria also suffer D1 photo-damage in high light. The replacement of the D1 protein by a high light resistant version is an interesting aspect of the PS II repair cycle in cyanobacteria. The degradation of the phycobilisome antenna and down-regulation of its synthesis further relieves the excitation pressure on PS II.

The stoichiometry of photosystems also changes in response to light intensity in both cyanobacteria and plants. The PS II/PS I ratio is close to unity in low light grown plants, while it is 2.2 for high light-acclimated plants (Chow *et al.*, 1988b). Only the PS II reaction centre seems to increase with increasing light intensity in plants. In cyanobacteria, however, regulation of the PS I content, rather than the PS II, forms the main component of the light quantity-driven photosystem stoichiometry changes (Murakami and Fujita, 1991). Down-regulation of electron flow to minimize free radical generation seems to be the main incentive for cyanobacterial photosystem stoichiometry adjustment in high light.

Signalling circuits underlying high light acclimation

The signalling pathways that mediate the high light response are not fully elucidated. Reactive Oxygen Species (ROS), generated during high light stress, is clearly involved in some responses (Li *et al.*, 2009). Some components of the ROS signalling pathway have been identified in algae and plants. The signalling pathway that mediates light quantity-driven plant photosystem stoichiometry adjustment is unclear, but may operate independent of the light quality-driven pathway. Due to differences in promoter strengths, the intrinsic transcriptional rate of PS II genes is higher than that of PS I genes. This, in conjunction with the progressive inhibition of chloroplast gene regulatory protein kinases PTK and CSK by the stromal reductants thioredoxin or glutathione in high light, may generate more PS II than PS I (Puthiyaveetil *et al.*, 2012).

The sensor kinase Hik33 and the response regulator RpaB are important for the cyanobacterial light quantity-responsive photosystem stoichiometry adjustment (Kappell and van Waasbergen, 2007). In low light, the RpaB binds to high light regulatory 1 (HLR1) elements in PS I and high light-inducible (*hli*) genes. This has a positive regulatory effect on PS I gene transcription and an inhibitory effect on *hli* genes. In high light the same system down-regulates PS I genes and up-regulates *hli* genes. It is likely that these differential effects in low and high light arise from a combinatorial control that uses the unique positioning of the HLR1 elements on the target genes and by RpaB phosphorylation level (Seino *et al.*, 2009). Hik33-RpaB system has survived in certain non-green algae, but whether its function is also conserved remains to be seen. The high light-dependent photosystem stoichiometry adjustment in cyanobacteria also requires PmgA, a transcriptional regulator (Hihara *et al.*, 1998). It remains to be clarified whether PmgA protein works with the Hik33-RpaB system or on its own. NblR is another one-component system that governs cyanobacteria's high light response (Kato *et al.*, 2014). It rapidly induces the accumulation of the NblA peptide that targets the phycobilisomes for degradation in high light and under nutrient stress, resulting in the characteristic bleaching response of cyanobacteria.

Nutrient deficiencies affect thylakoid membrane structure and function

Nutrient availabilities profoundly affect the structure and function of the photosynthetic apparatus. Plant growth and photosynthesis require micronutrients such as iron (Fe), copper (Cu), zinc (Zn), and manganese (Mn) (Palmer and Guerinot, 2009). Due to their ability to accept and donate electrons, that is, to exist in different redox states, Fe, Mn and Cu play crucial roles in the photosynthetic electron transport chain.

Influence of iron content on the thylakoid membrane and its protein complexes

Both nutrient deficiencies and excesses should be avoided, since many nutrients have deleterious effects when in surplus. Fe is one of the most abundant elements on earth. An excess of iron leads to a significantly lower photosynthetic efficiency in the aquatic plant *Spirodela polyrhiza* (Xing *et al.*, 2010). Nevertheless, iron deficiency is a prevalent problem in natural environments of plants, algae, and cyanobacteria as the bioavailability of iron is restricted. Chlorosis is a widespread symptom of iron deficiency in photosynthetic organisms (Moseley *et al.*, 2002; Spiller *et al.*, 1982). The main cause of chlorosis is severe degradation of PS I. As PS I exhibits a high iron content, containing 12 iron atoms per complex (Sandmann and Malkin, 1983), it is the primary target of iron deficiency in the photosynthetic electron transport chain. Iron deficiency causes particular degradation of the PS I subunits PsaC, PsaD, and PsaE. The synthesis of PsaC is compromised in this condition as it contains terminal 4Fe–4S clusters (Yadavalli *et al.*, 2012). PsaC protein in turn is required for the stable binding of PsaD and PsaE (Yu *et al.*, 1995), and in its absence they too will be degraded. A consequence of PS I degradation is a decreased PS I/PS II ratio. This has been reported in iron deficient *C. reinhardtii* (Naumann *et al.*, 2007) and cyanobacteria (Guikema and Sherman, 1983). In iron deficient *C. reinhardtii*, a remodelling and uncoupling of the LHC I complex results in an impaired transfer of excitation energy from LHC I to PS I. The first step in the iron deficiency-induced remodelling of LHC I complex is the N-terminal processing of Lhca3. A consequence of this remodelling process is a characteristic blue-shifted maximum fluorescence emission signal of PS I at 77K (Naumann *et al.*, 2005). Cyanobacteria have a different response to iron deficiency; they express the iron stress-induced A (*isiA*) gene (Leonhardt and Straus, 1994). *IsiA* codes for the iron stress induced protein A (ISIA). This protein has sequence similarity with the CP 43 subunit of PS II (Burnap *et al.*, 1993; Laudenbach and Straus, 1988). ISIA arranges itself into a belt, composed of up to 18 molecules, around the trimeric PS I and increases its light harvesting ability by ~ 70%, compared with PS I in iron-sufficient conditions (Bibby *et al.*, 2001). This was interpreted as a strategy to compensate for a reduction in light harvesting phycobilisomes (Guikema and Sherman, 1983) and PS I during iron deficiency (Leonhardt and Straus, 1994). The alga *Dunaliella salina* expresses the thylakoid protein of 45 kDa (Tidi) to increase PS I antenna size during iron limitation. Tidi protein has high sequence similarity with the LHC I chlorophyll *a*/*b*-binding proteins from higher plants (Varsano *et al.*, 2003, 2006).

Iron deficiency results in a severe decrease in the photosynthetic parameters Fv/Fm and PS II operating efficiency in light ($\phi_{PS\ II}$), which has been reported for sugar beet (Larbi *et al.*, 2004) and *C. reinhardtii* (Naumann *et al.*, 2007; Terauchi *et al.*, 2010). Iron deficiency leads not only to the reorganization of protein complexes but at the same time affects the ultrastructure of the thylakoid membrane. The formation of grana is hindered, and a reduction of grana and stroma lamella per chloroplast and the number of thylakoids per granum have been reported (Spiller and Terry, 1980). In iron deficient *C. reinhardtii,* decreased levels of MGDG and an increase in triacylglyceride (TAG) could be shown (Urzica *et al.*, 2013). This decrease in MGDG, which occurs also in higher plants during iron starvation (Nishio *et al.*, 1985), affects membrane stacking and the photo-stability of PS II (Wu *et al.*, 2013).

Influence of the copper content on the thylakoid membrane

Copper is another important micronutrient for photosynthesis. However, copper in excess leads to thylakoid swelling and a decrease in lipid content and changed lipid composition (Quartacci *et al.*, 2000). Copper deficiency has deleterious effects on photosynthesis, as PC, which contains copper, becomes less abundant. This leads to a decreased electron flow between cyt b_6f and PS I (Ayala and Sandmann, 1988). Furthermore, in sugar beet the internal membrane system of chloroplasts greatly decreases under copper deficiency (Henriques, 1989). Interestingly *C. reinhardtii* can survive copper deficiency by replacing PC with the cytochrome c_6 protein, which contains haem as the redox active cofactor (Merchant and Bogorad, 1986; Wood, 1978). Plants do contain a homologue of cytochrome c_6, but there are conflicting reports whether it can replace PC. Moseley and co-workers (Moseley *et al.*, 2000) further identified the *cdr1* gene, whose protein product is thylakoid-localized and seems to be required for the maintenance of PS I and LHC I in copper-deficient *C. reinhardtii.* This implies that copper is not only important for PC but also for the maintenance of chlorophyll-binding proteins (Moseley *et al.*, 2002).

Significance of the manganese content for thylakoid function

For *Vigna unguiculata,* a 250-fold increase in manganese concentration leads to a reduced abundance of the Mn-stabilizing subunit PsbO of the water splitting complex, the likely explanation for *Vigna*'s compromised electron transfer rate (Führs *et al.*, 2008). If the manganese concentration falls below 0.1 μM, no phototrophic growth occurs in *C. reinhardtii* (Merchant *et al.*, 2006). This growth defect is the result of a severe decrease in the D1 protein, which binds the Mn_4Ca cluster and the related proteins of the water splitting complex (Allen *et al.*, 2007). Along these lines, a freeze fracture micrograph of the manganese deficient spinach thylakoids reveals the specific loss of three-quarters of the appressed, PS II containing thylakoids (Simpson and Robinson, 1984).

Nitrogen and sulfur deprivation and their biotechnological significance

Sulfur deprivation, for instance in *Oryza sativa,* leads to a decline not only in chlorophyll content but also in the entire photosynthetic apparatus, with the PS II efficiency decreasing by about 30%, and the ability of PS I to reduce $NADP^+$ decreased by 60% (Lunde *et al.*, 2008). A similar trend is also seen for *C. reinhardtii.* Under sulfur deprivation the PS II shuts down oxygen evolution, generating an anaerobic condition (Melis *et al.*, 2000). This manipulation is now used to study (and tune) the hydrogen production through the oxygen-sensitive hydrogenase (Ghirardi *et al.*, 2000). In *C. reinhardtii,* nitrogen starvation is used for the biotechnological application of increasing the starch or lipid (TAG) content (Hu *et al.*, 2008), the accumulation of the latter may increase if the starch biosynthesis is inhibited or absent (Wang *et al.*, 2009; Work *et al.*, 2010). CHT7 is a key genetic switch that couples nutrient starvation with lipid accumulation in *C. reinhardtii* (Tsai *et al.*, 2014). In contrast to the severe PS II deactivation effect of sulfur deficiency, nitrogen deprivation leads to a specific degradation of the cyt b_6f complexes (Bulté and Wollman, 1992) through NO production from intracellular nitrites (Wei *et al.*, 2014).

Conclusions

In this chapter we discussed diverse strategies that plants and algae use to optimize photosynthetic efficiency in an ever changing environment. Fine-tuning of photosynthesis happens at different levels, which include changes in the architecture, organization and composition of thylakoid membranes. Compositional changes involve changes in protein and lipid abundance or their rearrangements in the thylakoid membrane. The interaction between proteins and lipids is also essential for membrane integrity and function. These acclimatory responses operate on a timescale of seconds to days. Underpinning many responses is the diffusional freedom of electron carriers, photo-protective pigments, and light harvesting complexes in a highly crowded thylakoid membrane. Optimizing light energy harvesting and conversion is also an overwhelming response under light and nutrient stresses. The redox state of the PQ pool seems to be an important signal for initiating many of the acclimatory responses and its signalling pathways are just beginning to be unravelled. It should be noted that many of the acclimatory responses discussed here are studied under controlled growth conditions in the laboratory. In fact, in the natural environment plants are exposed to a combination of different stresses such as fluctuating light, drought, salinity and biotic factors. This combination of stresses may result in complex plant responses, underlined by diverse signalling pathways, which could interact or inhibit each other (Suzuki *et al.*, 2014). Investigating these complex environments and their effects on photosynthesis should be an important goal of future investigations.

Future trends

The molecular mechanisms that give thylakoids their enormous flexibly will certainly be the focus of intense investigations in years to come. A complete understanding of the dynamic design principles of the thylakoid membrane will illuminate efforts to engineer crops and biofuel algae for increased photosynthetic efficiency. The design principles of thylakoids could also be incorporated in artificial photosynthetic membranes and systems. Since photosynthesis is a highly efficient process, with a quantum efficiency of nearly 100% in converting solar energy into chemical energy, mimicking the photosynthetic processes is of high interest (Blankenship, 2002; Blankenship *et al.*, 2011; Calkins *et al.*, 2013). In recent years, the conversion of light into electricity by artificial photosynthetic systems has emerged as an important pursuit. One early study was the design of artificial reaction centres embedded in a liposome bilayer (Steinberg-Yfrach *et al.*, 1997). More recently, artificial photosynthetic antenna complexes (Gust *et al.*, 2012) have been designed, with some of them containing photo-protective functions that are similar to NPQ. Further, the artificial oxidation of water with a catalyst similar to PS II has been the subject of intensive research (Young *et al.*, 2012). However, Calkins *et al.* (2013) emphasize the advantage of using intact thylakoid membranes or whole organelles for electrochemical applications. With a potential for photo-fuel production, they have developed the thylakoid-carbon nanotube with high photo-electrochemical activity. The 'artificial leaf' highlights the importance of mimicking natural photosynthesis and having the same overall energy storage efficiency (Nocera, 2012).

References

Abrahams, J.P., Leslie, A.G.W., Lutter, R., and Walker, J.E. (1994). Structure at 2.8 A resolution of F_1-ATPase from bovine heart mitochondria. Nature *370*, 621–628.

Albertsson, P. (2001). A quantitative model of the domain structure of the photosynthetic membrane. Trends Plant Sci. *6*, 349–358.

Alexciev, K., and Tullberg, A. (1997). Regulation of *petB* mRNA stability in pea chloroplasts by redox poise. Physiol. Plant. *99*, 477–485.

Allen, J.F. (2003). State transitions – a question of balance. Science *299*, 1530–1532.

Allen, J.F. (2015). Why chloroplasts and mitochondria retain their own genomes and genetic systems: colocation for redox regulation of gene expression. Proc. Natl. Acad. Sci. *112*, 10231–10238.

Allen, J.F., and Forsberg, J. (2001). Molecular recognition in thylakoid structure and function. Trends Plant Sci. *6*, 317–326.

Allen, J.F., Bennett, J., Steinback, K.E., and Arntzen, C.J. (1981). Chloroplast protein phosphorylation couples plastoquinone redox state to distribution of excitation-energy between photosystems. Nature *291*, 25–29.

Allen, M.D., Kropat, J., Tottey, S., Del Campo, J. a, and Merchant, S.S. (2007). Manganese deficiency in *Chlamydomonas* results in loss of photosystem II and MnSOD function, sensitivity to peroxides, and secondary phosphorus and iron deficiency. Plant Physiol. *143*, 263–277.

Amunts, A., Drory, O., and Nelson, N. (2007). The structure of a plant photosystem I supercomplex at 3.4 A resolution. Nature *447*, 58–63.

Anderson, J.M. (1986). Photoregulation of the Composition, Function, and Structure of Thylakoid Membranes. Annu. Rev. Plant Physiol. *37*, 93–136.

Anderson, J.M. (1999). Insights into the consequences of grana stacking of thylakoid membranes in vascular plants: a personal perspective. Funct. Plant Biol. *26*, 625–639.

Anderson, J.M., and Andersson, B. (1988). The dynamic photosynthetic membrane and regulation of solar energy conversion. Trends Biochem. Sci. *13*, 351–355.

Anderson, J., Chow, W., and Goodchild, D. (1988). Thylakoid Membrane Organisation in Sun/Shade Acclimation. Aust. J. Plant Physiol. *15*, 11–15.

Anderson, J., Chow, W., and Park, Y.-I. (1995). The grand design of photosynthesis: Acclimation of the

photosynthetic apparatus to environmental cues. Photosynth. Res. *46*, 129–139.
Anderson, J.M., Chow, W.S., and De Las Rivas, J. (2008). Dynamic flexibility in the structure and function of photosystem II in higher plant thylakoid membranes: the grana enigma. Photosynth. Res. *98*, 575–587.
Andersson, B., and Anderson, J.M. (1980). Lateral heterogeneity in the distribution of chlorophyll–protein complexes of the thylakoid membranes of spinach chloroplasts. Biochim. Biophys. Acta *593*, 427–440.
Armbruster, U., Hertle, A., Makarenko, E., Zühlke, J., Pribil, M., Dietzmann, A., Schliebner, I., Aseeva, E., Fenino, E., Scharfenberg, M., *et al.* (2009). Chloroplast proteins without cleavable transit peptides: rare exceptions or a major constituent of the chloroplast proteome? Mol. Plant *2*, 1325–1335.
Armbruster, U., Labs, M., Pribil, M., Viola, S., Xu, W., Scharfenberg, M., Hertle, A.P., Rojahn, U., Jensen, P.E., Rappaport, F., *et al.* (2013). *Arabidopsis* curvature thylakoid 1 proteins modify thylakoid architecture by inducing membrane curvature. Plant Cell *25*, 2661–2678.
Asada, K. (2000). The water–water cycle as alternative photon and electron sinks. Phil. Trans. R. Soc. Lond. B. Biol. Sci. *355*, 1419–1431.
Ayala, M.B., and Sandmann, G. (1988). Activities of Cu-containing proteins in Cu-depleted pea leaves. Physiol. Plant. *72*, 801–806.
Bailey, S., Mann, N.H., Robinson, C., and Scanlan, D.J. (2005). The occurrence of rapidly reversible non-photochemical quenching of chlorophyll a fluorescence in cyanobacteria. FEBS Lett. *579*, 275–280.
Barber, J. (1982). Influence of surface charges on thylakoid structure and function. Annu. Rev. Plant Physiol. Plant Mol. Biol. *33*, 261–295.
Baroli, I., and Melis, A. (1998). Photoinhibitory damage is modulated by the rate of photosynthesis and by the photosystem II light-harvesting chlorophyll antenna size. Planta *205*, 288–296.
Barzda, V., Garab, G., Gulbinas, V., and Valkunas, L. (1996). Evidence for long-range excitation energy migration in macroaggregates of the chlorophyll a/b light-harvesting antenna complexes. Biochim. Biophys. Acta – Bioenerg. *1273*, 231–236.
Bellafiore, S., Barneche, F., Peltier, G., and Rochaix, J.D. (2005). State transitions and light adaptation require chloroplast thylakoid protein kinase STN7. Nature *433*, 892–895.
Bendall, D.S., and Manasse, R.S. (1995). Cyclic photophosphorylation and electron transport. Biochim. Biophys. Acta – Bioenerg. *1229*, 23–38.
de Bianchi, S., Dall'Osto, L., Tognon, G., Morosinotto, T., and Bassi, R. (2008). Minor antenna proteins CP24 and CP26 affect the interactions between photosystem II subunits and the electron transport rate in grana membranes of *Arabidopsis*. Plant Cell *20*, 1012–1028.
Bibby, T.S., Nield, J., and Barber, J. (2001). Iron deficiency induces the formation of an antenna ring around trimeric photosystem I in cyanobacteria. Nature *412*, 743–745.
Blankenship, R.E. (2002). Molecular Mechanisms of Photosynthesis (Oxford: Blackwell Science Ltd).
Blankenship, R.E., Tiede, D.M., Barber, J., Brudvig, G.W., Fleming, G., Ghirardi, M., Gunner, M.R., Junge, W., Kramer, D.M., Melis, A., *et al.* (2011). Comparing photosynthetic and photovoltaic efficiencies and recognizing the potential for improvement. Science *332*, 805–809.
Boekema, E.J., Van Roon, H., Van Breemen, J.F., and Dekker, J.P. (1999). Supramolecular organization of photosystem II and its light-harvesting antenna in partially solubilized photosystem II membranes. Eur. J. Biochem. *266*, 444–452.
Boekema, E.J., van Breemen, J.F., van Roon, H., and Dekker, J.P. (2000). Arrangement of photosystem II supercomplexes in crystalline macrodomains within the thylakoid membrane of green plant chloroplasts. J. Mol. Biol. *301*, 1123–1133.
Bonardi, V., Pesaresi, P., Becker, T., Schleiff, E., Wagner, R., Pfannschmidt, T., Jahns, P., and Leister, D. (2005). Photosystem II core phosphorylation and photosynthetic acclimation require two different protein kinases. Nature *437*, 1179–1182.
Bonaventura, C., and Myers, J. (1969). Fluorescence and oxygen evolution from *Chlorella pyrenoidosa*. Biochim. Biophys. Acta *189*, 366–383.
Boyer, J.S. (1982). Plant productivity and environment. Science *218*, 443–448.
Bulté, L., and Wollman, F.A. (1992). Evidence for a selective destabilization of an integral membrane protein, the cytochrome b_6f complex, during gametogenesis in *Chlamydomonas reinhardtii*. Eur. J. Biochem. *204*, 327–336.
Burnap, R.L., Troyan, T., and Sherman, L.A. (1993). The highly abundant chlorophyll–protein complex of iron deficient *Synechococcus* sp. PCC7942 (CP43′) is encoded by the isiA gene. Plant Physiol. *103*, 893–902.
Cadoret, J.C., Demoulière, R., Lavaud, J., Van Gorkom, H.J., Houmard, J., and Etienne, A.L. (2004). Dissipation of excess energy triggered by blue light in cyanobacteria with CP43′ (isiA). Biochim. Biophys. Acta – Bioenerg. *1659*, 100–104.
Calkins, J.O., Umasankar, Y., O'Neill, H., and Ramasamy, R.P. (2013). High photo-electrochemical activity of thylakoid–carbon nanotube composites for photosynthetic energy conversion. Energy Environ. Sci. *6*, 1891–1900.
Cardol, P., Bailleul, B., Rappaport, F., Derelle, E., Béal, D., Breyton, C., Bailey, S., Wollman, F.A., Grossman, A., Moreau, H., *et al.* (2008). An original adaptation of photosynthesis in the marine green alga *Ostreococcus*. Proc. Natl. Acad. Sci. U.S.A. *105*, 7881–7886.
Choquet, Y., and Vallon, O. (2000). Synthesis, assembly and degradation of thylakoid membrane proteins. Biochimie *82*, 615–634.
Chow, W., Qian, L., Goodchild, D., and Anderson, J. (1988a). Photosynthetic acclimation of Alocasia macrorrhiza (L.) G. Don. Funct. Plant Biol. *15*, 107–122.
Chow, W.S., Anderson, J.M., and Hope, A.B. (1988b). Variable stoichiometries of photosystem II to photosystem I reaction centres. Photosynth. Res. *17*, 277–281.
Chow, W.S., Melis, A., and Anderson, J.M. (1990). Adjustments of photosystem stoichiometry in chloroplasts improve the quantum efficiency of photosynthesis. Proc. Natl. Acad. Sci. U.S.A. *87*, 7502–7506.

Consoli, E., Croce, R., Dunlap, D.D., and Finzi, L. (2005). Diffusion of light-harvesting complex II in the thylakoid membranes. EMBO Rep. *6*, 782–786.

Croce, R., and van Amerongen, H. (2014). Natural strategies for photosynthetic light harvesting. Nat. Chem. Biol. *10*, 492–501.

Daum, B., Nicastro, D., Austin, J., McIntosh, J.R., and Kühlbrandt, W. (2010). Arrangement of photosystem II and ATP synthase in chloroplast membranes of spinach and pea. Plant Cell *22*, 1299–1312.

Dekker, J.P., and Boekema, E.J. (2005). Supramolecular organization of thylakoid membrane proteins in green plants. Biochim. Biophys. Acta *1706*, 12–39.

Delosme, R., Olive, J., and Wollman, F.A. (1996). Changes in light energy distribution upon state transitions: an *in vivo* photoacoustic study of the wild type and photosynthesis mutants from *Chlamydomonas reinhardtii*. Biochim. Biophys. Acta – Bioenerg. *1273*, 150–158.

Depege, N., Bellafiore, S., and Rochaix, J.D. (2003). Role of chloroplast protein kinase Stt7 in LHCII phosphorylation and state transition in *Chlamydomonas*. Science *299*, 1572–1575.

Dietzel, L., Bräutigam, K., and Pfannschmidt, T. (2008). Photosynthetic acclimation: State transitions and adjustment of photosystem stoichiometry – Functional relationships between short-term and long-term light quality acclimation in plants. FEBS J. *275*, 1080–1088.

Dietzel, L., Bräutigam, K., Steiner, S., Schüffler, K., Lepetit, B., Grimm, B., Schöttler, M.A., and Pfannschmidt, T. (2011). Photosystem II supercomplex remodeling serves as an entry mechanism for state transitions in *Arabidopsis*. Plant Cell *23*, 2964–2977.

Dörmann, P., Hoffmann-Benning, S., Balbo, I., and Benning, C. (1995). Isolation and characterization of an *Arabidopsis* mutant deficient in the thylakoid lipid digalactosyl diacylglycerol. Plant Cell *7*, 1801–1810.

Drop, B., Sathish Yadav, K.N., Boekema, E.J., and Croce, R. (2014). Consequences of state transitions on the structural and functional organization of photosystem I in the green alga *Chlamydomonas reinhardtii*. Plant J. *78*, 181–191.

Dubertret, G., Gerard-Hirne, C., and Trémolières, A. (2002). Importance of trans-Δ^3-hexadecenoic acid containing phosphatidylglycerol in the formation of the trimeric light-harvesting complex in *chlamydomonas*. Plant Physiol. Biochem. *40*, 829–836.

Ellis, R.J. (2001). Macromolecular crowding : obvious but under appreciated. Trends Biochem. Sci. *26*, 597–604.

Escoubas, J.M., Lomas, M., LaRoche, J., and Falkowski, P.G. (1995). Light intensity regulation of cab gene transcription is signaled by the redox state of the plastoquinone pool. Proc. Natl. Acad. Sci. U.S.A. *92*, 10237–10241.

Fan, M., Li, M., Liu, Z., Cao, P., Pan, X., Zhang, H., Zhao, X., Zhang, J., and Chang, W. (2015). Crystal structures of the PsbS protein essential for photoprotection in plants. Nat. Struct. Mol. Biol. *22*, 729–735.

Finazzi, G., Rappaport, F., Furia, A., Fleischmann, M., Rochaix, J.D., Zito, F., and Forti, G. (2002). Involvement of state transitions in the switch between linear and cyclic electron flow in *Chlamydomonas reinhardtii*. EMBO Rep. *3*, 280–285.

Flachmann, R., and Kühlbrandt, W. (1995). Accumulation of plant antenna complexes is regulated by post-transcriptional mechanisms in tobacco. Plant Cell *7*, 149–160.

Foudree, A., Putarjunan, A., Kambakam, S., Nolan, T., Fussell, J., Pogorelko, G., and Rodermel, S. (2012). The mechanism of variegation in immutans provides insight into chloroplast biogenesis. Front. Plant Sci. *3*, 260.

Fristedt, R., and Vener, A.V. (2011). High light induced disassembly of photosystem II supercomplexes in *Arabidopsis* requires STN7-dependent phosphorylation of CP29. PLoS One *6*, e24565.

Fuhrmann, E., Bultema, J.B., Kahmann, U., Rupprecht, E., Boekema, E.J., and Schneider, D. (2009). The vesicle-inducing Protein 1 from *Synechocystis* sp. PCC 6803 organizes into diverse higher-ordered ring structures. Mol. Biol. Cell *20*, 4620–4628.

Führs, H., Hartwig, M., Molina, L.E.B., Heintz, D., Van Dorsselaer, A., Braun, H.P., and Horst, W.J. (2008). Early manganese-toxicity response in Vigna unguiculata L. – A proteomic and transcriptomic study. Proteomics *8*, 149–159.

Galka, P., Santabarbara, S., Khuong, T.T., Degand, H., Morsomme, P., Jennings, R.C., Boekema, E.J., and Caffarri, S. (2012). Functional analyses of the plant photosystem I–light-harvesting complex II supercomplex reveal that light-harvesting complex II loosely bound to photosystem II is a very efficient antenna for photosystem I in state II. Plant Cell *24*, 2963–2978.

Gan, F., Zhang, S., Rockwell, N.C., Martin, S.S., Lagarias, J.C., and Bryant, D.A. (2014). Extensive remodeling of a cyanobacterial photosynthetic apparatus in far-red light. Science *345*, 1312–1317.

Garab, G., and Musárdy, L. (2000). Role of LHCII-containing macrodomains in the structure, function and dynamics of grana. Aust. J. Plant Physiol. *27*, 649–658.

Garab, G., Kieleczawa, J., Sutherland, J.C., Bustamante, C., and Hind, G. (1991). Organisation of pigment–protein complexes into macrodomains in the thylakoid membranes of wild-type and chlorophyll less mutant of barley as revealed by circular dichroism. Photochem. Photobiol. *54*, 273–281.

Garab, G., Lohner, K., Laggner, P., and Farkas, T. (2000). Self-regulation of the lipid content of membranes by non-bilayer lipids: A hypothesis. Trends Plant Sci. *5*, 489–494.

Garber, M.P., and Steponkus, P.L. (1976). Alterations in Chloroplast Thylakoids during Cold Acclimation. Plant Physiol. *57*, 681–686.

Ghirardi, M.L., Zhang, L., Lee, J.W., Flynn, T., Seibert, M., Greenbaum, E., and Melis, A. (2000). Microalgae: A green source of renewable H_2. Trends Biotechnol. *18*, 506–511.

Goral, T.K., Johnson, M.P., Duffy, C.D.P., Brain, A.P.R., Ruban, A.V., and Mullineaux, C.W. (2012). Light-harvesting antenna composition controls the macrostructure and dynamics of thylakoid membranes in *Arabidopsis*. Plant J. *69*, 289–301.

Gruner, S.M. (1989). Stability of lyotropic phases with curved interfaces. J. Phys. Chem. *93*, 7562–7570.

Guikema, J. a, and Sherman, L. a (1983). Organization and Function of Chlorophyll in Membranes of Cyanobacteria during Iron Starvation. Plant Physiol. 73, 250–256.

Guo, J., Zhang, Z., Bi, Y., Yang, W., Xu, Y., and Zhang, L. (2005). Decreased stability of photosystem I in *dgd1* mutant of *Arabidopsis thaliana*. FEBS Lett. *579*, 3619–3624.

Guo, J., Wei, X., Li, M., Pan, X., Chang, W., and Liu, Z. (2013). Structure of the catalytic domain of a state transition kinase homolog from Micromonas algae. Protein Cell *4*, 607–619.

Guss, J.M., Harrowell, P.R., Murata, M., Norris, V.A., and Freeman, H.C. (1986). Crystal structure analyses of reduced (CuI) poplar plastocyanin at six pH values. J. Mol. Biol. *192*, 361–387.

Gust, D., Moore, T., A., and Moore, A.L. (2012). Realizing artificial photosynthesis. Faraday Discuss. *155*, 9–26.

Haferkamp, S., and Kirchhoff, H. (2008). Significance of molecular crowding in grana membranes of higher plants for light harvesting by photosystem II. Photosynth. Res. *95*, 129–134.

Haferkamp, S., Haase, W., Pascal, A. a, van Amerongen, H., and Kirchhoff, H. (2010). Efficient light harvesting by photosystem II requires an optimized protein packing density in Grana thylakoids. J. Biol. Chem. *285*, 17020–17028.

Harwood, J. (1998). Involvement of chloroplast lipids in the reaction of plants submitted to stress. In Lipids in Photosynthesis: Structure, Function and Genetics SE – 15, S. Paul-André, and M. Norio, eds. (Springer: Netherlands), pp. 287–302.

Hasan, S.S., and Cramer, W.A. (2012). Lipid functions in cytochrome bc complexes: an odd evolutionary transition in a membrane protein structure. Phil. Trans. R. Soc. Lond. B. Biol. Sci. *367*, 3406–3411.

Henriques, F.S. (1989). Effects of copper deficiency on the photosynthetic apparatus of sugar beet (*Beta vulgaris* L.). J. Plant Physiol. *135*, 453–458.

Herbstova, M., Tietz, S., Kinzel, C., Turkina, M.V., and Kirchhoff, H. (2012). Architectural switch in plant photosynthetic membranes induced by light stress. Proc. Natl. Acad. Sci. U.S.A. *109*, 20130–20135.

Hihara, Y., Sonoike, K., and Ikeuchi, M. (1998). A novel gene, pmgA, specifically regulates photosystem stoichiometry in the cyanobacterium *Synechocystis* species PCC 6803 in response to high light. Plant Physiol. *117*, 1205–1216.

Hill, R., and Bendall, F. (1960). Function of the two cytochrome components in chloroplasts – a working hypothesis. Nature *186*, 136–137.

Hohmann-Marriott, M.F., and Blankenship, R.E. (2011). Evolution of photosynthesis. Annu. Rev. Plant Biol. *62*, 515–548.

Horton, P. (1999). Hypothesis: Are grana necessary for regulation of light harvesting? Aust. J. Plant Physiol. *26*, 659–669.

Hu, Q., Sommerfeld, M., Jarvis, E., Ghirardi, M., Posewitz, M., Seibert, M., and Darzins, A. (2008). Microalgal triacylglycerols as feedstocks for biofuel production: perspectives and advances. Plant J. *54*, 621–639.

Ibrahim, I.M., Puthiyaveetil, S., and Allen, J.F. (2016). A two-component regulatory system in transcriptional control of photosystem stoichiometry: redox-dependent and sodium ion-dependent phosphoryl transfer from cyanobacterial histidine kinase Hik2 to response regulators Rre1 and RppA. Front. Plant Sci. *7*, 137.

Iwai, M., Takizawa, K., Tokutsu, R., Okamuro, A., Takahashi, Y., and Minagawa, J. (2010). Isolation of the elusive supercomplex that drives cyclic electron flow in photosynthesis. Nature *464*, 1210–1213.

Jahns, P., Latowski, D., and Strzalka, K. (2009). Mechanism and regulation of the violaxanthin cycle: The role of antenna proteins and membrane lipids. Biochim. Biophys. Acta – Bioenerg. *1787*, 3–14.

Johnson, M.P., Vasilev, C., Olsen, J.D., and Hunter, C.N. (2014). Nanodomains of cytochrome b_6f and photosystem II complexes in spinach grana thylakoid membranes. Plant Cell *26*, 3051–3061.

Junge, W., Sielaff, H., and Engelbrecht, S. (2009). Torque generation and elastic power transmission in the rotary F_0F_1-ATPase. Nature *459*, 364–370.

Kaftan, D., Brumfeld, V., Nevo, R., Scherz, A., and Reich, Z. (2002). From chloroplasts to photosystems: *in situ* scanning force microscopy on intact thylakoid membranes. EMBO J. *21*, 6146–6153.

Kappell, A.D., and van Waasbergen, L.G. (2007). The response regulator RpaB binds the high light regulatory 1 sequence upstream of the high-light-inducible *hliB* gene from the cyanobacterium *Synechocystis* PCC 6803. Arch. Microbiol. *187*, 337–342.

Kargul, J., Turkina, M.V., Nield, J., Benson, S., Vener, A.V., and Barber, J. (2005). Light-harvesting complex II protein CP29 binds to photosystem I of *Chlamydomonas reinhardtii* under State 2 conditions. FEBS J. *272*, 4797–4806.

Kato, H., Chibazakura, T., and Yoshikawa, H. (2014). NblR is a novel one-component response regulator in the cyanobacterium *Synechococcus elongatus* PCC 7942. Biosci. Biotechnol. Biochem. *72*, 1072–1079.

Kereïche, S., Kiss, A.Z., Kouril, R., Boekema, E.J., and Horton, P. (2010). The PsbS protein controls the macro-organisation of photosystem II complexes in the grana membranes of higher plant chloroplasts. FEBS Lett. *584*, 759–764.

Kim, J.H., Glick, R.E., and Melis, A. (1993). Dynamics of photosystem stoichiometry adjustment by light quality in chloroplasts. Plant Physiol. *102*, 181–190.

Kirchhoff, H., Horstmann, S., and Weis, E. (2000). Control of the photosynthetic electron transport by PQ diffusion microdomains in thylakoids of higher plants. Biochim. Biophys. Acta *1459*, 148–168.

Kirchhoff, H. (2008a). Molecular crowding and order in photosynthetic membranes. Trends Plant Sci. *13*, 201–207.

Kirchhoff, H. (2008b). Significance of protein crowding, order and mobility for photosynthetic membrane functions. Biochem. Soc. Trans. *36*, 967–970.

Kirchhoff, H. (2014). Diffusion of molecules and macromolecules in thylakoid membranes. Biochim. Biophys. Acta *1837*, 495–502.

Kirchhoff, H., Mukherjee, U., and Galla, H. (2002). Molecular architecture of the thylakoid membrane: lipid diffusion space for plastoquinone. Biochemistry *41*, 4872–4882.

Kirchhoff, H., Borinski, M., Lenhert, S., Chi, L., and Büchel, C. (2004a). Transversal and lateral exciton energy transfer in grana thylakoids of spinach. Biochemistry *43*, 14508–14516.

Kirchhoff, H., Tremmel, I., Haase, W., and Kubitscheck, U. (2004b). Supramolecular photosystem II organization in grana thylakoid membranes: evidence for a structured arrangement. Biochemistry *43*, 9204–9213.

Kirchhoff, H., Haase, W., Wegner, S., Danielsson, R., Ackermann, R., and Albertsson, P.A. (2007). Low-Light-induced formation of semicrystalline photosystem II arrays in higher plant chloroplasts. Biochemistry *46*, 11169–11176.

Kirchhoff, H., Haferkamp, S., Allen, J.F., Epstein, D.B.A., and Mullineaux, C.W. (2008). Protein diffusion and macromolecular crowding in thylakoid membranes. Plant Physiol. *146*, 1571–1578.

Kirchhoff, H., Hall, C., Wood, M., Herbstova, M., Tsabari, O., Nevo, R., Charuvi, D., Shimoni, E., and Reich, Z. (2011). Dynamic control of protein diffusion within the granal thylakoid lumen. Proc. Natl. Acad. Sci. U.S.A. *108*, 20248–20253.

Kirilovsky, D., and Kerfeld, C.A. (2013). The Orange Carotenoid Protein: a blue-green light photoactive protein. Photochem. Photobiol. Sci. *12*, 1135–1143.

Kouril, R., Dekker, J.P., and Boekema, E.J. (2012). Supramolecular organization of photosystem II in green plants. Biochim. Biophys. Acta *1817*, 2–12.

Kouřil, R., Wientjes, E., Bultema, J.B., Croce, R., Boekema, E.J., Kouril, R., Wientjes, E., Bultema, J.B., Croce, R., and Boekema, E.J. (2013). High-light vs. low-light: effect of light acclimation on photosystem II composition and organization in *Arabidopsis thaliana*. Biochim. Biophys. Acta *1827*, 411–419.

Kurisu, G., Zhang, H., Smith, J.L., and Cramer, W.A. (2003). Structure of the cytochrome b_6f complex of oxygenic photosynthesis: tuning the cavity. Science *302*, 1009–1014.

Larbi, A., Abadía, A., Morales, F., and Abadía, J. (2004). Fe resupply to Fe-deficient sugar beet plants leads to rapid changes in the violaxanthin cycle and other photosynthetic characteristics without significant *de novo* chlorophyll synthesis. Photosynth. Res. 79, 59–69.

Larsson, K. (1989). Cubic lipid–water phases : structures and biomembrane aspects. J. Phys. Chem. 93, 7304–7314.

Laudenbach, D.E., and Straus, N.A. (1988). Characterization of a cyanobacterial iron stress-induced gene similar to psbC. J. Bacteriol. *170*, 5018–5026.

Lavergne, J., Bouchard, J.P., and Joliot, P. (1992). Plastoquinone compartmentation in chloroplasts. II. Theoretical aspects. Biochim. Biophys. Acta Bioenerg. *1101*, 13–22.

Lee, A.G. (2000). Membrane lipids: It's only a phase. Curr. Biol. *10*, 377–380.

Lemeille, S., Willig, A., Depege-Fargeix, N., Delessert, C., Bassi, R., and Rochaix, J.D. (2009). Analysis of the chloroplast protein kinase Stt7 during state transitions. PLoS Biol 7, e45.

Leong, T.Y., and Anderson, J.M. (1984). Adaptation of the thylakoid membranes of pea chloroplasts to light intensities. II. Regulation of electron transport capacities, electron carriers, coupling factor (CF1) activity and rates of photosynthesis. Photosynth. Res. 5, 117–128.

Leonhardt, K., and Straus, N.A. (1994). Photosystem II genes isiA, psbDI and psbC in Anabaena sp. PCC 7120: cloning, sequencing and the transcriptional regulation in iron-stressed and iron-repleted cells. Plant Mol. Biol. *24*, 63–73.

Li, M., Semchonok, D.A., Boekema, E.J., and Bruce, B.D. (2014). Characterization and evolution of tetrameric photosystem I from the thermophilic cyanobacterium *Chroococcidiopsis* sp. TS-821. Plant Cell *26*, 1230–1245.

Li, X.P., Björkman, O., Shih, C., Grossman, A.R., Rosenquist, M., Jansson, S., and Niyogi, K.K. (2000). A pigment-binding protein essential for regulation of photosynthetic light harvesting. Nature *403*, 391–395.

Li, X.-P., Gilmore, A.M., Caffarri, S., Bassi, R., Golan, T., Kramer, D., and Niyogi, K.K. (2004). Regulation of photosynthetic light harvesting involves intrathylakoid lumen pH sensing by the PsbS protein. J. Biol. Chem. *279*, 22866–22874.

Li, Z., Wakao, S., Fischer, B.B., and Niyogi, K.K. (2009). Sensing and responding to excess light. Annu. Rev. Plant Biol. *60*, 239–260.

Lindahl, M., Yang, D.H., and Andersson, B. (1995). Regulatory proteolysis of the major light-harvesting chlorophyll a/b protein of photosystem II by a light-induced membrane-associated enzymic system. Eur. J. Biochem. *231*, 503–509.

Liu, Z.F., Yan, H.C., Wang, K.B., Kuang, T.Y., Zhang, J.P., Gui, L.L., An, X.M., and Chang, W.R. (2004). Crystal structure of spinach major light-harvesting complex at 2.72 angstrom resolution. Nature *428*, 287–292.

Lunde, C., Zygadlo, A., Simonsen, H.T., Nielsen, P.L., Blennow, A., and Haldrup, A. (2008). Sulfur starvation in rice: the effect on photosynthesis, carbohydrate metabolism, and oxidative stress protective pathways. Physiol. Plant. *134*, 508–521.

Macko, S., Wehner, A., and Jahns, P. (2002). Comparison of violaxanthin de-epoxidation from the stroma and lumen sides of isolated thylakoid membranes from *Arabidopsis*: Implications for the mechanism of de-epoxidation. Planta *216*, 309–314.

Mattoo, A.K., Hoffman-Falk, H., Marder, J.B., and Edelman, M. (1984). Regulation of protein metabolism: Coupling of photosynthetic electron transport to *in vivo* degradation of the rapidly metabolized 32-kilodalton protein of the chloroplast membranes. Proc. Natl. Acad. Sci. U.S.A. *81*, 1380–1384.

Mcdonnel, A., and Staehelin, L.A. (1980). Adhesion between liposomes mediated by the chlorophyll a/b light-harvesting complex isolated from chloroplast membranes. J. Cell Biol. *84*, 40–56.

Van De Meene, A.M.L., Hohmann-Marriott, M.F., Vermaas, W.F.J., and Roberson, R.W. (2006). The three-dimensional structure of the cyanobacterium *Synechocystis* sp. PCC 6803. Arch. Microbiol. *184*, 259–270.

Melis, A., Zhang, L., Forestier, M., Ghirardi, M.L., and Seibert, M. (2000). Sustained photobiological hydrogen gas production upon reversible inactivation of oxygen evolution in the green alga *Chlamydomonas reinhardtii*. Plant Physiol. *122*, 127–136.

Merchant, S., and Bogorad, L. (1986). Rapid degradation of apoplastocyanin in Cu(II)-deficient cells of

Chlamydomonas reinhardtii. J. Biol. Chem. *261*, 15850–15853.
Merchant, S.S., Allen, M.D., Kropat, J., Moseley, J.L., Long, J.C., Tottey, S., and Terauchi, A.M. (2006). Between a rock and a hard place: Trace element nutrition in *Chlamydomonas*. *1763*, 578–594.
Meyer, G., and Kloppstech, K. (1984). A rapidly light-induced chloroplast protein with a high turnover coded for by pea nuclear DNA. Eur. J. Biochem. *138*, 201–207.
Miller, K.R., Miller, G.J., and McIntyre, K.R. (1976). The light-harvesting chlorpohyll–protein complex of photosystem II. Its location in the photosynthetic membrane. J. Cell Biol. *71*, 624–638.
Minton, A.P. (2001). The influence of macromolecular crowding and macromolecular confinement on biochemical reactions in physiological media. J. Biol. Chem. *276*, 10577–10580.
Miyake, C., and Yokota, A. (2001). Cyclic flow of electrons within PSII in thylakoid membranes. Plant Cell Physiol. *42*, 508–515.
Morosinotto, T., Bassi, R., Frigerio, S., Finazzi, G., Morris, E., and Barber, J. (2006). Biochemical and structural analyses of a higher plant photosystem II supercomplex of a photosystem I-less mutant of barley. Consequences of a chronic over-reduction of the plastoquinone pool. FEBS J. *273*, 4616–4630.
Moseley, J., Quinn, J., Eriksson, M., and Merchant, S. (2000). The *Crd1* gene encodes a putative di-iron enzyme required for photosystem I accumulation in copper deficiency and hypoxia in *Chlamydomonas reinhardtii*. EMBO J. *19*, 2139–2151.
Moseley, J.L., Allinger, T., Herzog, S., Hoerth, P., Wehinger, E., Merchant, S., and Hippler, M. (2002). Adaptation to Fe-deficiency requires remodeling of the photosynthetic apparatus. EMBO J. *21*, 6709–6720.
Mullineaux, C.W. (2005). Function and evolution of grana. Trends Plant Sci. *10*, 521–525.
Mullineaux, C.W., and Allen, J.F. (1986). The state 2 transition in the cyanobacterium Synechococcus 6301 can be driven by respiratory electron flow into the plastoquinone pool. FEBS Lett. *205*, 155–160.
Murakami, A., and Fujita, Y. (1991). Regulation of photosystem stoichiometry in the photosynthetic system of the cyanophyte *Synechocystis* PCC 6714 in response to light-intensity. Plant Cell Physiol. *32*, 223–230.
Murakami, A., Kim, S.J., and Fujita, Y. (1997a). Changes in photosystem stoichiometry in response to environmental conditions for cell growth observed with the cyanophyte *Synechocystis* PCC 6714. Plant Cell Physiol. *38*, 392–397.
Murakami, A., Fujita, Y.A., Nemson, J., and Melis, A. (1997b). Chromatic regulation in *Chlamydomonas reinhardtii*: time course of photosystem stoichiometry adjustment following a shift in growth light quality. Plant Cell Physiol. *38*, 188–193.
Murata, N. (1969). Control of excitation transfer in photosynthesis. I. Light-induced changes of chlorophyll *a* fluorescence in *Porphyridium cruentum*. Biochim. Biophys. Acta *172*, 242–251.
Myers, J. (1971). Enhancement studies in photosynthesis. Annu. Rev. Plant Physiol. *22*, 289–312.
Nagy, G., Ünnep, R., Zsiros, O., Tokutsu, R., Takizawa, K., Porcar, L., Moyet, L., Petroutsos, D., Garab, G., Finazzi, G., *et al.* (2014). Chloroplast remodeling during state transitions in *Chlamydomonas reinhardtii* as revealed by noninvasive techniques in vivo. Proc. Natl. Acad. Sci. U.S.A. *111*, 5042–5047.
Nakamura, Y., Kaneko, T., Sato, S., Mimuro, M., Miyashita, H., Tsuchiya, T., Sasamoto, S., Watanabe, A., Kawashima, K., Kishida, Y., *et al.* (2003). Complete genome structure of *Gloeobacter violaceus* PCC 7421, a cyanobacterium that lacks thylakoids. DNA Res. *10*, 137–145.
Naumann, B., Stauber, E.J., Busch, A., Sommer, F., and Hippler, M. (2005). N-terminal processing of Lhca3 is a key step in remodeling of the photosystem I-light-harvesting complex under iron deficiency in *Chlamydomonas reinhardtii*. J. Biol. Chem. *280*, 20431–20441.
Naumann, B., Busch, A., Allmer, J., Ostendorf, E., Zeller, M., Kirchhoff, H., and Hippler, M. (2007). Comparative quantitative proteomics to investigate the remodeling of bioenergetic pathways under iron deficiency in *Chlamydomonas reinhardtii*. *7*, 3964–3979.
Nawrocki, W.J., Tourasse, N.J., Taly, A., Rappaport, F., and Wollman, F.A. (2015). The plastid terminal oxidase: its elusive function points to multiple contributions to plastid physiology. Annu. Rev. Plant Biol. *66*, 49–74.
Nelson, N., and Ben-Shem, A. (2004). The complex architecture of oxygenic photosynthesis. Nat. Rev. Mol. Cell Biol. *5*, 971–982.
Nickelsen, J., and Rengstl, B. (2013). Photosystem II assembly: from cyanobacteria to plants. Annu. Rev. Plant Biol. *64*, 609–635.
Nickelsen, J., and Zerges, W. (2013). Thylakoid biogenesis has joined the new era of bacterial cell biology. Front. Plant Sci. *4*, 458.
Nishio, J.N., Taylor, S.E., and Terry, N. (1985). Changes in thylakoid galactolipids and proteins during iron nutrition-mediated chloroplast development. Plant Physiol. *77*, 705–711.
Nocera, D.G. (2012). The artificial leaf. Acc. Chem. Res. *45*, 767–776.
Nordhues, A., Schöttler, M.A., Unger, A.K., Geimer, S., Schönfelder, S., Schmollinger, S., Rütgers, M., Finazzi, G., Soppa, B., Sommer, F., *et al.* (2012). Evidence for a role of VIPP1 in the structural organization of the photosynthetic apparatus in *Chlamydomonas*. Plant Cell *24*, 637–659.
Ohad, I., Kyle, D.J., and Arntzen, C.J. (1984). Membrane protein damage and repair: removal and replacement of inactivated 32-kilodalton polypeptides in chloroplast membranes. J. Cell Biol. *99*, 481–485.
Paget, M.S., and Helmann, J.D. (2003). The sigma70 family of sigma factors. Genome Biol. *4*, 203.
Palmer, C.M., and Guerinot, M.L. (2009). Facing the challenges of Cu, Fe and Zn homeostasis in plants. Nat. Chem. Biol. *5*, 333–340.
Peers, G., Truong, T.B., Ostendorf, E., Busch, A., Elrad, D., Grossman, A.R., Hippler, M., and Niyogi, K.K. (2009). An ancient light-harvesting protein is critical for the regulation of algal photosynthesis. Nature *462*, 518–521.
Pesaresi, P., Hertle, A., Pribil, M., Kleine, T., Wagner, R., Strissel, H., Ihnatowicz, A., Bonardi, V., Scharfenberg, M., Schneider, A., *et al.* (2009). *Arabidopsis* STN7

kinase provides a link between short- and long-term photosynthetic acclimation. Plant Cell *21*, 2402–2423.

Petrillo, E., Godoy Herz, M.A., Fuchs, A., Reifer, D., Fuller, J., Yanovsky, M.J., Simpson, C., Brown, J.W.S., Barta, A., Kalyna, M., *et al.* (2014). A chloroplast retrograde signal regulates nuclear alternative splicing. Science *344*, 427–430.

Petroutsos, D., Busch, A., Janssen, I., Trompelt, K., Bergner, S., V., Weinl, S., Holtkamp, M., Karst, U., Kudla, J., and Hippler, M. (2011). The Chloroplast Calcium Sensor CAS Is Required for Photoacclimation in *Chlamydomonas reinhardtii*. Plant Cell *23*, 2950–2963.

Pfannschmidt, T., Nilsson, A., and Allen, J.F. (1999). Photosynthetic control of chloroplast gene expression. Nature *397*, 625–628.

Pribil, M., Pesaresi, P., Hertle, A., Barbato, R., and Leister, D. (2010). Role of plastid protein phosphatase TAP38 in LHCII dephosphorylation and thylakoid electron flow. PLoS Biol *8*, e1000288.

Puthiyaveetil, S. (2011). A mechanism for regulation of chloroplast LHC II kinase by plastoquinol and thioredoxin. FEBS Lett. *585*, 1717–1721.

Puthiyaveetil, S., and Allen, J.F. (2008). Transients in chloroplast gene transcription. Biochem. Biophys. Res. Commun. *368*, 871–874.

Puthiyaveetil, S., and Allen, J.F. (2009). Chloroplast two-component systems: evolution of the link between photosynthesis and gene expression. Proc. Biol. Sci. *276*, 2133–2145.

Puthiyaveetil, S., and Kirchhoff, H. (2013). A phosphorylation map of the photosystem II supercomplex C2S2M2. Front. Plant Sci. *4*, 459.

Puthiyaveetil, S., Kavanagh, T.A., Cain, P., Sullivan, J.A., Newell, C.A., Gray, J.C., Robinson, C., van der Giezen, M., Rogers, M.B., and Allen, J.F. (2008). The ancestral symbiont sensor kinase CSK links photosynthesis with gene expression in chloroplasts. Proc. Natl. Acad. Sci. U.S.A. *105*, 10061–10066.

Puthiyaveetil, S., Ibrahim, I.M., Jelicić, B., Tomasić, A., Fulgosi, H., and Allen, J.F. (2010). Transcriptional control of photosynthesis genes: the evolutionarily conserved regulatory mechanism in plastid genome function. Genome Biol. Evol. *2*, 888–896.

Puthiyaveetil, S., Ibrahim, I.M., and Allen, J.F. (2012). Oxidation-reduction signalling components in regulatory pathways of state transitions and photosystem stoichiometry adjustment in chloroplasts. Plant Cell Environ. *35*, 347–359.

Puthiyaveetil, S., Ibrahim, I.M., and Allen, J.F. (2013). Evolutionary rewiring: a modified prokaryotic gene-regulatory pathway in chloroplasts. Phil. Trans. R. Soc. Lond. B. Biol. Sci. *368*, 20120260.

Puthiyaveetil, S., Tsabari, O., Lowry, T., Lenhert, S., Lewis, R.R., Reich, Z., and Kirchhoff, H. (2014). Compartmentalization of the protein repair machinery in photosynthetic membranes. Proc. Natl. Acad. Sci. U.S.A. *111*, 15839–15844.

Qin, X., Suga, M., Kuang, T., and Shen, J.R. (2015). Structural basis for energy transfer pathways in the plant PSI-LHCI supercomplex. Science *348*, 989–995.

Quartacci, M.F., Pinzino, C., Sgherri, C.L.M., Dalla, F., and Navari-izzo, F. (2000). Growth in excess copper induces changes in the lipid composition and fluidity of PSII-enriched membranes in wheat. Physiol. Plant. *108*, 87–93.

Rast, A., Heinz, S., and Nickelsen, J. (2015). Biogenesis of thylakoid membranes. Biochim. Biophys. Acta – Bioenerg. *1847*, 821–830.

Rintamaki, E., Martinsuo, P., Pursiheimo, S., and Aro, E.M. (2000). Cooperative regulation of light-harvesting complex II phosphorylation via the plastoquinol and ferredoxin-thioredoxin system in chloroplasts. Proc. Natl. Acad. Sci. U.S.A. *97*, 11644–11649.

Rippka, R., Waterbury, J., and Cohen-Bazire, G. (1974). A cyanobacterium which lacks thylakoids. Arch. Microbiol. *100*, 419–436.

Roach, T., Na, C.S., and Krieger-Liszkay, A. (2015). High light-induced hydrogen peroxide production in *Chlamydomonas reinhardtii* is increased by high CO_2 availability. Plant J. *81*, 759–766.

Ruban, A.V. (2015). Evolution under the sun: optimizing light harvesting in photosynthesis. J. Exp. Bot. *66*, 7–23.

Ruban, A.V., Wentworth, M., Yakushevska, A.E., Andersson, J., Lee, P.J., Keegstra, W., Dekker, J.P., Boekema, E.J., Jansson, S., and Horton, P. (2003). Plants lacking the main light-harvesting complex retain photosystem II macro-organization. Nature *421*, 648–652.

Sandmann, G., and Malkin, R. (1983). Iron-sulfur centers and activities of the photosynthetic electron transport chain in iron deficient cultures of the blue-green alga aphanocapsa. Plant Physiol. 73, 724–728.

Sato, N., Aoki, M., Maru, Y., Sonoike, K., Minoda, A., and Tsuzuki, M. (2003). Involvement of sulfoquinovosyl diacylglycerol in the structural integrity and heat-tolerance of photosystem II. Planta *217*, 245–251.

Saxton, M.J. (1993). Lateral diffusion in an archipelago. Dependence on tracer size. Biophys. J. *64*, 1053–1062.

Schöttler, M.A., and Tóth, S.Z. (2014). Photosynthetic complex stoichiometry dynamics in higher plants: environmental acclimation and photosynthetic flux control. Front. Plant Sci. *5*, 188.

Seino, Y., Takahashi, T., and Hihara, Y. (2009). The response regulator RpaB binds to the upstream element of photosystem I genes to work for positive regulation under low-light conditions in *Synechocystis* sp. Strain PCC 6803. J. Bacteriol. *191*, 1581–1586.

Semenova, G.A. (1995). Particle regularity on thylakoid fracture faces is influenced by storage conditions. Can. J. Bot. 73, 1676–1682.

Shapiguzov, A., Ingelsson, B., Samol, I., Andres, C., Kessler, F., Rochaix, J.D., Vener, A.V., and Goldschmidt-Clermont, M. (2010). The PPH1 phosphatase is specifically involved in LHCII dephosphorylation and state transitions in *Arabidopsis*. Proc. Natl. Acad. Sci. U.S.A. *107*, 4782–4787.

Shikanai, T. (2014). Central role of cyclic electron transport around photosystem I in the regulation of photosynthesis. Curr. Opin. Biotechnol. *26*, 25–30.

Shimizu, M., Kato, H., Ogawa, T., Kurachi, A., Nakagawa, Y., and Kobayashi, H. (2010). Sigma factor phosphorylation in the photosynthetic control of photosystem stoichiometry. Proc. Natl. Acad. Sci. U.S.A. *107*, 10760–10764.

Simidjiev, I., Barzda, V., Mustárdy, L., and Garab, G. (1997). Isolation of lamellar aggregates of the light-harvesting chlorophyll a/b protein complex of photosystem II with long-range chiral order and structural flexibility. Anal. Biochem. *250*, 169–175.

Simidjiev, I., Barzda, V., Mustárdy, L., and Garab, G. (1998). Role of thylakoid lipids in the structural flexibility of lamellar aggregates of the isolated light-harvesting chlorophyll a/b complex of photosystem II. Biochemistry *37*, 4169–4173.

Simidjiev, I., Stoylova, S., Amenitsch, H., Javorfi, T., Mustardy, L., Laggner, P., Holzenburg, A., and Garab, G. (2000). Self-assembly of large, ordered lamellae from non-bilayer lipids and integral membrane proteins *in vitro*. Proc. Natl. Acad. Sci. U.S.A. *97*, 1473–1476.

Simpson, D.J., and Robinson, S.P. (1984). Freeze-fracture ultrastructure of thylakoid membranes in chloroplasts from manganese-deficient plants. Plant Physiol. *74*, 735–741.

Snyders, S., and Kohorn, B.D. (1999). TAKs, thylakoid membrane protein kinases associated with energy transduction. J. Biol. Chem. *274*, 9137–9140.

Snyders, S., and Kohorn, B.D. (2001). Disruption of thylakoid-associated kinase 1 leads to alteration of light harvesting in *Arabidopsis*. J Biol. Chem. *276*, 32169–32176.

Spiller, S., and Terry, N. (1980). Limiting factors in photosynthesis: II. Iron stress diminishes photochemical capacity by reducing the number of of photosynthetic units. Plant Physiol. *65*, 121–125.

Spiller, S.C., Castelfranco, A.M., and Castelfranco, P.A. (1982). Effects of iron and oxygen on chlorophyll biosynthesis: I. *In vivo* observations on iron and oxygen deficient plants. Plant Physiol. *69*, 107–111.

Staehelin, L.A. (2003). Chloroplast structure : from chlorophyll granules to supra-molecular architecture of thylakoid membranes. Photosynth. Res. *76*, 185–196.

Staehelin, L.A., and van der Staay, G.M. (1996). Structure, composition, functional organization and dynamic properties of thylakoid membranes. In Oxygenic Photosynthesis: The Light Reactions SE – 2, D. Ort, C. Yocum, and I. Heichel, eds. (Springer Netherlands), pp. 11–30.

Steinberg-Yfrach, G., Liddell, P.A., Hung, S.C., Moore, A.L., Gust, D., and Moore, T.A. (1997). Conversion of light energy to proton potential in liposomes by artificial photosynthetic reaction centres. Nature *385*, 239–241.

Stock, A.M., Robinson, V.L., and Goudreau, P.N. (2000). Two-component signal transduction. Annu. Rev. Biochem. *69*, 183–215.

Stroebel, D., Choquet, Y., Popot, J.L., and Picot, D. (2003). An atypical haem in the cytochrome b(6)f complex. Nature *426*, 413–418.

Suga, M., Akita, F., Hirata, K., Ueno, G., Murakami, H., Nakajima, Y., Shimizu, T., Yamashita, K., Yamamoto, M., Ago, H., *et al.* (2014). Native structure of photosystem II at 1.95 Å resolution viewed by femtosecond X-ray pulses. Nature *517*, 99–103.

Suorsa, M., Rantala, M., Danielsson, R., Jarvi, S., Paakkarinen, V., Schroder, W.P., Styring, S., Mamedov, F., and Aro, E.M. (2013). Dark-adapted spinach thylakoid protein heterogeneity offers insights into the photosystem II repair cycle. Biochim. Biophys. Acta. *1837*, 1463–1471.

Suzuki, N., Rivero, R.M., Shulaev, V., Blumwald, E., and Mittler, R. (2014). Tansley review. Abiotic and biotic stress combinations. New Phytol. *203*, 32–43.

Sznee, K., Dekker, J.P., Dame, R.T., van Roon, H., Wuite, G.J.L., and Frese, R.N. (2011). Jumping mode atomic force microscopy on grana membranes from spinach. J. Biol. Chem. *286*, 39164–39171.

Takahashi, H., Iwai, M., Takahashi, Y., and Minagawa, J. (2006). Identification of the mobile light-harvesting complex II polypeptides for state transitions in *Chlamydomonas reinhardtii*. Proc. Natl. Acad. Sci. U.S.A. *103*, 477–482.

Takahashi, H., Clowez, S., Wollman, F.A., Vallon, O., and Rappaport, F. (2013). Cyclic electron flow is redox-controlled but independent of state transition. Nat. Commun. *4*, 1954.

Terauchi, A.M., Peers, G., Kobayashi, M.C., Niyogi, K.K., and Merchant, S.S. (2010). Trophic status of *Chlamydomonas reinhardtii* influences the impact of iron deficiency on photosynthesis. Photosynth. Res. *105*, 39–49.

Tietz, S., Puthiyaveetil, S., Enlow, H.M., Yarbrough, R., Wood, M., Semchonok, D.A., Lowry, T., Li, Z., Jahns, P., Boekema, E.J., *et al.* (2015). Functional implications of photosystem II crystal formation in photosynthetic membranes. J. Biol. Chem. *290*, 14091–14106.

Tikkanen, M., and Aro, E.M. (2012). Thylakoid protein phosphorylation in dynamic regulation of photosystem II in higher plants. Biochim. Biophys. Acta *1817*, 232–238.

Tikkanen, M., Nurmi, M., Kangasjarvi, S., and Aro, E.M. (2008). Core protein phosphorylation facilitates the repair of photodamaged photosystem II at high light. Biochim. Biophys. Acta *1777*, 1432–1437.

Tomizioli, M., Lazar, C., Brugière, S., Burger, T., Salvi, D., Gatto, L., Moyet, L., Breckels, L.M., Hesse, A.M., Lilley, K.S., *et al.* (2014). Deciphering thylakoid sub-compartments using a mass spectrometry-based approach. Mol. Cell. Proteomics *13*, 2147–2167.

Tremmel, I., Kirchhoff, H., Weis, E., and Farquhar, G. (2003). Dependence of plastoquinol diffusion on the shape, size, and density of integral thylakoid proteins. Biochim. Biophys. Acta – Bioenerg. *1607*, 97–109.

Trissl, H.W., and Wilhelm, C. (1993). Why do thylakoid membranes from higher plants form grana stacks? Trends Biochem. Sci. *18*, 415–419.

Tsabari, O., Nevo, R., Meir, S., Carrillo, L.R., Kramer, D.M., and Reich, Z. (2015). Differential effects of ambient or diminished CO_2 and O_2 levels on thylakoid membrane structure in light-stressed plants. Plant J. *81*, 884–894.

Tsai, C.H., Warakanont, J., Takeuchi, T., Sears, B.B., Moellering, E.R., and Benning, C. (2014). The protein Compromised Hydrolysis of Triacylglycerols 7 (CHT7) acts as a repressor of cellular quiescence in *Chlamydomonas*. Proc. Natl. Acad. Sci. U.S.A. *111*, 15833–15838.

Tsvetkova, N.M., Quinn, P.J., and Williams, W.P. (1995). Factors influencing PS II particle array formation in *Arabidopsis thaliana* chloroplasts and the relationship of such arrays to the thermostability of PS II. Biochim. Biophys. Acta *1228*, 201–210.

Umena, Y., Kawakami, K., Shen, J.R., and Kamiya, N. (2011). Crystal structure of oxygen-evolving photosystem II at a resolution of 1.9 angstrom. Nature *473*, 55–U65.

Ünlü, C., Drop, B., Croce, R., and van Amerongen, H. (2014). State transitions in *Chlamydomonas reinhardtii* strongly modulate the functional size of photosystem II but not of photosystem I. Proc. Natl. Acad. Sci. U.S.A. *111*, 3460–3465.

Urzica, E.I., Vieler, A., Hong-Hermesdorf, A., Page, M.D., Casero, D., Gallaher, S.D., Kropat, J., Pellegrini, M., Benning, C., and Merchant, S.S. (2013). Remodeling of Membrane Lipids in Iron-starved *Chlamydomonas*. J. Biol. Chem. *288*, 30246–30258.

Varsano, T., Kaftan, D., and Pick, U. (2003). Effects of Iron Deficiency on Thylakoid Membrane Structure and Composition in the Alga Dunaliella salina. J. Plant Nutr. *26*, 2197–2210.

Varsano, T., Wolf, S.G., and Pick, U. (2006). A chlorophyll a/b-binding protein homolog that is induced by iron deficiency is associated with enlarged photosystem I units in the eucaryotic alga Dunaliella salina. J. Biol. Chem. *281*, 10305–10315.

Vuthoori, S., Bowers, C.W., McCracken, A., Dombroski, A.J., and Hinton, D.M. (2001). Domain 1.1 of the sigma(70) subunit of Escherichia coli RNA polymerase modulates the formation of stable polymerase/promoter complexes. J. Mol. Biol. *309*, 561–572.

Wang, Z.T., Ullrich, N., Joo, S., Waffenschmidt, S., and Goodenough, U. (2009). Algal lipid bodies: stress induction, purification, and biochemical characterization in wild-type and starchless. Eukaryot. Cell *8*, 1856–1868.

Wei, L., Derrien, B., Gautier, A., Houille-Vernes, L., Boulouis, A., Saint-Marcoux, D., Malnoë, A., Rappaport, F., de Vitry, C., Vallon, O., *et al.* (2014). Nitric oxide-triggered remodeling of chloroplast bioenergetics and thylakoid proteins upon nitrogen starvation in *Chlamydomonas reinhardtii*. Plant Cell *26*, 353–372.

Wei, X., Guo, J., Li, M., and Liu, Z. (2015). Structural mechanism underlying the specific recognition between the *Arabidopsis* state-transition phosphatase TAP38/PPH1 and phosphorylated light-harvesting complex protein Lhcb1. Plant Cell *27*, 1113–1127.

Wientjes, E., van Amerongen, H., and Croce, R. (2013a). LHCII is an antenna of both photosystems after long-term acclimation. Biochim. Biophys. Acta *1827*, 420–426.

Wientjes, E., Drop, B., Kouřil, R., Boekema, E.J., and Croce, R. (2013b). During state 1 to state 2 transition in *Arabidopsis thaliana*, the photosystem II supercomplex gets phosphorylated but does not disassemble. J. Biol. Chem. *288*, 32821–32826.

Wood, P.M. (1978). Interchangeable copper and iron proteins in algal photosynthesis. Eur. J. Biochem. *87*, 9–19.

Work, V.H., Radakovits, R., Jinkerson, R.E., Meuser, J.E., Elliott, L.G., Vinyard, D.J., Laurens, L.M.L., Dismukes, G.C., and Posewitz, M.C. (2010). Increased lipid accumulation in the *Chlamydomonas reinhardtii* sta7-10 starchless isoamylase mutant and increased carbohydrate synthesis in complemented strains. Eukaryot. Cell *9*, 1251–1261.

Wu, D., Wright, D.A., Wetzel, C., Voytas, D.F., and Rodermel, S. (1999). The Immutans variegation locus of *Arabidopsis* defines a mitochondrial alternative oxidase homolog that functions during early chloroplast biogenesis. Plant Cell *11*, 43–56.

Wu, W., Ping, W., Wu, H., Li, M., Gu, D., and Xu, Y. (2013). Monogalactosyldiacylglycerol deficiency in tobacco inhibits the cytochrome b_6f-mediated intersystem electron transport process and affects the photostability of the photosystem II apparatus. Biochim. Biophys. Acta *1827*, 709–722.

Wunder, T., Liu, Q., Aseeva, E., Bonardi, V., Leister, D., and Pribil, M. (2013a). Control of STN7 transcript abundance and transient STN7 dimerisation are involved in the regulation of STN7 activity. Planta *237*, 541–558.

Wunder, T., Xu, W., Liu, Q., Wanner, G., Leister, D., and Pribil, M. (2013b). The major thylakoid protein kinases STN7 and STN8 revisited: effects of altered STN8 levels and regulatory specificities of the STN kinases. Front. Plant Sci. *4*, 417.

Xing, W., Huang, W., and Liu, G. (2010). Effect of excess iron and copper on physiology of aquatic plant Spirodela polyrrhiza (L.) Schleid. Environ. Toxicol. *25*, 103–112.

Yadavalli, V., Jolley, C.C., Malleda, C., Thangaraj, B., Fromme, P., and Subramanyam, R. (2012). Alteration of proteins and pigments influence the function of photosystem I under iron deficiency from *Chlamydomonas reinhardtii*. PLoS One *7*, e35084.

Yoshioka, M., Nakayama, Y., Yoshida, M., Ohashi, K., Morita, N., Kobayashi, H., and Yamamoto, Y. (2010). Quality control of photosystem II: FtsH hexamers are localized near photosystem II at grana for the swift repair of damage. J. Biol. Chem. *285*, 41972–41981.

Yoshioka-Nishimura, M., Nanba, D., Takaki, T., Ohba, C., Tsumura, N., Morita, N., Sakamoto, H., Murata, K., and Yamamoto, Y. (2014). Quality control of photosystem II: direct imaging of the changes in the thylakoid structure and distribution of FtsH proteases in spinach chloroplasts under light stress. Plant Cell Physiol. *55*, 1255–1265.

Young, K.J., Martini, L.A., Milot, R.L., Snoeberger, R.C., Batista, V.S., Schmuttenmaer, C.A., Crabtree, R.H., and Brudvig, G.W. (2012). Light-driven water oxidation for solar fuels. Coord. Chem. Rev. *256*, 2503–2520.

Yu, J., Smart, L., Jung, Y.S., Golbeck, J., and McIntosh, L. (1995). Absence of PsaC subunit allows assembly of photosystem I core but prevents the binding of PsaD and PsaE in *Synechocystis* sp. PCC6803. Plant Mol. Biol. *29*, 331–342.

Zito, F., Finazzi, G., Delosme, R., Nitschke, W., Picot, D., and Wollman, F.A. (1999). The Q_o site of cytochrome b_6f complexes controls the activation of the LHCII kinase. Embo J *18*, 2961–2969.

Distinct Energetics and Regulatory Functions of the Two Major Cyclic Electron Flow Pathways in Chloroplasts

4

Deserah D. Strand, Nicholas Fisher and David M. Kramer

Abstract

The output of the light reactions of photosynthesis, i.e. ATP and NADPH, must be finely controlled to meet the varying metabolic demands of the plant. Deleterious side reactions, including the production of reactive oxygen species (ROS), can occur if this balance is not maintained. In this chapter, we review recent advances in understanding of cyclic electron flow around photosystem I (CEF), a process that evolved to correct for such energy imbalances. CEF in higher plants has been proposed to function primarily through two pathways: the ferredoxin:plastoquinone reductase (FQR) and the NADPH dehydrogenase complex (NDH). Because these pathways appear to support the same function, they are often thought to be redundant. However, it was recently shown that the NDH complex is a type I proton pumping quinone reductase, making the complex a more efficient route for ATP generation via CEF than the FQR. In addition, these pathways are differentially regulated, the FQR through chloroplast redox status, and the NDH thermodynamically and, in part, by hydrogen peroxide. These observations imply that the two pathways are distinct in terms of their energetics and regulation and thus it is clear that the FQR and the NDH work under different conditions to provide with rapid or efficient energy balancing of the chloroplast. Specifically, we suggest a CEF model in which the FQR is rapidly activated to restore balance when an ATP deficit leads to a build-up of reducing power in the stroma, and the NDH activated when a prolonged deficit leads to closure of the PSI acceptor side with associated ROS production.

Introduction

Photosynthetic organisms obtain all their energetic requirements from sunlight. This energy is stored in two distinct forms, redox potential in the NADPH($+H^+$)/O_2 couple and the phosphorylation potential in the ATP/ADP+P_i couple, both of which are essential for photosynthesis and maintenance of cellular processes. These 'energy currencies' are used to drive endergonic biochemical reactions, but act in distinct ways in different reactions. The distribution of energy in the two forms must be finely and dynamically balanced to sustain efficient photosynthesis (Kramer and Evans, 2011; Eberhard *et al.*, 2008; Cruz *et al.*, 2005; Strand and Kramer, 2014). It is not clear how this balancing is accomplished, but it is likely that a process called cyclic electron flow around photosystem I (CEF) plays a critical role.

To understand CEF and its importance in energy balance, we must first start with the basic wiring of oxygenic photosynthesis, through a process called linear electron flow (LEF) [reviewed in Allen (2002), Ort and Yocum (1996) and Chapter 5]. In LEF, light energy is harvested by light harvesting pigments and transferred to the reaction centres of photosystem I (PSI) and photosystem II (PSII) where it drives the initial electron transfer reactions of photosynthesis. Excitation of PSII with a photon results in transfer of an electron from a reaction centre chlorophyll molecule (P_{680}) through a series of electron carriers to reduce plastoquinone (PQ) to plastoquinol (PQH_2) through a stabilized semi-quinone intermediate. The oxidized P_{680} is re-reduced by electrons extracted from water at the oxygen-evolving complex. PQH_2 is oxidized at the Q_o (quinol oxidation) site of the cytochrome *bf* complex (*bf*), with one electron reducing plastocyanin (PC) and the other reducing a chain of cytochromes leading to reduction of PQ at the Q_i (quinol reduction) site via the Q-cycle mechanism (reviewed in Kramer *et al.* 2009; Cape *et al.* 2006). Similarly, excitation of P_{700} promotes charge separation in PSI that results in oxidation of PC and reduction of ferredoxin (Fd). From Fd, electrons are fed into several metabolic pathways, including the Calvin–Benson–Bassham (CBB) cycle in the form of NADPH (Chapter 5), and the multiple thioredoxin-dependent processes in the chloroplast [Buchanan and Balmer (2005), Motohashi *et al.* (2001) and Chapter 9]. Completing the pathway, electrons from PQH_2 are transferred to oxidized PC by the cytochrome *bf* complex.

Certain electron transport reactions in LEF are coupled to proton deposition in the lumen. Excitation of PSII with four photons results in the extraction of four electrons (from water) with the concomitant deposition of four protons into the lumen. The electrons from water are transferred across the thylakoid membrane and directed to the PSII Q_B site where PQ is reduced, taking up two protons from the stromal side of the thylakoid. Then, at the *bf* complex, 2 protons are released into the lumen during PQH_2 oxidation at the Q_o site and taken up from the stroma during PQ reduction at the Q_i site. This redox loop results in two protons released into the lumen per electron transferred to PC. Overall, six proton are transferred to the lumen per pair of electrons transferred from water to Fd by the LEF pathway. The resulting electrochemical gradient of protons ($\Delta\tilde{\mu}_{H^+}$), or protonmotive force (Δp), drives ATP synthesis by the CF_OF_1-ATP synthase. Within the ATP synthase, a proton is thought to be translocated across the thylakoid membrane via a rotary ring composed of 14 c-subunits in higher plants (Seelert *et al.*, 2000). According to current models, one complete rotation of the c-ring requires 14 protons, resulting in the condensation of three ATP molecules within the $\alpha_3\beta_3$ F1 domain of the enzyme [i.e. the H^+:ATP ratio is 4.67 (Seelert *et al.*, 2000; von Ballmoos *et al.*, 2008)].

The need for chloroplast energy balance

The energy products of the light reactions are consumed by multiple metabolic processes, which have different demands for ATP and NADPH and are activated to different extents under different conditions. Since ATP and NADPH production are tightly coupled in LEF, the associated stoichiometry for output of ATP:NADPH should be fixed at ~1.3:1. This is a problem because any mismatch in the production and consumption can lead to metabolic congestion (i.e. the build-up of products or depletion of substrates leading to slowing of the overall process). For instance, if ATP:NADPH production were too high compared to the ratio of consumption, the chloroplast may accumulate excess ATP, leading to high ΔG_{ATP} and increased thylakoid Δp, with the resulting protonic backpressure leading to lumen over-acidification, inhibiting plastoquinol deprotonation and oxidation at the Q_o site of the *bf* complex or in extreme cases leading to acid-induced destruction of the photosynthetic apparatus (Krieger and Weis, 1993; Kramer *et al.*, 1999). On the other hand, if the light reactions produce a deficit of ATP, relative to NADPH, excess reductant will accumulate, while the light-driven Δp will decrease (Kramer *et al.*, 2003), leading to the production of potentially damaging reactive oxygen species (ROS) (Rutherford *et al.*, 2012).

Photosynthetic organisms have multiple mechanisms to avoid these types of energy imbalances through a series of alternative electron transfer pathways (Asada, 1999; Scheibe, 2004; Houille-Vernes *et al.*, 2011; Bendall and Manasse, 1995; Strand and Kramer, 2014; Eberhard *et al.*, 2008; Kramer and Evans, 2011; see also Chapter 5).

We focus here on a set of mechanisms termed cyclic electron flow around photosystem I (CEF), and specifically on new results that show that different CEF pathways likely serve different roles and are regulated by different signals.

Energy balance through activation of cyclic electron flow around photosystem I (CEF)

In its most basic form, CEF catalyses the transfer of electrons from the pool of soluble PSI acceptors in the stroma into the PQ/PQH_2 pool via a plastoquinone reductase (Fig. 4.1). This reduction allows photochemically excited electrons to pass through the PQ pool and back to PSI through the *bf* complex and plastocyanin. During this processes, protons are taken up from the stroma and deposited into the lumen via the Q_o site of the *bf* complex, generating

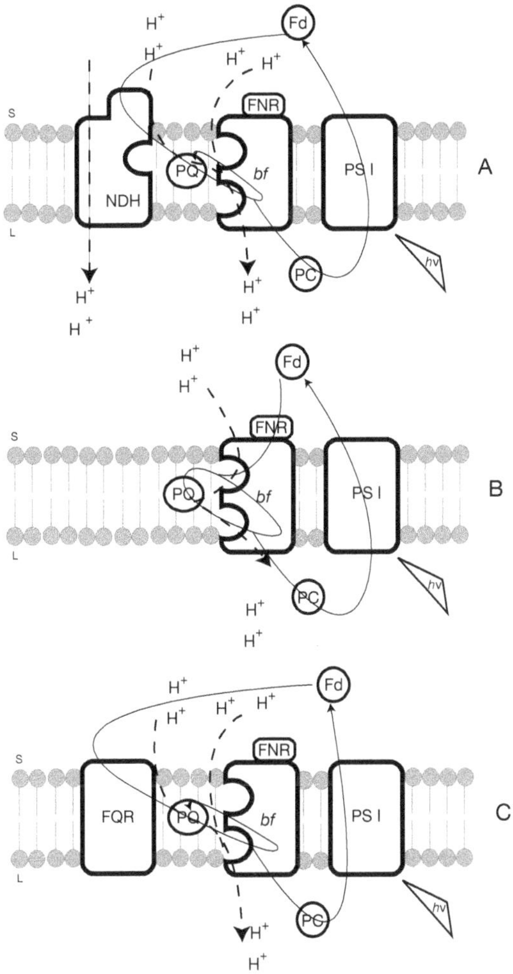

Figure 4.1 Cyclic electron flow around photosystem I. Electrons are transferred to plastoquinone (PQ) via one of several proposed routes. (A) a type I NAD(P)H dehydrogenase (NDH), (B) through the Q_i site of the *bf* complex via haem c_i, and (C) the ferredoxin-dependent, antimycin A-sensitive plastoquinone reductase (FQR) putatively comprised of a complex of PGR5 and PGRL1.

Δp, and therefore ATP. In contrast to LEF, CEF operates with no net generation of stromal reductants, thus increasing the ratio of ATP:NADPH. As will be discussed below, the efficiency of CEF is dependent on the nature of the plastoquinone reductase (PQR) utilized, the identity of which is still being debated.

The activity of CEF has been proposed to contribute to chloroplast energy balancing by increasing the ratio of ATP/NADPH produced by the light reactions (Kramer and Evans, 2011; Bendall and Manasse, 1995; Eberhard *et al.*, 2008). In addition, the Δp generated by CEF should also acidify the thylakoid lumen, leading to activation of the photoprotective q_E (exciton quenching) response, preventing build-up of electrons on the acceptor side of PSII (Kramer *et al.*, 2004a; Müller *et al.*, 2001).

Comparing the major pathways for CEF in vascular plants

There have been at least three proposed CEF pathways in higher plants (Fig. 4.1). In vascular plants, the most studied routes of PQ reduction include (1) a type I (analogous to complex I of mitochondria) NADPH dehydrogenase (NDH, Fig. 4.1A) (Sazanov *et al.*, 1998; Burrows *et al.*, 1998), (2) the Q_i site of *bf* through haem c_i (Fig. 4.1B) (Baniulis *et al.*, 2008), and (3) an antimycin A-sensitive, Fd-dependent plastoquinone reductase (FQR, Fig. 4.1C) (Tagawa *et al.* 1963; Bendall and Manasse 1995; Munekage *et al.* 2002; DalCorso *et al.* 2008). An additional CEF pathway is likely to be present in algae, catalysed by a single-subunit, non-protonmotive NADPH:PQ oxidoreductase (NDA2), homologous to the type II NDH dehydrogenases widespread in bacteria, and plant and fungal mitochondria (Melo *et al.*, 2004; Desplats *et al.*, 2009; Jans *et al.*, 2008). This latter CEF pathway is unlikely to be present in plant plastids, although a homologous enzyme [chloroplast-associated (non-protonmotive) NADPH:plastoquinone oxidoreductase (NDC)] has been proposed to participate in plastid plastoglobule metabolism (Ytterberg *et al.*, 2006).

In the following, we review recent work on NDH and FQR, specifically towards understanding why both CEF pathways are conserved in most vascular plants.

The FQR pathway

Much research has focused on the antimycin A-sensitive pathway involving the FQR (Tagawa *et al.* 1963; Arnon and Chain, 1977, 1979; Cleland and Bendall, 1992; Bendall and Manasse, 1995) that has yet to be unambiguously identified. In recent years two proteins have been postulated as participating in the FQR route of CEF, PGR5 (proton gradient regulation 5) and PGRL1 (PGR-like 1) (Fig. 4.1C) (Munekage *et al.*, 2002; DalCorso *et al.*, 2008; Hertle *et al.*, 2013). These were originally identified in plants, but have functional homologues in algae (Iwai *et al.*, 2010; Terashima *et al.*, 2012; Alric, 2014; Johnson *et al.*, 2014). The current model posits that PGR5 and PGRL1 form a complex capable of oxidizing Fd and reducing PQ in a non-protonmotive manner (Hertle *et al.*, 2013). While multiple lines of evidence point to the involvement of PGR5 and PGRL1 in CEF, the identification of this complex as a direct quinone reductase is problematic, as the acceptor used in the study by Hertle *et al.* (2013) was a highly soluble (non-physiological) quinone species, and the authors were unable to demonstrate steady-state Fd:plastoquinone oxidoreductase activity in the recombinant complex. In addition, the reaction was only shown to be inhibited with a concentration of antimycin A three orders of magnitude higher than has been shown previously, making interpretation of this experiment ambiguous. With these results in mind, the hypothesis that a PGR5–PGRL1 complex is the FQR requires further investigation. Further complicating FQR research is the increase of proton conductance of the thylakoid membrane in PGR5-deficient mutants, resulting in a large decrease in Δp and pH-dependent mechanisms of photoprotection (Suorsa *et al.*, 2012; Avenson *et al.*, 2005a). Interpretation of experiments using mutants must be considered carefully, as many have secondary effects. For example, the assertion that CEF is necessary for the accumulation of q_E is from data showing that the *pgr5* mutant is greatly diminished in q_E; however, this may also be due increased activation state of the ATP synthase (Avenson *et al.*, 2005a). In fact, flux analysis of *pgr5* shows a much larger decrease in the rate constant of Δp decay and overall Δp than the decrease in H^+/e^- (Avenson *et al.*, 2005a), indicating that while CEF may be decreased in this mutant, the ~10% loss of CEF does not account for the ~90% loss in q_E. This may be due to a prolonged ATP deficit due to the absence of CEF altering the ΔG_{ATP}, but an alternative hypothesis may be that PGR5 is a regulatory protein and may interact with the ATP synthase (Avenson *et al.*, 2005a,b; Kramer *et al.*, 2004b). Similarly, the recent report of increased PSI photodamage and insufficient lumen acidification in field-grown PGR5 can also be understood as due to altered membrane conductivity (Suorsa *et al.*, 2012).

Many of the *in vitro* studies used to investigate the PGR5 phenotype have utilized a chlorophyll fluorescence assay that is considered to reflect the non-photochemical reduction (i.e. F_{npr}) of the PQ pool on addition of Fd and NAD(P)H to thylakoid preparations (Munekage *et al.*, 2002, 2004; Courteille *et al.*, 2013). The sluggish kinetics of the F_{npr} (t½ ~ 20s) have proved contentious, and it has been argued that the observed rate is too slow to support a physiologically useful rate of CEF *in vivo* (see reviews in Kramer and Evans, 2011; Fisher and Kramer, 2014). A recent systematic re-examination of the nature of this fluorescence phenomenon (Fisher and Kramer, 2014) concluded that it arose from direct reduction of a low potential variant of Q_A within a subpopulation of PSII centres by Fd, and as such is not an appropriate assay for *in vitro* CEF activity. This conclusion was drawn from the following observations: (i) the F_{npr} is still observed in thylakoid preparations in which PSII variable fluorescence has been abolished by treatment with hydroxylamine and DCMU; (ii) quenching of the F_{npr} by far-red illumination is unaffected by saturating amounts of tridecylstigmatellin, a Q-cycle blocker; and (iii) the kinetics of the F_{npr} and rate of (FNR-mediated) reduction of the PQ pool by NADPH + Fd differ, indicative of redox disequilibrium between these two processes. As such, the F_{npr}-based studies of PGR5, and its assignment as a direct participant in CEF, are worthy of reassessment in the light of the new understanding of the biochemical basis of this fluorescence phenomenon.

Regulation of FQR

In recent years, evidence has been seen for regulation of both the FQR related proteins and formation of the PSI-*bf* supercomplex by chloroplast redox state (Breyton *et al.*, 2006; Hertle *et al.*, 2013; Takahashi *et al.*, 2013a). The FQR related PGRL1 has multiple redox-sensitive thiol groups that may signal for the formation of the PGRL1–PGR5 complex (Hertle *et al.*, 2013). More recently, FQR activity has recently been demonstrated to be redox labile and redox controlled in *Spinacia oleracea* chloroplast preparations (Strand *et al.*, 2016). It was shown that FQR activity is irreversibly lost when thylakoids are isolated in the absence of thiol reductant (in this case dithiothreitol at a concentration of 2 mM) (Strand *et al.*, 2016). This observation provides a partial solution to the long-standing problem of maintaining active CEF in chloroplast preparations for *in vitro* enzymatic investigation. A further, reversible regulation of FQR activity was observed in thylakoids with active CEF, wherein FQR activity requires the reduction of a two electron component with an apparent midpoint potential of ~ −310 mV, consistent with a thiol-disulfide exchange, possibly reflecting the redox potential of the NAD(P)H/NADP+ couple *in vivo* (Strand *et al.*, 2016).

The NDH pathway

The chloroplast NAD(P)H:plastoquinone oxidoreductase (NAD(P)H dehydrogenase, NDH; Fig. 4.1A) is a multisubunit complex, orthologous to respiratory complex I, encoded in both the nuclear and chloroplast genomes (Suorsa *et al.*, 2009). Despite extensive work on the characterization of NDH (Burrows *et al.* 1998; Hashimoto *et al.* 2003; Casano *et al.* 2004; Rumeau and Peltier 2005; Peng *et al.* 2009; Suorsa *et al.* 2009; Yamamoto *et al.* 2011), there are still many open questions, partly because of challenges in isolating sufficient amounts of the active complex in C_3 plants for physicochemical investigation, in which the content of this protein is low (i.e. the ratio of NDH:PSII in *Arabidopsis thaliana* has been estimated to be 1:20) (Peng *et al.*, 2009). However, NDH content has been shown to be greatly increased in C_4 plants (Takabayashi *et al.*, 2005), in which NDH thought to play a key role in producing ATP for the CO_2 concentrating mechanisms, as well as in mutants of *Arabidopsis* that show elevated CEF (Livingston *et al.*, 2010a).

The recent crystal structures of the respiratory Complex I from *Thermus thermophilus* and *Yarrowia lipolytica* offer mechanistic insights that may be used to formulate hypotheses about the function of plastid NDH (Kaila *et al.*, 2014; Efremov *et al.*, 2010; Efremov and Sazanov, 2011; Friedrich, 2014). Genetic and proteomic studies on the subunit composition of NDH have identified approximately 20 nuclear and plastid encoded genes, several of which are plant specific (Suorsa *et al.*, 2009). The plastid encoded subunits are sufficiently similar to those of Complex I to suggest conserved functions. Significantly, residues shown to be essential for protonmotive activity in respiratory Complex I are also conserved in higher plant plastid NDH (Strand, Fisher, and Kramer, submitted). Ongoing research by a number of groups continues to lend insight into assembly, association, and expression of NDH (Hashimoto *et al.*, 2003; Peng *et al.*, 2009; Zandueta-Criado and Bock, 2004).

Functional understanding of NDH has been hindered by the difficulty of measurement of the activity of the complex. Activity has been assessed variously in enriched thylakoid preparations and by ferricyanide reduction, in-gel reduction of nitroblue tetrazolium in the presence of NADH, and chlorophyll *a* fluorescence changes (Casano *et al.*, 2001; Sazanov *et al.*, 1998; Burrows *et al.*, 1998; Rumeau and Peltier, 2005), none of which are likely to fully represent activity under physiological conditions. NAD(P)H:quinone oxidoreductase assays with the partially purified complex have also been attempted using soluble quinones as the electron acceptor, but these are likely not physiologically relevant (Sazanov *et al.*, 1998). As such, the turnover rate of NDH is not known with certainty, however, rates of ~ 200 e^-/s are reported for respiratory complex I. Similar activity for the plastid NDH would be capable of supporting electron transport through CEF of ~10–13% of LEF required to maintain ATP balance (Avenson *et al.*, 2005a) (even less would be required with a protonmotive NDH, discussed below), despite the apparent low abundance of this enzyme.

Recently, we presented evidence that NDH is an authentic proton pumping enzyme, through (i) *in vivo* spectroscopic analysis of the Arabidopsis mutant *hcef1*, which displays a high CEF phenotype, and has been shown to be enriched in NDH, and (ii) systematic *in vitro* investigation of the post-illumination chlorophyll fluorescence rise (a phenomenon associated with NDH activity) in spinach, Arabidopsis and *Amaranthus hybridus* chloroplast preparations. (Strand, Fisher and Kramer, submitted) Consequentially, we demonstrated that NDH operates with a proton pumping stoichiometry of $2H^+$ for each electron transferred from ferredoxin to plastoquinone (i.e. $4H^+/2e^-$), double the efficiency of that proposed for the FQR pathway ($2H^+/2e^-$).

Regulation of NDH

The protonmotivity of NDH has profound consequences for the regulation of its activity as it is subject to (and constrained by) thermodynamic back-pressure from the proton gradient across the energized thylakoid membrane. The relationship at equilibrium between protons translocated (n) into the lumen against Δp per electron traversing a redox span of ΔE_h mV is given by equation (4.1), the 'gearing ratio' for protonmotivity:

$$\Delta E_h \geq n\Delta p \quad (4.1)$$

Assuming midpoint potentials of −430, −335 and +80 mV for the Fd^{3+}/Fd^{2+}, $NADP^+/NADPH$ and PQ/PQH_2 couples, respectively, at a stromal pH of 7.5, the protonmotive activity and directionality of a $2H^+/2e^-$ NDH would be unconstrained by physiological extents of Δp. Working from the assumption that the NADPH and Fd pools are in rapid redox equilibrium due to FNR activity, NDH operating with a $4H^+/2e^-$ pumping ratio would be constrained at Δp values $> \sim 180$ mV. This was demonstrated experimentally in spinach and *Amaranthus* chloroplast preparations where it was shown that the extent of the post illumination chlorophyll fluorescence rise was increased on addition of protonic uncouplers, alleviating the Δp-induced thermodynamic back-pressure on the system. Under conditions of high [such as may be found during photosynthetic induction, for example (Joliot and Johnson, 2011)], NDH would be expected to run in reverse as a plastoquinol:$NADP^+$ reductase, consuming and effectively interconverting ATP and NADPH. This is likely to be of considerable import for the fine-tuning of the energy budget within the chloroplast. It should be noted that the proposed $NADP^+$ reductase activity of NDH is not unprecedented, and has been widely observed in respiratory NDH homologues (Jackson and Crofts, 1968; Hirst, 2013).

In addition to thermodynamic control, multiple lines of evidence indicate a role for ROS, specifically hydrogen peroxide (H_2O_2), in the regulation of the NDH complex. Sabater and co-workers (Lascano *et al.*, 2003; Casano *et al.*, 2001) showed that both *in vitro* activity and accumulation of the complex increases after exposure to H_2O_2, and suggested that NDH is controlled by phosphorylation in response to a signal cascade likely involving Ca^{2+} (discussed further below). Recent *in vivo* work showed rapid activation of NDH-mediated CEF after exposure to H_2O_2 (Strand *et al.* 2015). In addition, mutants deficient in a type *m* thioredoxin had increased CEF through NDH and the FQR (Courteille *et al.*, 2013). It was thus proposed that TRX-M4 was a negative regulator of CEF, however, it was later shown these mutants accumulate significant levels of H_2O_2 (Wang *et al.*, 2013), suggesting CEF in these mutants may be a secondary, rather than a direct result of the loss of TRX-M4.

Calcium has been suggested as a signalling intermediate in the activation of NDH. In plants, Ca^{2+} could be an intermediate in the ROS signalling of NDH activation, as infiltration with a Ca^{2+} chelating agent inhibited the increase in activity of NDH in response to H_2O_2 (Lascano *et al.*, 2003). As with algae, the role of Ca^{2+} in regulating CEF in plants is unclear.

The CEF pathways in algae

Although this chapter is focused on the two major CEF pathways in vascular plant chloroplasts, it is worthwhile to cover recent advances in green algal CEF, partly because these systems are often considered to be models for plants. We emphasize here, though, that CEF in the model system *Chlamydomonas reinhardtii*, appears to be distinct from that seen in plants. For example, Chlamydomonas chloroplasts do not possess an NDH complex, but appear to rely on either a type II (non-proton pumping) NAD(P)H:plastoquinone oxidoreductase (NDA2) (Desplats *et al.*, 2009; Jans *et al.*, 2008) or a supercomplex of the *bf* complex, photosystem I, PGRL1 and other components (Iwai *et al.*, 2010). The role of NDA2 in chloroplast energy balance is still unclear, but it may be utilized for H_2 production under conditions of nutrient deprivation (Mignolet *et al.*, 2012). As a non-protonmotive enzyme, and in contrast to NDH, the plastoquinone reductase activity of NDA2 is irreversible.

The PSI-*bf* supercomplex was originally discovered under conditions that favoured a transition to state 2, where LHCII is associated with PSI (Iwai *et al.*, 2010). However, recent data suggest that a redox signals, rather than state transitions, are responsible for the signalling of PSI-*bf* supercomplex formation (Takahashi *et al.*, 2013a). It may be that the regulation of the PSI-*bf* supercomplex and FQR is the same switch, because the FQR associated protein, PGRL1, associates with the PSI-*bf* supercomplex (Iwai *et al.*, 2010), and experimental evidence for this has been presented *in vitro* (Strand *et al.*, 2016). Redox regulation of CEF would allow for a rapid response to fluctuations in metabolic demands, and may represent a short-term response.

Aside from the associated Q-cycle activity of the *bf* complex, FQR in itself is generally assumed to be non-protonmotive. As such, and in contrast to NDH, it is likely to operate unidirectionally and unconstrained by Δp. Nevertheless, FQR activity will be governed by the availability of oxidized PQ, which may become limiting under conditions of high ΔpH, slowing the PQ oxidase activity of the *bf* Q_o site.

Ca^{2+} has been proposed to be involved in CEF signalling in algae (Terashima *et al.*, 2012). Under anaerobic conditions in algae, which favours the formation of the PSI-*bf* supercomplex, both PGR5 and PGRL1 are necessary for full activation of CEF (Alric, 2014; Terashima *et al.*, 2012). The PSI-*bf* supercomplex further associates with a Ca^{2+} sensor (CAS) and an anaerobic response protein (ANR1) (Terashima *et al.*, 2012). While these proteins have been tied to the CEF response, their role in Ca^{2+} signalling is still not clearly defined.

Overall, like vascular plants, algae appear to possess multiple CEF pathways that involve separate complexes are independently regulated, possible to provide energy balancing over different time scales or under diverse physiological conditions (e.g. Lucker and Kramer, 2013).

The haem c_i pathway: possibilities and open questions

Relatively little research has been conducted specifically on the putative 'direct' mechanism of reduction of PQ by Fd oxidase activity of the *bf* complex, despite proposal that this mechanism is active in both plants and algae (Iwai *et al.*, 2010; Takahashi *et al.*, 2013b; Szymańska *et al.*, 2011). Structurally, haem c_i appears to be well placed to accept electrons from the stromal side of the complex (Stroebel *et al.*, 2003; Kurisu *et al.*, 2003).The model presented by Iwai *et al.* (2010) suggested that the CEF-supercomplex operated via this pathway, especially as flash activation of PSI resulted in transient reduction of a b-type cytochrome, suggesting that electrons flowed from Fd, through FNR, and eventually to cytochrome b_H. However, another possibility is that a fraction of supercomplexes contained reduced plastoquinone, which could be oxidized through the normal Q-cycle, resulting in oxidant-induced reduction of cytochrome *b*. There has been, to our knowledge, no direct test to discriminate between these possibilities. The Fd-dependent reduction of cytochrome *b* by NADPH was reported in early studies with higher plant thylakoid preparations (Chain and Malkin, 1979; Furbacher *et al.*, 1989), although this rate was observed to be extremely slow (Kramer, 1990), in apparent contradiction with CEF operating via the direct reduction of cytochrome c_i, because any electrons on haem c_i should equilibrate with cytochrome b_H Thus, under the conditions of these earlier works, haem c_i does not appear to be a kinetically viable route for CEF. However, the slow transfer of electron from Fd to haem b_H could also reflect regulatory processes that switch off this pathway in isolated thylakoids, as suggested by more recent work showing regulation by supercomplex formation (Iwai *et al.*, 2010), redox status (Takahashi *et al.*, 2013b; Strand *et al.*, 2016) and other factors such as Ca^{2+} activation (Terashima *et al.*, 2012). Clearly, further studies are needed to address these open questions.

Contribution of the CEF pathways to the plastid energy budget

The contribution of the NDH- and FQR CEF pathways to the chloroplast photosynthetic energy budget is not yet fully known. Net ATP generation from each proposed pathway is dependent on their relative contribution to Δp. FQR, if composed of a PGR5-PGRL1, and the PSI-*bf* supercomplex, would only contribute protonmotively by PQH_2 oxidation at Q_o at the *bf* complex, catalysing a CEF pathway with an overall H^+/e^- ratio of 4/2. In contrast, NDH, operating as a $4H^+/2e^-$ proton pump, would be capable of catalysing a CEF pathway with an overall H^+/e^- ratio of 8/2, a potential 2-fold increase in ATP yield per $2e^-$ compared to the FQR pathway (i.e. 1.71 ATP/$2e^-$ c.f. 0.86 ATP/$2e^-$).

These observations taken together suggest that a redundancy of pathways exists. Although sharing a common redox pool, these routes are expected to be energetically separated. As discussed above, the directionality of NDH activity is likely to be subject to thermodynamic constraints. We might expect under extreme metabolic imbalance, and high driving force, higher rates of the Mehler reaction due to the closure of the PSI acceptor side (and accumulation of the water–water cycle intermediate H_2O_2) (Asada, 1999), and therefore increased CEF through the NDH complex. However, application of exogenous H_2O_2 (Lascano *et al.*, 2003; Casano *et al.*, 2001; Strand *et al.*, 2015), or generation of H_2O_2 independent of energy imbalance (Strand *et al.*, 2015) led to rapid activation of NDH, indicating that H_2O_2 does indeed have a signalling role.

Conclusion: why are there multiple CEF pathways?

There is strong evidence that chloroplasts utilize multiple alternative electron pathways, some which directly contribute to the formation of thylakoid Δ*p*, including CEF, and others that do not (i.e. the water–water cycle and malate shunt) (reviewed in Eberhard *et al.*, 2008; Kramer and Evans, 2011; Strand and Kramer, 2014).

Of CEF, multiple seemingly redundant pathways exist, so which one is the most vital? Due to the large defect on q_E in *pgr5*, an observation not made in the NDH deficient mutants, the case has been made that the FQR is the main route of CEF (Munekage *et al.*, 2004, 2002). However, as discussed above, this mutant has a much larger defect on thylakoid proton conductivity, which could also explain the inability to maintain a ΔpH (Avenson *et al.*, 2005a). In early studies, cyclic photophosphorylation was completely abolished with antimycin A (Tagawa *et al.* 1963; Arnon and Chain 1975; Arnon and Chain, 1979; Bendall and Manasse, 1995), suggesting a major role for the FQR. The idea of a second route of CEF was only realized when NDH, a functional Complex I analogue (which had also been hypothesized to function in chlororespiratory pathways; Bendall and Manasse, 1995; Bennoun, 2002) was identified in chloroplast genome sequence data and shown to be enzymatically active (Burrows *et al.*, 1998; Sazanov *et al.*, 1998).

The seemingly redundant nature of the pathways may also be explained temporally. Most FQR observations were made during the induction of photosynthesis, or in the dark, which may differ significantly in metabolic demands from the steady state. Technology now exists, and is commercially available, to measure Δp in the steady state or pseudo steady state (Klughammer *et al.*, 2013). Using these *in vivo* methods to identify mutants with high rates of CEF, the NDH pathway has been the favoured route of all of this class of mutant (Strand *et al.*, 2015; Livingston *et al.*, 2010a,b; Gotoh *et al.*, 2010).

Additionally, these routes are likely different in their efficiencies for generating ATP. A PGR5-PGRL1-dependent pathway would only generate Δ*p* by quinol oxidation at the *bf* complex (Fig. 4.2A). Nevertheless, the importance of the FQR pathway for photosynthetic electron flow should not be underestimated. Aside from contributing to ATP generation, its utility is likely to lie in its capability for rapidly generating Δp for photoprotective down-regulation of antenna excitation. NDH is an independently protonmotive system (discussed above, Strand, Fisher, and Kramer, submitted), with double the proton-pumping stoichiometry of the FQR-CEF pathway (Fig. 4.2B). By virtue of the nature of its protonmotivity, the Δ*p*-dependent reverse (plastoquinol oxidase) activity of NDH would be expected to effectively interconvert ATP and NADPH, for further balancing of the system (Fig. 4.2C). Therefore, the two distinct routes of CEF, with different efficiencies, would give the chloroplast added flexibility in the metabolic output of the light reactions.

Future trends

In the seminal review of Bendall and Manasse (1995), cyclic electron flow is described as the 'Cinderella of chloroplast energetics', but the authors note that it has been gaining increased attention. Arguably, this description is still appropriate 20 years on. Although significant progress in CEF research has been achieved since, much remains to be determined. The identity of the FQR, and the redox effectors acting upon it, remain in question, likewise the mechanism of the signalling cascade for H_2O_2-mediated NDH activation. More generally, the importance of CEF to photosynthetic organisms experiencing fluctuating environmental conditions and challenges, as might be experienced in the field, as opposed to static laboratory conditions, requires attention.

Acknowledgements

We would like to thank Dr Jeffrey Cruz for helpful discussions. The authors are funded by Grant DE-FG02-11ER16220 from the Division of Chemical Sciences, Geosciences, and Biosciences, Office

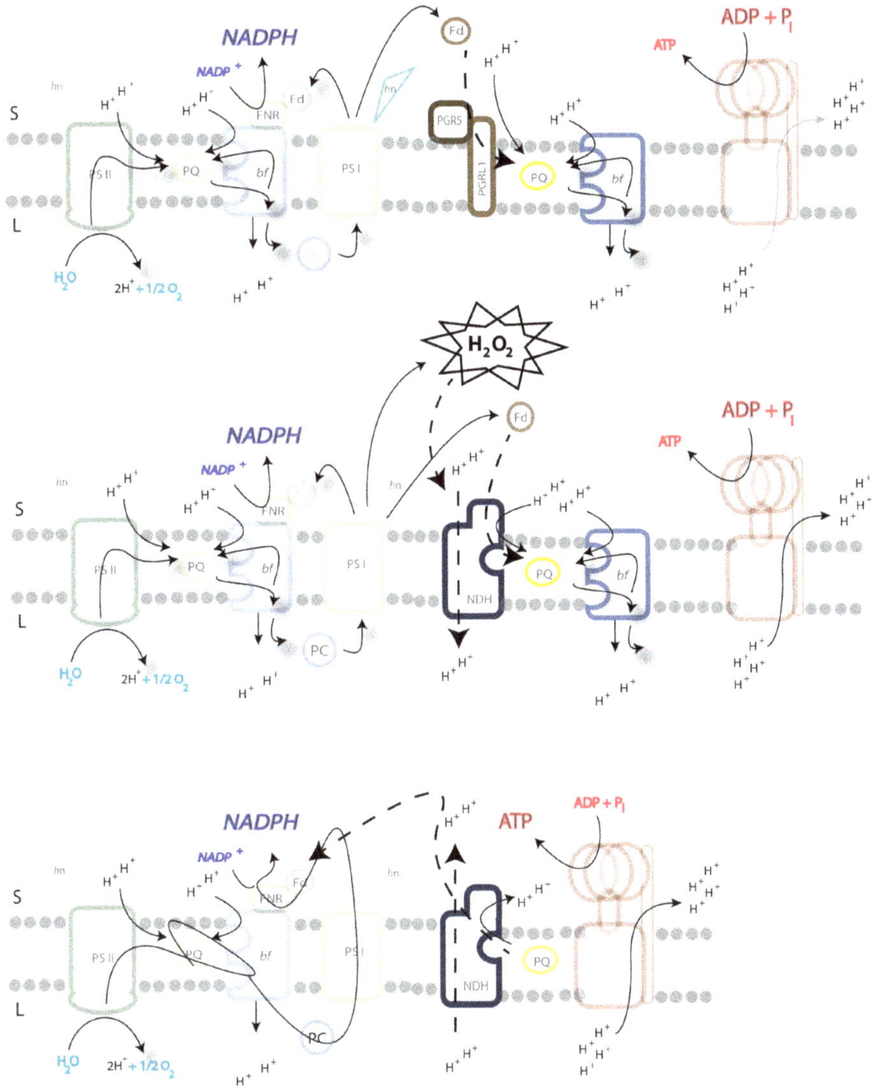

Figure 4.2 Alternative electron transfer pathways under diverse metabolic conditions. (A) An ATP deficit leads to a reduced stroma, and FQR is activated, (B) prolonged/extreme ATP deficit leads to the closure of the PSI acceptor side, and the subsequent generation of ROS, activating the NDH complex, (C) under conditions where there is an NADPH deficit, NDH consumes Δp to drive the quinol oxidase reaction.

of Basic Energy Sciences of the US Department of Energy.

References

Allen, J. (2002). Photosynthesis of ATP-electrons, proton pumps, rotors, and poise. Cell *110*, 273–276.

Alric, J. (2014). Redox and ATP control of photosynthetic cyclic electron flow in *Chlamydomonas reinhardtii*: (II) Involvement of the PGR5-PGRL1 pathway under anaerobic conditions. Biochim. Biophys. Acta *1837*, 825–834.

Arnon, D.I., and Chain, R.K. (1975). Regulation of ferredoxin-catalyzed photosynthetic phosphorylations. Proc. Natl. Acad. Sci. U.S.A. 72, 4961–4965.

Arnon, D.I., and Chain, R.K. (1979). Regulatory electron transport pathways in cyclic photophosphorylation: reduction in C-550 and cytochrome b6 by ferredoxin in the dark. FEBS Lett. *102*, 133–138.

Arnon, D.I., and Chain, R.K. (1977). Role of oxygen in ferredoxin-catalyzed cyclic photophosphorylations. FEBS Lett. *82*, 297–302.

Asada, K. (1999). The water–water cycle in chloroplasts: Scavenging of active oxygens and dissipation of excess photons. Annu. Rev. Plant Physiol. Plant Mol. Biol. *50*, 601–639.

Avenson, T.J., Cruz, J.A., Kanazawa, A., and Kramer, D.M. (2005a). Regulating the proton budget of higher plant photosynthesis. Proc. Natl. Acad. Sci. U.S.A. *102*, 9709–9713.

Avenson, T.J., Kanazawa, A., Cruz, J.A., Takizawa, K., Ettinger, W.E., and Kramer, D.M. (2005b). Integrating the proton circuit into photosynthesis: progress and challenges. Plant, Cell Environ. *28*, 97–109.

von Ballmoos, C., Cook, G.M., and Dimroth, P. (2008). Unique rotary ATP synthase and its biological diversity. Annu. Rev. Biophys. *37*, 43–64.

Baniulis, D., Yamashita, E., Zhang, H., Hasan, S.S., and Cramer, W.A. (2008). Structure-function of the cytochrome b_6f complex. Photochem. Photobiol. *84*, 1349–1358.

Bendall, D.S., and Manasse, R.S. (1995). Cyclic photophosphorylation and electron transport. Biochim. Biophys. Acta *1229*, 23–38.

Bennoun, P. (2002). The present model for chlororespiration. Photosynth. Res. *73*, 273–277.

Breyton, C., Nandha, B., Johnson, G.N., Joliot, P., and Finazzi, G. (2006). Redox modulation of cyclic electron flow around photosystem I in C3 plants. Biochemistry *45*, 13465–13475.

Buchanan, B.B., and Balmer, Y. (2005). Redox regulation: a broadening horizon. Annu. Rev. Plant Biol. *56*, 187–220.

Burrows, P.A., Sazanov, L.A., Svab, Z., Maliga, P., and Nixon, P.J. (1998). Identification of a functional respiratory complex in chloroplasts through analysis of tobacco mutants containing disrupted plastid ndh genes. EMBO J. *17*, 868–876.

Cape, J.L., Bowman, M.K., and Kramer, D.M. (2006). Understanding the cytochrome bc complexes by what they don't do. The Q-cycle at 30. Trends Plant Sci. *11*, 46–55.

Casano, L.M., Lascano, H.R., Martín, M., and Sabater, B. (2004). Topology of the plastid Ndh complex and its NDH-F subunit in thylakoid membranes. Biochem. J. *382*, 145–155.

Casano, L.M., Martín, M., and Sabater, B. (2001). Hydrogen peroxide mediates the induction of chloroplastic Ndh complex under photooxidative stress in barley. Plant Physiol. *125*, 1450–1458.

Chain, R.K., and Malkin, R. (1979). On the interaction of 2,5-dibromo-3-methyl-6-isopropylbenzoquinone (DBMIB) with bound electron carriers in spinach chloroplasts. Arch. Biochem. Biophys. *197*, 52–56.

Cleland, R., and Bendall, D. (1992). Photosystem I cyclic electron transport: measurement of ferredoxin-plastoquinone reductase activity. Photosynth. Res. *34*, 409–418.

Courteille, A., Vesa, S., Sanz-Barrio, R., Cazalé, A.-C., Becuwe-Linka, N., Farran, I., Havaux, M., Rey, P., and Rumeau, D. (2013). Thioredoxin m4 controls photosynthetic alternative electron pathways in *Arabidopsis*. Plant Physiol. *161*, 508–520.

Cruz, J.A., Avenson, T.J., Kanazawa, A., Takizawa, K., Edwards, G.E., and Kramer, D.M. (2005). Plasticity in light reactions of photosynthesis for energy production and photoprotection. J. Exp. Bot. *56*, 395–406.

DalCorso, G., Pesaresi, P., Masiero, S., Aseeva, E., Schünemann, D., Finazzi, G., Joliot, P., Barbato, R., and Leister, D. (2008). A complex containing PGRL1 and PGR5 is involved in the switch between linear and cyclic electron flow in *Arabidopsis*. Cell *132*, 273–285.

Desplats, C., Mus, F., Cuiné, S., Billon, E., Cournac, L., and Peltier, G. (2009). Characterization of Nda2, a plastoquinone-reducing type II NAD(P)H dehydrogenase in *Chlamydomonas* chloroplasts. J. Biol. Chem. *284*, 4148–4157.

Eberhard, S., Finazzi, G., and Wollman, F.-A. (2008). The dynamics of photosynthesis. Annu. Rev. Genet. *42*, 463–515.

Efremov, R.G., Baradaran, R., and Sazanov, L.A. (2010). The architecture of respiratory complex I. Nature *465*, 441–445.

Efremov, R.G., and Sazanov, L.A. (2011). Structure of the membrane domain of respiratory complex I. Nature *476*, 414–420.

Fisher, N., and Kramer, D.M. (2014). Non-photochemical reduction of thylakoid photosynthetic redox carriers *in vitro*: Relevance to cyclic electron flow around photosystem I? Biochim. Biophys. Acta. *1837*, 1944–1954.

Friedrich, T. (2014). On the mechanism of respiratory complex I. J. Bioenerg. Biomembr. *46*, 255–268.

Furbacher, P.N., Girvin, M.E., and Cramer, W. a (1989). On the question of interheme electron transfer in the chloroplast cytochrome b6 in situ. Biochemistry *28*, 8990–8998.

Gotoh, E., Matsumoto, M., Ogawa, K., Kobayashi, Y., and Tsuyama, M. (2010). A qualitative analysis of the regulation of cyclic electron flow around photosystem I from the post-illumination chlorophyll fluorescence transient in *Arabidopsis*: a new platform for the *in vivo* investigation of the chloroplast redox state. Photosynth. Res. *103*, 111–123.

Hashimoto, M., Endo, T., Peltier, G., Tasaka, M., and Shikanai, T. (2003). A nucleus-encoded factor, CRR2, is essential for the expression of chloroplast ndhB in *Arabidopsis*. Plant J. *36*, 541–549.

Hertle, A.P., Blunder, T., Wunder, T., Pesaresi, P., Pribil, M., Armbruster, U., and Leister, D. (2013). PGRL1 is the elusive ferredoxin-plastoquinone reductase in photosynthetic cyclic electron flow. Mol. Cell *49*, 511–523.

Hirst, J. (2013). Mitochondrial complex I. Annu. Rev. Biochem. *82*, 551–575.

Houille-Vernes, L., Rappaport, F., Wollman, F., Alric, J., and Johnson, X. (2011). Plastid terminal oxidase 2 (PTOX2) is the major oxidase involved in chlororespiration in *Chlamydomonas*. Proc. Natl. Acad. Sci. U.S.A. *108*, 20820–20825.

Iwai, M., Takizawa, K., Tokutsu, R., Okamuro, A., Takahashi, Y., and Minagawa, J. (2010). Isolation of the elusive supercomplex that drives cyclic electron flow in photosynthesis. Nature *464*, 1210–1213.

Jackson, J.B., and Crofts, A.R. (1968). Energy-linked reduction of nicotinamide adenine dinucleotides in cells of Rhodospirillum rubrum. Biochem. Biophys. Res. Commun. *32*, 908–915.

Jans, F., Mignolet, E., Houyoux, P.-A., Cardol, P., Ghysels, B., Cuiné, S., Cournac, L., Peltier, G., Remacle, C., and Franck, F. (2008). A type II NAD(P)H dehydrogenase mediates light-independent plastoquinone reduction in the chloroplast of *Chlamydomonas*. Proc. Natl. Acad. Sci. U.S.A. *105*, 20546–20551.

Johnson, X. *et al.* (2014). Proton gradient regulation 5-mediated cyclic electron flow under ATP- or redox-limited conditions: A study of ΔATPase pgr5 and ΔrbcL pgr5 mutants in the green alga *Chlamydomonas reinhardtii*. Plant Physiol. *165*, 438–452.

Joliot, P., and Johnson, G.N. (2011). Regulation of cyclic and linear electron flow in higher plants. Proc. Natl. Acad. Sci. U.S.A. *108*, 13317–13322.

Kaila, V.R.I., Wikström, M., and Hummer, G. (2014). Electrostatics, hydration, and proton transfer dynamics in the membrane domain of respiratory complex I. Proc. Natl. Acad. Sci. U.S.A. *111*, 6988–6993.

Klughammer, C., Siebke, K., and Schreiber, U. (2013). Continuous ECS-indicated recording of the proton-motive charge flux in leaves. Photosynth. Res. *117*, 471–487.

Kramer, D.M. (1990). Parallel instrumentational approaches to the study of the bioenergetics of photosynthesis in chloroplasts and intact plants. PhD thesis, Physiology and Biophysics, Urbana, Illinois: University of Illinois.

Kramer, D.M., Avenson, T.J., and Edwards, G.E. (2004a). Dynamic flexibility in the light reactions of photosynthesis governed by both electron and proton transfer reactions. Trends Plant Sci. *9*, 349–357.

Kramer, D.M., Avenson, T.J., and Edwards, G.E. (2004b). Response to Johnson: Controversy remains: regulation of pH gradient across the thylakoid membrane. Trends Plant Sci. *9*, 571–572.

Kramer, D.M., Cruz, J.A., and Kanazawa, A. (2003). Balancing the central roles of the thylakoid proton gradient. Trends Plant Sci. *8*, 27–32.

Kramer, D.M., and Evans, J.R. (2011). The importance of energy balance in improving photosynthetic productivity. Plant Physiol. *155*, 70–78.

Kramer, D.M., Nitschke, W., and Cooley, J.W. (2009).The cytochrome bc1 and related bc complexes: The Rieske/cytochrome b complex as the functional core of a central electron/proton transfer complex. In The Purple Phototrophic Bacteria, N. Hunter, F. Daldal, M.C. Thurnauer, and, J.T. Beatty, eds. (Springer), pp. 451–473.

Kramer, D.M., Sacksteder, C.A., and Cruz, J.A. (1999). How acidic is the lumen? Photosynth. Res. *60*, 151–163.

Krieger, A., and Weis, E. (1993). The role of calcium in the pH-dependent control of Photosystem II. Photosynth. Res. *37*, 117–130.

Kurisu, G., Zhang, H., Smith, J.L., and Cramer, W.A. (2003). Structure of the cytochrome b_6f complex of oxygenic photosynthesis: Tuning the cavity. Science *302*, 1009–1014.

Lascano, H.R., Casano, L.M., Martín, M., and Sabater, B. (2003). The activity of the chloroplastic Ndh complex is regulated by phosphorylation of the NDH-F subunit. Plant Physiol. *132*, 256–262.

Livingston, A.K., Cruz, J.A., Kohzuma, K., Dhingra, A., and Kramer, D.M. (2010a). An *Arabidopsis* mutant with high cyclic electron flow around photosystem I (hcef) involving the NADPH dehydrogenase complex. Plant Cell 22, 221–233.

Livingston, A.K., Kanazawa, A., Cruz, J.A., and Kramer, D.M. (2010b). Regulation of cyclic electron flow in C3 plants: differential effects of limiting photosynthesis at ribulose-1,5-bisphosphate carboxylase/oxygenase and glyceraldehyde-3-phosphate dehydrogenase. Plant. Cell Environ. *33*, 1779–1788.

Lucker, B., and Kramer, D.M. (2013). Regulation of cyclic electron flow in *Chlamydomonas reinhardtii* under fluctuating carbon availability. Photosynth. Res. *117*, 449–459.

Melo, A.M.P., Bandeiras, T.M., and Teixeira, M. (2004). New insights into type II NAD(P)H: quinone oxidoreductases new insights into type II NAD(P)H: quinone oxidoreductases. Microbiol. Mol. Biol. Rev. *68*, 603–616.

Mignolet, E., Lecler, R., Ghysels, B., Remacle, C., and Franck, F. (2012). Function of the chloroplastic NAD(P)H dehydrogenase Nda2 for H_2 photoproduction in sulphur-deprived *Chlamydomonas reinhardtii*. J. Biotechnol. *162*, 81–88.

Motohashi, K., Kondoh, A., Stumpp, M.T., and Hisabori, T. (2001). Comprehensive survey of proteins targeted by chloroplast thioredoxin. Proc. Natl. Acad. Sci. U.S.A. *98*, 11224–11229.

Müller, P., Li, X.P., and Niyogi, K.K. (2001). Non-photochemical quenching. A response to excess light energy. Plant Physiol. *125*, 1558–1566.

Munekage, Y., Hashimoto, M., and Miyake, C. (2004). Cyclic electron flow around photosystem I is essential for photosynthesis. Nature *429*, 579–582.

Munekage, Y., Hojo, M., Meurer, J., Endo, T., Tasaka, M., and Shikanai, T. (2002). PGR5 is involved in cyclic electron flow around photosystem I and is essential for photoprotection in *Arabidopsis*. Cell *110*, 361–371.

Ort, D.R., and Yocum, C.F. (1996). Electron transfer and energy transduction in photosynthesis: An overview. In, D.R. Ort and, C.F. Yocum eds. Oxygenic Photosynthesis: The Light Reactions. Advances in Photosynthesis. Volume 4. (Springer), pp. 1–9.

Peng, L., Fukao, Y., Fujiwara, M., Takami, T., and Shikanai, T. (2009). Efficient operation of NAD(P)H dehydrogenase requires supercomplex formation with photosystem I via minor LHCI in *Arabidopsis*. Plant Cell *21*, 3623–3640.

Rumeau, D., Becuwe-Linka, N., Beyley, A., Louwagie, M., Garin, J., and Peltier, G. (2005). New subunits NDH-M, -N, and -O, encoded by nuclear genes, are essential for plastid Ndh complex functioning in higher plants. Plant Cell. *17*, 219–232.

Rutherford, A.W., Osyczka, A., and Rappaport, F. (2012). Back-reactions, short-circuits, leaks and other energy wasteful reactions in biological electron transfer: redox tuning to survive life in O_2. FEBS Lett. *586*, 603–616.

Sazanov, L.A., Burrows, P.A., and Nixon, P.J. (1998). The plastid ndh genes code for an NADH-specific dehydrogenase: isolation of a complex I analogue from pea thylakoid membranes. Proc. Natl. Acad. Sci. U.S.A. *95*, 1319–1324.

Scheibe, R. (2004). Malate valves to balance cellular energy supply. Physiol. Plant. *120*, 21–26.

Seelert, H., Poetsch, A., Dencher, N.A., Engel, A., Stahlberg, H., and Müller, D.J. (2000). Proton-powered turbine of a plant motor. Nature *405*, 418–419.

Strand, D.D., Fisher, N., Davis, G.A., and Kramer, D.M. (2016). Redox regulation of the antimycin A sensitive pathway of cyclic electron flow around photosystem I in higher plant thylakoids. Biochim. Biophys. Acta. *1857*, 1–6.

Strand, D.D., and Kramer, D.M. (2014). Control of non-photochemical exciton quenching by the proton circuit of photosynthesis. In B. Demmig-Adams, G. Garab, W. Adams III, and Govindjee eds. Non-Photochemical Quenching and Energy Dissipation in Plants, Algae and Cyanobacteria. Advances in Photosynthesis and Respiration. Volume 40. (Springer), pp. 387–408.

Strand, D.D., Livingston, A.K., Satoh-cruz, M., Froehlich, J.E., and Maurino, V.G. (2015). Activation of cyclic electron flow by hydrogen peroxide in vivo. Proc. Natl. Acad. Sci. U.S.A. *112*, 5539–5544.

Stroebel, D., Choquet, Y., Popot, J., and Picot, D. (2003). An atypical haem in the cytochrome b_6f complex. Nature. *426*, 413–418.

Suorsa, M., Järvi, S., Grieco, M., Nurmi, M., Pietrzykowska, M., Rantala, M., Kangasjärvi, S., Paakkarinen, V., Tikkanen, M., Jansson, S., and Aro, E.-M. (2012). Proton Gradient Regulation 5 is essential for proper acclimation of *Arabidopsis* photosystem I to naturally and artificially fluctuating light conditions. Plant Cell. *24*, 2934–2948.

Suorsa, M., Sirpiö, S., and Aro, E.-M. (2009). Towards characterization of the chloroplast NAD(P)H dehydrogenase complex. Mol. Plant. 2, 1127–1140.

Szymańska, R., Dłuzewska, J., Ślesak, I., and Kruk, J. (2011). Ferredoxin:NADP+ oxidoreductase bound to cytochrome b_6f complex is active in plastoquinone reduction: Implications for cyclic electron transport. Physiol. Plant. *141*, 289–298.

Tagawa, K., Tsujimoto, H.Y., and Arnon, D.I. (1963). Separation by monochromatic light of photosynthetic phosphorylation from oxygen evolution. Proc. Natl. Acad. Sci. U.S.A. *50*, 544–549.

Takabayashi, A., Kishine, M., Asada, K., Endo, T., and Sato, F. (2005). Differential use of two cyclic electron flows around photosystem I for driving CO_2-concentration mechanism in C4 photosynthesis. Proc. Natl. Acad. Sci. U.S.A. *102*, 16898–16903.

Takahashi, H., Clowez, S., Wollman, F.-A., Vallon, O., and Rappaport, F. (2013). Cyclic electron flow is redox-controlled but independent of state transition. Nat. Commun. *4*, 1954.

Terashima, M., Petroutsos, D., Hüdig, M., Tolstygina, I., Trompelt, K., Gäbelein, P., Fufezan, C., Kudla, J., Weinl, S., Finazzi, G., and Hippler, M. (2012). Calcium-dependent regulation of cyclic photosynthetic electron transfer by a CAS, ANR1, and PGRL1 complex. Proc. Natl. Acad. Sci. U.S.A. *109*, 17717–17722.

Wang, P., Liu, J., Liu, B., Feng, D., Da, Q., Shu, S., Su, J., Zhang, Y., Wang, J., and Wang, H.-B. (2013). Evidence for a role of chloroplastic m-type thioredoxins in the biogenesis of photosystem II in *Arabidopsis*. Plant Physiol. 163: 1710–1728.

Yamamoto, H., Peng, L., Fukao, Y., and Shikanai, T. (2011). An Src homology 3 domain-like fold protein forms a ferredoxin binding site for the chloroplast NADH dehydrogenase–like complex in *Arabidopsis*. Plant Cell. *23*, 1480–1493.

Ytterberg, A.J., Peltier, J.-B., and van Wijk, K.J. (2006). Protein profiling of plastoglobules in chloroplasts and chromoplasts. A surprising site for differential accumulation of metabolic enzymes. Plant Physiol. *140*, 984–997.

Zandueta-Criado, A., and Bock, R. (2004). Surprising features of plastid ndhD transcripts: addition of non-encoded nucleotides and polysome association of mRNAs with an unedited start codon. Nucleic Acids Res. 32, 542–550.

Modelling Electron and Proton Transport in Chloroplasts

5

Alexander N. Tikhonov

Abstract

Mathematical modelling of photosynthesis provides a framework for in-depth analysis of light energy transduction and feedbacks in the network of plant cell metabolism. In this chapter, mathematical models of oxygenic photosynthesis are considered in the context of light-induced regulation of photosynthetic electron and proton transport in chloroplasts, the energy-transducing organelles of the plant cell. After a brief overview of electron and proton transport processes in chloroplasts and basic mechanisms of their regulation, general approaches to mathematical description of photosynthetic processes are outlined. As an example of computer modelling of oxygenic photosynthesis, a generalized mathematical model of electron and proton transport in chloroplasts is described. The model includes key stages of linear electron transport, alternative pathways of electron transfer around photosystem I, transmembrane proton transport and ATP synthesis. The model also takes into account pH-dependent mechanisms of the intersystem electron transport control and the light-induced activation of the Calvin–Benson cycle. Some peculiarities of diffusion-controlled reactions of electron and proton transport in chloroplasts are described within the framework of the model, which takes into account lateral heterogeneity of thylakoid membranes. At the end of the chapter, future trends in modelling electron and proton transport in chloroplasts are analysed.

Introduction

Photosynthesis represents a unique set of molecular events that unite different kinds of energy transducing processes: light capture and energy migration to photoreaction centres, charge separation in photoreaction centres, electron and ion transport, and biochemical reactions of organic substances synthesis from atmospheric CO_2 and water (Edwards and Walker, 1983; Ruban, 2012; Skulachev *et al.*, 2013). The light-induced processes of photosynthesis are initiated by light absorption in the light-harvesting antenna, followed by migration of energy and charge separation in photoreaction centres. Photosynthetic organisms of oxygenic type (higher plants, algae, cyanobacteria) have two multisubunit pigment–protein complexes, photosystem I (PSI) and photosystem II (PSII), which are interconnected via the cytochrome (Cyt) b_6f complex and mobile electron carriers, plastoquinone (PQ) and plastocyanin (Pc). Electron transport complexes are embedded into lamellar membranes of thylakoids, closed vesicles situated under the chloroplast envelope. Fig. 5.1 depicts a scheme of electron transport pathways in chloroplasts. The operation of PSII provides the cleavage of water by the water-oxidizing complex (WOC) and PQ reduction to plastoquinol (PQH_2). Decomposition of one H_2O molecule in the WOC leads to the release of two protons into the thylakoid lumen ($H_2O \rightarrow ½O_2 + 2H^+_{in} + 2e^-$); the light-induced formation of PQH_2 is accompanied by the uptake of two protons from stroma ($PQ + 2e^- + 2H^+_{out} \rightarrow PQH_2$). Diffusing in the thylakoid membrane, PQH_2 reaches the Cyt b_6f complex, which mediates the intersystem electron transport by oxidizing PQH_2 and reducing Pc. The oxidation of PQH_2 by the Cyt b_6f complex occurs at the quinone-binding centre Q_o on the luminal side of the thylakoid membrane. Oxidation of one PQH_2 molecule is accompanied by dissociation of two protons into the lumen. Photoexcitation of PSI provides the oxidation of Pc (or cytochrome

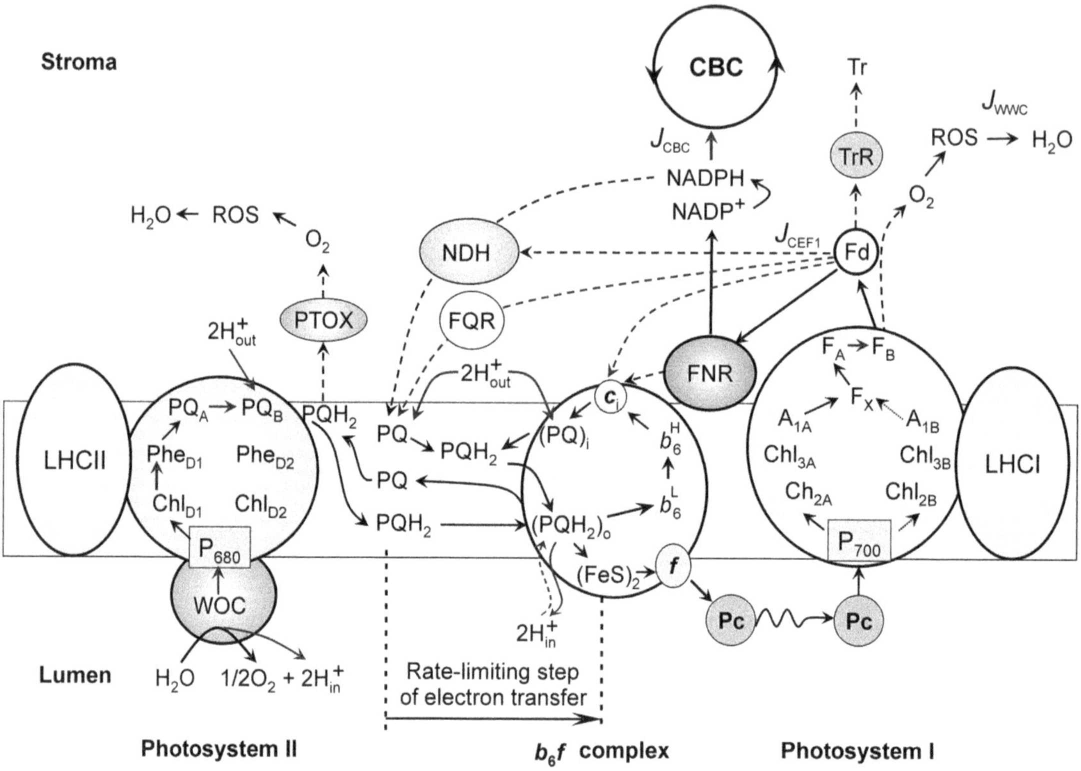

Figure 5.1 A scheme of photosynthetic electron transport pathways and the arrangement of protein complexes (Photosystem I, Photosystem II, cytochrome b_6f, LHCI, LHCII, FNR, and NDH) in the thylakoid membrane. Modified from Figure 1 in Tikhonov (2015). See the list of abbreviations and text for notations and other details.

c_6 in cyanobacteria) and reduction of ferredoxin (Fd), a mobile electron carrier on the stromal side of the membrane. Ferredoxin reduces $NADP^+$ to NADPH via the ferredoxin-NADP-oxidoreductase (FNR). Thus, acting in tandem, PSII and PSI provide electron transfer along the linear electron transport chain (ETC) from water to $NADP^+$, the terminal electron acceptor of PSI ($H_2O \rightarrow PSII \rightarrow PQ \rightarrow b_6f \rightarrow Pc \rightarrow PSI \rightarrow Fd \rightarrow FNR \rightarrow NADP^+$).

Electron transfer along the chloroplast ETC leads to generation of the transthylakoid difference in electrochemical potentials of protons, $\Delta\tilde{\mu}_{H^+}$, which serves as the proton motive force (*pmf*) used to drive ATP synthesis from ADP and P_i by the membrane-bound ATP synthase (Mitchell, 1966; Skulachev *et al.*, 2013). Two components of $\Delta\tilde{\mu}_{H^+}$, the pH difference ($\Delta pH = pH_{stroma} - pH_{lumen}$) and the transmembrane difference in electric potentials, $\Delta\psi = \psi_{lumen} - \psi_{stroma}$, are shown to be competent as the sources of energy for driving the ATP synthases (Gräber, 1982; Junesche and Gräber, 1991). In chloroplasts, steady-state ΔpH value is usually considered as a major component of the *pmf* (Johnson and Ruban, 2014), although under certain conditions one cannot ignore the contribution of $\Delta\psi$ (Cruz *et al.*, 2005). The products of the light-induced stages of photosynthesis, ATP and NADPH, are used in reductive biosynthetic reactions in the Calvin–Benson cycle (CBC) (Edwards and Walker, 1983).

Photosynthetic apparatus of chloroplasts represents a highly flexible system, which provides its efficient performance under variable natural conditions. Rapid response of photosynthetic ETC to fluctuations in the light environment and variations in atmospheric CO_2 should facilitate sustainable development of photosynthetic organisms and their survival under light stress and inclement environment conditions (for review, see Demmig-Adams *et al.*, 2012; Kaiser *et al.*, 2015). Elucidation of molecular mechanisms of the feedback control of electron transport and adaptation of photosynthetic apparatus to varying environment is a challenge to biochemistry and biophysics of photosynthesis.

A deep insight into the regulatory network of oxygenic photosynthesis is complicated by a large number of interacting components, entangled with numerous regulatory feedbacks that act over different time scales (Tikhonov, 2015). Mathematical modelling is increasingly recognized as an efficient tool to understand structural and functional organization of photosynthetic systems, providing an instrument for analysis of regulatory processes in the complex network of photosynthetic processes. Computer models has long been used for analysing the regulatory mechanisms of different photosynthetic processes (see, for example, Kukushkin *et al.*, 1973; Laisk and Walker, 1986; Karavaev and Kukushkin, 1993; Shinkarev, 1998; Fridlyand and Scheibe, 1999; Berry and Rumberg, 2000; Lazar, 2003, 2009; Laisk *et al.*, 2006, 2009; Nedbal *et al.*, 2007a,b; Lavergne, 2009; von Caemmerer *et al.*, 2009; Lazar and Schansker, 2009; Rubin and Riznichenko, 2009; Vershubskii *et al.* 2011; Zaks *et al.*, 2012; Xin *et al.*, 2013; Zhu *et al.*, 2013, and references therein). For the state-of-the-art in the field of mathematical modelling of photosynthetic processes, one can see the book *Photosynthesis in silico. Understanding Complexity from Molecules to Ecosystems* (Laisk, Nedbal, and Govindjee, eds., 2009) and Special issue of the *BioSystems* journal (Igamberdiev, ed., 2011). Recent reviews on computer modelling of electron and proton transport in chloroplasts has been presented in (Rubin and Riznichenko, 2014; Tikhonov and Vershubskii, 2014). The above-mentioned publications demonstrate how mathematical approaches can be used for understanding the regulatory feedbacks in complex photosynthetic systems. It should be noted that chloroplasts represent very attractive objects for mathematical modelling of metabolic processes, because they are largely separated from other subsystems of the plant cell. In the meantime, there may be complexity of theoretical description of photosynthetic processes in chloroplasts, which results from their complex topology. In particular, modelling electron and proton transport in chloroplasts is complicated by non-uniform distribution of electron transport and ATP synthase complexes between spatially segregated domains of laterally heterogeneous thylakoid membranes.

In this chapter, mathematical models of electron and proton transport in chloroplasts are considered in the context of light-induced regulation of oxygenic photosynthesis. After an overview of light-induced processes of photosynthesis, general approaches to mathematical description of photosynthetic electron transport are briefly outlined. In order to exemplify the use of mathematical approaches to the study of regulatory events in chloroplasts, a number of partial and general models of electron and proton transport have been described, including the models that take into account lateral heterogeneity of thylakoid membranes.

Photosynthetic electron transport and its regulation in chloroplasts

Let us consider some peculiarities of structural and functional organization of the chloroplast ETC that form the basis for constructing mathematical models of electron and proton transport processes and the feedback mechanisms of their regulation.

The intersystem electron transport

Fig. 5.1 schematically depicts the ETC of chloroplasts. Electron flow from H_2O to $NADP^+$ (linear electron flow, LEF) provides the formation of NADPH at the expense of electrons donated by PSII to PSI via the intersystem ETC. In chloroplasts, electron transport and ATP synthase complexes are distributed non-uniformly between the granal and stroma-exposed thylakoids. Closely stacked thylakoids of grana are enriched with PSII and the light-harvesting complex of PSII (LHCII); whereas most of PSI and ATP synthase complexes are localized in the membranes of unstacked stroma-exposed thylakoids, grana margins, and grana end membranes (Andersson and Anderson, 1980; Murphy, 1986; Albertsson, 2001; Dekker and Boekema, 2005). The Cyt b_6f complexes are distributed almost uniformly throughout all the domains of the thylakoid membrane (Albertsson, 2001).

There are two diffusion-controlled stages of long-range communication between PSII and PSI: (i) electron transport from PSII to the Cyt b_6f complex mediated by PQH_2 molecules diffusing in the thylakoid membrane, and (ii) electron transfer from the Cyt b_6f complex to PSI mediated by Pc diffusing within the thylakoid lumen. It is well-known fact that the rate of the intersystem electron transport is determined by PQ turnover as a shuttle connecting

PSII and Cyt b_6f complexes (Witt, 1979; Haehnel, 1984). The amount of PQ molecules is about 5–10 times higher than that of PSI or PSII (Stiehl and Witt, 1968, 1969; Siggel *et al.*, 1972). The rate of PQ turnover is determined by several events: (i) the reduction of the secondary quinone PQ_B to PQ_BH_2 ($PQ_B + 2e^- + 2H^+_{out} \rightarrow PQ_BH_2$), (ii) dissociation of PQ_BH_2 from PSII into the thylakoid membrane ($PQ_BH_2 \rightarrow PQH_2$), (iii) PQH_2 diffusion towards the Cyt b_6f complex, and (iv) PQH_2 oxidation at the Q_o-site of the Cyt b_6f complex.

The light-induced reduction of PQ to PQH_2 in PSII ($t_{½} \approx 0.6$ ms) occurs much more rapidly than PQH_2 oxidation ($t_{½} \geq 4$-20 ms). Thoroughly scrutinizing the light-induced turnover of PQ in spinach chloroplasts, Stiehl and Witt demonstrated that the electrons of the reduced plastoquinone (PQH_2) were accepted by P^+_{700} centres (via intermediates) with the half-time $t_{½} \approx 15$-17.5 ms (Stiehl and Witt, 1968, 1969). Note that the rate of the intersystem electron transport, as evaluated in terms of half-time for PQH_2 molecules oxidized per second, was almost the same at different capacities of the pool of reduced PQH_2 molecules. These works are of particular interest, because the redox transients of PQ were measured directly, by monitoring absorption changes of PQ in the UV region. Similar rates of PQH_2 oxidation and P^+_{700} reduction were reported by Haehnel, who also observed that the reduction of Cyt *f* ($PQH_2 \xrightarrow{e^-} \ldots \xrightarrow{e^-} \text{Cyt } f_{ox}$) occurred with the half-time $t_{½} \approx 17$-20 ms (Haehnel, 1973, 1976a,b). Further electron transfer from the Cyt b_6f complex to P^+_{700} ($\text{Cyt} f_{red} \xrightarrow{e^-} \text{Pc} \xrightarrow{e^-} P^+_{700}$) occurs rapidly as compared to PQH_2 oxidation. Characteristic times of electron transfer from Cyt *f* to Pc and from reduced Pc to P^+_{700} are ~35-350 µs and ~20-200 µs, respectively (Haehnel, 1984; Hope, 2000). Taken together, the oxidation of PQH_2 by the Cyt b_6f complex is the rate-limiting event in the chain of electron transport between PSII and PSI.

The question arises: which stage of PQH_2 oxidation determines the rate of the intersystem electron transport, either the lateral diffusion of PQH_2 in the membrane between spatially separated electron transport complexes or PQH_2 oxidation after its binding to the Cyt b_6f complex? There are indications that under certain experimental conditions the long-range lateral diffusion of PQH_2 restrains electron transfer between PSII and PSI (Lavergne and Joliot, 1991; Kirchhoff, 2008, 2014). Diffusion of PQH_2 may be retarded due to slow percolation of PQH_2 molecules through the lipid domains of the membrane over-crowded with densely packed protein complexes. However, the distribution of Cyt b_6f complexes among PSII supercomplexes located in granal membranes minimizes the average distance traversed by PQ molecules in the crowded membrane environment, providing rapid exchange of PQ between the Cyt b_6f and PSII complexes. Obstructed diffusion of Pc within the narrow thylakoid lumen may also restrict, under certain conditions, communication between the Cyt b_6f and PSI complexes (Kirchhoff *et al.*, 2011).

In the meantime, there are good reasons to believe that over a wide range of experimental conditions the light-induced formation PQH_2 in PSII and its diffusion within the membrane do not limit the overall rate of the intersystem electron transport. Although significant amounts of PSI and PSII complexes are laterally segregated, most of them are in close contact with the Cyt b_6f complexes (Albertsson, 2001). There is conclusive experimental evidence that PQH_2 formation in PSII and its lateral diffusion in the thylakoid membrane, as well as Pc movement within the lumen, should not limit the overall rate of electron transfer between PSII and PSI (Haehnel, 1976a; Tikhonov *et al.*, 1984). In isolated chloroplasts, over a wide range of pH, ionic strength, and temperature, the light-induced reduction of PQ to PQH_2, its dissociation from PSII, and PQH_2 diffusion towards to the Cyt b_6f complex occur much more rapidly than PQH_2 oxidation in the Cyt b_6f complex (for review, see Tikhonov, 2013, 2014). The conclusion that the major rate-limiting step in the intersystem chain of electron transport is not diffusion of PQH_2, but its oxidation in the Q-cycle, has been supported by the study of temporal kinetics of electron transport in intact leaves (Laisk *et al.*, 2014).

Interaction of PQH_2 with the Cyt b_6f complex

The rate of PQH_2 turnover is determined mainly by the redox processes taking place after its binding to the Q_o-centre of the Cyt b_6f complex. This complex is usually considered as a functional homodimer of multisubunit monomers (Fig. 5.2A). Each monomer consists of eight polypeptide subunits, including four major subunits: the iron–sulfur protein (ISP), the Cyt b_6 and Cyt *f* proteins, and subunit IV (for review, see Cramer *et al.*, 2006, 2011). The overall rate of PQH_2 oxidation can be influenced by several constituent reactions. According to the Q-cycle mechanism (Mitchell, 1976), the oxidation of PQH_2 in the Q_o-centre proceeds as two concerted reactions: one electron is transferred to a high-potential redox chain and the other to a low-potential chain (Fig. 5.1):

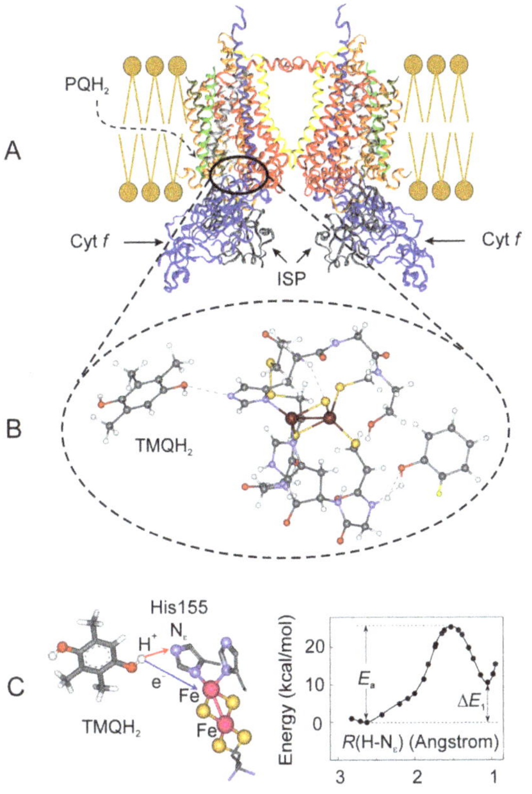

Figure 5.2 (A) Side view of the dimeric Cyt b_6f complex from *Chlamydomonas reinhardtii* (PDB entry 1Q90 (Stroebel *et al.*, 2003)). Figure was produced using Accelrys DV visualizer software package (http://www.accelrys.com). (B) Quantum chemical part of the model system, which includes trimethylquinol ($TMQH_2$, the analogue of PQH_2), the $(FeS)_2$ cluster, and surrounding them amino acid residues of the Cyt b_6f complex (Cys134–Thr135–His136–Leu137–Gly138–Cys139, Cys152, Cys154–His155–Gly156–Ser157, Tyr159) as described in (Frolov and Tikhonov, 2009). (C) Energy profile for the first step of proton-coupled oxidation of $TMQH_2$ by the ISP. The plot of energy *vs* the reaction coordinate $R(H\text{-}N_e)$ describes changes in the potential energy of the model system upon the displacement of the H atom from $TMQH_2$ to the N_e atom of His155. Modified from Figure 2 in Frolov and Tikhonov (2009) and Figure 12 in Tikhonov (2014).

$$PQH_2 + ISP_{ox} \rightarrow PQH^{\cdot} + ISP_{red} + H^+_{in} \quad (5.1)$$

$$PQH^{\cdot} + Cyt(b_6^L)_{ox} \rightarrow PQ + Cyt(b_6^L)_{red} + H^+_{in} \quad (5.2)$$

where $PQH^{\cdot}$ denotes the semiquinone species of PQ formed in reaction (5.1). The first electron is further transferred to Pc through the ISP and Cyt *f* proteins: $(PQH_2)_o \xrightarrow{e^-} ISP \xrightarrow{e^-} Cyt\ f \xrightarrow{e^-} Pc$. The second electron is directed through the low- and high-potential haems b_6^L and b_6^H to reduce PQ at the Q_i-site on the stromal side of the thylakoid membrane: $(PQH^{\cdot})_o \xrightarrow{e^-} b_6^L \xrightarrow{e^-} b_6^H \xrightarrow{e^-} (PQ)_i$. Since PQH_2 oxidation is accompanied by the release of two protons into the thylakoid lumen, the rate of this process appears to be under the control of the intrathylakoid pH_{in} (Rumberg and Siggel, 1969; Witt, 1979; Tikhonov *et al.*, 1984; Nishio and Whitmarsh, 1993).

The endergonic reaction (equation 5.1) has been identified as the rate-limiting step of quinol oxidation (for review, see Crofts, 2004; Crofts *et al.*, 2013; Tikhonov, 2014). In the Cyt b_6f complex, the ISP occupies a cleft between the large and small domains of the Cyt *f* protein (Kurisu *et al.*, 2003; Stroebel *et al.*, 2003). It is likely that after the reduction of ISP its mobile extrinsic domain, containing the redox cluster $(FeS)_2$, moves from the Q_o-site towards the haem *f*; thereby, the 'tethered' diffusion of the $(FeS)_2$ cluster enables electron transfer to Cyt *f*. There are several lines of evidence for a high mobility of the extrinsic domain of the ISP within the Cyt b_6f complex (Breyton, 2000; Yan and Cramer, 2003; de Vitry *et al.*, 2004; Hasan *et al.*, 2013). Therefore, the flip-flop movements of the $(FeS)_2$ cluster between the Q_o-site and the haem *f* may occur more rapidly than the immediate step of electron transfer from PQH_2 to $(FeS)_2$.

It should be noted that the literature data for the rates of partial reactions of electron transport in the Cyt b_6f complex are scattered, depending on the system investigated and its metabolic state. As noted above, Stiehl and Witt reported that a fast phase of post-illumination oxidation of PQH_2 in isolated spinach chloroplasts, which can be related to PQH_2 interaction with the Cyt b_6f complex, was about 15-17.5 ms (Stiehl and Witt, 1969). The half-times of post-illumination reduction of Cyt *f* in different species of intact leaves ranges from 20 ms to 28 ms for the entire range of light intensities up to 2800 μmol/m²/s (Kramer *et al.*, 1999). These values correspond to the rates of Cyt *f* reduction in isolated chloroplasts reported in earlier works (Velthuys, 1979; Nishio and Whitmarsh, 1993) and in the cyanobacterium *Synechocystis* PCC sp. 6803 (Schneider *et al.*, 2001). Otherwise, several authors have demonstrated that the rate of the Cyt b_6f complex turnover in intact leaves (measured on the basis of post-illumination reduction of P^+_{700}) becomes more rapid when illumination is increased (Harbinson and Hedley 1989; Laisk *et al.*, 2005). Relatively short apparent times of Cyt *f* and Cyt *b* reduction ($t_{½} \approx 3\text{-}6$ ms), indicating a rapid turnover of the Cyt b_6f complex, are typical of intact *C. reinhardtii* cells (Soriano *et al.*, 1996; Ponamarev and Cramer, 1998) and the cyanobacterium *Synechococcus* sp. PCC 7002 (Yan and Cramer, 2003).

There may be several reasons for the dispersion of kinetic data in terms of half-time, $t_{½}$. One of the reasons might be related to differences between the species. For instance, variable stoichiometry between PSII, Cyt b_6f, and PSI complexes (Schöttler *et al.*, 2015) might influence the apparent rate of post-illumination reduction of P^+_{700}. Variability of electron capacities of interacting donor and acceptor species may manifest itself in the kinetic behaviour of the system, exaggerating or underestimating the contributions of rapid and slow phases of electron transport reactions. Another reason for scattering kinetic data may be associated with the uncertainty of the redox status of electron carriers. For instance, in *C. reinhardtii* cells, the low redox poise ($E_h \approx -380$ mV) of the stroma suggests that the high-potential ISP centres are maintained in the reduced state even in the dark (Alric, 2014), thereby accelerating partial reactions of electron transport in the Cyt b_6f complex. Besides, transient kinetics of redox changes, when the oxidation of cytochromes take place concurrently with their reduction (Cyt f_{red} → Cyt f_{ox} → Cyt f_{red}, and Cyt b_{ox} → Cyt b_{red} → Cyt b_{ox}), will result to underestimation of half-times (Velthuys, 1979). This may explain why the apparent rates of partial reactions of PQH_2 oxidation in the Cyt b_6f complex may exceed the overall rate of the electron flow from PQH_2 to P^+_{700}. An accelerated post-illumination reduction of P^+_{700} may also result from the contribution of reduced Pc ($Pc^- + P^+_{700} \rightarrow Pc + P^+_{700}$ and/or due back reactions in PSI ($P^+_{700}F^-_X \rightarrow P_{700}F_X$, $P^+_{700}F^-_{A/B} \rightarrow P_{700}F_{A/B}$) with characteristic times ~2-80 ms (Brettel, 1997). These factors will manifest themselves upon the slowing down of electron drain from PSI. At any rate, however, we can safely state that the oxidation of PQH_2 in the Cyt b_6f complex comprises the rate-limiting step in the chain of electron transport between PSII and PSI.

Alternative pathways of electron transport

The feedback regulation of photosynthetic electron transport is associated with redox-dependent redistribution of electron fluxes, which provides flexibility of photosynthetic apparatus and its efficient interaction with other metabolic systems. Apart from the mainstream electron flow from PSI to $NADP^+$, electrons may be diverted to alternative pathways: (i) cyclic electron flux around PSI (CEF1) (for more details see Chapter 4); (ii) the water–water cycle (WWC), in which electrons from PSI reduce O_2 to H_2O (Mehler, 1951; Asada, 1999); (iii) the malate shunt, in which electrons from LEF are shuttled to oxidative phosphorylation in mitochondria (Scheibe, 2004). Proton pumping into the thylakoid lumen coupled to electron transport along the CEF1 and WWC routes supports the ATP synthesis without net reduction of NADPH, thereby increasing the ATP/NADPH ratio. Switching between alternative routes helps to sustain required ratio between ATP and NADPH (ATP/NADPH = 3/2) used for CO_2 accumulation in the CBC (Edward and Walker, 1983). Electrons from PSI may return into the ETC between PSII and PSI via illusive ferredoxin-quinone reductase (FQR) (Bendall and Manasse, 1995; Iwai *et al.*, 2010); this way will be called as a 'short' CEF1. Also, there may exist a 'long' route of CEF1, where electrons from NADPH return to the intersystem ETC via the NAD(P) oxidoreductase (NDH) (Shikanai, 2007; Johnson, 2011). Electron flow from PSI to O_2 represents additional channel of electron drain from PSI (Asada, 1999; Badger *et al.*, 2000; Heber, 2002). When O_2 rather than $NADP^+$ acts as a terminal electron acceptor in PSI, it is eventually reduced to water during the operation of pseudocyclic WWC ($H_2O \rightarrow PSII \rightarrow PSI \rightarrow O_2 \rightarrow H_2O$). Summing up, electron transfer along CEF1 and WWC pathways is accompanied by proton translocation into the lumen, supporting the ATP formation without the reduction of $NADP^+$ and providing an optimal stoichiometry of NADPH and ATP used in the CBC.

pH-dependent and redox-regulation of photosynthetic electron transport

Among the diversity of regulatory processes that serve to optimize electron and proton transport in chloroplasts, noteworthy are the feedbacks associated with the light-induced changes in the lumen and stroma pH. The intersystem electron transport is governed by the light-induced changes in the lumen pH_{in}. There are two main mechanisms of pH_{in}-dependent control of electron transport caused by lumen acidification (pH_{in}): (i) deceleration of PQH_2 oxidation, and (ii) attenuation of PSII activity. A decrease in pH_{in} causes the slowing down of PQH_2 oxidation by the Cyt b_6f complex, thereby reducing electron flow from PQH_2 to PSI (for review, see Tikhonov, 2012, 2013, 2014). At the same time, photochemical activity of PSII decreases with the light-induced generation of ΔpH. This occurs due to an enhancement of thermal dissipation of absorbed light energy in LHCII known as non-photochemical quenching (NPQ) of chlorophyll *a* excitation (for review, see Li *et al.* 2009; Demmig-Adams *et al.*, 2012; Horton, 2012; Ruban, 2012). The development of NPQ, caused by energization of thylakoid membranes (the so-called qE component of NPQ), induces conformational changes in LHCII, creating a quenching channel for dissipation of excess light energy. pH-Dependent mechanisms of electron transport control play important protective role, precluding too strong acidification of the lumen (for review, see Kramer *et al.*, 1999, 2003; Tikhonov, 2012, 2013, 2014) and decreasing the probability of damage to the photosynthetic apparatus under the solar stress conditions (Demmig-Adams *et al.*, 2012; Ruban, 2012).

In dark-adapted chloroplasts, activities of the CBC enzymes and FNR are low (Buchanan, 1980; Edwards and Walker, 1983; Woodrow and Berry, 1988). Immediately after the light onset, a low rate of NADPH consumption in the CBC will cause the over-reduction of electron carriers on the acceptor side of PSI. This would divert electrons from PSI to alternative routes of electron flow (CEF1 and WWC) and to thioredoxin (Tr). The light-induced activation of the CBC stimulates regeneration of $NADP^+$, promoting the outflow of electrons from PSI. Activation of the CBC enzymes is associated with the redox- and pH-dependent modulation of electron transport on the acceptor side of PSI (Werdan *et al.*, 1975; Mott and Berry, 1986; Andersson, 2008). Redox-dependent activation of photosynthetic enzymes occurs via the thioredoxin/thioredoxin reductase (TrR) system: at the excess of reductants on the acceptor side of PSI, the outflow of electron from PSI is diverted from Fd to Tr (via the TrR). Reduced Tr will activate, in turn, other enzymes, including those of the CBC (Motohashi and Hisabori, 2006; Dietz and Pfannschmidt, 2011; Michelet *et al.*, 2013).

Alkalization of the stroma is another factor of CBC activation. The light-induced increase in stromal pH_{out} (the rise of pH_{out} from 7.0-7.2 to 7.8-8.0; Heldt *et al.*, 1973; Robinson, 1985) is associated with the proton consumption upon the reduction of PQ ($PQ + 2e^- + 2H^+_{out} \rightarrow PQH_2$) and $NADP^+$ ($NADP^+ + 2e^- + H^+_{out} \rightarrow NADPH$). The light-induced rise of pH_{out} induces an increase in stromal concentration of Mg^{2+} (Barber, 1976), which is necessary for activation of Rubisco, the key enzyme of the CBC (Gardemann *et al.*, 1986).

The rate of electron flow in chloroplasts correlates with the chloroplast metabolic state characterized by the so-called 'phosphate potential', $P = [ATP]/([ADP] \times [P_i])$, where [ATP], [ADP] and $[P_i]$ are the concentrations of ATP, ADP, and P_i, respectively. Depending on the ADP/ATP ratio, the ATP synthase complexes (CF_0-CF_1) function either in the ATP synthase mode ($H^+_{in} \rightarrow H^+_{out}$, ATP formation) or in the ATPase mode (ATP hydrolysis, $H^+_{out} \rightarrow H^+_{in}$). According to the terminology coined by Chance and Williams, a state of chloroplasts upon intensive ATP synthesis (surplus of ADP and P_i, low P) should be termed as the 'metabolic state 3' (Chance and Williams, 1956). A state of photosynthetic control (exhausted pool of ADP or P_i, high P, when the net production of ATP is zero) would be termed as the 'metabolic state 4'. In state 3, the rate of the intersystem electron flow is usually high, because ATP synthesis is accompanied by stoichiometric drain of protons from the lumen to stroma, thus precluding too strong acidification of the lumen ($pH_{in} \approx 6$–6.2; for review, see Tikhonov, 2012, 2013). In this case, chloroplasts retain a high rate of electron transport, which is comparable with accelerated electron flow in uncoupled chloroplasts ($pH_{in} \approx pH_{out}$, 'metabolic state 5'). More significant acidification of the lumen ($pH_{in} < 6$) occurs in the state 4 (exhausted pool of ADP or P_i, and significant surplus of ATP, high P), when the overall proton flux through CF_0-CF_1 and ATP production virtually tend to zero. In state 4, the intersystem electron flow decelerates.

Light energy partitioning between PSI and PSII

Another mechanism of electron transport regulation in chloroplasts is associated with redistribution of light energy between PSI and PSII ('state transitions'; for review, see Lemeille and Rochaix, 2010; Minagawa, 2011; Tikkanen and Aro, 2012, 2014). The over-reduction of the PQ pool induces activation of a protein kinase that catalyses phosphorylation of LHCII, thereby initiating structural changes in photosynthetic apparatus and enhancing the light energy delivery to PSI at expense of PSII (state I → state II transition). These events are triggered by PQH_2 binding to the Q_o-site of the Cyt b_6f complex. According to traditional point of view, supported by biochemical and structural data (Kouril *et al.*, 2005; Hofmann, 2012), the loosely bound form of phosphorylated LHCII dissociates from the PSII–LHCII supercomplex and migrates to PSI, thereby increasing the light harvesting capacity of PSI. Novel insights into LHCII phosphorylation and state I ↔ state II transitions in higher plants have been suggested in (Tikkanen *et al.*, 2011; Tikkanen and Aro, 2012, 2014).

General principles of computer simulation of electron transport in chloroplasts

Deterministic approach to modelling electron transport in chloroplasts

Deterministic approach to mathematical description of molecular interactions in photosynthetic systems is based on the use of differential equations, which establish relationships between the interacting species. Classical approaches of chemical kinetics are based on the mass action law and equations of Michaelis–Menten kinetics. In the majority of traditional models of photosynthetic electron transport, concentrations of electron carriers are considered as the variables independent of spatial coordinates. However, in order to describe diffusion-controlled reactions of electron transport in spacious systems with the non-uniform distribution of electron transport complexes in the thylakoid membrane, differential equations with partial derivatives should be used. The question arises: is it correct to describe the interaction between distant electron transport complexes using the approximation based on the mass action law?

The application of the mass action law for diffusion-controlled processes can be approved if the pools of mobile electron carriers provide rapid interconnections between electron transport complexes (the so-called 'lake model'). This may be true for chloroplasts, if the mobile electron carriers (i.e. PQ and Pc) rapidly mediate electron transport between distant PSII, Cyt b_6f, and PSI complexes. Titration of chloroplasts by DCMU (an inhibitor which selectively blocks PSII activity) demonstrated that even ~10% of active (DCMU-untreated) PSII complexes were able to donate electrons to almost all Cyt b_6f and PSI complexes (Siggel *et al.*, 1972). This result suggests that spatially segregated PSII, Cyt b_6f, and PSI complexes are interconnected via the common pool of mobile electron carriers. By other words, PQ and PQH_2 molecules behave as a pool of freely diffusing mediators of electron transport dissolved in the lipid bilayer, which connect the PSII and Cyt b_6f complexes (at least within the microdomains where PQH_2 molecules are channelled from PSII to the Cyt b_6f complexes (Kirchhoff *et al.*, 2008)). The Cyt b_6f complexes, which are dispersed over the granal and stromal domains of the thylakoid lamellas, can communicate with PSI due to high mobility of Pc within the lumen. Taken together, the pool type behaviour of PQH_2 and Pc molecules sets the basis for the widely accepted random diffusion model of electron transport processes in chloroplasts. This model tacitly implies that integral protein complexes, including the basic electron transport complexes and ATP synthases, are randomly dispersed in the fluid domains (or microdomains) of the thylakoid membrane. The interactions between spatially segregated complexes are mediated by mobile electron carriers, PQH_2 and Pc.

If electron flow from PSII to the b_6f complex and further electron transfer to PSI are not limited by diffusion of PQH_2 and Pc molecules (the 'lake' type model), then one may use (as a first approximation) a conventional approach of chemical kinetics based on the mass action law. In this case, the reaction rate J_{ik} is proportional to the product of concentrations of reacting components, $c_i(t)$ and $c_j(t)$:

$$J_{ij}(t) \sim k_{ij} \times c_i(t) \times c_j(t) \tag{5.3}$$

where k_{ij} is the apparent rate constant. Rigorously speaking, concentrations of reagents, $c_i(t)$ and $c_j(t)$, should be considered as local concentrations of interacting reagents. However, in the vast majority of kinetic models, concentrations of electron carriers are considered as the variables independent of spatial coordinates. This approach may be valid for the 'lake' type models, if we assume that the processes considered are independent of the reaction vessel geometry.

In the meantime, for correct description of diffusion-dependent reactions in chloroplasts, one cannot ignore the spatial heterogeneity of the lamellar membranes. Integral protein complexes may present the obstacles that can cause significant decrease in the apparent coefficient of PQ diffusion (Kirchhoff, 2008, 2014). The long-range percolation of PQH_2 through the overcrowded areas of thylakoid membranes and slow diffusion of Pc within the narrow domains of the thylakoid lumen may hinder in the intersystem electron flow. Lateral diffusion of protons inside the narrow compartments confined by the thylakoid membranes may also be decelerated. In these cases, electron and proton processes should be described by differential equations with partial derivatives similar to heat-transfer equations:

$$\frac{\partial c_i(\vec{r},t)}{\partial t} = D_i \times \nabla^2 c_i(\vec{r},t) + \sum_k F_{ik}\left[c_i(\vec{r},t), c_k(\vec{r},t)\right] \tag{5.4}$$

where $c_i(\vec{r},t)$ stands for the local concentration of a species 'i' in the space with coordinate $\vec{r}$, D_i is the apparent diffusion coefficient, and $F_{ik}[c_i(\vec{r},t), c_k(\vec{r},t)]$ is the function, which describes the interaction of 'i' and 'k' components in the vicinity of spatial point $\vec{r}$. Applications of deterministic approach to computer simulation of diffusion-controlled processes of electron and proton transport in chloroplasts have recently been reviewed by Rubin and Riznichenko (2014) and Tikhonov and Vershubskii (2014). Another approach to modelling diffusion-controlled reactions is associated with the use of random walk models; this approach is outlined below.

Intersystem electron transport: random walk models of PQ and Pc diffusion

Complex topology of the lamellar system and non-uniform distribution of electron transport complexes in thylakoids pose a considerable challenge to everyone who applies traditional apparatus of mathematical physics for numerical description of light-induced processes in chloroplasts. Mathematical difficulties in modelling diffusion-controlled reactions in the systems with complex architecture may be obviated by using the Monte Carlo method for simulation of random walks of mobile molecules. This method became one of the efficient tools for analysis of diffusion-controlled processes in supramolecular biological structures with complex architecture (Blackwell and Whitmarsh, 1990; Drepper *et al.*, 1993; Blackwell *et al.*, 1994; Kirchhoff *et al.*, 2000, 2002, 2008; Shorten and Sneyd, 2009; Aliev and Tikhonov, 2011; Rubin and Riznichenko, 2014). Potential advantages of using the random walk simulations of electron transport reactions are evident; this method presents an efficient tool for modelling the diffusion-controlled processes in over-crowded domains of thylakoid membranes and can shed a new light on long-range communications between segregated electron transport complexes.

Let us consider electron transfer between PSII and Cyt b_6f complexes mediated by PQ. Rapid turnover of PQ as a shuttle in the intersystem electron transport chain implies sufficiently high lateral mobility of PQ in over-crowded thylakoid membranes. In the membranes of granal thylakoids, protein complexes occupy at least 70% of the total membrane area (Albertsson, 2001; Dekker and Boekema, 2005; Kirchhoff *et al.*, 2008). Collisions of PQ molecules with impermeable protein obstacles would restrict their lateral traffic in the membrane. The coefficient of obstructed PQ diffusion in photosynthetic membranes was found to be at least two or three orders of magnitude less than in pure lipid membranes free of proteins (Blackwell *et al.*, 1994). The question arises whether the restricted diffusion of PQ in the thylakoid membrane may provide an efficient long-range communication between PSII and Cyt b_6f complexes? According to percolation theory (Saxton, 1987, 1989; Kirchhoff, 2008, 2014), an apparent diffusion coefficient of low-molecular species significantly declines in the membrane crowded by impermeable obstacles. If an occupation of the thylakoid membrane by integral protein complexes is higher than critical value *cP* ('percolation threshold'), macromolecular obstacles will restrict the long-range diffusion of PQ, confining random walks of PQ within enclosed diffusion domains. The two-dimensional percolation threshold is defined as the surface fraction above which the long-range diffusion of small objects is restricted. In thylakoid membranes, integral proteins may occupy an area close to 70%, which is close to the percolation threshold ($cP \approx 0.7–0.75$). This suggests that rapid diffusion of PQ may be restricted to relatively small areas of the thylakoid membrane ('microdomains').

According to the microdomain concept of structural organization of thylakoid membranes (Lavergne and Joliot, 1991; Lavergne *et al.*, 1992, Kirchhoff *et al.*, 2000, Kirchhoff, 2014), rapid diffusion of PQ occurs within small lipid areas near active PSIIs surrounded by other protein complexes. The average distance between PSII and Cyt b_6f complexes within the microdomains of grana thylakoids has been evaluated as $r \sim 15\text{-}20$ nm (Tremmel *et al.*, 2003). Diffusion of PQH_2 and PQ within each microdomain does not limit electron transport between PSII and the Cyt b_6f complex. Typical diameter of grana falls in the range of 300-600 nm (Dekker and Boekema, 2005; Kirchhoff, 2014). PQ diffusion in the lipid areas is very fast. Therefore, if PQH_2 is not trapped inside the microdomain, it could migrate over a large area within PQ turnover time (~20 ms). Is it possible that PQH_2 molecules may be efficient in the long-range traffic between the grana thylakoid domains and stroma lamellae? Lateral diffusion coefficients for PQ-9, decyl PQ and PQ-2 in soybean phosphatidylcholine liposomes and in spinach thylakoid membranes have been estimated in the range of $(0.1\text{-}3) \times 10^{-9}$ cm²/s (Blackwell *et al.*, 1994). PQ turnover is determined by PQH_2 oxidation, which is characterized by the half-time $t_{1/2} \approx 20$ ms (Stiehl and Witt, 1969; Haehnel, 1984). For random walk motions in two-dimensional systems, the mean square displacement of particles may be evaluated using the Einstein equation: $\langle r^2 \rangle = 4D \times \Delta t$, where $\langle r^2 \rangle$ is the mean square of the particle displacement during the time interval Δt, and D is the diffusion coefficient. If $D_{PQ} \sim (0.1\text{-}3) \times 10^{-9}$ cm²/s, PQH_2 would traverse a distance about ~30-150 nm within the time interval $\Delta t = 20$ ms. It is important to note that numerical experiments (Tremmel *et al.*, 2003) suggest that mobility of macromolecular obstacles in the membrane increases the percolation threshold *cP*. In this case, the apparent diffusion coefficient for PQ should also increase, allowing the long-range diffusion of PQ within the overcrowded membrane. According to simulations by Tremmel *et al.* (2003), the apparent coefficient of the long-range diffusion of PQ in the thylakoid membrane approaches to $D_{PQ} \approx 2.1 \times 10^{-8}$ cm²/s. This means that PQ could travel farther than 400 nm in 20 ms, suggesting that PQ migration in the thylakoid membrane from PSII to the Cyt b_6f complex would not limit electron transfer between PSII and PSI.

Let us now consider Pc diffusion within the thylakoid lumen. Electron transfer from the Cyt *f* haem to Pc, and further from Pc^- to P^+_{700}, occurs rapidly. Characteristic times of these processes (~5–350 μs, and ~20–200 μs, respectively) are significantly shorter than PQ turnover time (~4-20 ms), suggesting that Pc diffusion between the Cyt b_6f and PSI complexes occurs rapidly and, therefore, should not limit the intersystem electron transport. How the latter statement could be reconciled with a relatively small volume for Pc diffusion within the over-crowded granal lumen? The face-to-face distance between the inner surfaces of the opposite thylakoid membranes is relatively short, $l_i \sim 10–20$ nm (Dekker and Boekema, 2005; Kirchhoff, 2014). The WOC domain of PSII protrudes from the membrane to the thylakoid lumen, thereby limiting the space available for Pc diffusion inside the lumen. According to Kirchhoff *et al.* (2011), diffusion of Pc within the granal thylakoid lumen of dark-adapted thylakoids is highly restricted because of relatively narrow interthylakoid gap. However, the grana lumen undergoes significant (~100%) expansion in the light, thereby considerably increasing the space available for Pc diffusion and providing conditions for facilitated lateral traffic of Pc. Takano *et al.* (1982) evaluated the coefficient of Pc diffusion in the thylakoid lumen as $D_{Pc} \approx 2 \times 10^{-9}$ cm²/s. This yields the Pc displacement $r \approx 130$ nm during $\Delta t = 20$ ms. Taking together, Pc movement inside the thylakoid lumen, as well as the lateral diffusion of PQH_2 in the thylakoid membrane, can influence the intersystem electron transport. However, under a wide range of experimental conditions, the diffusion-dependent steps of electron transport mediated by PQ and Pc may not be the limiting factors (for review, see Tikhonov 2013, 2014). In this case, the rate of electron transfer between PSII and PSI is determined predominantly by PQH_2 oxidation after its binding to the Cyt b_6f complex.

Computer modelling of electron transport on the basis of Brownian diffusion approximation has found its applications in biophysics of photosynthesis. Mobile electron transport proteins are usually considered as rigid bodies, which, diffusing in the lumen (Pc and Cyt c_6) or in the stroma (Fd), collide with immobilized multienzyme complexes, thus mediating electron transport in chloroplasts (see, for instance, Haddadian and Gross, 2006; Riznichenko *et al.* 2010; Kovalenko *et al.* 2011). Consideration of these models is beyond the frames of this chapter; for more information, see Rubin and Riznichenko (2014).

Modelling electron transport in multiprotein complexes

Electron transport reactions within each individual complex occur independently of the immediate states of other electron carriers in the ensemble of electron transport complexes. Therefore, for modelling electron transport in multienzyme complexes one cannot use traditional models of the 'lake' type based on the mass action law. Description of electron transfer processes within the multiprotein complexes is usually based on the 'master equation' method, when the catalytic cycle is described as a sequence of transitions between different states of the multiprotein complex (Rubin and Riznichenko, 2009, 2014). A complete number of states in each individual complex is determined by the numbers of electron carriers in the complex and possible states of each electron carrier (reduced or oxidized, excited or in the ground state, protonated or deprotonated). Let p_i is the probability of finding the multiprotein complex in the state 'i'$\{i = 1, 2 \ldots N\}$. Then the evolution of the system will be described by a set of ordinary differential equations (master equations):

$$\frac{dp_i}{dt} = \sum_{j=1}^{N} \left(k_{ji}p_j - k_{ij}p_i\right) \quad (5.5)$$

where k_{ij} is the rate constant of the transition from the state 'i' to state 'j'.

The 'master equation' approach has been successfully used for modelling electron transport and related processes in photosynthetic electron transport complexes. One of the first applications of the master equation theory to description of primary stages of photosynthesis belongs to Shinkarev and Venedictov (1977). Paillotin *et al.* (1983) successfully employed the master equation technique for modelling fluorescence induction, photochemical yield, and singlet-triplet exciton quenching in photosynthetic systems. Energy relaxation in the exciton transfer in photosynthetic antenna complexes have been studied theoretically using various models formulated in terms of generalized master equations (Singh and Brumer, 2012). The master equation approach has been widely used for detailed theoretical analysis of light-induced processes of electron transport in reaction centres of PSII and numerical description of the induction of chlorophyll *a* fluorescence in PSII (Stirbet *et al.* 1998; Lazar, 2003; Belyaeva *et al.*, 2008, 2011; Lazar and Schansker, 2009; Rubin and Riznichenko, 2009; Vredenberg and Prášil, 2009). For illustration of this approach, we consider Fig. 5.3, which shows a typical diagram of transitions between discrete states of the reaction centre of PSII. These states differ from each other with respect of redox states of electron carriers, protonation of the primary and secondary quinones, and excitation of chlorophyll molecules. Fitting the results of 'numerical' experiments to experimental data, it is possible to evaluate of rate constants of elementary steps of electron transfer, dissipative processes, and to calculate a yield of chlorophyll *a* fluorescence. Theoretical models based on this approach provide a rather well simulation of chlorophyll *a* fluorescence induction (the so-called OJIP curve), which characterizes the photochemical activity of PSII and performance of other photosynthetic processes (see for references Rubin and Riznichenko, 2014).

Quantum chemical modelling of proton-coupled electron transfer reactions

Contemporary methods of quantum chemistry provide an efficient tool for deep insight into the

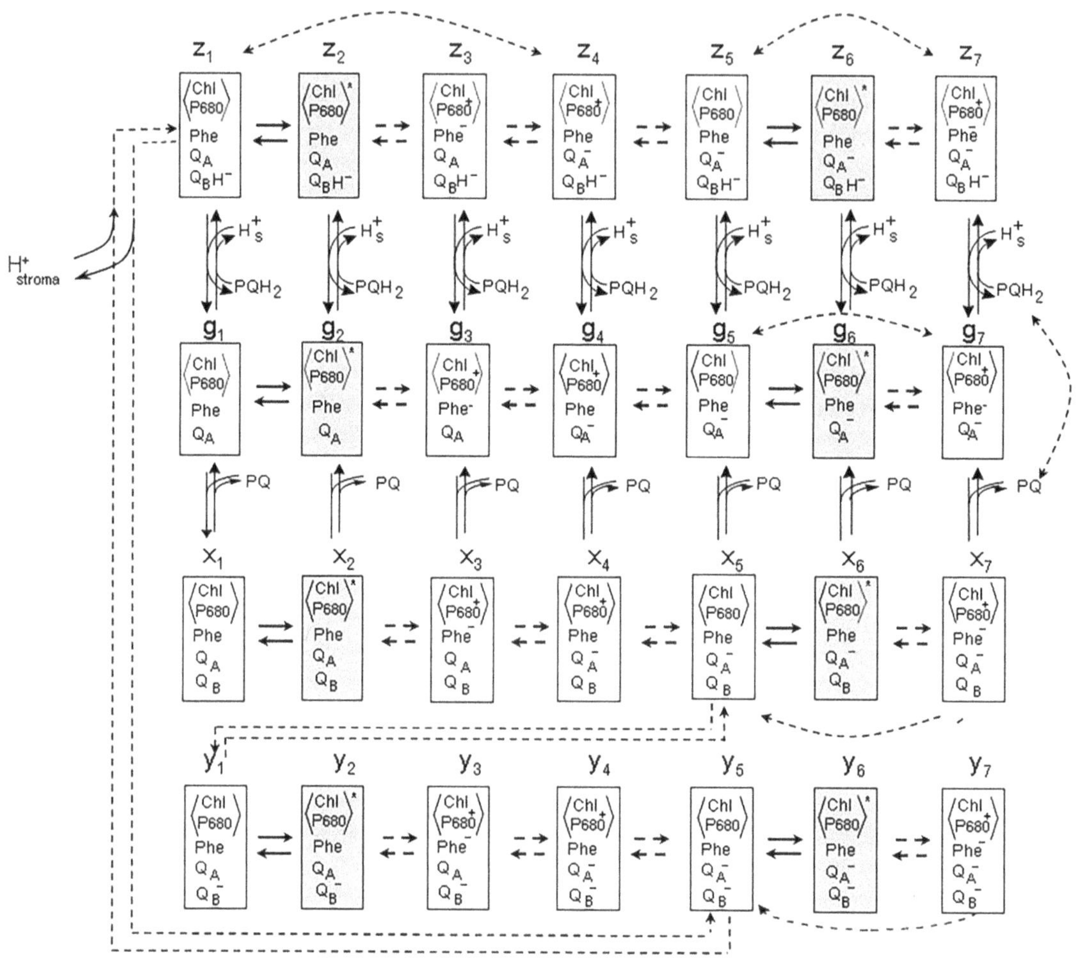

Figure 5.3 Diagram of catalytic cycle of Photosystem II, which depicts transitions between different states of the reaction centre associated with excitation of Chlorophyll (Ch1), oxidation/reduction of P_{700}, primary and secondary PQ molecules (Q_A and Q_B, respectively). Modified figure adapted from Belyaeva *et al.* (2008, 2011).

energy-transducing processes at molecular level. Quantum chemical methods have been used for in-depth analysis of proton-coupled electron transport in PSII and other photosynthetic processes (for review, see Sproviero *et al.*, 2008, 2009; Blomberg and Siegbahn, 2010). In this section, the use of quantum chemical approach to assay partial reactions of electron transport is exemplified for the reaction of PQH_2 oxidation in the Cyt b_6f complex, which is the rate-limiting step in the intersystem electron transport. There are convincing experimental data indicating that the rate of quinone turnover in the Cyt complexes of *bc* type is determined by the endergonic reaction of electron transfer from the quinol molecule (PQH_2 or ubiquinol, UQH_2) to the $(FeS)_2$ cluster of the ISP (for review, see Crofts, 2004; Crofts *et al.* 2013; Tikhonov, 2014). Quantum chemical modelling of PQH_2 oxidation in the Cyt b_6f complex (Frolov and Tikhonov, 2009) supports the notion that the first step of this reaction includes the endergonic step, which hampers the overall rate of the intersystem electron transport.

Electron transfer from PQH_2 to the ISP_{ox} is tightly coupled to proton transfer from PQH_2 to appropriate proton-accepting group at the Q_o-site. The N_e atom of the histidine residue (His155) liganding to the Fe atom of the $(FeS)_2$ cluster (Fig. 5.2C) is the prime candidate for the role of the primary recipient of the proton donated by

PQH_2 ($PQH_2 + ISP_{ox} \rightarrow PQH^{\cdot} + ISP^{\cdot}_{red}H^+$). The second step of PQH_2 oxidation is associated with electron transfer to the low-potential haem b_6^L. Thus, the overall process of PQH_2 oxidation to PQ includes two sequential electron transfer steps:

$$ISP_{ox} \ldots PQH_2 \ldots (b_6^L)_{ox} \rightarrow ISP^{\cdot}_{red}H^+ \ldots PQH^{\cdot} \ldots (b_6^L)_{ox} \rightarrow ISP^{\cdot}_{red}H^+ \ldots PQ \ldots (b_6^L)_{red} + H^+_{in} \quad (5.6)$$

Then, after oxidation of $ISP^{\cdot}_{red}H^+$ by Cyt f_{ox}, its affinity for the proton decreases, and the proton releases into the lumen (via a proton-conducting channel): $ISP^{\cdot}_{red}H^+ + \text{Cyt}\,f_{ox} \rightarrow ISP_{ox} + \text{Cyt}\,f_{red} + H^+_{in}$. The energy profile of the first step of PQH_2 oxidation (the transfer of the H atom from PQH_2 to the ISP_{ox}) has been calculated using the density functional theory (DFT) method (Frolov and Tikhonov, 2009). The model system used for DFT computations is shown in Fig. 5.2B. This system has two local minima of potential energy, which correspond to two positions of the H atom: one at the –OH group of the quinol molecule, and another at the imidazolate ring of His155 of the ISP (Fig. 5.2C). As one can see, the translocation of the H atom from the quinol molecule to the ISP_{ox} (the first step of quinol oxidation) is the endergonic reaction (~ 11 kcal/mol) with a rather high energy barrier E_a. The energy for realization of this energy-accepting process is regained due to the second (energy-donating) step of quinol oxidation, that is the electron transfer from plastosemiquinone (PQH·) to the low-potential haem b_6^L (Tikhonov, 2014). The overall balance of free energy changes for two steps of PQH_2 oxidation should be energy-favourable, $\Delta G_1 + \Delta G_2 \leq 0$. From biophysical point of view, the tight coupling of the energy-accepting ($\Delta G_1 > 0$) and energy-donating ($\Delta G_2 < 0$) reactions can be realized provided both reactions occur as concerted (almost simultaneous) processes (Blumenfeld and Tikhonov, 1994). Using the Moser–Dutton ruler for evaluation of the rate of electron tunnelling between redox centres (Page *et al.*, 1999), expanded by Crofts (2004) for proton-coupled electron transfer reactions, we evaluated the rate constant of PQH_2 oxidation in our model system as $k_1 \sim 40\text{–}170\,s^{-1}$. These values of the rate constant k_1 reasonably agree with the experimental data for electron transfer from PQH_2 to P^+_{700} ($t_{1/2} = \ln 2/k_1 = 4\text{–}17.5$ ms).

A generalized kinetic model of electron and proton transport in chloroplasts

For illustration of applications of deterministic approach to modelling electron transport in chloroplasts, consider a generalized model developed in our earlier works (for review, see Vershubskii *et al.*, 2011; Tikhonov and Vershubskii, 2014). The model mimics the typical patterns of electron transport kinetics, generation of ΔpH, ATP synthesis, and CO_2 uptake in chloroplasts. At first approximation, we have assumed that electron transport complexes and ATP synthases are randomly dispersed in the thylakoid membrane, neglecting the spatial heterogeneity of the chloroplast lamellar system. Peculiar effects of lateral heterogeneity of thylakoid membranes on electron and proton transport processes are considered below in the separate section.

Description of the model

A scheme of the processes modelled is depicted in Fig. 5.4. For numerical description of these processes, the following variables were considered: $[P^+_{700}]$ and $[P^+_{680}]$, concentrations of oxidized photoreaction centres P_{700} and P_{680} (electron donors in PSI and PSII, respectively); [Pc], concentration of oxidized electron donor for P^+_{700} (Pc in chloroplasts and/or Cyt c_6 in cyanobacteria); [PQ], concentration of oxidized plastoquinone; [Fd], concentration of oxidized ferredoxin; $[N^+]$ and [NH], concentrations of $NADP^+$ and NADPH; $[O_2]$, concentration of O_2. The amount of PQ molecules, which connect PSII with the Cyt b_6f complexes, is about 10 times higher than that of PSI or PSII. Variable [ATP] is the concentration of ATP; $[H^+_{in}]$ and $[H^+_{out}]$ are the concentrations of hydrogen ions in the lumen and in the stroma, respectively. The intensities of the light exciting photoreaction centres of PSI and PSII are described by parameters L_1 and L_2. The model comprises the set of eleven ordinary differential equations. Apparent rate constants for partial electron transport reactions have been found by fitting calculated kinetic curves to relevant experimental data as described in (Vershubskii *et al.*, 2001, 2004a,b, 2011).

For description of proton transport processes, we considered three compartments (Fig. 5.5A): the intrathylakoid lumen (pH_{in}), stroma (pH_{out}), and cytoplasm (pH_{cyt}). Three transmembrane fluxes

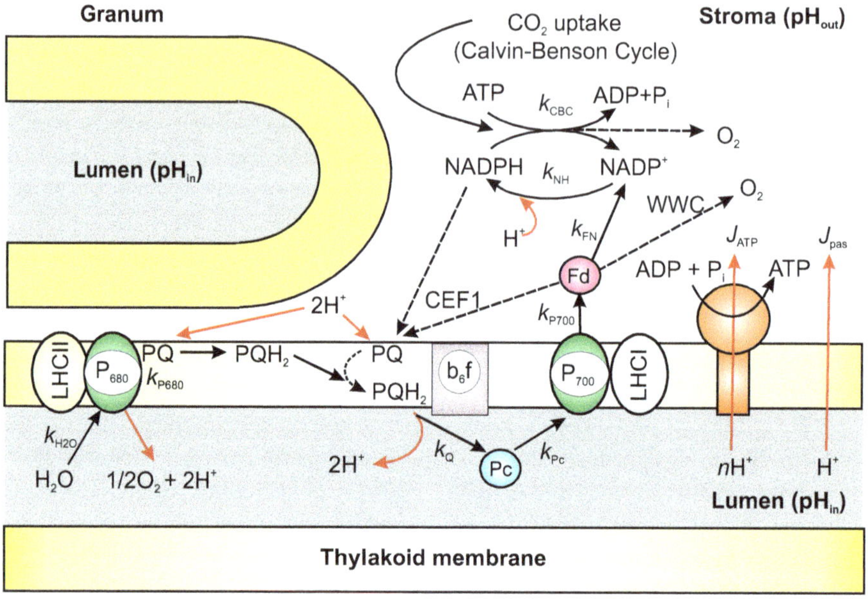

Figure 5.4 A scheme of electron and proton transport processes considered in the model. Two electrons extracted from the water molecule in PSII are transferred to PQ, reducing PQ to PQH_2. Electrons from PQH_2 are transferred to reduce plastocyanin (Pc). Oxidation of water molecule and PQH_2 is accompanied by dissociation of protons into the thylakoid lumen. PSI oxidizes Pc on the luminal side of the thylakoid membrane and reduces a mobile electron carrier ferredoxin (Fd) on the stromal side of the membrane. Reduced Fd molecules donate electrons to $NADP^+$. Reduced and protonated NADPH molecules are consumed in the CBC. Along with the linear electron flow from H_2O to $NADP^+$, the model takes into account alternative electron transport routes around PSI: NADPH- and Fd-mediated return of electrons to the PQ pool (CEF1) and O_2 reduction (WWC).

of protons were taken into account: (i) J_{ATP} (the proton flux through CF_0–CF_1 coupled to ATP synthesis), (ii) J_{pas} (passive drain of protons from the lumen to stroma) and (iii) J_{cyt} (exchange of protons between stroma and cytoplasm). Equations for J_{ATP}, J_{pas} and J_{cyt} were derived from the model for the transmembrane flow of protons as a set of proton-exchange processes through the membrane-bound acid-base groups (Fig. 5.5B; for details see Dubinskii and Tikhonov, 1995). Buffer capacities of the lumen and stroma compartments are taken into account by the pools of proton-accepting species B_{in} and B_{out}, respectively.

The rate of ATP synthesis, which is tightly coupled to the proton through CF_0-CF_1 (Fig. 5.5B), was calculated as the following function:

$$J_{ATP} = k_{ATP} \times ([ADP] \times [P_i]) \times \frac{[H^+_{out}] \times [10^{\Delta pH} - 1]}{\alpha + [H^+_{out}] \times [10^{\Delta pH} + \beta]} \quad (5.7)$$

Here, k_{ATP} is the normalizing coefficient, $\Delta pH = pH_{out} - pH_{in}$, [ADP] and $[P_i]$ are the concentration of ADP and P_i. The total concentration of adenine nucleotides is fixed, [ADP] + [ATP] = const; we assume, at first approximation, that P_i is present in large excess. The model implies that the ATP synthase is maintained in the activated state, because the processes of the redox-dependent regulation of photosynthetic enzymes were not considered within the framework of the model described. The last factor in equation 5.7 describes the proton flux through the membrane part of the ATP synthase (CF_0). Constants α and β are interrelated model parameters, $\alpha = 10^{-pK_a}(1 + \beta)$, where pK_a stands for an apparent pK value of the carboxyl group of the subunit *c* involved into the H^+ transfer through CF_0 (for more details see Vershubskii *et al.*, 2004a, 2011; Tikhonov and Vershubskii, 2014). This model of ATP synthase is based on the well-known fact that the carboxyl

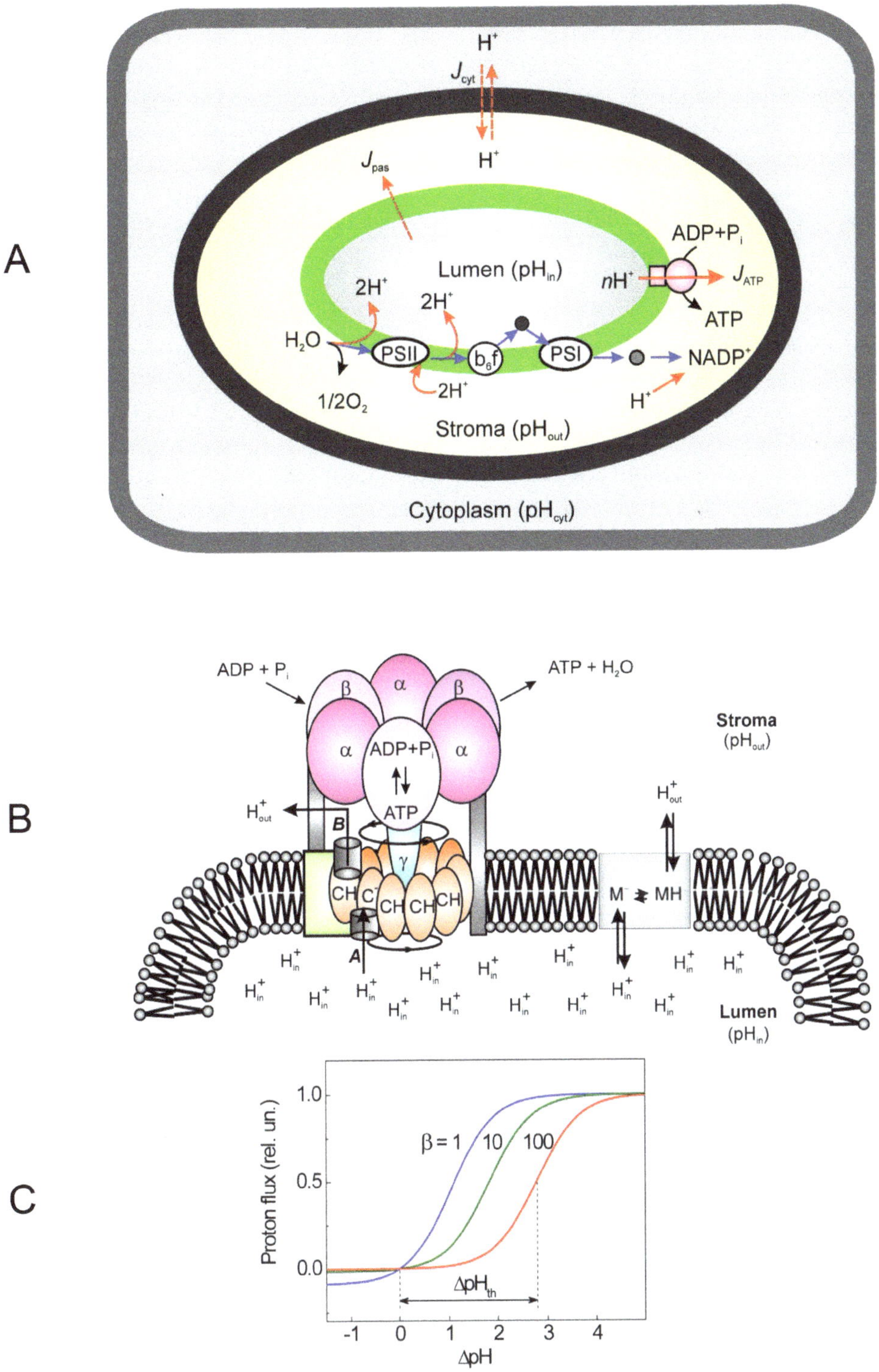

Figure 5.5 (A) A cartoon illustrating the modelled compartmentalization of hydrogen ions and transmembrane proton fluxes. (B) A sketch illustrating the mechanisms of proton transfer through the ATP synthase (flux J_{ATP}) and passive proton flow through the membrane (flux J_{pas}). (C) The dependence of the proton flux through the CF_0 fragment of ATP synthase *vs* the transthylakoid pH difference (ΔpH) calculated for three values of the model parameter β (adapted from Tikhonov and Vershubskii, 2014). See text for other details.

groups of n hydrophobic subunits c, which form the membrane-bound c_n-ring ($n = 13–15$ (Pogoryelov *et al.*, 2007)), are directly involved into the proton transfer through the membrane segment of the CF_0–CF_1 complex. The carboxyl group, which is exposed to the acidic reservoir (lumen) through the lumen-exposed 'semi-channel' A, becomes protonated ($–COO^- + H^+_{in} \rightarrow –COOH$), because its value $pK_a \approx 7.1–7.5$ (Assadi-Porter and Fillingame, 1995; Vollmar *et al.*, 2009) is higher than pH_{in} in energized thylakoids ($pH_{in} \leq 6.2–6.5$). The carboxyl group of the subunit c, which is exposed to the alkaline reservoir ($pH_{out} \approx 7.8–8.0$) through the stroma-exposed 'semi-channel' B, becomes deprotonated ($–COOH \rightarrow –COO– + H^+_{out}$). Rotating in the membrane, each subunit c periodically contacts either with the acid or with the alkaline reservoir, providing the translocation of the proton from the lumen to stroma: $–COO– + H^+_{in} \rightarrow –COOH \rightarrow –COO– + H^+_{out}$. The transmembrane pH difference (ΔpH) would support the directional rotation of the c_n-ring, provided the condition $pH_{in} \leq pK_a \leq pH_{out}$ is fulfilled. Rotating in the membrane, the c_n-ring will turn the subunit γ of CF_1, thereby driving the ATP formation from ADP and P_i. Despite its simplicity, the model can properly describe main peculiarities of ATP synthesis in chloroplasts. The bottom panel in Fig. 5.5 shows the force–flux relationships (the proton flux through CF_0 vs. ΔpH) calculated for $pK_a = 7.3$, $pH_{out} = 8.0$, and three values of the model parameter β. As one can see, the proton flux reveals the S-type dependence with a characteristic threshold (ΔpH_{th}), which is peculiar to experimental dependences of the rate of ATP synthesis *vs* ΔpH (Gräber, 1982; Junesche and Gräber, 1991). The model adequately describes the pH-dependence of ATP synthesis in chloroplasts (Versubskii *et al.*, 2004a, 2011).

Passive proton flux J_{pas} through the thylakoid membrane ($H^+_{lumen} \leftrightarrow H^+_{stroma}$) was described within the framework of a simple kinetic model (Dubinsky and Tikhonov, 1995) based on the assumption that the proton from the lumen first binds to the acid-base group M^- and then dissociates to the stromal compartment (Fig. 5.5B). The proton exchange between the stroma and cytoplasm (J_{cyt}) was described within the framework of the same model. The model parameters used for calculation of fluxes J_{ATP}, J_{pas} and J_{cyt} were fitted using experimental data on pH_{out} and pH_{in} values measured in stroma and in the lumen in metabolic states 3 and 4 as described earlier (Vershubskii *et al.*, 2004a, 2011).

For description of pH_{in}-dependent oxidation of PQH_2 we used a semiempirical function $k_Q([Q],[Pc],[H^+_i]) = 1/\tau_Q$, where τ_Q is a characteristic time of electron transfer from PQH_2 to Pc (Dubinskii and Tikhonov, 1997; Vershubskii *et al.*, 2001, 2011). For simulation of non-photochemical losses of energy in PSII, we treated the model parameter L_2 (a number of light quanta exciting P_{680} centres per time unit) as a function which decreases with the lumen acidification (Kuvykin *et al.*, 2009). The value of $pK_{NPQ} = 6.0$ was taken on the basis of experimental data (Pfündel and Dilley, 1993). pH-dependent activation of the CBC was modelled as described in (Frolov and Tikhonov, 2007).

Induction events and alternative pathways of electron transport

Let us consider few examples illustrating the applicability of the model for simulation of induction events in chloroplasts. A general term 'induction events' usually relates to transient photosynthetic processes observed after a dark-to-light transition (Edwards and Walker, 1983). Fig. 5.6A shows typical patterns of experimental and calculated kinetics of P_{700} photooxidation in dark-adapted chloroplasts. Immediately after switching on illumination, a concentration of oxidized centres P^+_{700} rapidly rises to the transient level A, and then declines to the intermediate level B, demonstrating a certain lag-phase prior to significant rise of $[P^+_{700}]$ towards the steady-state level C. A relatively low level of P^+_{700} at the initial phase of the induction curve may be explained by two main factors: (i) slow outflow of electrons from PSI to the CBC that causes the over-reduction of the pool of electron carriers on the acceptor side of PSI, and (ii) recycling of electrons from PSI to the intersystem ETC (CEF1). The model predicts (Fig. 5.6B) that immediately after switching the light on, during the lag period (phase A-B), the acceptor side of PSI becomes over-reduced, because the efflux of electrons from PSI is limited owing to slow consumption of NADPH in the CBC. The PQ pool also becomes over-reduced. The light-induced changes in relative concentrations of PQH_2, Fd, and NADPH follow non-monotonic kinetic patterns: after initial reduction, these carriers reoxidize due to activation of

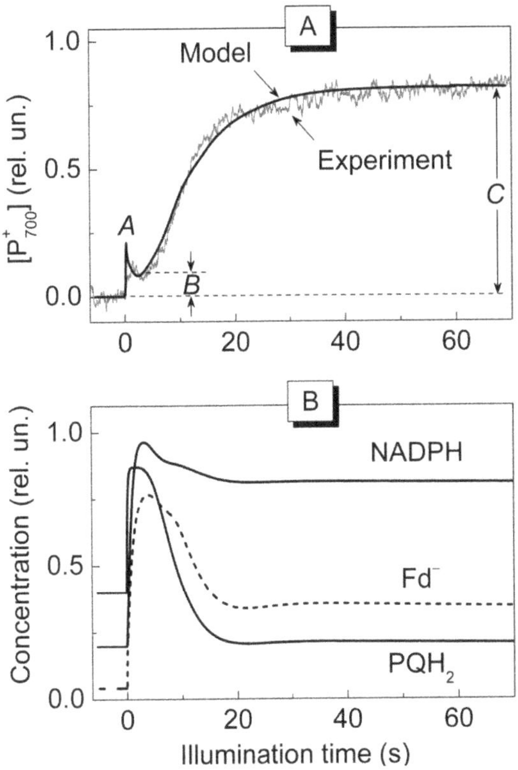

Figure 5.6 (A) Comparison of computed time-course of P_{700} photooxidation with experimental kinetics of P^{+}_{700} induction in dark-adapted (10 min) *Hibiscus rosa-sinensis* leaves. Modified from Figure 11 in Kuvykin *et al*., 2011). (B) Computer simulations of redox transitions of NADP, Fd, and PQ. Modified from Figure 6A inTikhonov and Vershubskii (2014).

the CBC. This stimulates the efflux of electrons from PSI and the subsequent growth of $[P^{+}_{700}]$ to the steady state level C (Fig. 5.6A). Acceleration of electron efflux from PSI to the CBC, induced by alkalization of stroma (Fig. 5.7A), is accompanied by temporal acceleration of LEF and concomitant decrease in CEF1 (Fig. 5.7B). The deceleration of electron fluxes with illumination represents a general trend, which can be explained by attenuation of PSII photochemical activity due to NPQ and slowing down of PQH_2 oxidation caused by the lumen acidification.

Alternative pathways of electron flow on the acceptor side of PSI may be essential for balancing the ATP/NADPH output ratio. Although electron flows through cyclic and pseudocyclic chains do not reduce $NADP^+$, they generate *pmf* needed to drive the ATP synthase complexes. Thus, the stoichiometric ATP/NADPH ratio of 3:2 required for the CBC is attained. Fig. 5.7B compares computed kinetics of electron flux through PSII (J_{PSII}) and electron fluxes on the acceptor side of PSI (J_{LEF}, J_{CEF1}, and J_{WWC}). The model predicts that after significant initial jump, the overall electron flux from PSII to the PQ pool (J_{PSII}) decays non-monotonically to a steady state level. A decrease in J_{PSII} can be explained by down-regulation of PSII activity due to enhanced NPQ in the result of acidification of the lumen ($pH_{in}\downarrow$). The light-induced decrease in pH_{in} also causes deceleration of PQH_2 oxidation by the b_6f complex. To add, the light-induced alkalization of stroma ($pH_o\uparrow$) may hinder protonation of PQ reduced by PSII ($PQ + 2e^- + 2H^+_{out} \rightarrow PQH_2$), thereby further reducing electron flux form PSII to PSI.

A substantial redistribution of electron fluxes occurs in the induction period. On the initial stage, there is a noticeable contribution of alternative electron fluxes J_{CEF1} (cyclic electron flow around PSI) and J_{WWC} (the Mehler reaction) to the efflux of electrons from PSI. Electron fluxes J_{CEF1} and

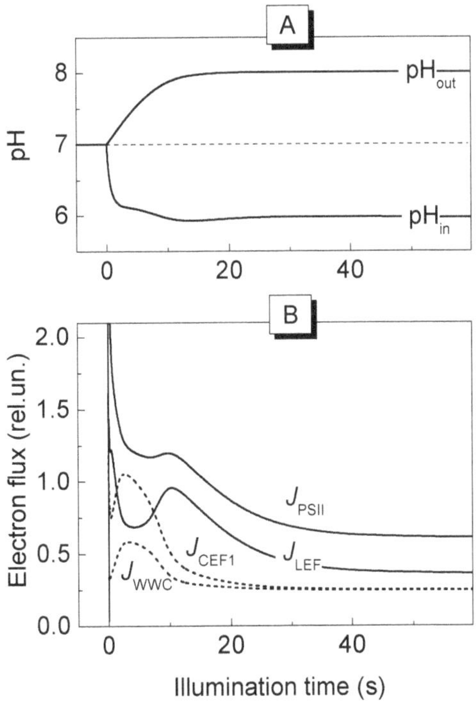

Figure 5.7 (A) Computed light-induced changes in the lumen and stromal pH. Modified from Fig. 6B in Tikhonov and Vershubskii (2014). (B) Computed time-courses of electron fluxes J_{PSII}, J_{LEF}, J_{CEF}, and J_{WWC}. Modified from Figure 18 in Tikhonov (2015).

J_{WWC} show their maximal values when the linear electron flow to the CBC (J_{LEF}) is restricted owing to over-reduction of $NADP^+$ at low activity of the CBC enzymes. This is consistent with the current notion that CEF1 and the Mehler reaction provide a bypass to avoid 'over-reduction' on the acceptor side of PSI. After a certain lag, when the CBC activates and the linear electron flux J_{LEF} increases, the relative contributions of alternative fluxes J_{CEF} and J_{WWC} gradually decline.

Taken together, the kinetic model described here can portray some general features of induction events in photosynthesis. In particular, it reproduces the appearance of the lag-phase in the kinetics of P_{700} photooxidation and redistribution of electron fluxes between alternative pathways (linear, cyclic, and pseudocyclic electron fluxes) after the dark-to-light transition.

Effects of CO_2 and O_2 on photosynthetic electron transport

Direct influence of atmospheric gases (CO_2 and O_2) on photosynthetic electron transport is determined by their interaction with Rubisco, the key enzyme of the CBC. Mathematical modelling of the CBC reactions has a long story. For more than 50 years ago, Chernavskaya and Chernavskii (1961) presented a rather simple model that predicted oscillations of metabolite concentrations in the CBC. Oscillations of gas-exchange reactions and chlorophyll *a* fluorescence in leaves are well-documented experimental facts (Walker, 1992; Rovers and Giersch, 1995).

The most general description of the CBC reactions includes the stages of carboxylation, reduction and regeneration (Hahn, 1984; Pettersson and Ryde-Pettersson, 1988; Karavaev and Kukushkin, 1993; Fridlyand and Scheibe, 1999; Igamberdiev and Lea, 2006; Laisk *et al.*, 1989, 2006, 2009; von Caemmerer *et al.*, 2009; Dubinsky *et al.*, 2010; Igamberdiev and Kleczkowski, 2011), which can be divided into more elementary processes, down to the level of single reactions (Farquhar *et al.*, 1980; Farazdaghi, 2011). A comprehensive critical review of the existing mathematical models of the CBC has been presented by Arnold and Nikoloski (2011). They scrutinized 15 models and identified those models that provided quantitatively accurate predictions for the levels of CBC intermediates and could be used in metabolic engineering and in the design of synthetic metabolic pathways for improved carbon fixation, growth, and yield. Detailed analysis of mathematical models of the CBC is beyond the scope of this chapter.

For illustration of the regulatory events associated with activation of the CBC, let us consider the results of simulations performed within the framework of the model outlined above. The consumption of NADPH and ATP in the CBC reactions was calculated using a phenomenological approach described in (Frolov and Tikhonov, 2007; Kuvykin *et al.*, 2009). Assimilation of CO_2 was modelled by the following function:

$$f_{gas}([CO_2],[O_2]) = \frac{[CO_2]-\Gamma}{[CO_2]+K_C(1+[O_2]/K_O)} \qquad (5.8)$$

This formula describes the consumption of CO_2 in the CBC (photosynthesis) and the uptake of O_2 due to oxygenase activity of Rubisco (photorespiration), taking into account the competition between CO_2 and O_2 for Rubisco. Parameter Γ corresponds to CO_2 concentration at the compensation point ($\Gamma = 35$ ppm); parameters K_C and K_O stand for the Michaelis constants of carboxylation and oxygenation reactions catalysed by Rubisco, respectively (Sharkey *et al.* 2007; von Caemmerer *et al.*, 2009).

The model provides a reasonably good simulation of CO_2 consumption and photorespiration. For instance, the model reproduces with fair accuracy effects of CO_2 on photosynthesis and photorespiration reported by André (2011) for wheat leaves (Fig. 5.8). Assimilation of CO_2 increases with the rise of CO_2 concentration, saturating at high levels of CO_2; photorespiration decreases with the rise of CO_2. Due to photorespiration, at atmospheric concentration of O_2 (21%) the yield of photosynthesis is lower than under the oxygen-deficient conditions (2% O_2). Thus, the model adequately portrays the Warburg effect (Turner and Brittain, 1962): deaeration increases photosynthesis and decreases the oxygen uptake by chloroplasts.

The model describes the effects of CO_2 and O_2 on electron transport in chloroplasts. At low level of CO_2, when the Rubisco turnover limits the rate of electron efflux from PSI, molecular oxygen plays the role of electron sink supporting the operation of PSI. Fig. 5.9 demonstrates that deoxygenation causes a marked decrease in the levels

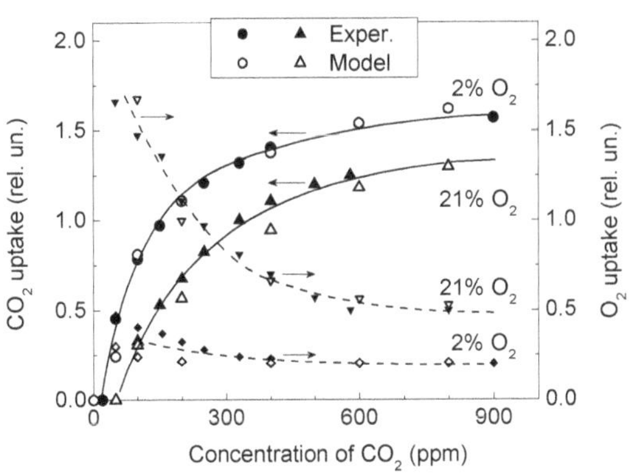

Figure 5.8 CO_2 consumption (photosynthesis) and O_2 uptake (photorespiration) at two concentrations of atmospheric O_2. Results of numerical experiments (open symbols) are compared with experimental data (solid symbols) obtained on wheat plant (data points are taken from Andre, 2011). Modified from Figure 8 in Tikhonov and Vershubskii (2014).

of P^+_{700} in fair agreement with experimental data. At sub-saturating concentrations of CO_2, just as in experiment, depletion of O_2 also causes a decrease in J_{PSII} (Kuvykin *et al.*, 2011). The inhibitory effect of oxygen depletion (2% O_2) disappears at saturating concentrations of CO_2. These results agree with experimental data (Munekage *et al.*, 2002), suggesting that the outflow of electrons from PSI to O_2 stimulates the overall electron flux through PSII.

Summing up, the model considered adequately describes the effects of CO_2 and O_2 on photosynthetic electron transport in chloroplasts. Down-regulation of LEF with lowering atmospheric CO_2 and depletion of O_2 can be explained by several reasons: (i) impediments to electron transfer on the acceptor side of PSI, (ii) deceleration of PQH_2 oxidation caused by the lumen acidification, and (iii) pH-dependent enhancement of energy losses in PSII (Kuvykin *et al.* 2011).

Modelling the proton transport processes in chloroplasts

Topological aspects of proton transport in chloroplasts

Structural peculiarities of the lamellar system may have an impact on photosynthetic capacity of chloroplasts. As noted above, PSI, PSII, and CF_0–CF_1 complexes are distributed non-uniformly between the grana stacks and stroma lamellae. Significant amounts of PSI and PSII complexes are laterally segregated, whereas the Cyt b_6f complexes are distributed almost uniformly along the granal and stromal membranes (Albertsson, 2001). These peculiarities of structural organization of photosynthetic apparatus raise two awkward questions about electron and proton transport in chloroplasts: (*i*) how may the obstructed diffusion of PQH_2 and restricted mobility of Pc affect the intersystem electron transport? and (*ii*) what is the influence of restricted diffusion of protons within the lumen and in the narrow gap between adjacent grana thylakoids on the operation of the chloroplast ETC?

Here we will consider the proton transport processes associated with PQ reduction to PQH_2 in PSII. The formation of PQH_2 requires two electrons donated by PSII and two protons taken from the chloroplast stroma ($PQ_B + 2e^- + 2H^+_{out} \rightarrow PQ_BH_2$). Under the normal physiological conditions, the light-induced raise of pH_s in the bulk of stroma does not exceed $pH_s \approx 7.8$–8.0 (Heldt *et al.*, 1973; Robinson, 1985). Such a moderate elevation of pH_s should not effect on the rate of PQ reduction to PQH_2. However, the formation of PQH_2 may be strongly affected by more significant rise of pH in the local vicinity of PSII complexes located in

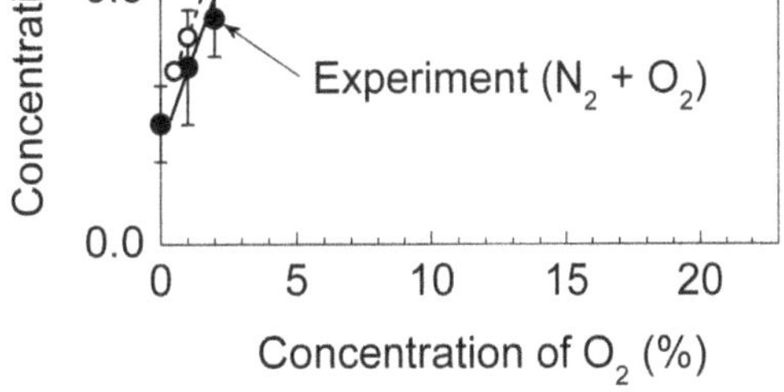

Figure 5.9 Effects of atmospheric O_2 on the steady-state level of P^+_{700} in *Hibiscus rosa-sinensis* leaves (solid circles) and calculated for two concentrations of CO_2 (open symbols) within the framework of the model. Modified from Figure 5 in Kuvykin *et al.* (2011).

granal thylakoids, as it predicts the model described by Vershubskii *et al.* (2001, 2004b, 2011). Most of PSII complexes are spatially separated from the stroma, their outer parts protrude from the membrane into the partition between appressed thylakoids of grana. The gap between the adjacent thylakoids is ~2.5-3.5 nm; the protein molecules protruding from oppositely located membranes are in close contact with each other. The interthylakoid partition contains a lot of buffering groups, mostly of the large proteins which cover ~75-80% of the thylakoid membrane area (Murphy, 1986; Kirchhoff, 2008; Kirchhoff *et al.*, 2008, 2011). The protons percolating through the interthylakoid gap from stroma to PSII will interact with the buffering groups. It is not surprising, therefore, that an efficient coefficient of proton diffusion within the gap may be significantly smaller than in the bulk water (Polle and Junge, 1986, 1989; Junge and Polle, 1986; Junge and McLauphlin, 1987). Slow diffusion of protons within the narrow gap would retard to pH levelling in the gap and in the stroma. Therein lies the reason why the proton consumption upon the PQ reduction in PSII might cause significant alkalization of the gap ($pH_{gap} > pH_{stroma}$).

The light-induced alkalization of the interthylakoid partition may turn to disadvantage in the formation of PQH_2. This is the case when pH_{gap} exceeds pK_a values of acid-base groups involved in protonation of the secondary quinol ($PQ_B + 2e^- + 2H^+_{gap} \rightarrow PQ_BH_2$). Analysis of experimental data reported in the literature suggests that pK_a values of the acid–base groups involved in PQ_BH_2 formation are about 8.5-9.0 (for references, see Tikhonov and Vershubskii, 2014). Therefore, upon sufficiently strong alkalization of the gap ($pH_{gap} \geq pK_a$) the reduction of PQ to PQH_2 should be retarded.

Lateral profiles of proton potential in chloroplasts

The existence of non-uniform lateral profiles of ΔpH has been discussed in the literature for a long time (Haraux *et al.*, 1983; Tikhonov and Blumenfeld, 1985) in the context of alternative ('local' or 'non-local') mechanisms of the proton delivery from the proton pumps to ATP synthases (Kell, 1979; Westerhoff *et al.*, 1984; Dilley, 1991; Blumenfeld and Tikhonov, 1994). Mathematical analysis of the problem predicts that obstructed lateral diffusion of protons in the lumen (Dubinskii and Tikhonov, 1997) or in the invaginations of mitochondrial membranes (Kara-Ivanov, 1983) might be one of the reasons for non-uniform lateral profiles of *pmf* in energy-transducing organelles. Direct measurement of pH profiles in small compartments is a very arduous experimental task. In the absence of reliable experimental data about the lateral profiles of pH_{gap} and pH_{in}, computer modelling becomes a useful tool for quantitative analysis of pH distribution along the chloroplast lamellae. Mathematical models, which take into account the lateral heterogeneity of thylakoids, predict the establishment of non-uniform lateral profiles of pH and significant alkalization of the interthylakoid gap. The latter might be one of the factors of down-regulation of electron transport in chloroplasts (for review, see Vershubskii *et al.*, 2011; Tikhonov and Vershubskii, 2014).

Below we will consider the results of computer simulation of proton transfer within the framework of our model, which takes into account non-uniform partitioning of PSI and PSII in thylakoids. Fig. 5.10A depicts the spatial arrangement of electron transport complexes in the membranes of granal and stromal thylakoids in the model system. Granal thylakoids are modelled by flattened cylinders of radius a; stroma-exposed thylakoids are modelled as cylinders of radius b, which protrude from the thylakoids of grana. The distances between the inner surfaces of the thylakoid, l_i, and the partition between the neighbouring thylakoids of grana, l_o, are geometrical parameters of the model. PSI complexes are localized exclusively in the stromal domain, PSII complexes are localized in granal regions of the thylakoid lamellae. The Cyt b_6f complexes are distributed uniformly along the membrane. ATP synthase complexes (CF_0–CF_1) are localized in the stroma-exposed membranes. Mobile electron carriers, PQ and Pc, can diffuse along the membrane and inside the thylakoid lumen, respectively. Since the lateral mobility of electron transport complexes in the thylakoid membrane is of three orders of magnitude lower than that of PQ (Kirchhoff, 2008; Kirchhoff *et al.*, 2008; Kaňa, 2013), PSI, PSII, Cyt b_6f, and CF_0–CF_1 complexes are assumed to be fixed in the membrane.

Restrictions to proton diffusion inside the narrow compartments of chloroplasts could influence the lateral profiles of variables $[H^+_{in}(\vec{r})]$ and $[H^+_{out}(\vec{r})]$.

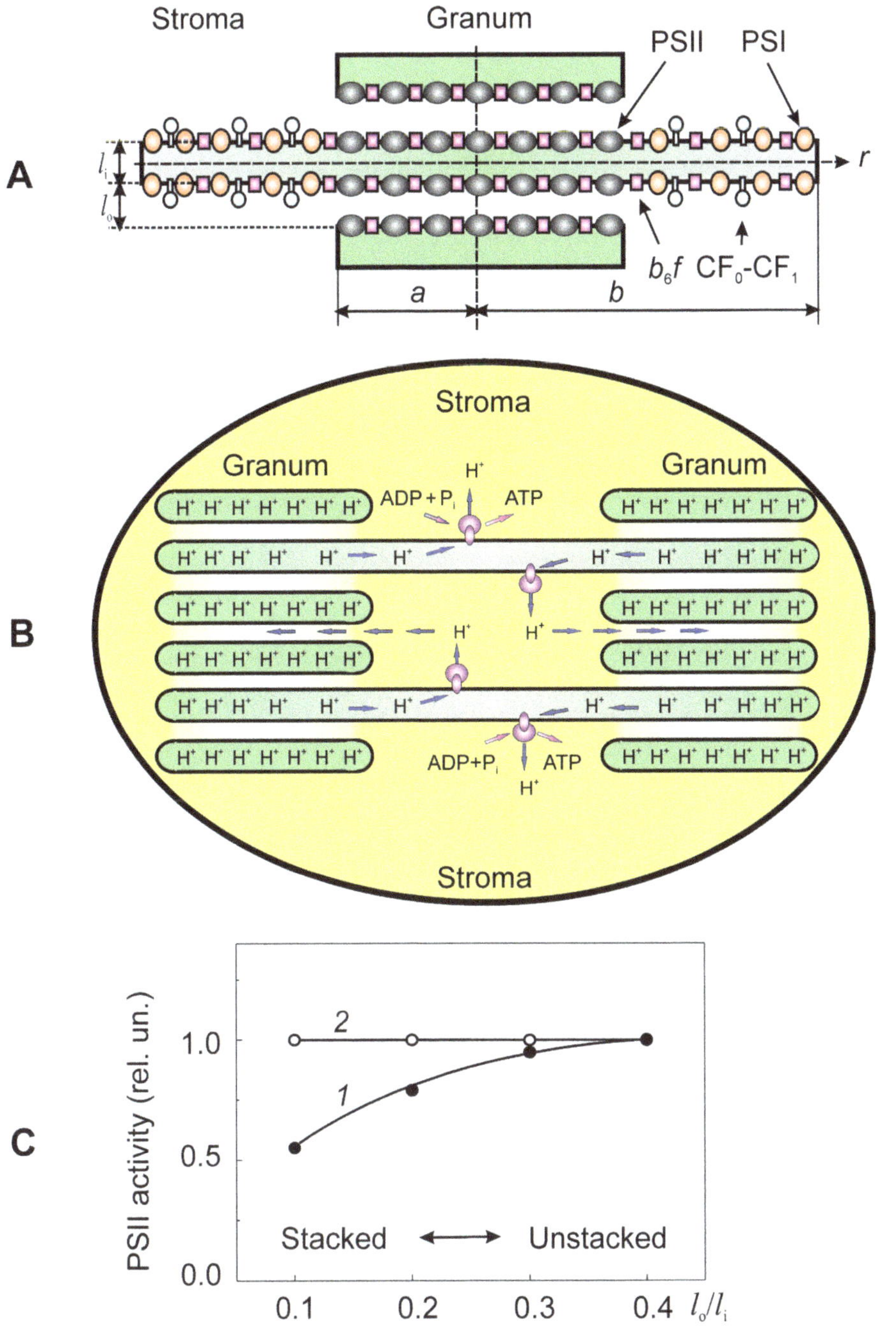

Figure 5.10 (A) A scheme of non-uniform distribution of electron transport (PSI, PSII, and b_6f) and the ATP synthase complexes between granal and stroma-exposed thylakoids. (B) Cartoon illustrating non-uniform distribution on pH in different compartments of the chloroplast. Since most of the ATP synthase complexes are localized in stroma-exposed thylakoids, an efficient drain of protons from the lumen via the ATP synthases may lead to the non-uniform profile of the intrathylakoid pH ($pH_{in}^{grana} < pH_{in}^{stroma}$). Significant alkalization of the interthylakoid gap ($pH_{out}^{grana} > pH_{out}^{stroma}$) is caused by limitations in the proton diffusion along the narrow interthylakoid gap. Non-uniform distribution of pH may cause circulation of proton fluxes: a 'proton pressure' developed in granal regions of the lumen will divert the protons to CF_0-CF_1, while the consumption of protons by PSII will support the proton flux from the stroma to the interthylakoid gap. Modified from Figure 14 in Tikhonov and Vershubskii (2014). (C) The influence of the gap width (variations of parameter l_0 at l_i = const) on calculated activity of PSII. Curves 1 and 2 correspond to different values of diffusion coefficient for protons, $D_{H^+} = 0.02D_0$ (curve 1) and $D_{H^+} = 0.1D_0$ (curve 2), where $D_0 = 10^{-5}$ cm^2/s. Modified from Figure 5 in Vershubskii *et al.* (2004b).

There are good reasons to believe that mobility of protons and low-molecular electron carriers nearby the thylakoid membrane may be significantly lower than in the aqueous bulk phase (Nesbitt and Berg, 1982; Junge and Polle, 1986; Polle and Junge, 1989; Trubitsin and Tikhonov, 2003). Slow percolation of protons through the gap between appressed thylakoids of grana was demonstrated by Polle and Junge (1986). They observed that stacking/unstacking of granal thylakoids strongly affected the rate of proton uptake by thylakoids. The rate of proton diffusion within the interthylakoid gap increased dramatically after unstacking of granal thylakoids. Comparison of theoretical and experimental data suggests that the apparent coefficient of proton diffusion in the intrathylakoid lumen and in the narrow partitions between grana thylakoids may decrease by a factor of ~500 as compared to proton mobility in the aqueous bulk phase (for more details see Vershubskii *et al.*, 2004b, 2011).

One of the most interesting theoretical predictions concerns significant alkalization of the interthylakoid gap ($pH_{gap} \geq 9.5$-10) caused by the proton consumption upon the PQ reduction to PQH_2. Restricted diffusion of protons from the stroma to PSII ($H^+_{stroma} \rightarrow H^+_{gap}$) cannot compensate a significant rise of pH in the partition between appressed granal thylakoids. Fig. 5.11 shows lateral profiles of pH inside the lumen and in the interthylakoid gap. In state 4 (when the overall proton flux through CF_0-CF_1 is virtually zero), pH_{lumen} decreases more significantly than in state 3 (when protons can escape from the lumen via active CF_0–CF_1 complexes). In state 3, there may establish non-uniform profile of pH_{lumen}, which shape depends on the model parameter l_0. In the case of a narrow gap (stacked thylakoids), the lumen in the centre of granum is more acidic than at the periphery enriched with CF_0-CF_1 complexes. Destacking of granal thylakoids (widening the gap) influences the pH-profiles: pH_{gap} decreases, the lumen becomes more acidic and the lateral profile of pH_{lumen} straightens (data not shown; see Vershubskii *et al.*, 2011). Significant alkalization of the gap should hinder to protonation of double-reduced form of PQ ($PQ_B^{2-} + 2H^+_{gap} \rightarrow PQ_BH_2$), thus slowing down the operation of PSII. Acceleration of proton diffusion inside the gap, either due to destacking of granal thylakoids (as modelled by the rise of parameter l_0) or due to an increase in the diffusion coefficient D_{H^+}, accelerates electron transport from PSII to PSI, thereby magnifying the proton pumping into the lumen. These results are explained by acceleration of PSII turnover and more intensive proton pumping into the lumen caused by rapid diffusion of protons to PSII in unstacked thylakoids.

Let us discuss energetic and regulatory aspects of non-uniform lateral profiles of proton potentials in chloroplasts. Since most of CF_0-CF_1 complexes are localized in stroma-exposed thylakoids, the efficient drain of protons from the lumen via CF_0-CF_1 would lead to the non-uniform lateral profile of ΔpH in state 3, when ΔpH difference across the stromal membrane is lower than ΔpH in grana. Note that the steady state value of $\Delta pH \approx 1.8$-2.5 across the stromal thylakoids (Fig. 5.11) is sufficient to support efficient ATP synthesis (for review, see Tikhonov, 2012, 2013). It is also interesting to note that there may be lateral circulation of proton fluxes (Fig. 5.10B): the 'proton pressure' developed inside the granal thylakoids ($pH^{grana}_{in} < pH^{stroma}_{in}$) will divert the protons to the lumen of stroma-exposed thylakoids, where CF_0-CF_1 complexes act as a sink for protons; otherwise, the operation of PSII supports the influx of protons from the bulk of stroma to the interthylakoid gap ($pH^{grana}_{out} > pH^{stroma}_{out}$).

Apart from its energy role in ATP synthesis, ΔpH plays important regulatory role: the light-induced decrease in the lumen pH decelerates PQH_2 oxidation by the Cyt b_6f complex and induces the attenuation of PSII activity due to the NPQ mechanism of energy dissipation in LHCII. It is interesting to note that the model described also predicts another mechanism of down-regulation of PSII activity. This mechanism is associated with the light-induced alkalization of the interthylakoid gap, which may hinder in protonation of fully reduced plastoquinol, PQ_B^{2-}, thereby slowing down the formation of PQH_2. The model predicts that unstacking of granal thylakoids, which releases the constraints to protons diffusion towards PSII, would stimulate the turnover of PSII (Fig. 5.10C). This prediction is in a good agreement with experimental data on the influence of osmotic conditions on electron transport in chloroplasts. Kirchhoff *et al.* (2000) reported that in destacked tobacco chloroplasts nearly all PQ molecules became reduced by a 120 ms light pulse (about 11 electrons were stored

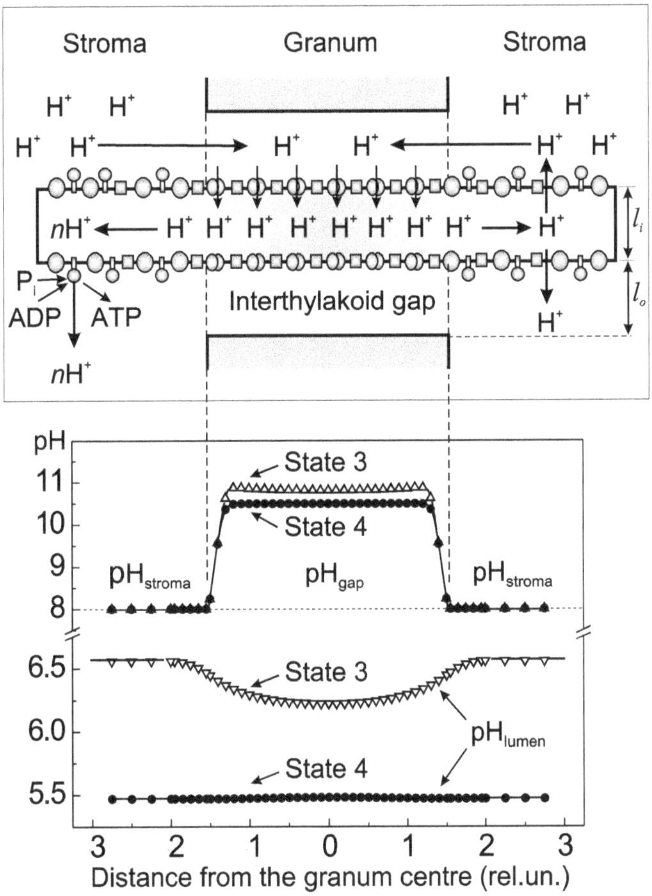

Figure 5.11 Modelled pathways of proton transfer (top panel) and calculated lateral profiles of pH inside the thylakoid lumen and in the interthylakoid gap in steady states 3 and 4 (bottom panel). Stromal pH was assumed to be constant, pH_s 8. Modified from Figures 12 and 13 in Tikhonov and Vershubskii *et al.* (2014).

in the ETC, equivalent to 5–6 PQH_2 molecules). Otherwise, in stacked thylakoids, a large fraction of PQ molecules (about 40-50%) remained oxidized after the light pulse. The authors concluded that their observations could be explained by inability of PQ to migrate rapidly throughout the membrane. Results of computer simulations suggest alternative explanation of experimental data: in stacked membranes the light-induced formation of PQH_2 may be limited by slow diffusion of protons throughout the narrow interthylakoid gap; unstacking of granal thylakoids stimulates diffusion of protons, thereby promoting the light-induced formation of PQH_2. Summing up, computer experiments suggest that along with the NPQ mechanism of attenuation of PSII activity and deceleration of PQH_2 oxidation by the Cyt b_6f complex, the intersystem electron transport may be down-regulated due to the light-induced alkalization of the interthylakoid partition.

Transthylakoid proton transport coupled to ATP synthesis

Transthylakoid transfer of protons through the membrane-bound ATP synthase (CF_0-CF_1) represents one of the last but not least events on the stage of photosynthetic performance collectively termed as the 'light-induced stages of photosynthesis'. It is not surprising that Paul Boyer, who was first to suggest that the ATP synthase operates as the macromolecular device with the γ shaft rotating in the central part of the F_1 fragment of the enzyme, wrote: '*All enzymes are beautiful, but the ATP synthase is one of the most beautiful as well as one of the most unusual and important*' (Boyer,

1997). Below, some of the works on mathematical modelling of the transthylakoid proton transport through the ATP synthase complex are briefly outlined.

Dynamic models of proton ATP synthases as rotating molecular machines

The ATP synthase of the F_o-F_1 type represents a reversible macromolecular machine, which comprises two rotary motors in one. The F_o motor, using an energy stored up in the form of *pmf*, generates a rotary torque, providing the operation of the enzyme in the ATP synthesis mode. The F_1 motor, using the energy released upon the ATP hydrolysis, can generate a mechanical torque, which provides turning the γ shaft in the opposite direction and proton pumping from the stroma to lumen. Biochemical and structural studies of ATP synthases of F_o-F_1 type give a wealth of experimental material for mathematical modelling of this complex. Mathematical modelling is a convenient tool for the analysis of various scenarios underlying the work of ATP synthase and the assessment of their energetic and dynamic properties (for review, see Romanovsky and Tikhonov, 2010).

The first comprehensive models for adequate description of ATP synthase functioning were suggested by Oster and collaborators, who constructed dynamic models for simulations of rotations of F_o and F_1 motors (Elston *et al.*, 1998; Wang and Oster, 1998; reviewed in Oster and Wang, 2000). Their works are based on the use of the stochastic Langevin approach for modelling dynamic systems. The most general form of the Langevin equation for numerical description of the rotor rotations in ATP synthase is the following:

$$\zeta \frac{d\theta}{dt} = F_E(\theta,s) + F_M(\theta,s) - \tau + f(t) \qquad (5.9)$$

where θ is the angle characterizing the rotation of the rotor (c_n-ring), ζ is the drag coefficient, $F_E(\theta,s)$ is the electrostatic force moment acting on the rotor from the membrane-bound subunit *a*, $F_M(\theta,s)$ is the 'hydrophobic' force moment acting on the c_n-ring from the membrane, τ is the torque acting on the subunit γ from the $a_3\beta_3$ hexamer, $f(t)$ is the moment of random force with the 'amplitude' $\sqrt{2\zeta k_B T}$, symbol *s* characterizes discrete states of the system, which differ with respect of occupation of the proton-binding sites of the membrane part of the F_o motor. The models constructed in (Elston *et al.*, 1998; Wang and Oster, 1998) describe rotations of both rotors: the ring c_n of *n* subunits *c* buried into the lipid membrane, which unidirectional rotation is driven by *pmf*, and rotations of the subunit γ revolving inside the central cavity of the hexamer $a_3\beta_3$ (see cartoon in Fig. 5.5B). The model suggested by Elston *et al.* (1998) predicts that biased rotational diffusion, augmented by electrostatic forces, does indeed generate sufficient torque to account for ATP production. The motor shows reversibility: ATP hydrolysis in the F_1 fragment generates a torque, which converts the motor into an efficient proton pump. The model described in (Wang and Oster, 1998) accounts for mechanochemical behaviour of the enzyme in both modes, rotations of the F_1 motor in the hydrolysing or synthesizing directions. The authors conclude that the F_1 motor achieves high mechanical torque and almost 100% efficiency because it converts the free energy of ATP binding into elastic strain.

The reader can find in the literature other mathematical models of ATP synthases (Cherepanov *et al.*, 1999; Pänke and Rumberg, 1999; Pänke *et al.*, 2001; Pogrebnaya *et al.*, 2005; Nartsissov and Mashkovtseva, 2006), which describe the unidirectional revolutions of F_o-F_1 molecular motors driven by the transmembrane *pmf* (see for references Romanovsky and Tikhonov, 2010). A rather simple model for the ATP synthase, which depicts how the rate of ATP synthesis depends on ΔpH (Fig. 5.5C), is outlined above. Our general model of electron and proton transport, supplemented by the ATP synthase module, properly describes induction events in chloroplasts upon the dark-to-light transitions (Vershubskii *et al.* 2011; Tikhonov and Vershubskii, 2014).

Small bioenergetic systems: methodological aspects of their thermodynamic description

At the end of this chapter, it would be reasonable to point out to some peculiarities of thermodynamic description of small bioenergetic systems related to proton transport processes. First of all, it should be noted that complex multienzyme systems, including energy-transducing organelles, usually function in metabolic states which are far from thermodynamic equilibrium (Nelson and Cox,

2012). Consider, for instance, the energy coupling between the chloroplast *pmf* (the driving force for ATP synthesis) and 'phosphorylation potential' $P = [ATP]/([ADP] \times [P_i])$, which determines the energy needed for ATP formation. The P and *pmf* ($\Delta\tilde{\mu}_{H^+}$) values are interconnected. In order to provide efficient ATP synthesis, the relationship $n \times \Delta\tilde{\mu}_{H^+} \geq \Delta G_{ATP}$ must be fulfilled. Here, $\Delta G_{ATP} = \Delta G^0_{ATP} + RT \times \ln P$ is the Gibbs free energy change for ATP synthesis, $\Delta\tilde{\mu}_{H^+}$ is the transthylakoid difference of electrochemical potentials of protons, and n is the H^+/ATP ratio. There is experimental evidence, however, that the *pmf* generated in intact chloroplasts is out of equilibrium with the 'phosphorylation potential' P (Kobayasi *et al.*, 1979a, 1979b). For systems which are far from thermodynamic equilibrium, conventional approaches of equilibrium and non-equilibrium thermodynamics to numerical description of 'force–flux' relationships should be used with precautions (Westerhoff and Van Dam, 1987). In this case, the use of kinetic models may have an obvious advantage, because kinetic description can be applied to bioenergetic systems that are far from equilibrium.

Thermodynamic description of bioenergetic systems in terms of the proton potential difference, ΔpH, raises a question about the physical meaning of the hydrogen ion concentration (activity) inside small vesicles (Blumenfeld *et al.*, 1991; Blumenfeld and Tikhonov, 1994). Is it correct to speak about pH inside small vesicles in conventional terms of physical chemistry when we consider, for instance, hydrogen ions inside a single thylakoid? The very essence of this question becomes clear when we calculate a number of free hydrogen ions in the bulk phase of the vesicle. Trivial estimates show that at moderate pH_{in} a number of free (unbound) protons within a single thylakoid of internal volume V_{in} may be only a few. For instance, $N_{in} = V_{in} \times 10^{-pHin} \sim 0.6$–3.6 at $pH_{in} = 6.0$ (Tikhonov and Blumenfeld, 1985). Is it correct to speak about pH_{in} if the mean numbers of free (unbound) hydrogen ions N_{in} are only a few or a probability of detecting even of one unbound hydrogen ion is < 1? Scrutinizing a reaction of the $PQ \leftrightarrow P + Q$ type (an analogue of the dissociation reaction $H_2O \leftrightarrow H^+ + OH^-$) by the methods of statistical thermodynamics, Blumenfeld *et al.* (1991) concluded that fluctuations in a number N_P of unbound particles P would sharply increase if internal volume V_{in} of the vesicle becomes sufficiently small, when $N_P \sim 1$. Due to strong fluctuations in N_P, the law of mass action may be violated dramatically. It should be stressed, however, that the Nernst equation for the free energy change ΔG_P, related to particle P transfer from the vesicle to the external volume, remains valid, if we deal with the mean number of free particles, $<N_P>$, or their mean concentration $<C_{in}>$. Regardless of a number of free particles inside the vesicle, $\Delta G_P = -k_B T \ln(<C_{in}>/C_{out})$, where k_B is the Boltzmann constant, T is temperature, $<C_{in}> = <N_P>/V_{in}$ is the mean concentration of free particles P within a vesicle, C_{out} is the external concentration of free particles P. Note that the presence of a large number of buffer groups has to damp fluctuations in a number of free protons inside the thylakoids. In chloroplasts, almost all protons taken up by illuminated chloroplasts (~99%) are bound to proton-accepting groups of the thylakoid membrane and buffer groups of the molecules dissolved in the lumen (Masarova and Tikhonov, 1989).

Conclusions

Mathematical modelling of photosynthetic electron and proton transport provides a framework for in-depth analysis of feedbacks in the network of plant cell metabolism and creates a platform for investigation of regulatory mechanisms that provide plasticity and optimal functioning of photosynthetic apparatus. The reader can find in the literature a number of excellent examples of computer models that describe a variety of photosynthetic processes, from light capture to sucrose synthesis (see, for instance, Laisk *et al.*, 2006, 2009; Nedbal *et al.*, 2007a,b; Zaks *et al.*, 2012; Zhu *et al.*, 2013; Rubin and Riznichenko, 2014; and references therein). In this chapter, as an example of computer modelling of oxygenic photosynthesis, a generalized model of electron and proton transport in chloroplasts has been described. The model includes key stages of the linear electron transport, alternative pathways of electron transfer around PSI, transmembrane proton transport and ATP synthesis. The model also takes into account pH-dependent regulatory processes: modulation of the intersystem electron transport, down-regulation of PSII activity due to the NPQ mechanism, and the light-induced activation of the CBC. Results of numerical experiments performed within the framework of this model show that the model adequately reproduces a variety of experimental data on induction events in intact chloroplasts observed under different experimental conditions (variations of CO_2 and O_2 concentrations in atmosphere), including a complex kinetics of P^+_{700} induction, CO_2 consumption and photorespiration. Simulation of diffusion-controlled photosynthetic reactions of electron and proton transport in laterally heterogeneous thylakoids suggests that along with the NPQ mechanism of attenuation of PSII activity and deceleration of PQH_2 oxidation by the Cyt b_6f complex, the intersystem electron transport may be down-regulated due to the light-induced alkalization of the narrow partition between adjacent thylakoids of grana. The model considered may provide a basic module for a comprehensive model of oxygenic photosynthesis.

Future trends in modelling electron and proton transport in chloroplasts

The models of electron and proton transport in chloroplasts, a few examples of which have been outlined above, might be integrated into comprehensive models of photosynthesis. Further development of these models will provide reliable and workable platforms for profound analysis of photosynthetic energy conversion in nature ranging from seconds to seasons. Putative prospects of mathematical modelling photosynthetic processes in chloroplasts, as one can expect, will be associated with two general trends: first, the expansion, and, second, the refinement of current models of photosynthesis at molecular level. Chloroplasts are the organelles integrated into the complex network of metabolic processes in the plant cell. The construction of comprehensive models of photosynthetic electron transport should take into account interactions of chloroplasts with mitochondria and various metabolic cycles. In particular, it will be necessary to consider the interaction of chloroplasts with mitochondria via common metabolites, which are closely related to photophosphorylation and electron transport in chloroplasts. Generalized models should also describe the influence of variable environment conditions (fluctuations of light intensity, atmosphere gases, nutrition, seasonal and round-the-clock variations of temperature) on oxygenic photosynthesis (Kaiser *et al.*, 2015). These demands will need the integration of current models of photosynthetic electron transport with the computer models of other metabolic processes. Putative ways of expansion and refinement of current models of oxygenic photosynthesis are briefly outlined below.

Refinement and generalization of the models of electron transport processes in chloroplasts

The refinement of molecular models of partial reactions of electron transport may be important for realistic modelling of electron transport in chloroplasts. For instance, development of advanced models of PSI might shed new light on the problem concerning the contribution of A- and B-branches to electron transport in PSI. Another problem concerns the nature of the rate-limiting step in the chloroplasts ETC. As noted above, the rate of electron transfer from PSII to the Cyt b_6f complex is determined by two factors, formation of PQH_2 in PSII and its diffusion in the thylakoid membrane, on the one hand, and direct interaction of PQH_2 with the Cyt b_6f complex, on the other hand. Alterations in structural organization of the membrane (over-crowding of the membrane with protein complexes, fluidity of the lipid domains) can affect the PQH_2 mobility and its ability in the long-range percolation through the membrane domains overcrowded with densely packed protein complexes. An adequate description of long-range diffusion of mobile electron carriers (PQ, as well as Pc and Fd) based on the use of deterministic approach faces with very complex mathematical problems, because of sophisticated architecture of chloroplasts and non-uniform distribution of electron transport complexes. The problem becomes more complex in view of light-induced structural changes in chloroplasts (see Chapter 3). The use of the Brownian diffusion approach to simulation of interactions of mobile electron carriers with electron transport complexes may offer several advantages for modelling the long-range percolation of mobile electron carriers through the overcrowded membrane (diffusion of PQ) or within the narrow thylakoid lumen (diffusion of Pc). Taking into account realistic topology of chloroplasts, the random walk models will shed new light on regulation of photosynthetic electron transport in plants.

Significant progress in the computer science and technology inspires further development of quantum chemical and molecular dynamics simulations of photosynthetic processes (see, for example, Blomberg and Siegbahn, 2010; Postila *et al.*, 2013; Milanovsky *et al.*, 2014). Detailed analysis of electron and proton transfer reactions in the Cyt b_6f complex with the use of combined quantum chemistry/molecular mechanics (molecular dynamics) methods will clarify the mechanisms of the rate-limiting step of the intersystem electron transfer and pH-dependent control of electron transport in chloroplasts. Another interesting trend in theoretical study of oxygenic photosynthetic is associated with the development of quantum chemical models of proton-coupled electron transport reactions in PSII, including the operation of the WOC (Sproviero *et al.*, 2008, 2009).

Efficient regulation of photosynthesis is associated with the redox-dependent modulation of electron transport processes mediated by Tr reduced by PSI (for instance, activation of the FNR and CBC enzymes, triggering the state I ↔ state II transitions, activation/deactivation of CEF1). The consideration of redox-dependent regulation of photosynthetic electron transport and related metabolic cycles will be of fundamental importance for the construction of comprehensive e-models of photosynthesis.

Architectural dynamics of chloroplasts, remodelling thylakoids and protein–pigment complexes

Architectural switches in chloroplasts induced by dark-to-light transitions and variations of metabolic conditions (see Chapter 3) may have a dramatic impact on electron transport mediated by mobile carriers (diffusion of PQ in the thylakoid membrane, Pc mobility within the lumen, and Fd diffusion in the stromal volume). A further development of current models of electron and proton transport would need to update them by taking into account the light-induced remodelling of thylakoids (swelling/shrinkage of thylakoids, lateral translocation of electron transport and LHCII complexes). Appropriate consideration of structural changes in chloroplasts seems to be important task for creation of comprehensive models of photosynthetic electron transport.

Proton and ion transport in chloroplasts

Advanced models of oxygenic photosynthesis should include a module of ion transport processes. However, the non-uniform partitioning of the proton- and ion-accepting groups between the membrane and the aqueous bulk phases inside and outside the thylakoids adds complexity to realistic description of ion transport processes in chloroplasts. In particular, there may be sequestered buffering groups and different pathways for the proton sink to ATP synthases related to the so-called 'localized' circuits of proton transfer to ATP synthases (Kell, 1979; Westerhoff *et al.*, 1984; Dilley, 1991). Switches between 'delocalized' and 'localized' routes of proton transport should be incorporated into advanced models of chloroplasts. Related problem of proton transport in chloroplasts concerns the peculiarities of kinetic and thermodynamic description of energy transduction in the systems of small sizes (Blumenfeld *et al.*, 1991). Realistic modelling of the ion transport processes, which would take into account local/nonlocal mechanisms of proton transport, non-uniform distribution of proton-accepting groups, and peculiarities of energy transduction within small systems, is a serious challenge to biophysicists who would like to solve the task of creating a comprehensive model of the chloroplast.

References

Albertsson, P.A. (2001). A quantitative model of the domain structure of the photosynthetic membrane. Trends Plant Sci. *6*, 349–354.

Aliev, M.K., and Tikhonov, A.N. (2011). Obstructed metabolite diffusion within skeletal muscle cells *in silico*. Mol. Cellular Biochem. *358*, 105–119.

Alric, J. (2014). Redox and ATP control of photosynthetic cyclic electron flow in *Chlamodomonas reinhardtii*. (II) Involvement of the PGR5-PGRL1 pathway under anaerobic conditions. Biochim. Biophys. Acta *1837*, 825–834.

Andersson, B., and Anderson, J.M. (1980). Lateral heterogeneity in the distribution of chlorophyll–protein complexes of the thylakoid membranes of spinach. Biochim. Biophys. Acta *593*, 427–440.

Andersson, I. (2008). Catalysis and regulation in Rubisco. J. Exp. Botany *59*, 1555–1568.

André, M.J. (2011). Modelling $^{18}O_2$ and $^{16}O_2$ unidirectional fluxes in plants: I. Regulation of pre-industrial atmosphere. BioSystems *103*, 239–251.

Arnold, A., and Nikoloski, Z. (2011). A quantitative comparison of Calvin–Benson cycle models. Trends Plant Sci. *16*, 676–683.

Asada, K. (1999). The water–water cycle in chloroplasts: scavenging of active oxygen and dissipation of excess photons. Annu. Rev. Plant Physiol. Plant Mol. Biol. *50*, 601–639.

Assadi-Porter, F.M., and Fillingame, R.H. (1995). Proton-translocating carboxyl of subunit *c* of F_1F_o H^+-ATP synthase: the unique environment suggested by the pK_a determined by ^1H NMR. Biochemistry *34*, 16186–16193.

Badger, M.R., von Caemmerer, S., Ruuska, S., and Nakano, H. (2000). Electron flow to oxygen in higher plants and algae: rates and control of direct photoreduction (Mehler reaction) and rubisco oxygenase. Philos. Trans. R. Soc., London B *355*, 1433–1446.

Barber, J. (1976). Ionic regulation in intact chloroplasts and its effects on primary photosynthetic processes. In Topics in Photosynthesis, In: Barber, J., ed. The Intact Chloroplasts, vol. 1. (Amsterdam, Netherlands: Elsevier), pp. 88–134.

Belyaeva, N.E., Schmitt, F.-J., Steffen, R., Paschenko, V.Z., Riznichenko, G.Y., Chemeris, Y.K., Renger, G., and Rubin, A.B. (2008). PSII model-based simulations of single turnover flash-induced transients of fluorescence yield monitored within the time domain of 100 ns-10 s on dark adapted *Chlorella pyrenoidosa* cells. Photosynth. Res. *98*, 105-119.

Belyaeva, N.E., Schmitt, F.-J., Paschenko, V.Z., Riznichenko, G.Y., Rubin, A.B., and Renger, G. (2011). PSII model based analysis of transient fluorescence yield measured on whole leaves of *Arabidopsis thaliana* after excitation with light flashes of different energies. BioSystems *103*, 188–195.

Bendall, D.S., and Manasse, R.S. (1995). Cyclic photophosphorylation and electron transport. Biochim. Biophys. Acta *1229*, 23–38.

Berry, S., and Rumberg, B. (2000). Kinetic modeling of the photosynthetic electron transport. Bioelectrochemistry *53*, 35–53.

Blackwell, M.F., and Whitmarsh, J. (1990). Effect of integral membrane proteins on the lateral mobility of plastoquinone in phosphatidylcholine proteoliposomes. Biophys. J. *28*, 1259–1271.

Blackwell, M., Gibas, C., Gygax, S., Roman, D., and Wagner, B. (1994). The plastoquinone diffusion coefficient in chloroplasts and its mechanistic implications. Biochim. Biophys. Acta *1183*, 533–543.

Blomberg, M.R.A., and Siegbahn, P.E.M. (2010). Quantum chemistry as a tool in bioenergetics. Biochim. Biophys. Acta *1797*, 129–142.

Blumenfeld, L.A., Grosberg, A.Y., and Tikhonov, A.N. (1991). Fluctuations and mass action law breakdown in statistical thermodynamics of small systems. J. Chem. Phys. *95*, 7541–7547.

Blumenfeld, L.A., and Tikhonov, A.N. (1994). Biophysical Thermodynamics of Intracellular Processes. Molecular Machines of the Living Cell. (New York, USA: Springer).

Boyer, P.D. (1997). The ATP synthase – a splendid molecular machine. Annu. Rev. Biochem. *66*, 717–749.

Brettel, K. (1997). Electron transfer and arrangement of the redox cofactors in photosystem I. Biochim. Biophys. Acta *1318*, 322–373.

Breyton, C. (2000). Conformational changes in the cytochrome b_6f complex induced by inhibitor binding. J. Biol. Chem. *275*, 13195-13201.

Buchanan, B.B. (1980). Role of light in the regulation of chloroplast enzymes. Annu. Rev. Plant Physiol. *31*, 341–374.

Buchanan, B.B. (1991). Regulation of CO_2 assimilation in oxygenic photosynthesis: the ferredoxin/thioredoxin system. Perspective on its discovery, present status, and future development. Arch. Biochem. Biophys. *288*, 1–9.

Chance, B., and Williams, G.R. (1956). The respiratory chain and oxidative phosphorylation. Adv. Enzymol. *17*, 65–134.

Cherepanov, D., A., Mulkidjanian, A., Y., and Junge, W. (1999). Transient accumulation of elastic energy in proton translocating ATP synthase. FEBS Lett. *449*, 1–6.

Chernavskaya, N.M., and Chernavskii, D.S. (1961). Periodic phenomena in photosynthesis. Sov. Phys. Uspekhi *4*, 850–865.

Cramer, W.A., Zhang, H., Yan, J., Kurisu, G., and Smith, J.L. (2006). Transmembrane traffic in the cytochrome b_6f complex. Annu. Rev. Biochem. *75*, 769-790.

Cramer, W.A., Hasan, S.S., and Yamashita, E. (2011). The Q cycle of cytochrome *bc* complexes: a structure perspectives. Biochim. Biophys. Acta *1807*, 788-802.

Crofts, A.R. (2004). Proton-coupled electron transfer at the Q_o-site of the bc_1 complex controls the rate of ubihydroquinone oxidation. Biochim. Biophys. Acta *1655*, 77–92.

Crofts, A.R., Hong, S., Wilson, C., Burton, R., Victoria, D., Harrison, C., and Schulten, K. (2013). The mechanism of ubihydroquinone oxidation at the Q_o-site of the cytochrome bc_1 complex. Biochim. Biophys. Acta *1827*, 1362–1377.

Cruz, J.A., Kanazawa, A., Treff, N., and Kramer, D.M. (2005). Storage of light-driven transthylakoid proton motive force as an electric field ($\Delta\psi$) under steady-state conditions in intact cells of *Chlamydomonas reinhardtii*. Photosynth. Res. *85*, 221–233.

de Vitry, C., Ouyang, Y., Finazzi, G., Wollman, F.-A., and Kallas, T. (2004). The chloroplast Rieske iron–sulfur protein: at the crossroad of electron transport and signal transduction. J. Biol. Chem. *279*, 44621-44627.

Dekker, J.P., and Boekema, E.J. (2005). Supramolecular organization of thylakoid membrane proteins in green plants. Biochim. Biophys. Acta *1706*, 12–39.

Demmig-Adams, B., Cohu, C.M., Muller, O., and Adams, W.W. (2012). Modulation of photosynthetic energy conversion efficiency in nature: from seconds to seasons. Photosynth. Res. *113*, 75–88.

Dietz, K.-J., and Pfannschmidt, T. (2011). Novel regulators in photosynthetic redox control of plant metabolism and gene expression. Plant Physiol. *155*, 1477–1485.

Dilley, R.A. (1991). Energy coupling in chloroplasts: a calcium-gated switch controls proton fluxes between localized and delocalized proton gradients. Curr. Topics Bioenerg. *16*, 265–318.

Drepper, F., Carlberg, I., Andersson, B., and Haehnel, W. (1993). Lateral diffusion of an integral membrane protein: Monte Carlo analysis of the migration of phosphorylated light-harvesting complex II in the thylakoid membrane. Biochemistry *32*, 11915–11922.

Dubinsky, A. Yu., and Tikhonov, A.N. (1995). Mathematical simulation of the light-induced uptake of protons by chloroplasts upon various mechanisms of proton leak through the thylakoid membrane. Biophysics *40*, 365–371.

Dubinsky, A. Yu., and Tikhonov, A.N. (1997). Mathematical model of thylakoid as the distributed heterogeneous system of electron and proton transport. Biophysics *42*, 644–660.

Dubinsky, A. Yu., Ivlev, A.A., and Igamberdiev, A.U. (2010). Theoretical analysis of the possibility of existence of oscillations in photosynthesis. Biophysics *55*, 55–58.

Edwards, G., and Walker, D.A. (1983). C_3, C_4: Mechanisms, and Cellular and Environmental Regulation of Photosynthesis. (Oxford, UK: Blackwell).

Elston, T., Wang, H., and Oster, G. (1998). Energy transduction in ATP synthase. Nature *391*, 510–513.

Farazdaghi, H. (2011). The single-process biochemical reaction of Rubisco: A unified theory and model with the effects of irradiance, CO_2 and rate-limiting step on the kinetics of C_3 and C_4 photosynthesis from gas exchange. BioSystems *103*, 265–284.

Farquhar, G.D., von Caemmerer, S., and Berry, J.A. (1980). A biochemical model of photosynthetic CO_2 assimilation in leaves of C_3 plants. Planta *149*, 78–90.

Fridlyand, L.E., and Scheibe, R. (1999). Regulation of the Calvin cycle for CO_2 fixation as an example for general control mechanisms in metabolic cycles. BioSystems *51*, 79–93.

Frolov, A.E., and Tikhonov, A.N. (2007). Influence of light-induced changes in stromal and luminal pH on electron transport kinetics in chloroplasts: mathematical modeling. Biophysics *52*, 398–405.

Frolov, A.E., and Tikhonov, A.N. (2009). The oxidation of plastoquinol by the cytochrome b_6f complex: a density functional theory study. Russian J. Phys. Chem. A, *83*, 506-508.

Gardemann, A., Schimkat, D., and Heldt, H.W. (1986). Control of CO_2 fixation regulation of stromal fructose-1,6-bisphosphatase in spinach by pH and Mg^{2+} concentration. Planta *168*, 536–545.

Golding, A.J., Joliot, P., and Johnson, G.N. (2005). Equilibration between cytochrome *f* and P700 in intact leaves. Biochim. Biophys. Acta *1706*, 105–109.

Gräber, P. (1982). Phosphorylation in chloroplasts: ATP synthesis driven by $\Delta\psi$ and by ΔpH of artificial or light-generated origin. Curr. Topics Membr. Transp. *16*, 215–245.

Haddadian, E.J., and Gross, E.L. (2006). A Brownian dynamics study of the interaction of the luminal domains of the cytochrome b_6f complex with plastocyanin and cytochrome c_6: the effects of the Rieske FeS protein on the interactions. Biophys. J. *91*, 2589–2600.

Haehnel, W. (1976a). The reduction kinetics chlorophyll a_I as indicator for proton uptake between the light reactions in chloroplasts. Biochim. Biophys. Acta *440*, 506–521.

Haehnel, W. (1976b). The ratio of the two light reactions and their coupling in chloroplasts. Biochim. Biophys. Acta *423*, 499–509.

Haehnel, W. (1984). Photosynthetic electron transport in higher plants. Annu. Rev. Plant Physiol. Plant Mol. Biol. *35*, 659–693.

Hahn, B. (1984). A mathematical model of leaf Calvin cycle. Annals of Botany *54*, 325–339.

Haraux, F., Sigalat, C., Moreau, A., and de Kouchkovsky, Y. (1983). The efficiency of energized protons for ATP synthesis depends on the membrane topography in thylakoids. FEBS Lett. *155*, 248-252.

Harbinson, J., and Hedley, C.L. (1989). The kinetics of P-700$^+$ reduction in leaves: a novel *in situ* probe of thylakoid functioning. Plant, Cell and Environ. *12*, 357–369.

Harbinson, J., and Hedley, C.L. (1993). Changes in P-700 oxidation during the early stages of the induction of photosynthesis. Plant Physiol. *103*, 649-660.

Hasan, S.S., Stofleth, J.T., Yamashita, E., and Cramer, W.A. (2013). Lipid-induced conformational changes within the cytochrome b_6f complex of oxygenic photosynthesis. Biochemistry 52, 2649-2654.

Heber, U. (2002). Irrungen, wirrungen? The Mehler reaction in relation to cyclic electron transport in C3 plants. Photosynth. Res. 73, 223–231.

Heldt, H.W., Werdan, K., Milivancev, M., and Geller, G. (1973). Alkalization of the chloroplast stroma caused by light-dependent proton flux into the thylakoid space. Biochim. Biophys. Acta *314*, 224–241.

Hofmann, N.R. (2012). A refined model of state transitions in plant thylakoid membranes. Plant Cell *24*, 2708.

Hope, A.B. (2000). Electron transfers amongst cytochrome *f*, plastocyanin and photosystem I: kinetics and mechanisms. Biochim. Biophys. Acta *1456*, 5–26.

Horton, P. (2012). Optimization of light harvesting and photoprotection: molecular mechanisms and physiological consequences. Phil. Trans. R. Soc. B *367*, 3455-3465.

Igamberdiev, A.U., ed. (2011). Computational models of photosynthesis. BioSystems *103*, pp. 113–314.

Igamberdiev, A.U., and Lea, P.J. (2006). Land plants equilibrate O_2 and CO_2 concentrations in the atmosphere. Photosynth. Res. *87*, 177–194.

Igamberdiev, A.U., and Kleczkowski, L.A. (2011). Optimization of CO_2 fixation in photosynthetic cells via thermodynamic buffering. BioSystems *103*, 224–229.

Iwai, M., Takizawa, K., Tokutsu, R., Okamuro, A., Takahashi, Y., and Minagawa, J. (2010). Isolation of the elusive supercomplex that drives cyclic electron flow in photosynthesis. Nature *464*, 1210–1214.

Järvi, S., Gollan, P.J., and Aro, E.-M. (2013). Understanding the roles of the thylakoid lumen in photosynthetic regulation. Front. Plant Sci. 4, Article 434. doi: 10.3389/fpls.2013.00434.

Johnson, G.N. (2011). Physiology of PSI cyclic electron transport in higher plants. Biochim. Biophys. Acta *1807*, 906–911.

Johnson, M.P., and Ruban, A.V. (2014). Rethinking the existence of a steady-state $\Delta\psi$ component of the proton motive force across plant thylakoid membranes. Photosynth. Res. *119*, 233–242.

Junesche, U., and Gräber, P. (1991). The rate of ATP synthesis as a function of ΔpH and $\Delta\psi$ catalysed by the active, reduced _H$^+$-ATPase from chloroplasts. FEBS Lett. *294*, 275–278.

Junge, W., and Polle, A. (1986). Theory of proton flow along appressed thylakoid membranes under both non-stationary and stationary conditions. Biochim. Biophys. Acta *848*, 265–273.

Junge, W., and McLaughlin, S. (1987). The role of fixed and mobile buffers in the kinetics of proton movement. Biochim. Biophys. Acta *890*, 1–5.

Kaiser, E., Morales, A., Harbinson, J., Kromdijk, J., Heuvelink, E., and Marcelis, L.F.M. (2015). Dynamic photosynthesis in different environmental conditions. J. Exp. Botany *66*, 2415–2426.

Kaňa, R. (2013). Mobility of photosynthetic proteins. Photosynth. Res. *116*, 465–479.

Kara-Ivanov, M. (1983). Nonuniform lateral distribution of transmembrane ΔpH in a coupling membrane as related to its curvature. J. Bioenerg. Biomembr. *15*, 111–119.

Karavaev, V.A., and Kukushkin, A.K. (1993). Theoretical model of the light and dark stages of photosynthesis: the regulation problem. Biophysics *38*, 958–975.

Kell, D. (1979). On the functional proton current pathway of electron transport phosphorylation. Biochim. Biophys. Acta *549*, 55–99.

Kirchhoff, H. (2008). Significance of protein crowding, order and mobility for photosynthetic membrane functions. Biochem. Soc. Trans. *36*, 967–970.

Kirchhoff, H. (2014). Diffusion of molecules and macromolecules in thylakoid membranes. Biochim. Biophys. Acta *1837*, 495–502.

Kirchhoff, H., Horstmann, S., and Weis, E. (2000). Control of the photosynthetic electron transport by PQ diffusion in microdomains in thylakoids of higher plants. Biochim. Biophys. Acta *1459*, 148–168.

Kirchhoff, H., Mukherjee, U., and Galla, H.J. (2002). Molecular architecture of the thylakoid membrane: lipid diffusion space for plastoquinone. Biochemistry *41*, 4872–4882.

Kirchhoff, H., Tremmel, I., Haase, W., and Kubitscheck, U. (2004). Supramolecular photosystem II organization in grana thylakoid membranes: evidence for a structured arrangement. Biochemistry 43, 9204–9213.

Kirchhoff, H., Haferkamp, S., Allen, J.F., Epstein, D.B.A., and Mulineaux, C.W. (2008). Protein diffusion and macromolecular crowding in thylakoid membranes. Plant Physiol. *146*, 1571–1578.

Kirchhoff, H., Hall, C., Wood, M., Herbstová, M., Tsabari, O., Nevo, R., Charuvi, D., Shimoni, E., and Reich, Z. (2011). Dynamic control of protein diffusion within the granal thylakoid lumen. Proc. Natl. Acad. Sci. U.S.A. *108*, 20248–20253.

Kobayasi, Y., Inoue, Y., Shibata, K., and Heber, U. (1979a). Control of electron flow in intact chloroplasts by the intrathylakoid pH, not by phosphorylation potential. Planta *146*, 481–486.

Kobayasi, Y., Inoue, Y., Furuya, F., Shibata, K., and Heber, U. (1979b). Regulation of adenylate level in intact spinach chloroplasts. Planta *147*, 69–75.

Kouril, R., Zygadlo, A., Arteni, A.A., DeWit, C.D., Dekker, J.P., Jensen, P.E., Scheller, H.V., and Boekema, E.J. (2005). Structural characterization of a complex of photosystem I and light-harvesting complex II of *Arabidopsis thaliana*. Biochemistry *44*, 10935–10940.

Kovalenko, I.B., Abaturova, A.M., Riznichenko, G.Y., and Rubin, A.B. (2011). Computer simulation of interaction of photosystem 1 with plastocyanin and ferredoxin. BioSystems *103*, 180–187.

Kramer, D.M., Sacksteder, C.A., and Cruz, J.A. (1999). How acidic is the lumen? Photosynth. Res. *60*, 151–163.

Kramer, D.M., Sacksteder, C.A., and Cruz, J.A. (2003). Balancing the central roles of the thylakoid proton gradient. Trends Plant Sci. *8*, 27–32.

Kramer, D.M., Avenson, T.J., and Edwards, G.E. (2004). Dynamic flexibility in the light reactions of photosynthesis governed by both electron and proton transfer reactions. Trends Plant. Sci. *9*, 349–357.

Kukushkin, A.K., Tikhonov, A.N., Blumenfeld, L.A., and Ruuge, E.K. (1973). Theoretical investigation of primary processes of photosynthesis of higher plants and algae. Doklady Akademii Nauk USSR *211*, 718-721.

Kurisu, G., Zhang, H., Smith, J.L., and Cramer, W.A. (2003). Structure of the cytochrome b_6f complex of oxygenic photosynthesis: tuning the cavity. Science *302*, 1009-1014.

Kuvykin, I.V., Vershubskii, A.V., Ptushenko, V.V., and Tikhonov, A.N. (2008). Oxygen as an alternative acceptor in the photosynthetic electron transfer chain in C_3 plants. Biochemistry (Moscow) *73*, 1063–1075.

Kuvykin, I.V., Vershubskii, A.V., Priklonskii, V.I., and Tikhonov, A.N. (2009). Computer simulation study of pH-dependent regulation of electron transport in chloroplasts. Biophysics *54*, 455-464.

Kuvykin, I.V., Ptushenko, V.V., Vershubskii, A.V., and Tikhonov, A.N. (2011). Regulation of electron transport in C_3 plant chloroplasts *in situ* and *in silico*. Short-term effects of atmospheric CO_2 and O_2. Biochim. Biophys. Acta *1807*, 336–347.

Laisk, A., and Walker, D.A. (1986). Control of phosphate turnover as a rate-limiting factor and possible cause of oscillations in photosynthesis: a mathematical model. Proc. R. Soc. Lond. B Biol. Sci. *227*, 281–302.

Laisk, A., Eichelmann, H., Oja, V., and Peterson, R.B. (2005). Control of cytochrome b_6f at low and high light intensity and cyclic electron transport in leaves. Biochim. Biophys. Acta *1708*, 79–90.

Laisk, A., Eichelmann, Y., and Oja, V. (2006). C_3 photosynthesis *in silico*. Photosynth. Res. *90*, 45–66.

Laisk, A., Eichelmann, Y., and Oja, V. (2009). Leaf C_3 photosynthesis *in silico*: Integrated carbon/nitrogen metabolism. In Photosynthesis *in silico*: Understanding Complexity from Molecules to Ecosystems, A. Laisk, L. Nedbal and Govindjee, eds. (Dordrecht: Springer), pp. 295–322.

Laisk, A., Eichelmann, H., Oja, V., Eatherall, A., and Walker, D.A. (1989). A mathematical model of the carbon metabolism in photosynthesis. Difficulties in explaining oscillations by fructose 2,6-bisphosphate regulation. Proc. R. Soc. Lond. B 237, 389–415.

Laisk, A., Nedbal, N., and Govindjee, eds. (2009). Photosynthesis *in silico*. Understanding complexity from molecules to ecosystems (Dordrecht: Springer).

Laisk, A., Oja, V., Eichelmann, H., and Dall'Osto, L. (2014). Action spectra of photosystems II and I and quantum yield of photosynthesis in leaves in State 1. Biochim. Biophys. Acta *1837*, 315-325.

Laverne, J. (2009). Clustering of electron transport components: kinetic and thermodynamic consequences. In Photosynthesis *in silico*: Understanding Complexity from Molecules to Ecosystems, A. Laisk, L. Nedbal and Govindjee, eds. (Dordrecht: Springer), pp. 177–205.

Lavergne, J., and Joliot, P. (1991). Restricted diffusion in photosynthetic membranes. Trends Biochem. Sci. *16*, 129–134.

Lavergne, J., Bouchaud, J.-P., and Joliot, P. (1992). Plastoquinone compartmentation in chloroplasts. II. Theoretical aspects. Biochim. Biophys. Acta *1101*, 13–22.

Lazar, D. (2003). Chlorophyll *a* fluorescence rise induced by high light illumination of dark-adapted plant tissue studied by means of a model of photosystem II and considering photosystem II heterogeneity. J. Theor. Biol. *220*, 469–503.

Lazar, D., and Schansker, G. (2009). Models of chlorophyll a fluorescence transients. In Photosynthesis *in silico*: Understanding Complexity from Molecules to Ecosystems, A. Laisk, L. Nedbal and Govindjee, eds. (Springer, Dordrecht), pp. 85–123.

Lazar, D. (2009). Modelling of light-induced chlorophyll *a* fluorescence rise (O-J-I-P transient) and changes in 820 nm-transmittance signal of photosynthesis. Photosynthetica *47*, 483–498.

Lemeille, S., and Rochaix, J.-D. (2010). State transitions at the crossroad of thylakoid signaling pathways. Photosyn. Res. *106*, 33–46.

Li, Z., Wakao, S., Fischer, B.B., and Niyogi, K.K. (2009). Sensing and responding to excess light. Annu. Rev. Plant. Biol. *60*, 239–260.

Masarova, M., and Tikhonov, A.N. (1989). Effects of the intrathylakoid buffer capacity on the light-induced uptake of protons and photophosphorylaton rate in chloroplasts. Biophysics *34*, 142–143.

Mehler, A.H. (1951). Studies on reactions of illuminated chloroplasts. I. Mechanisms of the reduction of oxygen and other Hill reagents. Arch. Biochem. Biophys. *33*, 65–77.

Michelet, L., Zaffagnini, M., Morisse, S., Sparla, F., Pérez-Pérez, M.E., Francia, F., Danon, A., Marchand, C.H., Fermani, S., Trost, P., and Lemaire, S.D. (2013). Redox regulation of the Calvin–Benson cycle: something old, something new. Front. Plant Sci. *4*, Article 470. doi: 10.3389/fpls.2013.00470.

Milanovsky, G.E., Ptushenko, V.V., Golbeck, J.H., Semenov, A.Y., and Cherepanov, D.A. (2014). Molecular dynamics study of the primary charge separation reactions in Photosystem I: effect of the replacement of the axial ligands to the electron acceptor A_0. Biochim. Biophys. Acta *1837*, 1472–1483.

Minagawa, J. (2011). State transitions – the molecular remodeling of photosynthetic supercomplexes that controls energy flow in the chloroplast. Biochim. Biophys. Acta *1807*, 897–905.

Mitchell, P. (1966). Chemiosmotic coupling in oxidative and photosynthetic phosphorylation. Biol. Rev. *41*, 445–502.

Mitchell, P. (1976). Possible molecular mechanisms of the protonmotive function of cytochrome systems. J. Theor. Biol. *62*, 327–367.

Mitchell, R., Spillmann, A., and Haehnel, W. (1990). Plastoquinol diffusion in linear photosynthetic electron transport. Biophys. J. *58*, 1011–1024.

Motohashi, K., and Hisabori, T. (2006). HCF164 receives reducing equivalents from stromal thioredoxin across the thylakoid membrane and mediates reduction of target proteins in the thylakoid lumen. J. Biol. Chem. *281*, 35039–35047.

Mott, K.A., and Berry, J.A. (1986). Effects of pH on activity and activation of ribulose 1,5-biphosphate carboxylase at air level CO_2. Plant Physiol. *82*, 77–82.

Murphy, D.J. (1986). The molecular organization of the photosynthetic membranes of higher plants. Biochim. Biophys. Acta *864*, 33–94.

Munekage, Y., Hojo, M., Meurer, J., Endo, T., Tasaka, M., and Shikanai, T. (2002). PGR5 is involved in cyclic electron flow around photosystem I and is essential for photoprotection in *Arabidopsis*. Cell *110*, 361–371.

Nartsissov, Y.R., and Mashkovtseva, E.V. (2006). Application of rigid body mechanics to theoretical description of rotation within F_0F_1-ATP synthase. J. Theor. Biol. *242*, 300–308.

Nedbal, L., Červený, J., Rascher, U., and Schmidt, H. (2007a). E-photosynthesis: a comprehensive modeling approach to understand chlorophyll fluorescence transients and other complex dynamic features of photosynthesis in fluctuating light. Photosynth. Res. *93*, 223–234.

Nedbal, L., Červený, J., and Schmidt, H. (2007b). Scaling and integration of kinetic models of photosynthesis: Towards comprehensive e-photosynthesis. In: A. Laisk, L. Nedbal and Govindjee, eds. In Photosynthesis *in silico*: Understanding Complexity from Molecules to Ecosystems, pp. 17–29.

Nelson, D.L., and Cox, M.M. (2012). Lehninger Principles of Biochemistry, Sixth Edition. (Freeman, W. H. & Company).

Nesbitt, D.M., and Berg, S.P. (1982). The influence of spinach thylakoid lumen volume and membrane proximity on the rotational motion of spin label tempoamine. Biochim. Biophys. Acta *679*, 169–174.

Nishio, J.N., and Whitmarsh, J. (1993). Dissipation of the proton electrochemical potential in intact chloroplasts II. The pH gradient monitored by cytochrome *f* reduction kinetics. Plant Physiol. *101*, 89–96.

Oster, G., and Wang, H. (2000). Reverse engineering a protein: the mechanochemistry of ATP synthase. Biochim. Biophys. Acta *1458*, 482–510.

Page, C.C., Moser, C.C., Chen, X., and Dutton, P.L. (1999). Natural engineering principles of electron tunnelling in biological oxidation-reduction. Nature *402*, 47–52.

Paillotin, G., Geacintov, N., E., and Breton, J. (1983). A master equation theory of fluorescence induction, photochemical yield, and singlet-triplet exciton quenching in photosynthetic systems. Biophys J. *44*, 65–77.

Pänke, O., and Rumberg, B. (1999). Kinetic modeling of rotary CF_0F_1-ATP synthase: storage of elastic energy during energy transduction. Biochim. Biophys. Acta *1412*, 118–128.

Pänke, O., Cherepanov, D.A., Gumbiowski, K., Engelbrecht, S., and Junge, W. (2001). Viscoelastic dynamics of actin filaments coupled to rotary F-ATPase: torque profile of the enzyme. Biophys. J. *81*, 1220–1233.

Pettersson, G., and Ryde-Pettersson, U. (1988). A mathematical model of the Calvin photosynthesis cycle. Eur. J. Biochem. *175*, 661–672.

Pfündel, E.E., and Dilley, R.A. (1993). The pH dependence of violaxanthin deepoxidation in isolated pea chloroplasts. Plant Physiol. *101*, 65–71.

Pogoryelov, D., Reichen, C., Klyszejko, A.L., Brunisholz, R., Muller, D.J., Dimroth, P., and Meier, T. (2007). The oligomeric state of *c* rings from cyanobacterial F-ATP synthases varies from 13 to 15. J. Bacteriol. *189*, 5895–5902.

Pogrebnaya, A., Romanovsky, Y., and Tikhonov, A. (2005). Rotation of F_1-ATPase: the stochastic model. Fluct. Noise Lett. *5*, L217-L224.

Polle, A., and Junge, W. (1986). The slow rise of the flash-light-induced alkalization by photosystem II of the suspending medium of thylakoids is reversibly related to thylakoid stacking. Biochim. Biophys. Acta *848*, 257–264.

Polle, A., and Junge, W. (1989). Proton diffusion along the membrane surface of thylakoids is not enhanced over that in bulk water. Biophys. J. *56*, 27–31.

Ponamarev, M.V., and Cramer, W.A. (1998). Perturbation of the internal water chain in cytochrome *f* of oxygenic photosynthesis: loss of concerted reduction of cytochromes *f* and *b*. Biochemistry *37*, 17199-17208.

Postila, P.A., Kaszuba, K., Sarewicz, M., Osyczka, A., Vattulainen, I., and Róg, T. (2013). Key role of water in proton transfer at the Q_o-site of the cytochrome bc_1 complex predicted by atomistic molecular dynamics simulations. Biochim. Biophys. Acta *1827*, 761–768.

Riznichenko, G.Y., Kovalenko, I.B., Abaturova, A.M., Diakonova, A.N., Ustinin, D.M., Grachev, E.A., and Rubin, A.B. (2010). New direct dynamic models of protein interactions coupled to photosynthetic electron transport reactions. Biophys. Rev. *2*, 101–110.

Robinson, S.P. (1985). The involvement of stromal ATP in maintaining the pH gradient across the chloroplast envelope in the light. Biochim. Biophys. Acta *806*, 187–194.

Romanovsky, Y.M., and Tikhonov, A.N. (2010). Molecular energy transducers of the living cell. Proton ATP synthase: a rotating molecular motor. Physics-Uspekhi *53*, 893–914.

Rovers, W., and Giersch, C. (1995). Photosynthetic oscillations and the interdependence of photophosphorylation and electron transport as studied by a mathematical model. BioSystems *35*, 63–73.

Ryzhikov, S.B., and Tikhonov, A.N. (1988). Regulation of electron transfer in photosynthetic membranes of higher plants. Biophysics *33*, 642-646.

Ruban, A. (2012). The Photosynthetic Membrane: Molecular Mechanisms and Biophysics of Light Harvesting (Oxford, UK: Wiley-Blackwell).

Rubin, A., and Riznichenko, G. (2009). Modeling of the primary processes in a photosynthetic membrane. In Photosynthesis *in silico*: Understanding Complexity from Molecules to Ecosystems. A. Laisk, L. Nedbal and Govindjee, eds. (Dordrecht, The Netherlands: Springer) pp. 151–176.

Rubin, A., and Riznichenko, G. (2014). Mathematical Biophysics. Series: Biological and Medical Physics, Biomedical Engineering, XV.

Saxton, M.J. (1987). Lateral diffusion in an archipelago. The effect of mobile obstacles. Biophys. J. *52*, 989–997.

Saxton, M.J. (1989). Lateral diffusion in an archipelago. Distance dependence of the diffusion coefficient. Biophys. J. *56*, 615–622.

Scheibe, R. (2004). Malate valves to balance cellular energy supply. Physiol. Plant. *120*, 21–26.

Schneider, D., Berry, S., Rich, P., Seidler, A., and Rögner, M. (2001). A regulatory role of the PetM subunit in a cyanobacterial b_6f complex. J. Biol. Chem. *276*, 16780-16785.

Schöttler, M.A., Tóth, S.Z., Boulouis, A., and Kahlau, S. (2015). Photosynthetic complex stoichiometry dynamics in higher plants: biogenesis, function, and turnover of ATP synthase and the cytochrome b_6f complex. J. Exp. Bot. *66*, 2373–2400.

Schreiber, U., Klughammer, C., and Neubauer, C. (1988). Measuring P700 absorbance changes around 830 nm with a new type of pulse modulation system. Z. Naturforsch. *43c*, 686–698.

Sharkey, T.D., Bernacchi, C.J., Farquhar, G.D., and Singsaas, E.L. (2007). Fitting photosynthetic carbon dioxide response curves for C_3 leaves. Plant Cell Environ. *30*, 1035–1040.

Shikanai, T., 2007. Cyclic electron transport around photosystem I: Genetic approaches. Annu. Rev. Plant Biol. *58*, 199–217.

Shinkarev, V.P., and Venedictov, P.S. (1977). Probability description of the processes of electron transport in multimolecular complexes. Biophysics *22*, 413–418.

Shinkarev, V.P. (1998). The general kinetic model of electron transfer in photosynthetic reaction centers activated by multiple flashes. Photochem. Photobiol. *67*, 683–699.

Shorten, P.R., and Sneyd, J. (2009). A mathematical analysis of obstructed diffusion within skeletal muscle. Biophys. J. *96*, 4764–4778.

Siggel, U., Renger, G., Stiehl, H.H., and Rumberg, B. (1972). Evidence for electronic and ionic interaction between electron transport chains in chloroplasts. Biochim. Biophys. Acta *256*, 328–335.

Singh, N., and Brumer, P. (2012). Efficient computational approach to the non-Markovian second order quantum master equation: electronic energy transfer in model photosynthetic systems. Molecular Physics *110*, 1815–1828.

Sivak, M.N., and Walker, D.A. (1987). Oscillations and other symptoms of limitation of *in vivo* photosynthesis by inadequate phosphate supply to the chloroplast. Plant Physiol. Biochem. *25*, 635-648.

Skulachev, V.P., Bogachev, A.V., and Kasparinsky, F.O. (2013). Principles of Bioenergetics (Heidelberg: Springer-Verlag).

Soriano, G.M., Ponamarev, M.V., Tae, G.-S., and Cramer, W.A. (1996). Effect of the interdomain basic region of cytochrome *f* on its redox reactions *in vivo*. Biochemistry *35*, 14590-14598.

Sproviero, E.M., Gascón, J.A., McEvoy, J.P., Brudvig, G.W., Batista, E.R., and Batista, V.S. (2008). Computational studies of the O_2-evolving complex of photosystem II and biomimetic oxomanganese complexes. Coord. Chem. Rev. *252*, 395–415.

Sproviero, E.M., Newcomer, M.B., Gascón, J.A., Batista, E.R., Brudvig, G.W., and Batista, V.S. (2009). The MoD-QM/MM methodology for structural refinement of photosystem II and other biological macromolecules. Photosynth. Res. *102*, 455–470.

Stiehl, H.H., and Witt, H.T. (1968). Die kurzzeitigen ultravioletten Differenzspektren bei der Photosynthese. Z. Naturforsch. *23b*, 220–224.

Stiehl, H.H., and Witt, H.T. (1969). Quantitative treatment of the function of plastoquinone in photosynthesis. Z. Naturforsch. *24b*, 1588–1598.

Stirbet, A., Govindjee, Strasser, B.J., and Strasser, R. (1998). Chlorophyll *a* fluorescence induction in higher plants: modelling and numerical simulation. J. Theor. Biol. *193*, 131–151.

Stroebel, D., Choquet, Y., Popot, J.-L., and Picot, D. (2003). An atypical heam in the cytochrome b_6f complex. Nature *426*, 413–418.

Takano, M., Takahashi, M.-A., and Asada, K. (1982). Reduction of photosystem I reaction center, P700, by plastocyanin in stroma thylakoids from spinach: lateral diffusion of plastocyanin. Arch. Biochem. Biophys. *218*, 369–375.

Tikhonov, A.N. (2012). Energetic and regulatory role of proton potential in chloroplasts. Biochemistry (Moscow) *77*, 956–974.

Tikhonov, A.N. (2013). pH-Dependent regulation of electron transport and ATP synthesis in chloroplasts. Photosynth. Res. *116*, 511-534.

Tikhonov, A.N. (2014). The cytochrome b_6f complex at the crossroad of photosynthetic electron transport pathways. Plant Physiol. Biochem. *81*, 163–183.

Tikhonov, A.N. (2015). Induction events and short-term regulation of electron transport in chloroplasts: an overview. Photosynth. Res. *125*, 65–94.

Tikhonov, A.N., and Blumenfeld, L.A. (1985). Concentration of hydrogen ions inside the subcellular particles: physical meaning and determination methods. Biophysics *30*, 527–537.

Tikhonov, A.N., and Vershubskii, A.V. (2014). Computer modeling of electron and proton transport in chloroplasts. BioSystems *121*, 1-21.

Tikhonov, A.N., Khomutov, G.B., and Ruuge, E.K. (1984). Electron transport control in chloroplasts. Effects of magnesium ions on the electron flow between two photosystems. Photobiochem. Photobiophys. *8*, 261–269.

Tikkanen, M., and Aro, E.-M. (2012). Thylakoid protein phosphorylation in dynamic regulation of photosystem II in higher plants. Biochim. Biophys. Acta *1817*, 232–238.

Tikkanen, M., and Aro, E.-M. 2014. Integrative regulatory network of plant thylakoid energy transduction. Trends Plant Sci. *19*, 10–17.

Tikkanen, M., Grieco, M., and Aro, E.-M. (2011). Novel insights into plant light-harvesting complex II phosphorylation and 'state transitions'. Trends Plant Sci. *16*, 126–131.

Tremmel, I.G., Kirchhoff, H., Weis, E., and Farquhar, G.D. (2003). Dependence of plastoquinol diffusion on the shape, size, and density of integral thylakoid proteins. Biochim. Biophys. Acta *1607*, 97–109.

Trubitsin, B.V., and Tikhonov, A.N. (2003). Determination of a transmembrane pH difference in chloroplasts with a spin label Tempamine. J. Magn. Reson. *163*, 257–269.

Turner, J.E., and Brittain, E.G. (1962). Oxygen as a factor of photosynthesis. Biol. Rev. *37*, 130–170.

Velthuys, B.R. (1979). Electon flow through plastoquinone and cytochromes b_6 and f in chloroplasts. Proc. Natl. Acad. Sci. U.S.A. *76*, 2765-2769.

Vershubskii, A.V., Priklonskii, V.I., and Tikhonov, A.N. (2001). Electron and proton transport in chloroplasts: a mathematical model constructed with regard for the lateral heterogeniety of thylakoids. Biophysics *46*, 448–457.

Vershubskii, A.V., Priklonskii, V.I., and Tikhonov, A.N. (2004a). Mathematical modeling of electron and proton transport coupled with ATP synthesis in chloroplasts. Biophysics *49*, 52–65.

Vershubskii, A.V., Priklonskii, V.I., and Tikhonov, A.N. (2004b). Effects of diffusion and topological factors on the efficiency of energy coupling in chloroplasts with heterogeneous partitioning of protein complexes in thylakoids of grana and stroma. A mathematical model. Biochemistry (Moscow) *69*, 1016–1024.

Vershubskii, A.V., Kuvykin, I.V., Priklonskii, V.I., and Tikhonov, A.N. (2011). Functional and topological aspects of pH-dependent regulation of electron and proton transport in chloroplasts *in silico*. BioSystems *103*, 164–179.

Vollmar, M., Schlieper, D., Winn, M., Büchner, C., and Groth, G. (2009). Structure of the c_{14} rotor ring of the proton translocating chloroplast ATP synthase. J. Biol. Chem. *284*, 18228–18235.

von Caemmerer, S., Farquhaer, G., and Berry, J. (2009). Biochemical model of C_3 photosynthesis. In Photosynthesis *in silico*: Understanding Complexity from Molecules to Ecosystems. A. Laisk, L. Nedbal, and Govindjee, eds. (Dordrechr, The Netherlands: Springer), pp. 209–230.

Vredenberg, W., and Prášil, O. (2009). Modeling of chlorophyll *a* fluorescence kinetics in plant cells: derivation of a descriptive algorithm. In Photosynthesis *in silico*: Understanding Complexity from Molecules to Ecosystems, A. Laisk, L. Nedbal, and Govindjee, eds. (Dordrecht: Springer), pp. 125–149.

Walker, D. (1992). Concerning oscillations. Photosynth. Res. *34*, 387–395.

Wang, H., and Oster, G. (1998). Energy transduction in the F_1 motor of ATP synthase. Nature *396*, 279–282.

Werdan, K., Heldt, H.W., and Milovancev, M. (1975). The role of pH in the regulation of carbon fixation in the chloroplast stroma. Studies on CO_2 fixation in the light and dark. Biochim. Biophys. Acta *396*, 276–292.

Westerhoff, H.V., and Van Dam, K. (1987). Thermodynamics and Control of Biological Free Energy Transduction. (Amsterdam: Elsevier).

Westerhoff, H.V., Melandri, B.A., Venturoli, G., Azzone, G.F., and Kell, D.B. (1984). A minimal hypothesis for membrane-linked free-energy transduction. The role of independent, small coupling units. Biochim. Biophys. Acta *768*, 257–292.

Witt, H.T. (1979). Energy conversion in the functional membrane of photosynthesis. Analysis by light pulse and electric pulse methods. Biochim. Biophys. Acta *505*, 355–427.

Woodrow, I.E., and Berry, J.A. (1988). Enzymatic regulation of photosynthetic CO_2 fixation in C_3 plants. Annu. Rev. Plant Physiol. Plant Mol. Biol. *39*, 533–594.

Xin, C.P., Yang, J., and Zhu, X.G. (2013). A model of chlorophyll *a* fluorescence induction kinetics with explicit description of structural constraints of individual photosystem II units. Photosynth. Res. *117*, 339–354.

Yan, J., and Cramer, W.A. (2003). Functional insensitivity of the cytochrome b_6f complex to structure changes in the hinge region of the Rieske iron–sulfur protein. J. Biol. Chem. *278*, 20926–20933.

Zaks, J., Amarnath, K., Kramer, D.M., Niyogi, K.K., and Fleming, G.R. (2012). A kinetic model of rapidly reversible nonphotochemical quenching. Proc. Natl. Acad. Sci. U.S.A. *109*, 15757–15762.

Zhu, X.G., Wang, Y., Ort, D.R., and Long, S.P. (2013). e-Photosynthesis: a comprehensive dynamic mechanistic model of C3 photosynthesis: from light capture to sucrose synthesis. Plant Cell Environ. *36*, 1711–1727.

New Players for Photoprotection and Light Acclimation

6

Denis Jallet, Michael Cantrell and Graham Peers

Abstract

Light is the source of energy for oxygenic photosynthesis. Large variations in light availability expose photoautotrophic organisms to low fluxes that limit net photosynthesis and high fluxes that are in great excess of photosynthetic capacity. Plants, algae and cyanobacteria need to adjust their photosynthetic apparatus over the time scales of light variability – from seconds to seasons – to balance the energy needs of the organism and to avoid the photooxidative damage associated with excess light energy absorption (photoprotection). This chapter reviews recent discoveries regarding the light acclimation and photoprotective processes within the chloroplast and cyanobacteria. We focus on energy dissipation pathways and the signalling pathways associated with the response to excess light stress.

Introduction

The predictable but also mercurial nature of light in natural environments offers many challenges to oxygenic photoautotrophs. On the one hand, natural environments present slow cycles of changing light (Demmig-Adams *et al.*, 2012). These are represented by the yearly progression of seasons and the associated changing photoperiod. The 24 hour cycle of night and day also exposes organisms to darkness and full sunlight in a regular periodic pattern. On the other hand, rapid changes in the local environment can cause unpredictable and substantial fluctuations in light fluxes (Demmig-Adams *et al.*, 2012; Falkowski and Raven, 2013). These rapid changes can be due to a diverse set of factors including change in cloud cover and variations associated with the immediate setting of a plant, algal or cyanobacterial cell, such as sun flecks underneath a leaf canopy or the rapid mixing of turbid water column. Combining these two categories results in a highly dynamic light environment, which places distinct demands on photosynthetic organisms (Fig. 6.1).

Oxygenic photosynthesis takes place in thylakoid membranes. It relies on two photosystems (PSII, PSI) connected by cytochrome b_6/f and mobile electron transporters. Each photosystem contains pigment-protein inner antennae for light harvesting plus a reaction centre where electron transport related cofactors are localized. Additionally, accessory antennae associate with the photosystems, forming large pigment-protein supercomplexes. Light energy is collected by antennae pigments and funnelled towards the reaction centres that perform the first step of energy conversion. Photochemistry in the PS reaction centre liberates electrons from water and ultimately they are delivered to $NADP^+$. This also generates a proton gradient across the thylakoid membrane that is used to generate adenosine triphosphate (ATP). NADPH and ATP produced through these reactions are then used to drive cellular metabolism, including the Calvin–Benson–Bassham cycle.

The light-dependent reactions of photosynthesis are precarious metabolic processes. The light excited states of photosynthetic pigments and the highly reducing (and oxidizing) intermediates of electron transport are necessary for the removal of electrons from water and their ultimate delivery to $NADP^+$. However, these same high-energy states and intermediates potentially react with oxygen to create reactive oxygen species (ROS) that can damage photosynthetic components.

The probability of damage increases when the absorption of light energy is greater than the capacity for electron transport or for utilization of

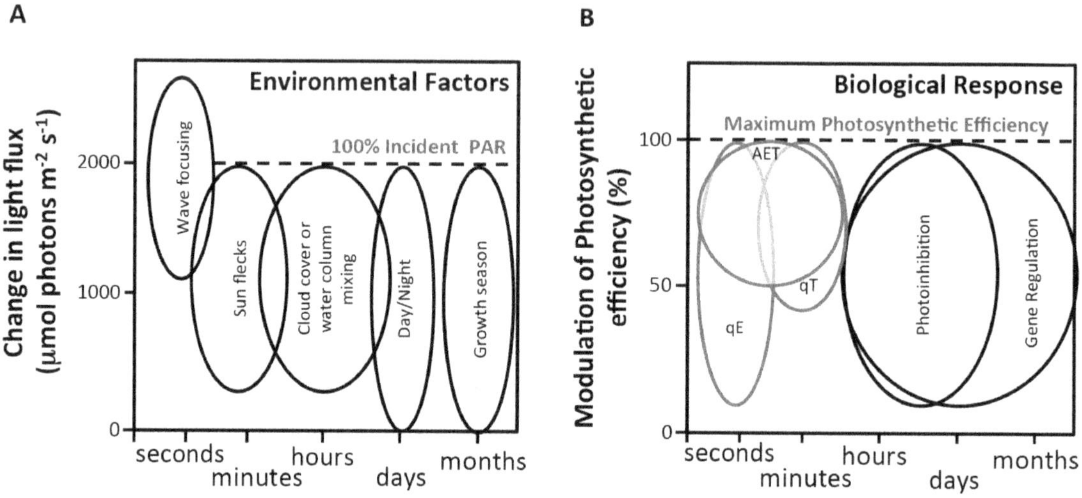

Figure 6.1 Illustration showing the complexity of environmental factors that affect light supply to photosynthetic organisms (A) and the complexity of the biological responses to changes in light (B) that are discussed in this chapter. Note that time scales, changes in light flux and modulation of photosynthetic efficiency (as % absorbed photons converted to biomass) are all approximate. Abbreviations are defined in the main text.

reducing equivalents/ATP. Singlet oxygen ($^1O_2$) appears when the triplet states of antennae chlorophylls or P680 of the PSII reaction centre pass their energy to molecular oxygen (Krieger-Liszkay 2005). This reactive species can readily damage lipids, proteins and pigments within the chloroplast (Laloi and Havaux, 2015) and it is the primary ROS generated by photosynthesis in excess light (Triantaphylides *et al.*, 2008). Superoxide (O_2^-), hydrogen peroxide (H_2O_2) and hydroxyl radicals (HO·) are also produced in excess light conditions (Apel and Hirt, 2004; Asada, 2006; Foyer and Shigeoka, 2011).

The photosynthetic apparatus has many strategies for reducing the formation of ROS. These will be discussed in more detail within the main body of this chapter, such as the rapidly inducible dissipation of excess light excitation as heat (qE, Derks *et al.*, 2015), the redistribution of antenna complexes between the photosystems to maintain a balanced excitation pressure (state transitions; Minagawa, 2011) and alternative electron transfer routes that remove excess reductant from the chloroplast (Peltier *et al.*, 2010) (Fig. 6.1B). These processes all reduce the probability of forming ROS, but these first lines of defence against ROS can be overwhelmed and oxidative stress can occur.

Chloroplasts and cyanobacteria contain a large diversity of antioxidant molecules to scavenge ROS. Specific enzymes play key roles in ROS detoxification including the superoxide dismutases (SODs) and ascorbate peroxidases (APXs) in the chloroplast (Mallick and Mohn, 2000; Mittler, 2002). The genes encoding these stress-related enzymes are in the nucleus and their transcription level is controlled by retrograde signalling pathways from the chloroplast to nucleus (Foyer and Shigeoka, 2011; Mittler *et al.*, 2011). A variety of metabolites also contribute to the antioxidant system including carotenoids (Ramel *et al.*, 2012c), tocopherols (Falk and Munne-Bosch, 2010), ascorbate and glutathione (Foyer and Noctor, 2011).

Collectively all of these processes are considered important aspects of photoprotection – the avoidance of oxidative stress due to light-induced photodamage. In many cases, after exposure to extended excess light fluxes, many plants, algae and cyanobacteria exhibit a reduction of carbon fixation/uptake or oxygen evolution. This is termed photoinhibition. This term is also used to describe damage to PSII, as measured by the Fv/Fm parameter of chlorophyll fluorescence.

The efficiencies of light energy capture versus excess light energy dissipation must be balanced to maintain high fitness in a natural light environment.

Plants, algae and cyanobacteria are able to regulate this balance at the physiological and at the genetic level. Light acclimation is the process by which plants respond to the changing light environment in an attempt to balance photosynthesis and photoprotection. This acclimation is reflective of the light history of the organism and is queued by changes in gene expression.

Given the detrimental effects of excess light to the overall fitness and productivity of plants, algae or cyanobacteria, it is not surprising that the majority of studies investigating light acclimation thus far have focused on the stress aspects of shifting from low light to excess light. These studies continue to reveal novel components of photoprotection, stress mitigation and signalling processes that modulate photosynthesis.

This chapter reviews and summarizes recent additions to our understanding of the light acclimation process and protection from oxidative damage associated with excess light. Particularly, we focus on proteins and regulators discovered since 2011. We also will expand on the recent discoveries that help explain, or change our view of, well-characterized aspects of photoacclimation and photoprotection. There are several chapters contained within this volume that overlap with this subject matter. These include chapters on the structure/function of thylakoids, the photosystem II repair cycle and chloroplast redox regulation – all of which are involved in photoprotection and photoacclimation processes. The reader is referred to these chapters for specifics regarding these topics.

Light-harvesting antennae-based mechanisms of photoprotection

qE

Simplistically, light energy that is absorbed by the antenna complex of PSII can be either passed on to the reaction centre where this energy is used for photochemistry, it can be lost as fluorescence or it can be dissipated as heat.

The phenomena associated with excitation-dependent quenching of light energy as heat (qE) have received the most attention over the past few decades. Efficient qE is required for plant and algal fitness in rapidly changing light regimes (Kulheim *et al.*, 2002; Peers *et al.*, 2009). qE is quickly induced upon exposure to excess light and also relaxes within minutes in the dark. Cyanobacterial based qE mechanisms are described beginning in the 'Orange carotenoid protein' section. Plants and algae that contain chlorophyll *a/b* or *a/c*-based light-harvesting complex (LHC) antenna systems require acidification of the chloroplast lumen to induce qE. It also requires the activation of a xanthophyll cycle that accumulates de-epoxidized xanthophylls including zeaxanthin (Zea), antheraxanthin, lutein and diatoxanthin. The accumulating xanthophyll depends on the species of eukaryote. Readers interested in greater detail about the diversity of non-photochemical quenching (NPQ) processes and particularly qE mechanisms are referred to an excellent recent review on this subject (Derks *et al.*, 2015). However, here we note that the exact biophysical mechanism associated with quenching in LHCII has been a matter of great debate. It may involve the transfer of energy from chlorophyll to a carotenoid in one (or more) of the LHCII components or it may involve the internal conversion of energy by chlorophyll within LHCIIs (Derks *et al.*, 2015). Some recent single particle analyses suggest that there could be multiple dissipative pathways within the LHCII (Kruger *et al.*, 2012; Schlau-Cohen *et al.*, 2015). Regardless of the quenching model of qE, it is clear that there is a conformational change associated with the thylakoid membranes and the proteins therein that switches PSII–LHCII complexes from a light-harvesting state to a photoprotective, quenching state.

PsbS

PsbS is required for qE in plants and mosses (Alboresi *et al.*, 2010; Li *et al.*, 2000; Niyogi and Truong, 2013). It is an unusual member of the LHC family of proteins in that it contains four membrane-spanning helixes and does not appear to bind pigments *in vivo* (Dominici *et al.*, 2002). It is highly unlikely that PsbS is the direct site of qE-related excitation quenching (Ruban *et al.*, 2012). PsbS does have a demonstrated role in sensing luminal pH. This facilitates changes in the conformation of the thylakoids and the proteins located therein (Brooks *et al.*, 2014). Goral *et al.* (2012) found that the absence of PsbS in the *npq4* mutant of *Arabidopsis* reduced the mobility of LHCs within the grana. Two other recent studies have provided evidence that PsbS facilitates qE quenching. Wilk *et al.* (2013) found that efficient quenching of LHCs in reconstituted proteoliposomes required the presence of both PsbS and Zea. They also found that LCHIIs could form heterodimers with PsbS, suggesting this interaction could occur *in vivo*. Sylak-Glassman *et al.* (2014) performed fluorescence lifetime experiments on whole leaves of wild-type (WT), *npq4* and Zea-accumulating *npq1 Arabidopsis* plants. They found that the presence/absence of PsbS alone did not influence the rate of excited chlorophyll relaxation. But its presence changes the magnitude of quenching and the speed with which it was activated. These data all suggest that PsbS is essential for modifying the thylakoidal membrane structure and allowing the PSII–LHCII supercomplex to enter into a quenching conformation.

LHCSR/LHCX

Most eukaryotic algae (and moss) share one distinct family of LHC relatives (Neilson and Durnford, 2010) despite the large evolutionary distances between these eukaryotic supergroups. This family of LHCs falls into a clade that includes LHCSR and LHCX which have been shown to play a role in qE in *Chlamydomonas* (Peers *et al.*, 2009), *Physcomitrella* (Alboresi *et al.*, 2010) and the diatom *Phaeodactylum* (Bailleul *et al.*, 2010). LHCSR binds to the PSII–LHCII supercomplex isolated from excess light-grown *Chlamydomonas* (Tokutsu and Minagawa, 2013). These isolated complexes had reduced fluorescent lifetimes at a pH of 5.5 and this pH effect was abolished by inhibitor-mediated blocking of acidic residues and in PSII-complexes isolated from the LHCSR3 mutant *npq4*. This suggested a direct role for LHCSR in sensing and transducing the luminal pH change to energy quenching. Recombinant LHCSR3 has been reconstituted with chlorophylls and xanthophylls. These proteins display fast fluorescence lifetime components of less than 100 ps compared to 3.5–4.5 ns for other LHC proteins, suggesting they are indeed involved in energy dissipation versus energy transfer (Bonente *et al.*, 2011). Further experiments on LHCSR3 revealed that it forms a transient lutein cation radical that quenches chlorophyll energy transfer (Bonente *et al.*, 2011). This is distinct from LHCs from higher plants, where a zeaxanthin appears to play the role of energy quencher (Ahn *et al.*, 2008). Finally, the C-terminal domain of LHCSR appears to be the pH sensor. Replacement of approximately 20 amino acids near the C-terminus with neutral residues in a synthetic LHCSR reduced quenching capacity *in vitro* (Liguori *et al.*, 2013). All *LHCSR* transcripts are accumulated in *Chlamydomonas* in response to excess light (Peers *et al.*, 2009). LHCSR3.1 and -3.2 share near identical promoter sequences and their transcriptional induction is suppressed by DCMU (leading to an oxidized plastoquinone (PQ) pool) and also by a calmodulin inhibitor (Maruyama *et al.*, 2014). These stimuli did not affect the transcription of LHCSR1, which instead appears to be influenced by a singlet oxygen accumulation signal (Wakao *et al.*, 2014). Other components of the *Chlamydomonas* LCHII antenna appear to be required for efficient qE in addition to LHCSR. These include the Lhcbm5 peptide (Elrad *et al.*, 2002) and now Lhcbm9 appears to be required for effective NPQ as well. Lhcbm9 is up-regulated in response to nutrient stress and it also appears to play a role in PSII stabilization and NPQ (Grewe *et al.*, 2014). The exact role of these LHCs has not yet been determined.

Orange carotenoid protein

Cyanobacteria and red algae possess unique accessory antennae, the phycobilisomes (PBSs). PBSs appear as highly organized complexes made of phycobiliproteins associated with colourless linker polypeptides. They typically present a hemidiscoidal arrangement, where rods of phycocyanin (PC) radiate from an allophycocyanin (APC) core. Located on thylakoid membranes, they are optimized for a directional energy transfer towards the photosystem reaction centres (Adir, 2005; Watanabe and Ikeuchi, 2013). They can absorb orange and green lights that are poorly absorbed by chlorophyll containing antennae. PBSs play important roles in photoprotection and photoacclimation processes.

Because the phycobilisomes and LHC antenna systems are quite different it is therefore not surprising that their mechanisms of excitation quenching are also quite distinctive. qE in cyanobacteria does not depend on a pH gradient or the xanthophyll cycle. Instead, qE in well-characterized cyanobacteria relies on a soluble Orange Carotenoid Protein (OCP) (Wilson *et al.*, 2006). The sequencing of 54 cyanobacterial strains in 2013 revealed that most of them possess *ocp* orthologues (Kirilovsky and Kerfeld, 2012; Shih *et al.*, 2013). OCP non-covalently attaches a ketocarotenoid – either echinenone or hydroxyechinenone in *Synechocystis* (Wilson *et al.*, 2010). OCP primarily exists in the dark as an orange, stable form (OCP^o) and upon illumination with strong blue-green light, it gets photoconverted to a red, metastable form (OCP^r) (Wilson *et al.*, 2008). OCP^r – but not OCP^o – can bind to PBS cores, triggering the quenching of excess excitation energy and a drop in fluorescence emission (Gwizdala *et al.*, 2011). Energy transfer from PBSs to both PSI and PSII decreases upon binding of OCP^r (Rakhimberdieva *et al.*, 2010; Tian *et al.*, 2011). Accordingly, OCP-deficient *Synechocystis* mutant strains appear more sensitive to photoinhibition (Wilson *et al.*, 2006).

The molecular mechanism for how the conversion from OCP^o to OCP^r allows binding to PBSs has only lately been elucidated. Hydrogen bonds situated at the N- and C- terminal domains interface in OCP^o might get disrupted upon OCP^r formation (Wilson *et al.*, 2012). Several aspartate and glutamate residues situated at the same interface also become more accessible to solvent (Liu *et al.*, 2014). The N-terminal domain was isolated following tryptic digestion and it retained a bound carotenoid and appeared red. It was able to interact with PBSs *in vitro* and induced fluorescence quenching (Leverenz *et al.*, 2014). The crystal structure of this protein domain revealed a major shift of carotenoid position compared to OCP^o N-terminal domain (Leverenz *et al.*, 2015). These results indicate that photoconversion to OCP^r causes a conformation change that exposes amino acids on the N-terminal domain and that this facilitates OCP^r binding to PBSs.

In the dark, fluorescence recovery protein (FRP) triggers the reconversion of OCP^r to OCP^o (Boulay *et al.*, 2010) making OCP-related NPQ reversible. FRP's 3D structure was recently solved (Sutter *et al.*, 2013). Site-directed mutagenesis coupled to co-immunoprecipitation studies showed that FRP interacts with OCP's C-terminal domain which could be available for interaction when OCP^r is bound to PBSs. More has to be done to characterize the molecular mechanism associated with FRP in order to understand how OCP-mediated NPQ is reversed.

The exact site of OCP-related NPQ inside PBSs remains uncertain. Spectrally resolved picosecond fluorescence spectroscopy indicated that it could be one of the major allophycocyanin (APC) forms (ApcA, ApcB), with maximum emission at 660 nm (Tian *et al.*, 2011). *Synechocystis* mutant strains deficient in some of the minor APC forms (ApcD, ApcF) retained normal OCP-related NPQ induction upon high light exposure (Jallet *et al.*, 2012; Stadnichuk *et al.*, 2012). Cross-linking of OCP^r to PBSs followed by LC-MS/MS analysis showed that ApcB and ApcE might participate in OCP binding (Zhang *et al.*, 2014). These data position OCP's carotenoid in close vicinity to ApcB's bilin, designating ApcB as the quenching site.

The mechanism allowing heat dissipation of excess excitation energy has not been solved. Tian and co-workers derived a quenching rate of 240 ± 60/fs using spectrally resolved picosecond fluorescence on whole cells of *Synechocystis*. Their modelling suggested that quenching was mediated by charge transfer between a bilin in the APC core and the OCP carotenoid (Tian *et al.*, 2011). Another model, based on *in vitro* transient absorption spectroscopy on isolated OCP, suggests that there is direct energy transfer from APC to OCP (Berera *et al.*, 2012). However, neither mechanism has been validated.

Other systems that may reveal new qE mechanisms

Physiological studies of algae with diverse light harvesting antenna systems suggest that there may be slight variations on the themes discussed above. The cryptomonad alga *Rhodomonas salina* has an unusual antenna system in that it has both phycobiliproteins and chlorophyll *a/c* binding LHCs. Kana *et al.* (2012) found a rapidly inducible and relaxing NPQ in this alga that is induced by ΔpH with no change in xanthophyll cycle pigments. It appeared that quenching responding to low pH occurred in purified LHCs and not in the phycobillisomes. The exact nature and location of this quenching has not been elucidated. *Chromera velia* is an alveolate alga related to the apicomplexan parasites. Induction of NPQ in this organism is correlated with a rapid formation of Zea, with no detectible accumulation of antheraxanthin (Kotabova *et al.*, 2011) suggesting a novel type of violoxanthin de-epoxidase. Interestingly, violaxanthin has a light-harvesting role in this alga, which is not the case for our more established model photosynthetic organisms.

Symbiodinium, a dinoflagellate known for its symbiotic associations with corals and cnidairans, also possesses a potential novel mechanism of NPQ. Many dinoflagellates, including *Symbiodinium*, contain unique antennae made up of water-soluble perdinin–chlorophyll *a* proteins and chlorophyll *a*–chlorophyll *c2*–peridinin–protein complexes (acpPC, Maruyama *et al.*, 2015). The mechanism for their NPQ is also unresolved as yet, but it appears that the diatoxanthin/diadinoxanthin cycle and lumen pH change lead to quenching in the acpPC complex (Kanazawa *et al.*, 2014). The intriguing observation that an AcpPC-encoding transcript related to LHCSR increases in high light grown *Symbiodinium* suggests this protein may also play a role in NPQ in the dinoflagellates (Xiang *et al.*, 2015).

Additionally, some PBS containing cyanobacteria possess neither OCP nor FRP orthologues, for example *Synechococcus elongatus* PCC7942 or *Thermosynechococcus elongatus* (Kirilovsky and Kerfeld, 2012) and their photoprotective capacity has not been studied. Explorations of light energy dissipation mechanisms in diverse algae and cyanobacteria over the next decade will likely reveal new mechanisms for the dissipation of excess light.

Ions

Trafficking of inorganic ions into the chloroplast stroma and lumen is required to maintain efficient photosynthetic activity and chloroplast homeostasis (Finazzi *et al.*, 2014). Ion transport activity turns out to be particularly important for modulating the proton motive force associated with qE. Mutants of *Arabidopsis* without a stromal lamellae-localized K^+ transporter (TPK3) are unable to export K^+ from the lumen. This leads to an imbalance in the electrical potential across the thylakoid membranes during light-driven electron transport and reduces the capacity to build a pH gradient. This results in plants that are deficient in qE and also have decreased photosynthetic rates (Carraretto *et al.*, 2013). Furthermore, the thylakoid-localized KEA3 H+/K+ antiporter of *Arabidopsis* may play an important role in the dark/low light recovery of qE and increased light harvesting efficiency by facilitating the re-equilibrium of these ions upon the release of stress (Armbruster *et al.*, 2014; Kunz *et al.*, 2014).

Miyaji *et al.* (2015) have recently discovered a chloroplast envelope ascorbate transporter in *Arabidopsis* (AtPHT4;4), that is a member of the SLC17 family of transporters. Knockout of this protein decreases the accumulation of Zea and antheraxanthin in high light, because violaxanthin de-epoxidase requires ascorbate as a co-factor. This phenotype correlates with a decrease in NPQ capacity when compared to WT. The investigators propose that AtPHT4;4 is a required component of the ascorbate transport from its point of synthesis (the mitochondria) into the chloroplast.

The phosphate anion is required in the chloroplast for extensive post-translation modifications of photosynthetic enzymes in addition to its essential role in the production of ATP. The *Arabidopsis* PHT 4;1 transporter, localized in the thylakoids, appears to have an essential role for maintaining high ATP synthesis rates, but its absence has little effect on the phosphorylation status of PSII, LHCII and the PSII repair cycle (Karlsson *et al.*, 2015). Interestingly, the differences seen in thylakoid proton conductivity between the mutant and WT in low light lessens in high light. This suggests that there is a complementary mechanism for phosphate transport that is not yet discovered.

Antenna-based quenching of photosystem I

While considerable attention has been paid to the modulation of photoprotection and light harvesting of PSII, very little has been paid to the same processes in PSI. PSI shows little fluorescence at room temperature suggesting a low probability that triplet states of chlorophyll and singlet oxygen will be formed. However, PSI photoinhibition is observed when high light stress is coupled with cold temperatures (Sonoike, 2011). This damage is caused by O_2^- and H_2O_2 produced by reduction of oxygen by PSI. Alboresi *et al.* (2009) found that photoprotection of PSI required a complete PSI–LHCI complex *in vitro* and violaxanthin and Zea were already known to be associated with the LHCI antenna proteins (Wehner *et al.*, 2004). Ballottari *et al.* (2014) utilized the Zea-accumulating *npq2* mutant of *Arabidopsis* to show that Zea also plays a role in regulating functional antenna size of PSI. Furthermore, they also showed that the Lhca4 protein with Zea was capable of generating a carotenoid radical cation that could explain chlorophyll excitation quenching in the PSI antenna.

Yokono *et al.* (2015) recently have suggested that light energy can be effectively transferred from PSII directly to PSI and that this is mediated by megacomplexes of the two photosystems. First, they used delayed fluorescence spectroscopy to show that energy released by charge recombination at PSII can be transferred to PSI at −196°C. Then they isolated a megacomplex containing both PSI and PSII from *Arabidopsis*. It appears that half the PSII are functionally attached to PSI such that light energy that cannot be effectively used by PSII is transferred to PSI. This connection appears to increase in high light conditions and reverts back to a low light state within a minute of transfer to low light. The phenomenon likely interacts with PSI quenching mechanisms described above and may provide a distinct path for light energy dissipation independent of state transitions. This research suggests that the assumptions of chlorophyll fluorometry methods that typically assume PSI fluorescence yields remain constant should be refined (Giovagnetti *et al.*, 2015; Pfuendel *et al.*, 2013).

State transitions – flexible light harvesting capability between PSII or PSI

State transitions associated with LHC antenna systems

State transitions in plants and green algae are a phosphorylation dependent alteration of the PSII-associated light harvesting antenna (LHCII) that results in a change in the functional absorption cross section of the photosystems. These changes occur to balance the flux of energy through electron transport in response to changing light quantity and quality (Allen, 1992; Minagawa, 2011). A chloroplast in state 1 has an oxidized PQ pool and the relative light harvesting capacity of PSII is increased while a chloroplast in state 2 has a reduced PQ pool and an increase in the relative light harvesting capability of PSI. This behaviour is typically observed as an increase or decrease in PSII chlorophyll fluorescence and it was believed to represent changes of 15–25% of light-harvesting capacity in plants (Allen 1992) and 80% in *Chlamydomonas* (Delosme *et al.*, 1996). It is well established that the PQ pool and cytochrome b_6f are essential players in this process (Zito *et al.*, 1999). Phosphorylation state of LHCII also dictates the migration of antennae between the photosystems. The identity of the kinase (such as STN7, Bellafiore *et al.*, 2005) and the phosphatase (e.g. PPH1/TAP38, Pribil *et al.*, 2010; Shapiguzov *et al.*, 2010) that control this process in *Arabidopsis* are also known.

The traditional model of state transitions in *Chlamydomonas* involved a large migration of LHCII complexes to PSI in state 2 conditions (Minagawa, 2011). However, we now know that this may not reflect the true rearrangement of the PSII supercomplex. A new model of state transitions predicts that phosphorylated LHCIIs remain associated with PSII or are detached from either photosystem (Minagawa and Tokutsu, 2015; Nagy *et al.*, 2014). Phosphorylation would repel the membrane faces within the granal stacks, moving the thylakoids and photosynthetic complexes apart. This is supported by data from small angle neutron scattering and circular dichroism observations of cells in state 1 and state 2. Only ~10% of the LHCII actually migrate to increase the absorption cross section of PSI, similar to what is observed in plants (Drop *et al.*, 2014; Minagawa and Tokutsu, 2015; Nagy *et al.*, 2014; Unlu *et al.*, 2014).

There has also been progress in our understanding of the role of individual components of the LHCII trimers in state transitions. LHCII in *Arabidopsis* are made up of three closely related proteins – Lhcb1, -2 and -3. Lhcb2 is phosphorylated to 70% of maximum within 10 s of illumination favouring state 2 conditions and the PSI-supercomplex becomes enriched in phosphorylated Lhcb2 (Leoni *et al.*, 2013). The generation of specific antibodies for the phosphorylated forms of Lhcb2 and Lhcb1 were essential for resolving these changes. Furthermore, mutant plants lacking Lhcb2 do not perform state transitions at all, despite hyperphosphorylation of Lhcb1, showing the necessity of this one LHCII component for this process (Pietrzykowska *et al.*, 2014). Meanwhile, Lhcb1 is required for the proper ultrastructural rearrangements associated with state transitions (Pietrzykowska *et al.*, 2014; Pribil *et al.*, 2010). Each of these LHCII components play complementary roles in this process.

The role of LHCII components for state transitions in *Chlamydomonas* seems to be less specific. Biochemical analysis of purified PSI-LHCI-LHCII supercomplexes isolated in state 2 *Chlamydomonas* showed that the CP29 inner antenna complex and all the classes of the PSII associated LhcbMs can be also be found with PSI (Drop *et al.*, 2014). But only LhcbMI, -II and IV classes appeared to be phosphorylated. Single particle analysis via electron microscopy suggested that these large PSI–LHCI–LCHII complexes contain one monomer (perhaps CP29) and two LHCII trimers and this represents the largest PSI supercomplex observed in eukaryotes thus far (Drop *et al.*, 2014). Thus, the heterogeneity of the protein composition of the LHCII trimers appears to play an important role in how light harvesting is regulated in plants and green algae.

The studies described above also highlight how a conserved strategy of photoprotection is manifested by slightly different mechanisms within the embryophytes and chlorophytes. However, there is one major difference. The detached phosphorylated LHCII complexes observed in state 2 *Chlamydomonas* have a shortened excited state lifetime, suggesting that they are effectively quenched (Unlu *et al.*, 2014). State transitions have been clearly shown to be an efficient mechanism to protect PSII from overexcitation during a rapid change in light levels (Allorent *et al.*, 2013) and the quenching described by Unlu *et al.* (2014) may be the mechanism of photoprotection.

State transitions associated with phycobilisomes

In cyanobacteria, state transitions rely both on PBS movement and photosystem rearrangements to balance the redox state of the PQ pool (Kirilovsky, 2014; Kirilovsky *et al.*, 2014; Mullineaux and Emlyn-Jones, 2005). Regulator of phycobilisome association C (RpaC) is required for the PBS-related component (Emlyn-Jones *et al.*, 1999; Mullineaux and Emlyn-Jones, 2005). PBSs remained attached to PSII (State I) in *Synechocystis* Δ*rpaC* strains, even under conditions that would normally trigger their partial reallocation to PSI (state II, Emlyn-Jones *et al.*, 1999). The *rpaC* mutants grew normally under high light conditions, showing that cyanobacterial state transitions do not play a photoprotective role under saturating illumination. However, the *rpaC* mutants had a 30% longer doubling time than WT under 2 μmol photons/m^2/s white light and a 40% longer doubling time under 2 μmol photons/m^2/s yellow light, absorbed preferentially by PBSs (Emlyn-Jones *et al.*, 1999). Unlike LHC-based state transitions, that appear to be involved in photoprotection, PBS-based state transitions play a physiological role at very low light to balancing the excitation pressure on PSI and PSII.

Current research focuses on describing the molecular interactions between PSI/PSII and PBSs. Liu and co-workers combined protein cross-linking and LC-MS/MS to identify subunits responsible for PBS-photosystem contacts in *Synechocystis* (Liu *et al.*, 2013). They found that ApcD, situated in the PBS core, can interact with the PsaA subunit of PSI. ApcE, another PBS core component, appeared to interact with PsbC, PsbD and PsbI subunits of PSII. Furthermore, they showed that excitation energy from the PBS could effectively be transferred to either photosystem. This leads to a variation on the state transition model whereby the PBS may facilitate the formation of a megacomplex containing both PSI and PSII. Further work on both the structural aspects of state transitions and the physiological importance of these in natural light regimes should be extended to red algae (rhodophytes) to examine the diversity of PBS related state transitions or quenching (e.g. Kana *et al.*, 2014).

Alternative electron transport

NPQ represents the dominant pathway by which oxygenic photosynthesis can dissipate excess energy. But this can still be insufficient to avoid ROS generation and associated photodamage. Alternative electron transport (AET) pathways remove excess reductant from the chloroplast, thus preventing over-reduction of the inter-photosystems transport chain. AET mutants show increased sensitivity to high light intensities, as exemplified by flavodiiron (Flv)-deficient strains of the cyanobacterium *Synechocystis* (Allahverdiyeva *et al.*, 2011, 2013) or plastid terminal oxidase (PTOX) knockout lines of the green alga *Chlamydomonas* (Houille-Vernes *et al.*, 2011). Recent advances have revealed the diversity of AET components, capacities and physiological roles in oxygenic photosynthesis.

Flavodiiron proteins

Flavodiiron proteins (Flv) are present in anaerobic organisms where they act as O_2 detoxifiers. They contain a core made of a metallo β-lactamase-like domain as well as a flavodoxin domain. Homologous proteins are found in cyanobacteria (Helman *et al.*, 2003) and some photosynthetic eukaryotes

(Peltier *et al.*, 2010). In these organisms, Flvs present a supplementary C-terminal flavin reductase domain that potentially catalyses NADPH oxidation (Zhang *et al.*, 2009). It has recently been shown that Flvs play photoprotective roles in photosynthetic organisms.

Flvs are best characterized in the model cyanobacterium *Synechocystis* PCC6803, which has four homologous genes encoding for Flvs. Flv1 and Flv3 form heterodimers (Allahverdiyeva *et al.*, 2011). Recombinant Flv3 oxidizes NADH and reduces O_2 to water *in vitro* (Vicente *et al.*, 2002). *In vivo*, Flv1Flv3 participates in a Mehler-like reaction that uses the reducing power generated by PSI (NADPH) to convert O_2 to water (Helman *et al.*, 2003). This helps avoiding an over-reduction of the inter-photosystem electron transfer chain, particularly under conditions of inorganic carbon (C_i) deprivation or fluctuating light conditions (Allahverdiyeva *et al.*, 2011, 2013). In contrast to the Mehler reaction, no reactive oxygen species are generated by Flv1Flv3. Flv1/Flv3 homologues are found in all sequenced cyanobacterial genomes, plus green algae and mosses where they potentially catalyse similar Mehler-like reactions (Dang *et al.*, 2014; Peltier *et al.*, 2010). The reason why embryophytes aside from bryophytes have lost the *Flv* gene family remains an open question.

Flv2 and Flv4 form heterodimers accumulated under conditions of C_i deprivation, whereas they are barely detected under high C_i conditions (Zhang *et al.*, 2009). Orthologues of this group are only found in PBSs containing cyanobacteria where they appear to play a novel role in photoprotection compared to the Flv1/3 group.

Studies of *flv4* and *flv2* mutants revealed that the heterodimer plays a specific role in PSII photoprotection (Zhang *et al.*, 2009, 2012). These strains displayed a lower PSII photochemical yield as well as a higher sensitivity to photoinhibition under C_i deprivation. Flash-induced chlorophyll fluorescence kinetics studies also indicated that they maintained a more reduced PQ pool than the WT (Bersanini *et al.*, 2014; Zhang *et al.*, 2012). This suggests that Flv2Flv4 acts in an alternative electron route, which takes electrons either from the acceptor side of PSII or directly from PQH_2 to reduce a yet-to-identify acceptor. Current research focuses on unravelling the underlying mechanism. Bersanini *et al.* (2014) used artificial electron acceptors to show that Flv2Flv4 can receive electrons directly at the Q_B-binding pocket of PSII. Recently, it has been shown that recombinant Flv4 can transfer electrons from NADH to O_2 *in vitro* (Shimakawa *et al.*, 2015). *In vivo*, the Flv2Flv4 heterodimer could then alleviate over-reduction of the PQ pool by creating an AET route from PSII to O_2. Moreover, thermoluminescence studies revealed a modified redox potential for Q_B in the Flv2/4 deficient mutant under Ci deprivation (Chukhutsina *et al.*, 2014). The Flv2Flv4 heterodimer, when binding to the Q_B pocket, might also modify the redox properties of PSII to avoid possible charge recombinations leading to $^1O_2$ production.

Further support for the roles of Flv2Flv4 came from studies on *Synechococcus* PCC7942, which lacks *Flv2* and *Flv4* orthologues. Shimikawa *et al.* (2015) found that alternative electron flow under C_i deprivation conditions was mediated by Flv2Flv4 in *Synechocystis*, agreeing with the model presented above, but that *Synechococcus* PCC7942 was highly photoinhibited during low CO_2 stress. They interpret these data as support for the role of Flv proteins and associated AET in protecting the photosynthetic apparatus.

Flv2, Flv4 and Sll0218 are encoded within one operon in *Synechocystis* (Zhang *et al.*, 2012). Eisenhut *et al.* (2012) found that an antisense RNA directed against *flv4* helps modulate the accumulation of FLV4 during a low CO_2 shift. AET is essential for photoprotection in long term C_i deprivation stress, but it appears that the activation of this pathway is suppressed during a transient response to avoid overall reductions in carbon fixation rates.

Interestingly, the oxygen-consuming function of Flvs also appears to play a role in nitrogen fixation. Nitrogenase requires an anaerobic environment and the specialized N_2-fixing cells of *Anabaena*, the heterocysts, must maintain low oxygen for its proper function. *Flv1a* and *Flv3b* are exclusively expressed in the heterocysts (Ermakova *et al.*, 2013) and Δ*flv3b* mutants were unable to grow efficiently in oxic environments, where they displayed nitrogen-deficient characteristics and had reduced levels of nitrogenase (Ermakova *et al.*, 2014). This shows that respiratory oxygen consumption is unable to compensate for the reduced levels of AET associated with Flvs in these cells.

Plastiquinol terminal oxidase

The clear photoprotective role of AET in cyanobacteria has not been shown in plants. However, the IMMUTANS protein has PTOX function, maintaining an oxidized PQ pool during chlororespiration (Carol *et al.* 1999), and it was proposed that PTOX could also have a photoprotective function in excess light. Despite some contrasting results in the literature, it appears as though the presence/absence of PTOX does not significantly influence photoprotection in *Arabidopsis* (Rosso *et al.*, 2006) but its role in releasing excitation pressure on photosystem II in other species is still a matter of open debate. For instance, the role of PTOX becomes more prevalent under temperature stress in a wide variety of plants and can function as a significant sink of electrons in mutants unable to fix CO_2 (McDonald *et al.*, 2011).

However, the role of PTOX in relieving photoinhibition in the green alga *Chlamydomonas* is much clearer. Houille-Vernes *et al.* (2011) isolated a knockout mutant of PTOX2 (PTOX1 was still intact) by screening for mutants that illustrated a post-illumination rise in chlorophyll fluorescence yields. This suggests that, in the absence of light-driven linear photosynthesis, the PQ pool cannot be oxidized by chlororespiration. Most importantly, the *ptox2* mutant had reduced fitness in moderate light levels compared to WT cells showing that AET is important for photoprotection in this alga. Interestingly, because the *ptox2* mutant maintains a more reduced PQ pool in the dark, they also found that cells remained in State 2.

PTOX activity may become more important when electron transport is unbalanced. *gun4* mutants of *Chlamydomonas* show a reduced PSI:PSII ratio and an apparent uncoupling of linear electron transport compared to WT (Formighieri *et al.*, 2012). PTOX was estimated to account for 20% of electron transport versus 2% in WT. This was determined by comparing measurements of the quantum yield of PSII under actinic illumination, with and without the addition of the PTOX inhibitor propylgallate. In this case, PTOX activity was important to reduce the excitation pressure on PSII.

The role of PTOX or Flv in other eukaryotes is currently unknown. Genes encoding PTOX homologues are found in most lineages of oxygenic photoautotrophs including cyanobacteria, plants and algae that evolved from secondary endosymbiosis such as diatoms (McDonald *et al.*, 2011). Genes encoding Flv homologues have been found in Chlorophytes, Prasinophytes and mosses. Many algae exhibit significant light dependent oxygen consumption in high light conditions (Helman *et al.*, 2005; Waring *et al.*, 2010; Weger *et al.*, 1990), suggesting that AET may be an important mechanism to release excess electrons in the chloroplast.

Other terminal oxidases

Lea-Smith *et al.* (2013) made a mutant of *Synechocystis* without the thylakoid membrane-located cytochrome c oxidase (COX) and quinol oxidase (Cyd). These mutants were most sensitive to rapid fluctuations in light (square wave light/dark cycles or rapid on/off cycles) but tolerated sine-wave changes in irradiance, albeit with reduced growth rates. Mutants displayed greater photoinhibition, more oxidative stress damage products and a reduced ability to utilize glycogen. This suggests that these oxidases play an important role not only in the light, but also in dark metabolism.

Malate valve

The malate valve is proposed to be a significant mechanism for chloroplasts to dissipate excess electrons. NADP-malate dehydrogenase (NADP-MDH) consumes NADPH to reduce oxaloacetate to malate. The malate is then exported from the chloroplast to the cytoplasm through a dicarboxylate transporter (Scheibe, 2004), where it can be oxidized by a cytosolic NAD-MDH. NADP-MDH knockout mutants of *Arabidopsis* show the importance of this pathway for maintaining redox balance in the chloroplast. They display reduced growth and accumulate more NADPH in the chloroplast compared to WT in excess light (Heyno *et al.*, 2014). Earlier studies found little change in photosynthesis and oxidative stress parameters in this mutant (Hebbelmann *et al.*, 2012), which raised the issues of where the excess reductant could be shunted. Deeper investigations of metabolism found increases in photorespiration capacity, the NTRC/2-Cys peroxiredoxin system and an increase in proline content, all of which could consume excess reductant (Hebbelmann *et al.*, 2012). Increases in chlorophyll fluorescence after illumination also suggested that NADPH can be oxidized by the NADPH dehydrogenase complex of chlororespiration, reducing the PQ pool instead of oxaloacetate (Heyno *et al.* 2014). In whole, these results show the multiple strategies that plants have to consume excess reductant. It also suggests a significant amount of electrons can flow through the malate valve in C_3 plants. Additionally, Heyno *et al.* (2014) proposed that the malate valve plays an important signalling role between the chloroplast and mitochondria.

The transporter component of the malate valve has only recently been discovered. A plastidic 2-oxoglutarate/malate transporter (AtpOMT1) had been previously shown to have a role in linking carbon and nitrogen metabolism. Kinoshita *et al.* (2011) found that isolated chloroplasts from the *AtpOMT1* mutant of *Arabidopsis* were unable to import oxaloacetate to be used as a substrate for NADP-MDH. The plants, in turn, had greater photoinhibition in high light. However, these data should not be interpreted as due to the lack of a malate valve alone. The lack of 2-oxoglutarate transport also explains the observed decreases in amino acid biosynthesis and photorespiration seen in this mutant. It is likely that the bifunctionality of this transporter creates these phenotypes (see below).

Photorespiration

Ribulose-1,5-bisphosphate carboxylase/oxygenase (RuBisCO) can generate 2-phosphoglycolate (2-PG) because of its dual activity as an oxygenase (Bauwe *et al.*, 2010). 2-PG inhibits several metabolic reactions and needs to be detoxified through photorespiration. This process consumes excess reducing power generated by photosynthesis thus contributing to photoprotection (Niyogi, 2000). It also regenerates substrate for the Calvin–Benson–Bassham cycle. The underlying enzymatic steps are well described, taking place in various cellular compartments: chloroplast, cytoplasm, peroxisome and mitochondria (Bauwe *et al.*, 2010; Eisenhut *et al.*, 2013). However, the molecular transporters that allow intermediates to transit from one compartment to another remain largely unidentified (Eisenhut *et al.*, 2013, 2014).

A recent study in *Arabidopsis* showed that chloroplastic 2-oxoglutarate malate transporter 1 (AtpDMP1) acts as both a 2-oxoglutarate/malate transporter and as a malate/oxaloacetate transporter (Kinoshita *et al.*, 2011), which is discussed further above. The plastidic glycolate glycerate transporter 1 (AtPLGG1) was identified in a forward genetics screen and its activity is required for photorespiration in *Arabidopsis* (Pick *et al.*, 2013).

Transporters might differ from one organism to another. For example, diatoms possess plastids that arose from a secondary endosymbiosis event and are surrounded by four membranes instead of two (Prihoda *et al.*, 2012). Specific, undiscovered transporters might thus be required to allow metabolite traffic between chloroplasts and cytosol in diatoms (Weber and Linka, 2011). Transport of metabolites between organelles may represent important control points for metabolism (Tegeder and Weber, 2008; Weber and Linka, 2011) and a full understanding on the controls of carbon flow out of and into the chloroplast is an essential part of understanding its biology.

New proteins that scavenge or reduce the formation of $^1O_2$

The mechanisms described above are effective at reducing the excitation pressure exerted on photosynthesis in excess light, but, even so, $^1O_2$ is still produced. Several carotenoid-binding, water soluble proteins appear to contribute to $^1O_2$ quenching. For instance, the cyanobacterial OCP acts as an excellent $^1O_2$ quencher *in vivo* and *in vitro*, independently of its photoactivation state (Sedoud *et al.*, 2014). Another example is in the microalga *Scenedesmus*. It produces a soluble protein called AstaP which binds astaxanthin under stress conditions (Kawasaki *et al.*, 2013). AstaP shows no sequence homology to OCP (discussed above). *In vitro*, AstaP quenches $^1O_2$ but its subcellular localization and quencher activity *in vivo* remain to be assessed. In order to be effective $^1O_2$ quenchers, these proteins need to be located immediately next to the site of $^1O_2$ generation, but outside of the thylakoid membranes.

Synechocystis produces several high light-inducible proteins (HLIPs), namely HliA-D (Komenda and Sobotka, 2015). HLIPs are single-helix transmembrane proteins that share sequence similarity with eukaryotic LHCs. HLIPs are proposed to play an important role in the early steps of PSII assembly, binding chlorophyll *a* and transporting it to its incorporation site on newly synthesized reaction centres (Knoppova *et al.*, 2014; Komenda and

Sobotka, 2015). HliD was recently shown to bind both chlorophyll *a* and β-carotene *in vivo* (Staleva *et al.*, 2015). Interestingly, transient absorption spectroscopy revealed that the excited states of the bound chlorophyll *a* are physically quenched by β-carotene *in vitro*, minimizing $^1O_2$ formation. This mechanism probably reduces oxidative stress during the synthesis of the photosystems.

Sensing and responding to light quality

Photoreceptors allow living organisms to perceive light intensity and quality. They can transcriptionally activate or repress target genes, thus driving metabolic changes in response to illumination status (Galvão and Fankhauser, 2015; Kianianmomeni and Hallmann, 2014). New insights into how photosynthetic organisms sense light suggest that various families of photoreceptors are important players in this process.

Phytochromes absorb light through their covalently bound bilins, consequently triggering signalling cascades involving phosphorylation or dephosphorylation events of response regulators. In plants, they primarily exist as inter-convertible red/far-red light absorbing species. Recent advances include the discovery that several post-translational modifications, including phosphorylation and SUMO-ylation (small ubiquitin-like modifier), modulate phytochrome activity in *Arabidopsis* (Medzihradszky *et al.*, 2013; Nito *et al.*, 2013; Sadanandom *et al.*, 2015). The potential effects on photoprotection and/or photoacclimation have yet to be assessed.

Phytochromes are also present in algae and cyanobacteria (Duanmu *et al.*, 2014; Kehoe and Grossman, 1996), which might appear surprising because red light is strongly attenuated under water. However, it was recently demonstrated that isolated algal and cyanobacterial phytochromes display absorption maxima spanning the whole visible spectrum (Kehoe, 2010; Rockwell *et al.*, 2014). Algal and cyanobacterial phytochromes thus constitute powerful molecular tools for the perception of changes in light quality and quantity.

Cyanobacteriochromes (cyanobacterial phytochromes) have been shown to control complementary chromatic adaptation (CCA). CCA is a molecular process where cyanobacteria modify the proportion of red absorbing (phycocyanin, PC) and/or green absorbing (phycoerythrin, PE) phycobiliproteins in their phycobilisomes as a function of light quality (Demarsac, 1977; Kehoe and Gutu, 2006). It adjusts the action spectrum of photosynthesis to incident light. The deletion of *rcaE*, encoding for a cyanobacteriochrome, led to intermediate PC/PE accumulation both under green and red illumination in *Fremyella diplosiphon* (Terauchi *et al.*, 2004). A similar phenotype was observed in *Nostoc punctiforme* when the gene encoding for CcaS, another cyanobacteriochrome, was deleted (Hirose *et al.*, 2010). CcaS exists as a red light-absorbing form (Pr) with kinase activity and a green light-absorbing one (Pg) with phosphatase activity, as revealed by *in vitro* assays (Hirose *et al.*, 2008, 2010). CcaS Pr, present under green illumination, phosphorylates the response regulator CcaR that subsequently activates PE related genes transcription and the appropriate light quality adaptation.

Interestingly, CcaS homologues are found in many cyanobacteria that do not perform obvious chromatic adaptation, including *Synechocystis* (Hirose *et al.*, 2008). In this species, CcaS activates transcription of the CpcG2 encoding gene under green light illumination. CpcG2-containing PBSs associate preferentially with PSI and might help maintaining a balanced excitation pressure between PSII and PSI (Watanabe *et al.*, 2014).

A new kind of chromatic adaptation was described in the filamentous cyanobacterium *Leptolyngbia* sp. JSC-1 (Gan *et al.*, 2014b). Upon far-red light illumination, cells extensively remodel their photosynthetic apparatus and they produce alternative photosystems subunits that can bind chlorophyll *d* and *f*. This remodelling process also produces alternative PBSs subunits, with red-shifted absorbance maxima. The whole process, called far-red light photoacclimation (FaRLiP), improves the absorption and utilization of far-red light. It results from the transcriptional activation of at least two gene clusters. One of these clusters contains genes encoding for the cyanobacteriochrome RfpA and the response regulators RfpC/RfpB. These probably regulate FaRLiP but more characterization has to be done. Interestingly, FaRLiP might be present in strains from all sections of the cyanobacterial phylum (Gan *et al.*, 2014a).

As highlighted above, cyanobacteriochromes have shifted absorbance maxima compared to plant phytochromes. They play a key role in controlling chromatic adaptation. Some algal phytochromes displays similar shifts (Rockwell *et al.*, 2014) but their potential modulatory roles in photoprotection or photoacclimation responses remain largely uncharacterized.

Cryptochromes are photoreceptors that bind flavin adenine dinucleotide (FAD) as a cofactor and absorb blue light (Chaves *et al.*, 2011; Galvão and Fankhauser, 2015). They appear closely related to photolyases, enzymes used for the repair of DNA after UV exposure, but have essentially lost this repair function (Chaves *et al.*, 2011). Cryptochromes modulate diverse processes such as the establishment of circadian rhythms or the priming of developmental transitions. They are present in plants, animals, fungi, algae and prokaryotic organisms (Chaves *et al.*, 2011).

Recent studies have shown that cryptochromes also regulate the expression of photosynthesis related genes. In the diatom *Phaeodactylum tricornutum*, overexpression of *PtCPF1* led to the differential expression of various LHC encoding genes and chlorophyll/carotenoid biosynthesis genes under blue light (Coesel *et al.*, 2009). Interestingly, this cryptochrome maintains some DNA repair activity as well.

A few more studies show the importance of cryptochromes in regulating the photosynthetic apparatus in diverse organisms. aCRY is an animal-type cryptochrome found in *Chlamydomonas* that is capable of sensing blue, red and yellow light (Beel *et al.*, 2012). *aCRY* mutants have reduced accumulation of a diverse set of transcripts involved in photosynthesis, such as *Lhcbm6* and various chlorophyll and carotenoid biosynthesis genes (Beel *et al.*, 2012). In the purple bacterium *Rhodobacter sphaeroides*, a *cryB* insertional mutant displayed lower accumulation of pigment binding proteins under blue illumination (Hendrischk *et al.*, 2009). Altogether, these results indicate that cryptochromes can elicit changes in the photosynthetic apparatus according to incident light quality. The underlying signalling pathways remain largely unknown and might partially overlap or compete with the phytochrome related signals.

There are other kinds of photoreceptors that might modulate photoacclimative or photoprotective processes, such as blue light sensing phototropins in plants and green algae or aureochromes in diatoms (Galvão and Fankhauser, 2015; Kianianmomeni and Hallmann, 2014; Lepetit and Dietzel, 2014).

Retrograde signals from plastid to nucleus

Chloroplasts are responsible not only for photosynthesis and the biosynthesis of many essential plant and algal compounds, but also for the perception of and acclimation to various environmental stresses such as high irradiance (Dietz, 2015; Neuhaus and Emes, 2000). The chloroplast organelle of Archaeplastidae (including plants) is the remnant of an ancient endosymbiosis between a cyanobacterium and a heterotrophic eukaryote. Plastids of other algae found in the eukaryotic tree of life are the result of independent and different endosymbiosis where separate heterotrophs enslaved a phototrophic eukaryote (Keeling, 2013). This complex evolutionary history has led to a complex coordination of chloroplast and extra-chloroplast metabolism. This is complicated by the chloroplasts' dependence on the nucleus for, on average, over 95% of their proteins (Barbrook *et al.*, 2006). In plants and algae, this coordination is achieved by intracellular signalling from the chloroplast to the nucleus, so-called retrograde signalling (Chi *et al.*, 2013; Dietz, 2015; Pogson *et al.*, 2008). Recent advances in our understanding of retrograde signalling have led to an emerging appreciation of the sheer diversity of signals involved in triggering light acclimation.

Plastoquinone pool-dependent signalling

The redox state of the PQ pool, the electron acceptor of PSII, has historically been considered a major initiation point in transcriptional regulation of photoacclimation. It is easily manipulated experimentally by adding the inhibitors DCMU (3-(3,4-dichlorophenyl)-1,1-dimethylurea) to induce an oxidized PQ pool and DBMIB (2,5-dibromo-3-methyl-6-isopropyl-p-benzoquinone) to reduce the PQ pool. The application of these inhibitors has been instrumental in unequivocally linking the expression of *petE, lhcb,* and *apx* to the redox state of the PQ pool (Escoubas *et al.*, 1995; Fey *et al.*, 2005). Manipulation of PQ pool has demonstrated that some LHCs required for qE, LHCSR3 from *Chlamydomonas* and LHCX1 and LHCX2 from *Phaeodactylum*, are induced in response to PQ reduction (Lepetit *et al.*, 2013; Maruyama *et al.*, 2014).

The PQ redox signal controls more than just genes involved directly in photosynthesis. Alternative splicing is a form of post-transcriptional regulation that involves processing a single pre-mRNA into multiple mRNAs, which can either encode a different protein isoform or lead degradation of mRNA through nonsense mediated decay (McGlincy and Smith, 2008). Petrillo *et al.* (2014) used DCMU/DBMIB to illustrate that the *Arabidopsis* nuclear serine arginine rich splicing factor, *AtRS31*, undergoes alternative splicing in response to a reduced PQ pool. *AtRS31* belongs to a large family of serine arginine splicing factors involved in constitutive and alternative splicing (Reddy *et al.*, 2013) and the mRNA encoding this gene was already known to be alternatively spliced in response to temperature stress (Palusa *et al.*, 2007). Several other light-dependent splicing events were observed under a light/dark treatment. These spliced transcripts mainly encoded proteins involved in RNA-binding and processing as well as DNA binding proteins (Petrillo *et al.*, 2014). Interestingly, RuBisCO activase was the only photosynthetic protein shown to undergo alternative splicing in response to a light/dark regime. Surprisingly, light-dependent *AtRS31* splicing was observed in both leaves and roots, with roots showing at least a 2-hour time delay response (Petrillo *et al.* 2014). This suggests that an uncharacterized signal reporting the plastoquinone pool redox state can be transmitted across plant tissues. These observations highlight the complexity and critical importance of alternative splicing in light response within the whole plant. Future studies focused on the impact of these splicing events on the restructuring of the chloroplast proteome under different light regimes could reveal their overall importance to metabolism.

Systems-level responses to PQ manipulation, measured by RNAseq and proteomics focused on posttranslational modification, will undoubtedly lead to the identification of genes under the control of the PQ pool, and possibly even some of the transcription factors that control gene expression.

Reactive oxygen species

The build-up of excitation pressure on PSII by the over reduction of the electron transport chain increases the probability of $^1O_2$ production by excited chlorophyll at PSII (Krieger-Liszkay, 2005) and superoxide and subsequent hydrogen peroxide production through oxygen reduction by PSI (Asada, 2006). $^1O_2$ is proposed to be the major ROS generated in excess light (Gonzalez-Perez *et al.*, 2011), but both $^1O_2$ and H_2O_2 act as potent signals for retrograde signalling. This mediates acclimation to light stress through the expression of genes controlling the production of ROS scavenging pigments, proteins, and metabolites or by instigating programmed cell death (Chi *et al.*, 2013).

New components of the excess light stress-sensing pathway have been isolated from mutants that show unusual responses to exogenous $^1O_2$ exposure. The singlet oxygen-resistant 1 (*sor1*) mutant was isolated from a UV-mutagenized population of *Chlamydomonas* that was screened for constitutive tolerance to the $^1O_2$-photosensitizer Rose Bengal (Fischer *et al.*, 2012). The *SOR1* gene encodes a putative basic leucine zipper domain transcription factor. In the *sor1* line, a gain of function mutation led to constitutively higher expression of both *SOR1* and oxidative stress response genes. Intriguingly, an 8-bp palindromic sequence element was enriched in many of these genes. This element was shown to be required for the transcriptional induction of the σ-class glutathione-S-transferase gene *GSTS1* in response to increased levels of reactive electrophile species such as oxylipins (Fischer *et al.*, 2012). These findings implicate SOR1 in regulating genes associated with detoxification of $^1O_2$-induced lipid peroxidation and that these genes may also play a part in acclimation to $^1O_2$ (see discussion of oxylipins in 'Integrating complex retrograde signals').

SAK1 is another putative basic leucine zipper domain transcription factor involved in the $^1O_2$ response in *Chlamydomonas* (Wakao *et al.*, 2014). In WT cells, the SAK1 protein is accumulated 6- to 10-fold in the cytosol and is phosphorylated in response to $^1O_2$. *sak1* mutants show increased sensitivity to $^1O_2$ exposure (Wakao *et al.*, 2014). *sak1* mutants showed a reduction in transcript accumulation of two important high light stress-related proteins; glutathione peroxidase (GPX5), which responds specifically to $^1O_2$ challenges (Ledford *et al.*, 2007) and LHCSR1, which is involved in qE response (Truong *et al.* unpublished). All these pieces together point towards SAK1 acting as an intermediary in the $^1O_2$-dependent retrograde signal.

Methylene blue sensitivity (MBS) is a conserved zinc finger protein that is also required for singlet oxygen-dependent retrograde signalling (Shao *et al.*, 2013). *mbs* mutants in *Chlamydomonas* and *Arabidopsis* are hypersensitive to photooxidative stress while overexpression in plants leads to increased tolerance. MBS accumulates in distinct RNA granules in the cytosol during exposure to oxidative stress. Perhaps MBS affects mRNA stability, translation of a specific mRNA or another intermediary factor controlling $^1O_2$ acclimation (Shao *et al.*, 2013).

Brzezowski *et al.* (2012) isolated a GPX5 regulation mutant of *Chlamydomonas* that had increased sensitivity to exogenously generated $^1O_2$. This mutant was missing a paralogue of the PSBP1 protein (PSBP2) (Brzezowski *et al.*, 2012). PSBP1 is involved in the water-splitting reactions of the oxygen-evolving complex (Ifuku *et al.*, 2008). The PSPB2 peptide may be targeted to the chloroplast but its role in $^1O_2$ detection or signal transduction remains unclear at this time.

Unlike $^1O_2$, H_2O_2 generated in the chloroplast can diffuse to the cytosol (Mubarakshina *et al.*, 2010). While no responsive element has been identified in the H_2O_2 retrograde signalling cascade, H_2O_2 accumulation in the cytosol also appears to regulate excess light stress responses. Work in *Chlamydomonas* has shown that H_2O_2 transiently accumulates during the first 10 min following a shift from dark to high light and this accumulation occurs concurrently with a decrease in catalase activity (Michelet *et al.*, 2013). This transient rise in H_2O_2 was shown to activate the expression of the nuclear encoded genes for the antioxidant enzymes ascorbate peroxidase (APX) and GPX, enzymes involved in chloroplast H_2O_2 defence response (Michelet *et al.*, 2013).

Pigment-related metabolites participating in retrograde signalling

Contemporary studies of retrograde signalling have also highlighted the importance of ROS-dependent oxidative products, chlorophyll biosynthetic intermediates and chloroplast metabolites as potential signalling molecules.

Carotenoids have received extensive study for their role in quenching $^1O_2$ generated by PSII (Telfer, 2014). The reaction of $^1O_2$ with β-carotene results in the generation of a number of volatile short-chain compounds, including β-cyclocitral, and these are accumulated in light-stressed *Arabidopsis* (Ramel *et al.*, 2012a). Five different $^1O_2$ stress marker transcripts were up-regulated in *Arabidopsis* leaves after exposure to exogenous β-cyclocitral. Specific markers for H_2O_2 stress were not affected by exogenous β-cyclocitral (Ramel *et al.*, 2012b). These results suggest that oxidative damage products derived from carotenoids can also serve as a retrograde signal.

Metabolic intermediates involved in tetrapyrrole biosynthesis exert control over the transcription of photosynthetic nuclear genes during dark to light transitions. Four classes of tetrapyrroles – chlorophyll, haem, sirohaem, and phytochromobilin – are generated in the chloroplast from a common biosynthetic pathway (Tanaka and Tanaka, 2007). The genome uncoupled 4 (GUN4) mutants have been instrumental for the characterization of tetrapyrroles in retrograde signalling. Lately, Brzezowski *et al.* (2014) proposed a modified model for the role of GUN4 in retrograde signalling in *Chlamydomonas* whereby it is the protein itself, and not a metabolite alone, that senses the accumulation of chlorophyll biosynthesis intermediates and mediates the singlet oxygen signal to the retrograde pathway.

Haem has also been shown to be a retrograde signal in *Chlamydomonas* and *Arabidopsis* (von Gromoff *et al.*, 2008; Woodson *et al.*, 2011). Haem can be converted to linear bilins through the action of two enzymes, a haem oxygenase (HMOX1) and a ferredoxin-dependent bilin reductase (PCYA). These bilins are then integrated into phytochromes in plants (Quail, 2007). Phytochromes are absent in *Chlamydomonas,* but the production of bilins occurs nonetheless. They are required for efficient growth in the light and are proposed to be a retrograde signal (Duanmu *et al.*, 2013) A multifactor RNAseq experiment found a set of genes that were induced during a dark to light shift were also repressed by exogenously added bilins. These included high light stress response genes such as

LHCSR1, *PSBS1*, *ELI3*, *GPX*, *HSP70* and others (Duanmu *et al.*, 2013). Bilins clearly play some role in dampening the excess light response, but it is not clear how.

Other chloroplast metabolites involved in retrograde signalling

Finally, recent studies have suggested a series of unexpected chloroplastic metabolites may also serve as signals for retrograde gene regulation in the nucleus. The first is dihydroxyacetone phosphate (DHAP), an intermediate of the Calvin–Benson–Bassham cycle that is exported to the cytosol by the triose phosphate/phosphate translocators. Vogel *et al.* (2014) proposed a model where mitogen-activated protein kinase 6 is activated by the exported DHAP in excess light. Within 10–45 min, the kinase influences the transcription of high light stress acclimation Apetala2/ethylene response factors (Dietz *et al.*, 2010; Lata and Prasad, 2011; Vogel *et al.*, 2014).

The second signal appears to be an isoprenoid precursor (MEcPP, methylerythritol cyclodiphosphate) produced in the chloroplast. This, and other isoprenoids, accumulate in leaves during light and heat stress (Xiao *et al.*, 2012). A *ceh1* mutant that accumulates MEcPP along with pharmacological manipulations of MEcPP levels convincingly demonstrated that the molecule increases nuclear transciption of a high light stress reponse gene encoding a plastid localized hydroperoxide lyase. Interstingly, MEcPP did not affect any of the gun responsive genes, indicating independent signalling pathways (Xiao *et al.*, 2012).

Finally, the nucleotide precursor metabolite, 3′-phosphoadenosine 5′-phosphate (PAP), regulates retrograde signalling in high light and drought stress through the likely inhibition of a nuclear 5′ to 3′ exoribonuclease (XRNs, Estavillo *et al.*, 2011). PAP is produced by sulfation reactions in the chloroplast and high light induces a 20-fold accumulation of PAP, which can move out of the chloroplast (Estavillo *et al.*, 2011). This accumulation correlated with the increased expression of the high light responsive gene ELIP2 and drought tolerance. PAP in yeast is known to inhibit XRNs (Dichtl *et al.*, 1997) and Estavillo *et al.* (2011) showed that mutants lacking XRNs showed similar nuclear gene expression to mutants with altered PAP levels. Clearly, these three examples, along with the examples provided above, reveal an incredible complexity in the signals that influence light acclimation state.

Calcium

Calcium is an essential nutrient that plays structural roles in photosynthetic proteins and cell walls as well as signalling functions in both development and environmental stress response (Gilroy *et al.*, 2014; Hochmal *et al.*, 2015). Calcium binding proteins like calmodulin are important actors in the calcium-dependent signalling cascades involved in activating transcriptional regulation. Recent work in *Chlamydomonas* on the chloroplast localized, calcium sensor CAS demonstrated that both CAS and Ca^{2+} are involved in the nuclear regulation of LHCSR3 during acclimation to high light (Petroutsos *et al.*, 2011). CAS-RNAi knockdown lines displayed reduced LHCSR3 protein accumulation, which was rescued by a 10-fold increase in Ca^{2+} concentration. This together with the repression of LHCSR3 protein accumulation and gene expression by calmodulin inhibitors indicates that Ca^{2+} play a central role in the light-dependent retrograde regulation of LHCSR3 (Maruyama *et al.*, 2014). The role of this possible retrograde signalling pathway in other photoacclimation responses remains uncharacterized.

Integrating complex retrograde signals

Efficient high light acclimation requires a range of temporally separated responses that occur across multiple signalling pathways (Dietz, 2015). Systems wide characterization of retrograde signals responding to high light stress observed some of the metabolites described above in addition to the oxylipins oxophytodeinoic acid and jasmonic acid, and redox cues as critical components in light-dependent retrograde signalling (Alsharafa *et al.*, 2014; Oelze *et al.*, 2012). Their accumulation and effects are temporally separated. The overlap of these retrograde signalling pathways agrees with the ‘cross tolerance hypothesis’ where multiple retrograde signals lead to synergistic co-activation of non-specific and specific stress responsive pathways enabling tolerance to both biotic and abiotic stresses (Bartoli *et al.*, 2013). This hypothesis is supported by a meta-analysis of retrograde signalling research on *Arabidopsis* that found a core response module consisting of 36 transcripts (Gläßer *et al.*, 2014). This core module consists of a variety of pathways including genes involved in transcriptional regulation, biotic defence response, cell wall establishment and cell wall synthesis. There was a distinct lack of chloroplast generated signals in regulating general plant physiology. Contrary to the findings of Alsharafa *et al.* (2014), this analysis found that auxin-based signalling, and not oxylipins, is the key secondary component integrating plastidial signals in the response cascade. This contradiction highlights the complexity of determining the specific signalling mechanisms involved in high light acclimation.

Conclusions

Photoprotection and light acclimation respond to excess light stress over a wide variety of time scales. This is fitting for a resource whose supply can vary by orders of magnitude within seconds or be slowly modulated on and off over a day. The staggeringly complex physiological coordination of photoprotection strategies while maintaining efficient carbon fixation represents one of biology's most complex problems. Many of the new players mentioned in this review play a role in regulating this complexity and we are only just beginning to understand how the system is coordinated by post-translational changes within the thylakoids, chloroplast or whole cell. Our recent advances in understanding how chloroplast biology controls genetic regulation in the nucleus suggests that a vast number of signals communicate light, energetic status and stress.

Future prospects

While assembling our ideas for this chapter, one thing became abundantly clear: we know very little about the mechanisms associated with low light acclimation and how photoprotective and light acclimation strategies behave in natural light systems. There is a clear physiological and genetic regulation of photosynthesis associated with growing in these environments (Demmig-Adams *et al.*, 2012). We will need to think 'out of the chloroplast' and consider how the ecophysiology of an entire organism influences the essential, but integrated, metabolism of the cell (see also Demmig-Adams *et al.*, 2014).

However, there are still obviously more important players waiting to be discovered. Forward and reverse genetic screens in single-shift experiments will continue to reveal new regulators of metabolism, increasing our catalogue of alternate energy dissipation pathways and the transcriptional and post-translational factors that regulate them. We believe that systems-based inquiries into whole organism responses to light will provide unexpected insights into photosynthetic metabolism in changing light. An excellent example of this is the recent work investigating the systems response of *Chlamydomonas* during a low light to moderate light shift (Mettler *et al.*, 2014). They found short-term (< 0.5 h) changes in catabolic metabolism fuelled increased growth, independent of photosynthetic rate. Importantly, many proteins associated with photosynthesis changed independently of transcript abundance and post-translational modifications likely played a strong role in metabolism changes. Applying this same approach to some of the mutants described in this chapter and the meta-analyses of these data will draw us closer to a full understanding of the regulation of light harvesting in a variable light environment.

References

Adir, N. (2005). Elucidation of the molecular structures of components of the phycobilisome: reconstructing a giant. Photosynth. Res. *85*, 15–32.

Ahn, T.K., Avenson, T.J., Ballottari, M., Cheng, Y.-C., Niyogi, K.K., Bassi, R., and Fleming, G.R. (2008). Architecture of a charge-transfer state regulating light harvesting in a plant antenna protein. Science *320*, 794–797.

Alboresi, A., Ballottari, M., Hienerwadel, R., Giacometti, G.M., and Morosinotto, T. (2009). Antenna complexes protect Photosystem I from Photoinhibition. BMC Plant Biol. 9.

Alboresi, A., Gerotto, C., Giacometti, G.M., Bassi, R., and Morosinotto, T. (2010). *Physcomitrella patens* mutants affected on heat dissipation clarify the evolution of photoprotection mechanisms upon land colonization. Proc. Natl. Acad. Sci. U.S.A. *107*, 11128–11133.

Allahverdiyeva, Y., Ermakova, M., Eisenhut, M., Zhang, P., Richaud, P., Hagemann, M., Cournac, L., and Aro, E.-M. (2011). Interplay between flavodiiron proteins and photorespiration in *Synechocystis* sp. PCC 6803. J. Biol. Chem. *286*, 24007–24014.

Allahverdiyeva, Y., Mustila, H., Ermakova, M., Bersanini, L., Richaud, P., Ajlani, G., Battchikova, N., Cournac, L., and Aro, E.-M. (2013). Flavodiiron proteins Flv1 and Flv3 enable cyanobacterial growth and photosynthesis under fluctuating light. Proc. Natl. Acad. Sci. U.S.A. *110*, 4111–4116.

Allen, J.F. (1992). Protein-hosphorylation in regulation of photosynthesis Biochim. Biophys. Acta *1098*, 275–335.

Allorent, G., Tokutsu, R., Roach, T., Peers, G., Cardol, P., Girard-Bascou, J., Petroutsos, D., Kuntz, M., Franck, F., Niyogi, K.K., *et al.* (2013). A dual strategy to cope with high light in *Chlamydomonas reinhardtii*. Plant Cell *25*, 545–557.

Alsharafa, K., Vogel, M.O., Oelze, M.-L., Moore, M., Stingl, N., Koenig, K., Friedman, H., Mueller, M.J., and Dietz, K.-J. (2014). Kinetics of retrograde signalling initiation in the high light response of *Arabidopsis thaliana*. Phil. Trans. R. Soc. B *26*, 1151–1165.

Apel, K., and Hirt, H. (2004). Reactive oxygen species: Metabolism, oxidative stress, and signal transduction. Annu. Rev. Plant Biol. *55*, 373–399.

Armbruster, U., Carrillo, L.R., Venema, K., Pavlovic, L., Schmidtmann, E., Kornfeld, A., Jahns, P., Berry, J.A., Kramer, D.M., and Jonikas, M.C. (2014). Ion antiport accelerates photosynthetic acclimation in fluctuating light environments. Nat Comm *5*, 5439.

Asada, K. (2006). Production and scavenging of reactive oxygen species in chloroplasts and their functions. Plant Physiol. *141*, 391–396.

Aspinall-O'Dea, M., Wentworth, M., Pascal, A., Robert, B., Ruban, A., and Horton, P. (2002). *In vitro* reconstitution of the activated zeaxanthin state associated with energy dissipation in plants. Proc. Natl. Acad. Sci. U.S.A. *99*, 16331–16335.

Bailleul, B., Rogato, A., de Martino, A., Coesel, S., Cardol, P., Bowler, C., Falciatore, A., and Finazzi, G. (2010). An atypical member of the light-harvesting complex stress-related protein family modulates diatom responses to light. Proc. Natl. Acad. Sci. U.S.A. *107*, 18214–18219.

Ballottari, M., Alcocer, M.J.P., D'Andrea, C., Viola, D., Ahn, T.K., Petrozza, A., Polli, D., Fleming, G.R., Cerullo, G., and Bassi, R. (2014). Regulation of photosystem I light harvesting by zeaxanthin. Proc. Natl. Acad. Sci. U.S.A. *111*, 2431–2438.

Barbrook, A.C., Howe, C.J., and Purton, S. (2006). Why are plastid genomes retained in non-photosynthetic organisms? Trends Plant Sci. *11*, 101–108.

Bartoli, C.G., Casalongué, C.A., Simontacchi, M., Marquez-Garcia, B., and Foyer, C.H. (2013). Interactions between hormone and redox signalling pathways in the control of growth and cross tolerance to stress. Environ. Exp. Bot. *94*, 73–88.

Bauwe, H., Hagemann, M., and Fernie, A.R. (2010). Photorespiration: players, partners and origin. Trends Plant Sci. *15*, 330–336.

Beel, B., Prager, K., Spexard, M., Sasso, S., Weiss, D., Mueller, N., Heinnickel, M., Dewez, D., Ikoma, D., Grossman, A.R., *et al.* (2012). A flavin binding cryptochrome photoreceptor responds to both blue and red light in *Chlamydomonas reinhardtii*. Plant Cell *24*, 2992–3008.

Bellafiore, S., Barneche, F., Peltier, G., and Rochaix, J.D. (2005). State transitions and light adaptation require chloroplast thylakoid protein kinase STN7. Nature *433*, 892–895.

Berera, R., van Stokkum, I.H.M., Gwizdala, M., Wilson, A., Kirilovsky, D., and van Grondelle, R. (2012). The photophysics of the orange carotenoid protein, a light-powered molecular switch. J. Phys. Chem. B *116*, 2568–2574.

Bersanini, L., Battchikova, N., Jokel, M., Rehman, A., Vass, I., Allahverdiyeva, Y., and Aro, E.-M. (2014). Flavodiiron protein Flv2/Flv4-related photoprotective mechanism dissipates excitation pressure of psii in cooperation with phycobilisomes in cyanobacteria. Plant Physiol. *164*, 805–818.

Bonente, G., Ballottari, M., Truong, T.B., Morosinotto, T., Ahn, T.K., Fleming, G.R., Niyogi, K.K., and Bassi, R. (2011). Analysis of lhcsr3, a protein essential for feedback de-excitation in the green alga *Chlamydomonas reinhardtii*. PLoS Biol. *9*, e1000577.

Boulay, C., Wilson, A., D'Haene, S., and Kirilovsky, D. (2010). Identification of a protein required for recovery of full antenna capacity in OCP-related photoprotective mechanism in cyanobacteria. Proc. Natl. Acad. Sci. U.S.A. *107*, 11620–11625.

Brooks, M., Jansson, S., and Niyogi, K. (2014). PsbS-dependent non-photochemical quenching. In Non-Photochemical Quenching and Energy Dissipation in Plants, Algae and Cyanobacteria, B. Demmig-Adams, G. Garab, W. Adams III, and Govindjee, eds. (Dordrecth, The Netherlands: Springer Netherlands), pp. 297–314.

Brzezowski, P., Wilson, K.E., and Gray, G.R. (2012). The PSBP2 protein of *Chlamydomonas reinhardtii* is required for singlet oxygen-dependent signaling. Planta *236*, 1289–1303.

Brzezowski, P., Schlicke, H., Richter, A., Dent, R.M., Niyogi, K.K., and Grimm, B. (2014). The GUN4 protein plays a regulatory role in tetrapyrrole biosynthesis and chloroplast-to-nucleus signalling in *Chlamydomonas reinhardtii*. Plant J. *79*, 285–298.

Carol, P., Stevenson, D., Bisanz, C., Breitenbach, J., Sandmann, G., Mache, R., Coupland, G., and Kuntz, M. (1999). Mutations in the *Arabidopsis* gene immutans cause a variegated phenotype by inactivating a chloroplast terminal oxidase associated with phytoene desaturation. Plant Cell *11*, 57–68.

Carraretto, L., Formentin, E., Teardo, E., Checchetto, V., Tomizioli, M., Morosinotto, T., Giacometti, G.M., Finazzi, G., and Szabó, I. (2013). A thylakoid-located two-pore k+ channel controls photosynthetic light utilization in plants. Science *342*, 114–118.

Chaves, I., Pokorny, R., Byrdin, M., Hoang, N., Ritz, T., Brettel, K., Essen, L.-O., van der Horst, G.T.J., Batschauer, A., and Ahmad, M. (2011). The cryptochromes: blue light photoreceptors in plants and animals. Annu. Rev. Plant Biol. *62*, 335–364.

Chi, W., Sun, X., and Zhang, L. (2013). Intracellular signaling from plastid to nucleus. Annu. Rev. Plant Biol. *64*, 559–582.

Chukhutsina, V., Bersanini, L., Aro, E.-M., and van Amerongen, H. (2015). Cyanobacterial flv4–2 operon-encoded proteins optimize light harvesting and charge separation in photosystem II. Mol. Plant *8*, 747–761.

Coesel, S., Mangogna, M., Ishikawa, T., Heijde, M., Rogato, A., Finazzi, G., Todo, T., Bowler, C., and Falciatore, A. (2009). Diatom PtCPF1 is a new cryptochrome/photolyase family member with DNA repair and transcription regulation activity. EMBO Rep. *10*, 655–661.

Dang, K.-V., Plet, J., Tolleter, D., Jokel, M., Cuine, S., Carrier, P., Auroy, P., Richaud, P., Johnson, X., Alric, J., *et al.* (2014). Combined increases in mitochondrial cooperation and oxygen photoreduction compensate for deficiency in cyclic electron flow in *Chlamydomonas reinhardtii.* Plant Cell *26,* 3036–3050.

Delosme, R., Olive, J., and Wollman, F.A. (1996). Changes in light energy distribution upon state transitions: An *in vivo* photoacoustic study of the wild type and photosynthesis mutants from *Chlamydomonas reinhardtii.* Biochim. Biophys. Acta Bioenerg. *1273,* 150–158.

Demarsac, N.T. (1977). Occuarence and nature of chromatic adaptation in cyanobacteria. J. Bacteriol. *130,* 82–91.

Demmig-Adams, B., Cohu, C.M., Muller, O., and Adams III, W.W. (2012). Modulation of photosynthetic energy conversion efficiency in nature: from seconds to seasons. Photosynth. Res. *113,* 75–88.

Demmig-Adams, B., Stewart, J.J., and Adams, W.W. (2014). Multiple feedbacks between chloroplast and whole plant in the context of plant adaptation and acclimation to the environment. Phil. Trans. R. Soci. B 369.

Derks, A., Schaven, K., and Bruce, D. (2015). Diverse mechanisms for photoprotection in photosynthesis. Dynamic regulation of photosystem II excitation in response to rapid environmental change. Biochim. Biophys. Acta Bioenerg. *1847,* 468–485.

Dichtl, B., Stevens, A., and Tollervey, D. (1997). Lithium toxicity in yeast is due to the inhibition of RNA processing enzymes. The EMBO Journal *16,* 7184–7195.

Dietz, K.-J., Vogel, M., and Viehhauser, A. (2010). AP2/EREBP transcription factors are part of gene regulatory networks and integrate metabolic, hormonal and environmental signals in stress acclimation and retrograde signalling. Protoplasma *245,* 3–14.

Dietz, K.J. (2015). Efficient high light acclimation involves rapid processes at multiple mechanistic levels. J. Exp. Bot. *66,* 2401–2414.

Dominici, P., Caffarri, S., Armenante, F., Ceoldo, S., Crimi, M., and Bassi, R. (2002). Biochemical properties of the PsbS subunit of photosystem II either purified from chloroplast or recombinant. J. Biol. Chem. *277,* 22750–22758.

Drop, B., Yadav, S.K.N., Boekema, E.J., and Croce, R. (2014). Consequences of state transitions on the structural and functional organization of Photosystem I in the green alga *Chlamydomonas reinhardtii.* Plant J. *78,* 181–191.

Duanmu, D., Casero, D., Dent, R.M., Gallaher, S., Yang, W., Rockwell, N.C., Martin, S.S., Pellegrini, M., Niyogi, K.K., Merchant, S.S., *et al.* (2013). Retrograde bilin signaling enables *Chlamydomonas* greening and phototrophic survival. Proc. Natl. Acad. Sci. U.S.A. *110,* 3621–3626.

Duanmu, D., Bachy, C., Sudek, S., Wong, C.-H., Jimenez, V., Rockwell, N.C., Martin, S.S., Ngan, C.Y., Reistetter, E.N., van Baren, M.J., *et al.* (2014). Marine algae and land plants share conserved phytochrome signaling systems. Proc. Natl. Acad. Sci. U.S.A. *111,* 15827–15832.

Eisenhut, M., Georg, J., Klaehn, S., Sakurai, I., Mustila, H., Zhang, P., Hess, W.R., and Aro, E.-M. (2012). The Antisense RNA As1_flv4 in the cyanobacterium *Synechocystis* sp PCC 6803 prevents premature expression of the flv4-2 operon upon shift in inorganic Carbon supply. J. Biol. Chem. *287,* 33153–33162.

Eisenhut, M., Pick, T.R., Bordych, C., and Weber, A.P.M. (2013). Towards closing the remaining gaps in photorespiration – the essential but unexplored role of transport proteins. Plant Biol. *15,* 676–685.

Eisenhut, M., Hocken, N., and Weber, A.P.M. (2014). Plastidial metabolite transporters integrate photorespiration with carbon, nitrogen, and sulfur metabolism. Cell Calcium *58,* 98–104.

Elrad, D., Niyogi, K.K., and Grossman, A.R. (2002). A major light-harvesting polypeptide of photosystem II functions in thermal dissipation. Plant Cell *14,* 1801–1816.

Emlyn-Jones, D., Ashby, M.K., and Mullineaux, C.W. (1999). A gene required for the regulation of photosynthetic light harvesting in the cyanobacterium *Synechocystis* 6803. Mol. Microbiol. *33,* 1050–1058.

Ermakova, M., Battchikova, N., Allahverdiyeva, Y., and Aro, E.-M. (2013). Novel heterocyst-specific flavodiiron proteins in Anabaena sp PCC 7120. FEBS Lett. *587,* 82–87.

Ermakova, M., Battchikova, N., Richaud, P., Leino, H., Kosourov, S., Isojarvi, J., Peltier, G., Flores, E., Cournac, L., Allahverdiyeva, Y., *et al.* (2014). Heterocyst-specific flavodiiron protein Flv3B enables oxic diazotrophic growth of the filamentous cyanobacterium Anabaena sp. PCC 7120. Proc. Natl. Acad. Sci. U.S.A. *111,* 11205–11210.

Escoubas, J.M., Lomas, M., Laroche, J., and Falkowski, P.G. (1995). Light-intensity regulation of CAB gene-transcription is signaled by the redox state of the plastoquinone pool. Proc. Natl. Acad. Sci. U.S.A. *92,* 10237–10241.

Estavillo, G.M., Crisp, P.A., Pornsiriwong, W., Wirtz, M., Collinge, D., Carrie, C., Giraud, E., Whelan, J., David, P., Javot, H., *et al.* (2011). Evidence for a SAL1-PAP chloroplast retrograde pathway that functions in drought and high light signaling in *Arabidopsis.* Plant Cell *23,* 3992–4012.

Falk, J., and Munne-Bosch, S. (2010). Tocochromanol functions in plants: antioxidation and beyond. J. Exp. Bot. *61,* 1549–1566.

Falkowski, P.G., and Raven, J.A. (2013). Aquatic Photosynthesis, 2nd edn (Princeton, NJ: Princeton University Press).

Fey, V., Wagner, R., Brautigam, K., Wirtz, M., Hell, R., Dietzmann, A., Leister, D., Oelmuller, R., and Pfannschmidt, T. (2005). Retrograde plastid redox signals in the expression of nuclear genes for chloroplast proteins of *Arabidopsis thaliana.* J. Biol. Chem. *280,* 5318–5328.

Finazzi, G., Petroutsos, D., Tomizioli, M., Flori, S., Sautron, E., Villanova, V., Rolland, N., and Seigneurin-Berny, D. (2015). Ions channels/transporters and chloroplast regulation. Cell Calcium *58,* 86–97.

Fischer, B.B., Ledford, H.K., Wakao, S., Grace, S., Casero, D., and Pellegrini, M., *et al.* (2012). SINGLET OXYGEN RESISTANT 1 links reactive electrophile signaling to singlet oxygen acclimation in *Chlamydomonas reinhardtii.* Proc. Natl. Acad. Sci. U.S.A. *109,* E1302-E1311.

Formighieri, C., Ceol, M., Bonente, G., Rochaix, J.-D., and Bassi, R. (2012). Retrograde signaling and photoprotection in a gun4 mutant of *Chlamydomonas reinhardtii.* Mol. Plant *5,* 1242–1262.

Foyer, C.H., and Noctor, G. (2011). Ascorbate and glutathione: the heart of the redox hub. Plant Physiol. *155,* 2–18.

Foyer, C.H., and Shigeoka, S. (2011). Understanding oxidative stress and antioxidant functions to enhance photosynthesis. Plant Physiol. *155*, 93–100.

Galvão, V.C., and Fankhauser, C. (2015). Sensing the light environment in plants: photoreceptors and early signaling steps. Curr. Opin. Neurobiol. *34*, 46–53.

Gan, F., Shen, G., and Bryant, D.A. (2014a). Occurrence of far-red light photoacclimation (FaRLiP) in diverse cyanobacteria. Life 5, 4–24.

Gan, F., Zhang, S., Rockwell, N.C., Martin, S.S., Lagarias, J.C., and Bryant, D.A. (2014b). Extensive remodeling of a cyanobacterial photosynthetic apparatus in far-red light. Science *345*, 1312–1317.

Gilroy, S., Suzuki, N., Miller, G., Choi, W.G., Toyota, M., Devireddy, A.R., and Mittler, R. (2014). A tidal wave of signals: calcium and ROS at the forefront of rapid systemic signaling. Trends Plant Sci. *19*, 623–630.

Giovagnetti, V., Ware, M., and Ruban, A. (2015). Assessment of the impact of photosystem I chlorophyll fluorescence on the pulse-amplitude modulated quenching analysis in leaves of *Arabidopsis thaliana*. Photosynth. Res. doi: 10.1007/s11120-015-0087-z, 1–11.

Gläßer, C., Haberer, G., Finkemeier, I., Pfannschmidt, T., Kleine, T., Leister, D., Dietz, K.-J., Häusler, R.E., Grimm, B., and Mayer, K.F.X. (2014). Meta-analysis of retrograde signaling in *Arabidopsis thaliana* reveals a core module of genes embedded in complex cellular signaling networks. Mol. Plant 7, 1167–1190.

Gonzalez-Perez, S., Gutierrez, J., Garcia-Garcia, F., Osuna, D., Dopazo, J., Lorenzo, O., Revuelta, J.L., and Arellano, J.B. (2011). Early transcriptional defense responses in *Arabidopsis* cell suspension culture under high-light conditions. Plant Physiol. *156*, 1439–1456.

Goral, T.K., Johnson, M.P., Duffy, C.D.P., Brain, A.P.R., Ruban, A.V., and Mullineaux, C.W. (2012). Light-harvesting antenna composition controls the macrostructure and dynamics of thylakoid membranes in *Arabidopsis*. Plant J. *69*, 289–301.

Grewe, S., Ballottari, M., Alcocer, M., D'Andrea, C., Blifernez-Klassen, O., Hankamer, B., Mussgnug, J.H., Bassi, R., and Kruse, O. (2014). Light-harvesting complex protein LHCBM9 is critical for photosystem ii activity and hydrogen production in *Chlamydomonas reinhardtii*. Plant Cell 26, 1598–1611.

Gwizdala, M., Wilson, A., and Kirilovsky, D. (2011). *In vitro* reconstitution of the cyanobacterial photoprotective mechanism mediated by the orange carotenoid protein in *Synechocystis* PCC 6803. Plant Cell 23, 2631–2643.

Hebbelmann, I., Selinski, J., Wehmeyer, C., Goss, T., Voss, I., Mulo, P., Kangasjarvi, S., Aro, E.-M., Oelze, M.-L., Dietz, K.-J., *et al.* (2012). Multiple strategies to prevent oxidative stress in *Arabidopsis* plants lacking the malate valve enzyme NADP-malate dehydrogenase. J. Exp. Bot. *63*, 1445–1459.

Helman, Y., Tchernov, D., Reinhold, L., Shibata, M., Ogawa, T., Schwarz, R., Ohad, I., and Kaplan, A. (2003). Genes encoding a-type flavoproteins are essential for photoreduction of O-2 in cyanobacteria. Curr. Biol. *13*, 230–235.

Helman, Y., Barkan, E., Eisenstadt, D., Luz, B., and Kaplan, A. (2005). Fractionation of the three stable oxygen isotopes by oxygen-producing and oxygen-consuming reactions in photosynthetic organisms. Plant Physiol. *138*, 2292–2298.

Hendrischk, A.K., Fruhwirth, S.W., Moldt, J., Pokorny, R., Metz, S., Kaiser, G., Jager, A., Batschauer, A., and Klug, G. (2009). A cryptochrome-like protein is involved in the regulation of photosynthesis genes in *Rhodobacter sphaeroides*. Mol. Microbiol. *74*, 990–1003.

Heyno, E., Innocenti, G., Lemaire, S.D., Issakidis-Bourguet, E., and Krieger-Liszkay, A. (2014). Putative role of the malate valve enzyme NADP-malate dehydrogenase in H_2O2 signalling in *Arabidopsis*. Phil. Trans. R. Soc. B *369*, 20130228.

Hirose, Y., Shimada, T., Narikawa, R., Katayama, M., and Ikeuchi, M. (2008). Cyanobacteriochrome CcaS is the green light receptor that induces the expression of phycobilisome linker protein. Proc. Natl. Acad. Sci. U.S.A. *105*, 9528–9533.

Hirose, Y., Narikawa, R., Katayama, M., and Ikeuchi, M. (2010). Cyanobacteriochrome CcaS regulates phycoerythrin accumulation in Nostoc punctiforme, a group II chromatic adapter. Proc. Natl. Acad. Sci. U.S.A. *107*, 8854–8859.

Hochmal, A.K., Schulze, S., Trompelt, K., and Hippler, M. (2015). Calcium-dependent regulation of photosynthesis. Biochim. Biophys. Acta Bioenerg. *1847*, 993–1003.

Houille-Vernes, L., Rappaport, F., Wollman, F.-A., Alric, J., and Johnson, X. (2011). Plastid terminal oxidase 2 (PTOX2) is the major oxidase involved in chlororespiration in *Chlamydomonas*. Proc. Natl. Acad. Sci. U.S.A. *108*, 20820–20825.

Ifuku, K., Ishihara, S., Shimamoto, R., Ido, K., and Sato, F. (2008). Structure, function, and evolution of the PsbP protein family in higher plants. Photosynth. Res. *98*, 427–437.

Jallet, D., Gwizdala, M., and Kirilovsky, D. (2012). ApcD, ApcF and ApcE are not required for the Orange Carotenoid Protein related phycobilisome fluorescence quenching in the cyanobacterium *Synechocystis* PCC 6803. Biochim. Biophys. Acta Bioenerg. *1817*, 1418–1427.

Kana, R., Kotabova, E., Sobotka, R., and Prasil, O. (2012). Non-photochemical quenching in cryptophyte alga rhodomonas salina is located in chlorophyll a/c antennae. PLoS One 7, e29700.

Kana, R., Kotabova, E., Lukes, M., Papacek, S., Matonoha, C., Liu, L.N., Prasil, O., and Mullineaux, C.W. (2014). Phycobilisome mobility and its role in the regulation of light harvesting in red algae. Plant Physiol. *165*, 1618–1631.

Kanazawa, A., Blanchard, G.J., Szabo, M., Ralph, P.J., and Kramer, D.M. (2014). The site of regulation of light capture in Symbiodinium: Does the peridinin-chlorophyll alpha-protein detach to regulate light capture? Biochim. Biophys. Acta Bioenerg. *1837*, 1227–1234.

Karlsson, P.M., Herdean, A., Adolfsson, L., Beebo, A., Nziengui, H., Irigoyen, S., Uennep, R., Zsiros, O., Nagy, G., Garab, G., *et al.* (2015). The *Arabidopsis* thylakoid transporter PHT4;1 influences phosphate availability for ATP synthesis and plant growth. Plant J. *84*, 99–110.

Kawasaki, S., Mizuguchi, K., Sato, M., Kono, T., and Shimizu, H. (2013). A novel astaxanthin-binding photooxidative stress-inducible aqueous carotenoprotein from a eukaryotic microalga isolated from asphalt in midsummer. Plant Cell Physiol. *54*, 1027–1040.

Keeling, P.J. (2013). The number, speed, and impact of plastid endosymbioses in eukaryotic evolution. Annu. Rev. Plant Biol. *64*, 583–607.

Kehoe, D.M., and Grossman, A.R. (1996). Similarity of a chromatic adaptation sensor to phytochrome and ethylene receptors. Science *273*, 1409–1412.

Kehoe, D.M., and Gutu, A. (2006). Responding to color: The regulation of complementary chromatic adaptation. Annu. Rev. Plant Biol. *57*, 127–150.

Kehoe, D.M. (2010). Chromatic adaptation and the evolution of light color sensing in cyanobacteria. Proc. Natl. Acad. Sci. U.S.A. *107*, 9029–9030.

Kianianmomeni, A., and Hallmann, A. (2014). Algal photoreceptors: *in vivo* functions and potential applications. Planta *239*, 1–26.

Kinoshita, H., Nagasaki, J., Yoshikawa, N., Yamamoto, A., Takito, S., Kawasaki, M., Sugiyama, T., Miyake, H., Weber, A.P.M., and Taniguchi, M. (2011). The chloroplastic 2-oxoglutarate/malate transporter has dual function as the malate valve and in carbon/nitrogen metabolism. Plant J. *65*, 15–26.

Kirilovsky, D., and Kerfeld, C.A. (2012). The orange carotenoid protein in photoprotection of photosystem II in cyanobacteria. Biochim. Biophys. Acta Bioenerg. *1817*, 158–166.

Kirilovsky, D., Kaňa, R., and Prášil, O. (2014). Mechanisms modulating energy arriving at reaction centers in cyanobacteria. In Non-Photochemical Quenching and Energy Dissipation in Plants, Algae and Cyanobacteria, B. Demmig-Adams, G. Garab, W. Adams III, and Govindjee, eds. (Dordrecht,: Springer Netherlands), pp. 471–501.

Kirilovsky, D. (2015). Modulating energy arriving at photochemical reaction centers: orange carotenoid protein-related photoprotection and state transitions. Photosynth. Res. *126*, 3–17.

Knoppova, J., Sobotka, R., Tichy, M., Yu, J., Konik, P., Halada, P., Nixon, P.J., and Komenda, J. (2014). Discovery of a chlorophyll binding protein complex involved in the early steps of photosystem ii assembly in Synechocystis. Plant Cell *26*, 1200–1212.

Komenda, J., and Sobotka, R. (2015). Cyanobacterial high-light-inducible proteins – Protectors of chlorophyll–protein synthesis and assembly. Biochim. Biophys. Acta Bioenerg. doi:10.1016/j.bbabio.2015.08.011.

Kotabova, E., Kana, R., Jaresova, J., and Prasil, O. (2011). Non-photochemical fluorescence quenching in *Chromera velia* is enabled by fast violaxanthin de-epoxidation. FEBS Lett. *585*, 1941–1945.

Krieger-Liszkay, A. (2005). Singlet oxygen production in photosynthesis. J. Exp. Bot. *56*, 337–346.

Kruger, T.P.J., Ilioaia, C., Johnson, M.P., Ruban, A.V., Papagiannakis, E., Horton, P., and van Grondelle, R. (2012). Controlled disorder in plant Light-Harvesting Complex II explains its photoprotective role. Biophys. J. *102*, 2669–2676.

Kulheim, C., Agren, J., and Jansson, S. (2002). Rapid regulation of light harvesting and plant fitness in the field. Science *297*, 91–93.

Kunz, H.-H., Gierth, M., Herdean, A., Satoh-Cruz, M., Kramer, D.M., Spetea, C., and Schroeder, J.I. (2014). Plastidial transporters KEA*1*, -*2*, and -3 are essential for chloroplast osmoregulation, integrity, and pH regulation in *Arabidopsis*. Proc. Natl. Acad. Sci. U.S.A. *111*, 7480–7485.

Laloi, C., and Havaux, M. (2015). Key players of singlet oxygen-induced cell death in plants. Frontiers Plant Sci. *6*, 39.

Lata, C., and Prasad, M. (2011). Role of DREBs in regulation of abiotic stress responses in plants. J. Exp. Bot. *62*, 4731–4748.

Lea-Smith, D.J., Ross, N., Zori, M., Bendall, D.S., Dennis, J.S., Scott, S.A., Smith, A.G., and Howe, C.J. (2013). Thylakoid terminal oxidases are essential for the cyanobacterium *synechocystis* sp pcc 6803 to survive rapidly changing light intensities. Plant Physiol. *162*, 484–495.

Ledford, H.K., Chin, B.L., and Niyogi, K.K. (2007). Acclimation to singlet oxygen stress in *Chlamydomonas reinhardtii*. Eukaryot. Cell *6*, 919–930.

Leoni, C., Pietrzykowska, M., Kiss, A.Z., Suorsa, M., Ceci, L.R., Aro, E.-M., and Jansson, S. (2013). Very rapid phosphorylation kinetics suggest a unique role for Lhcb2 during state transitions in *Arabidopsis*. Plant J. *76*, 236–246.

Lepetit, B., Sturm, S., Rogato, A., Gruber, A., Sachse, M., Falciatore, A., Kroth, P.G., and Lavaud, J. (2013). High light acclimation in the secondary plastids containing diatom phaeodactylum tricornutum is triggered by the redox state of the plastoquinone pool. Plant Physiol. *161*, 853–865.

Lepetit, B., and Dietzel, L. (2015). Light signaling in photosynthetic eukaryotes with 'green'and 'red'chloroplasts. Environ. Exp. Bot. *114*, 30–47.

Leverenz, R.L., Jallet, D., Li, M.D., Mathies, R.A., Kirilovsky, D., and Kerfeld, C.A. (2014). Structural and functional modularity of the orange carotenoid protein: distinct roles for the n- and c-terminal domains in cyanobacterial photoprotection. Plant Cell *26*, 426–437.

Leverenz, R.L., Sutter, M., Wilson, A., Gupta, S., Thurotte, A., de Carbon, C.B., Petzold, C.J., Ralston, C., Perreau, F., Kirilovsky, D., *et al.* (2015). A 12 angstrom carotenoid translocation in a photoswitch associated with cyanobacterial photoprotection. Science *348*, 1463–1466.

Li, X.P., Bjorkman, O., Shih, C., Grossman, A.R., Rosenquist, M., Jansson, S., and Niyogi, K.K. (2000). A pigment-binding protein essential for regulation of photosynthetic light harvesting. Nature *403*, 391–395.

Liguori, N., Roy, L.M., Opacic, M., Durand, G., and Croce, R. (2013). Regulation of light harvesting in the green alga *Chlamydomonas reinhardtii*: The C-terminus of LHCSR is the knob of a dimmer switch. J. Am. Chem. Soc. *135*, 18339–18342.

Liu, H., Zhang, H., Niedzwiedzki, D.M., Prado, M., He, G., Gross, M.L., and Blankenship, R.E. (2013). Phycobilisomes Supply Excitations to Both Photosystems in a Megacomplex in Cyanobacteria. Science *342*, 1104–1107.

Liu, H.J., Zhang, H., King, J.D., Wolf, N.R., Prado, M., Gross, M.L., and Blankenship, R.E. (2014). Mass spectrometry footprinting reveals the structural rearrangements of cyanobacterial orange carotenoid protein upon light activation. Biochim. Biophys. Acta Bioenerg. *1837*, 1955–1963.

Mallick, N., and Mohn, F.H. (2000). Reactive oxygen species: response of algal cells. J. Plant Physiol. *157*, 183–193.

Maruyama, S., Tokutsu, R., and Minagawa, J. (2014). Transcriptional regulation of the stress-responsive light harvesting complex genes in *Chlamydomonas reinhardtii*. Plant Cell Physiol. 55, 1304–1310.

Maruyama, S., Shoguchi, E., Satoh, N., and Minagawa, J. (2015). Diversification of the light-harvesting complex gene family via intra- and intergenic duplications in the coral symbiotic alga symbiodinium. Plos One *10*, e0119406.

McDonald, A.E., Ivanov, A.G., Bode, R., Maxwell, D.P., Rodermel, S.R., and Huenter, N.P.A. (2011). Flexibility in photosynthetic electron transport: The physiological role of plastoquinol terminal oxidase (PTOX). Biochim. Biophys. Acta Bioenerg. *1807*, 954–967.

McGlincy, N.J., and Smith, C.W. (2008). Alternative splicing resulting in nonsense-mediated mRNA decay: what is the meaning of nonsense? Trends Biochem. Sci. *33*, 385–393.

Medzihradszky, M., Bindics, J., Adam, E., Viczian, A., Klement, E., Lorrain, S., Gyula, P., Merai, Z., Fankhauser, C., Medzihradszky, K.F., *et al.* (2013). Phosphorylation of phytochrome b inhibits light-induced signaling via accelerated dark reversion in arabidopsis. Plant Cell 25, 535–544.

Mettler, T., Muhlhaus, T., Hemme, D., Schottler, M.A., Rupprecht, J., Idoine, A., Veyel, D., Pal, S.K., Yaneva-Roder, L., Winck, F.V., *et al.* (2014). Systems analysis of the response of photosynthesis, metabolism, and growth to an increase in irradiance in the photosynthetic model organism *chlamydomonas reinhardtii*. Plant Cell *26*, 2310–2350.

Michelet, L., Roach, T., Fischer, B.B., Bedhomme, M., Lemaire, S.D., and Krieger-Liszkay, A. (2013). Down-regulation of catalase activity allows transient accumulation of a hydrogen peroxide signal in *Chlamydomonas reinhardtii*. Plant Cell Environ. *36*, 1204–1213.

Minagawa, J. (2011). State transitions-The molecular remodeling of photosynthetic supercomplexes that controls energy flow in the chloroplast. Biochim. Biophys. Acta Bioenerg. *1807*, 897–905.

Minagawa, J., and Tokutsu, R. (2015). Dynamic regulation of photosynthesis in *Chlamydomonas reinhardtii*. Plant J. *82*, 413–428.

Mittler, R. (2002). Oxidative stress, antioxidants and stress tolerance. Trends Plant Sci. 7, 405–410.

Mittler, R., Vanderauwera, S., Suzuki, N., Miller, G., Tognetti, V.B., Vandepoele, K., Gollery, M., Shulaev, V., and Van Breusegem, F. (2011). ROS signaling: the new wave? Trends Plant Sci. *16*, 300–309.

Miyaji, T., Kuromori, T., Takeuchi, Y., Yamaji, N., Yokosho, K., Shimazawa, A., Sugimoto, E., Omote, H., Ma, J.F., Shinozaki, K., *et al.* (2015). AtPHT4;4 is a chloroplast-localized ascorbate transporter in *Arabidopsis*. Nat. Commun. *6*, 5928.

Mubarakshina, M.M., Ivanov, B.N., Naydov, I.A., Hillier, W., Badger, M.R., and Krieger-Liszkay, A. (2010). Production and diffusion of chloroplastic H_2O2 and its implication to signalling. J. Exp. Bot. *61*, 3577–3587.

Mullineaux, C.W., and Emlyn-Jones, D. (2005). State transitions: an example of acclimation to low-light stress. J. Exp. Bot. *56*, 389–393.

Nagy, G., Uennep, R., Zsiros, O., Tokutsu, R., Takizawa, K., Porcar, L., Moyet, L., Petroutsos, D., Garab, G., Finazzi, G., *et al.* (2014). Chloroplast remodeling during state transitions in *Chlamydomonas reinhardtii* as revealed by noninvasive techniques in vivo. Proc. Natl. Acad. Sci. U.S.A. *111*, 5042–5047.

Neilson, J.A.D., and Durnford, D.G. (2010). Evolutionary distribution of light-harvesting complex-like proteins in photosynthetic eukaryotes. Genome *53*, 68–78.

Neuhaus, H.E., and Emes, M.J. (2000). Nonphotosynthetic metabolism in plastids. Annu. Rev. Plant Physiol. Plant Mol. Biol. *51*, 111–140.

Nito, K., Wong, C.C.L., Yates, J.R., and Chory, J. (2013). Tyrosine phosphorylation regulates the activity of phytochrome photoreceptors. Cell Reports 3, 1970–1979.

Niyogi, K.K. (2000). Safety valves for photosynthesis. Curr. Opin. Plant Biol. 3, 455–460.

Niyogi, K.K., and Truong, T.B. (2013). Evolution of flexible non-photochemical quenching mechanisms that regulate light harvesting in oxygenic photosynthesis. Curr. Opin. Plant Biol. *16*, 307–314.

Oelze, M.-L., Vogel, M.O., Alsharafa, K., Kahmann, U., Viehhauser, A., Maurino, V.G., and Dietz, K.-J. (2012). Efficient acclimation of the chloroplast antioxidant defence of *Arabidopsis thaliana* leaves in response to a 10- or 100-fold light increment and the possible involvement of retrograde signals. J. Exp. Bot. *63*, 1297–1313.

Palusa, S.G., Ali, G.S., and Reddy, A.S. (2007). Alternative splicing of pre-mRNAs of *Arabidopsis* serine/arginine-rich proteins: regulation by hormones and stresses. Plant J. *49*, 1091–1107.

Peers, G., Truong, T.B., Ostendorf, E., Busch, A., Elrad, D., Grossman, A.R., Hippler, M., and Niyogi, K.K. (2009). An ancient light-harvesting protein is critical for the regulation of algal photosynthesis. Nature *462*, 518–528.

Peltier, G., Tolleter, D., Billon, E., and Cournac, L. (2010). Auxiliary electron transport pathways in chloroplasts of microalgae. Photosynth. Res. *106*, 19–31.

Petrillo, E., Godoy Herz, M.A., Fuchs, A., Reifer, D., Fuller, J., Yanovsky, M.J., Simpson, C., Brown, J.W.S., Barta, A., Kalyna, M., *et al.* (2014). A chloroplast retrograde signal regulates nuclear alternative splicing. Science *344*, 427–430.

Petroutsos, D., Busch, A., Janssen, I., Trompelt, K., Bergner, S.V., Weinl, S., Holtkamp, M., Karst, U., Kudla, J., and Hippler, M. (2011). The chloroplast calcium sensor cas is required for photoacclimation in *Chlamydomonas reinhardtii*. Plant Cell *23*, 2950–2963.

Pfuendel, E.E., Klughammer, C., Meister, A., and Cerovic, Z.G. (2013). Deriving fluorometer-specific values of relative PSI fluorescence intensity from quenching of F-0 fluorescence in leaves of *Arabidopsis thaliana* and Zea mays. Photosynth. Res. *114*, 189–206.

Pick, T.R., Braeutigam, A., Schulz, M.A., Obata, T., Fernie, A.R., and Weber, A.P.M. (2013). PLGG1, a plastidic glycolate glycerate transporter, is required for photorespiration and defines a unique class of metabolite transporters. Proc. Natl. Acad. Sci. U.S.A. *110*, 3185–3190.

Pietrzykowska, M., Suorsa, M., Semchonok, D.A., Tikkanen, M., Boekema, E.J., Aro, E.M., and Jansson, S. (2014). The light-harvesting chlorophyll a/b binding proteins Lhcb1 and Lhcb2 play complementary roles during state transitions in arabidopsis. Plant Cell *26*, 3646–3660.

Pogson, B.J., Woo, N.S., Förster, B., and Small, I.D. (2008). Plastid signalling to the nucleus and beyond. Trends Plant Sci. *13*, 602–609.

Pribil, M., Pesaresi, P., Hertle, A., Barbato, R., and Leister, D. (2010). Role of plastid protein phosphatase TAP38 in LHCII dephosphorylation and thylakoid electron flow. PLoS Biol. *8*, e1000288.

Prihoda, J., Tanaka, A., de Paula, W.B.M., Allen, J.F., Tirichine, L., and Bowler, C. (2012). Chloroplast–mitochondria cross-talk in diatoms. J. Exp. Bot. *63*, 1543–1557.

Quail, P.H. (2007). Phytochrome-regulated gene expression. J. Integr. Plant Biol. *49*, 11–20.

Rakhimberdieva, M.G., Elanskaya, I.V., Vermaas, W.F.J., and Karapetyan, N.V. (2010). Carotenoid-triggered energy dissipation in phycobilisomes of *Synechocystis* sp PCC 6803 diverts excitation away from reaction centers of both photosystems. Biochim. Biophys. Acta Bioenerg. *1797*, 241–249.

Ramel, F., Birtic, S., Cuine, S., Triantaphylides, C., Ravanat, J.L., and Havaux, M. (2012a). Chemical quenching of singlet oxygen by carotenoids in plants. Plant Physiol. *158*, 1267–1278.

Ramel, F., Birtic, S., Ginies, C., Soubigou-Taconnat, L., Triantaphylides, C., and Havaux, M. (2012b). Carotenoid oxidation products are stress signals that mediate gene responses to singlet oxygen in plants. Proc. Natl. Acad. Sci. U.S.A. *109*, 5535–5540.

Ramel, F., Birtic, S., Ginies, C., Soubigou-Taconnat, L., Triantaphylidès, C., and Havaux, M. (2012c). Carotenoid oxidation products are stress signals that mediate gene responses to singlet oxygen in plants. Proc. Natl. Acad. Sci. U.S.A. *109*, 5535–5540.

Reddy, A.S., Marquez, Y., Kalyna, M., and Barta, A. (2013). Complexity of the alternative splicing landscape in plants. Plant Cell *25*, 3657–3683.

Rockwell, N.C., Duanmu, D., Martin, S.S., Bachy, C., Price, D.C., Bhattacharya, D., Worden, A.Z., and Lagarias, J.C. (2014). Eukaryotic algal phytochromes span the visible spectrum. Proc. Natl. Acad. Sci. U.S.A. *111*, 3871–3876.

Rosso, D., Ivanov, A.G., Fu, A., Geisler-Lee, J., Hendrickson, L., Geisler, M., Stewart, G., Krol, M., Hurry, V., Rodermel, S.R., *et al.* (2006). IMMUTANS does not act as a stress-induced safety valve in the protection of the photosynthetic apparatus of arabidopsis during steady-state photosynthesis. Plant Physiol. *142*, 574–585.

Ruban, A.V., Johnson, M.P., and Duffy, C.D.P. (2012). The photoprotective molecular switch in the photosystem II antenna. Biochim. Biophys. Acta Bioenerg. *1817*, 167–181.

Sadanandom, A., Adam, E., Orosa, B., Viczian, A., Klose, C., Zhang, C.J., Josse, E.M., Kozma-Bognar, L., and Nagy, F. (2015). SUMOylation of phytochrome-B negatively regulates light-induced signaling in *Arabidopsis thaliana*. Proc. Natl. Acad. Sci. U.S.A. *112*, E5447-E5448.

Scheibe, R. (2004). Malate valves to balance cellular energy supply. Physiol. Plant. *120*, 21–26.

Schlau-Cohen, G.S., Yang, H.Y., Kruger, T.P.J., Xu, P.Q., Gwizdala, M., van Grondelle, R., Croce, R., and Moerner, W.E. (2015). Single-molecule identification of quenched and unquenched states of LHCII. J. Phys. Chem. Lett. *6*, 860–867.

Sedoud, A., Lopez-Igual, R., Rehman, A.U., Wilson, A., Perreau, F., Boulay, C., Vass, I., Krieger-Liszkay, A., and Kirilovsky, D. (2014). The cyanobacterial photoactive orange carotenoid protein is an excellent singlet oxygen quencher. Plant Cell *26*, 1781–1791.

Shao, N., Duan, G.Y., and Bock, R. (2013). A mediator of singlet oxygen responses in *Chlamydomonas reinhardtii* and arabidopsis identified by a luciferase-based genetic screen in algal cells. Plant Cell *25*, 4209–4226.

Shapiguzov, A., Ingelsson, B., Samol, I., Andres, C., Kessler, F., Rochaix, J.-D., Vener, A.V., and Goldschmidt-Clermont, M. (2010). The PPH1 phosphatase is specifically involved in LHCII dephosphorylation and state transitions in *Arabidopsis*. Proc. Natl. Acad. Sci. U.S.A. *107*, 4782–4787.

Shih, P.M., Wu, D., Latifi, A., Axen, S.D., Fewer, D.P., Talla, E., Calteau, A., Cai, F., de Marsac, N.T., Rippka, R., *et al.* (2013). Improving the coverage of the cyanobacterial phylum using diversity-driven genome sequencing. Proc. Natl. Acad. Sci. U.S.A. *110*, 1053–1058.

Shimakawa, G., Shaku, K., Nishi, A., Hayashi, R., Yamamoto, H., Sakamoto, K., Makino, A., and Miyake, C. (2015). FLAVODIIRON2 and FLAVODIIRON4 proteins mediate an oxygen-dependent alternative electron flow in *Synechocystis* sp PCC 6803 under CO_2-limited conditions. Plant Physiol. *167*, 472–732.

Sonoike, K. (2011). Photoinhibition of photosystem I. Physiol. Plant. *142*, 56–64.

Stadnichuk, I.N., Yanyushin, M.F., Maksimov, E.G., Lukashev, E.P., Zharmukhamedov, S.K., Elanskaya, I.V., and Paschenko, V.Z. (2012). Site of non-photochemical quenching of the phycobilisome by orange carotenoid protein in the cyanobacterium *Synechocystis* sp PCC 6803. Biochim. Biophys. Acta Bioenerg. *1817*, 1436–1445.

Staleva, H., Komenda, J., Shukla, M.K., Šlouf, V., Kaňa, R., Polívka, T., and Sobotka, R. (2015). Mechanism of photoprotection in the cyanobacterial ancestor of plant antenna proteins. Nat. Chem. Biol. *11*, 287–291.

Sutter, M., Wilson, A., Leverenz, R.L., Lopez-Igual, R., Thurotte, A., Salmeen, A.E., Kirilovsky, D., and Kerfeld, C.A. (2013). Crystal structure of the FRP and identification of the active site for modulation of OCP-mediated photoprotection in cyanobacteria. Proc. Natl. Acad. Sci. U.S.A. *110*, 10022–10027.

Sylak-Glassman, E.J., Malnoe, A., De Re, E., Brooks, M.D., Fischer, A.L., Niyogi, K.K., and Fleming, G.R. (2014). Distinct roles of the photosystem II protein PsbS and zeaxanthin in the regulation of light harvesting in plants revealed by fluorescence lifetime snapshots. Proc. Natl. Acad. Sci. U.S.A. *111*, 17498–17503.

Tanaka, R., and Tanaka, A. (2007). Tetrapyrrole biosynthesis in higher plants. Annu. Rev. Plant Biol. *58*, 321–346.

Tegeder, M., and Weber, A.P.M. (2008). 4 Metabolite transporters in the control of plant primary metabolism. In Control of Primary Metabolism in Plants, M.T.

McManus and W. Plaxton, eds. (Oxford, UK: Blackwell Publishing Ltd.), pp. 85–120.
Telfer, A. (2014). Singlet oxygen production by PSII under light stress: mechanism, detection and the protective role of beta-carotene. Plant Cell Physiol. *55*, 1216–1223.
Terauchi, K., Montgomery, B.L., Grossman, A.R., Lagarias, J.C., and Kehoe, D.M. (2004). RcaE is a complementary chromatic adaptation photoreceptor required for green and red light responsiveness. Mol. Microbiol. *51*, 567–577.
Tian, L.J., van Stokkum, I.H.M., Koehorst, R.B.M., Jongerius, A., Kirilovsky, D., and van Amerongen, H. (2011). Site, rate, and mechanism of photoprotective quenching in cyanobacteria. J. Am. Chem. Soc. *133*, 18304–18311.
Tokutsu, R., and Minagawa, J. (2013). Energy-dissipative supercomplex of photosystem II associated with LHCSR3 in *Chlamydomonas reinhardtii*. Proc. Natl. Acad. Sci. U.S.A. *110*, 10016–10021.
Triantaphylides, C., Krischke, M., Hoeberichts, F.A., Ksas, B., Gresser, G., Havaux, M., Van Breusegem, F., and Mueller, M.J. (2008). Singlet oxygen is the major reactive oxygen species involved in photooxidative damage to plants. Plant Physiol. *148*, 960–968.
Unlu, C., Drop, B., Croce, R., and van Amerongen, H. (2014). State transitions in *Chlamydomonas reinhardtii* strongly modulate the functional size of photosystem II but not of photosystem I. Proc. Natl. Acad. Sci. U.S.A. *111*, 3460–3465.
Vicente, J.B., Gomes, C.M., Wasserfallen, A., and Teixeira, M. (2002). Module fusion in an A-type flavoprotein from the cyanobacterium *Synechocystis* condenses a multiple-component pathway in a single polypeptide chain. Biochem. Biophys. Res. Commun. *294*, 82–87.
Vogel, M.O., Moore, M., Koenig, K., Pecher, P., Alsharafa, K., Lee, J., and Dietz, K.-J. (2014). Fast retrograde signaling in response to high light involves metabolite export, Mitogen-Activated Protein Kinase 6, and AP2/ERF transcription factors in *Arabidopsis*. Plant Cell *26*, 1151–1165.
von Gromoff, E.D., Alawady, A., Meinecke, L., Grimm, B., and Beck, C.F. (2008). Heme, a plastid-derived regulator of nuclear gene expression in *Chlamydomonas*. Plant Cell *20*, 552–567.
Wakao, S., Chin, B.L., Ledford, H.K., Dent, R.M., Casero, D., Pellegrini, M., Merchant, S.S., and Niyogi, K.K. (2014). Phosphoprotein SAK1 is a regulator of acclimation to singlet oxygen in *Chlamydomonas reinhardtii*. eLife 3, e02286.
Waring, J., Klenell, M., Bechtold, U., Underwood, G.J.C., and Baker, N.R. (2010). Light-induced responses of oxygen photoreduction, reactive oxygen species production and scavenging in two diatom species. J. Phycol. *46*, 1206–1217.
Watanabe, M., and Ikeuchi, M. (2013). Phycobilisome: architecture of a light-harvesting supercomplex. Photosynth. Res. *116*, 265–276.
Watanabe, M., Semchonok, D.A., Webber-Birungi, M.T., Ehira, S., Kondo, K., Narikawa, R., Ohmori, M., Boekema, E.J., and Ikeuchi, M. (2014). Attachment of phycobilisomes in an antenna-photosystem I supercomplex of cyanobacteria. Proc. Natl. Acad. Sci. U.S.A. *111*, 2512–2517.
Weber, A.P.M., and Linka, N. (2011). Connecting the plastid: transporters of the plastid envelope and their role in linking plastidial with cytosolic metabolism. Ann. Rev. Plant Biol., *62*, 53–77.
Weger, H.G., Chadderton, A.R., Lin, M., Guy, R.D., and Turpin, D.H. (1990). Cytochrome and alternative pathway respiration during transient ammonium assimilation by N-limited *Chlamydomonas reinhardtii*. Plant Physiol. *94*, 1131–1136.
Wehner, A., Storf, S., Jahns, P., and Schmid, V.H.R. (2004). De-epoxidation of violaxanthin in light-harvesting complex I proteins. J. Biol. Chem. *279*, 26823–26829.
Wilk, L., Grunwald, M., Liao, P.-N., Walla, P.J., and Kuehlbrandt, W. (2013). Direct interaction of the major light-harvesting complex II and PsbS in nonphotochemical quenching. Proc. Natl. Acad. Sci. U.S.A. *110*, 5452–5456.
Wilson, A., Ajlani, G., Verbavatz, J.M., Vass, I., Kerfeld, C.A., and Kirilovsky, D. (2006). A soluble carotenoid protein involved in phycobilisome-related energy dissipation in cyanobacteria. Plant Cell *18*, 992–1007.
Wilson, A., Punginelli, C., Gall, A., Bonetti, C., Alexandre, M., Routaboul, J.M., Kerfeld, C.A., van Grondelle, R., Robert, B., Kennis, J.T.M., *et al.* (2008). A photoactive carotenoid protein acting as light intensity sensor. Proc. Natl. Acad. Sci. U.S.A. *105*, 12075–12080.
Wilson, A., Kinney, J.N., Zwart, P.H., Punginelli, C., D'Haene, S., Perreau, F., Klein, M.G., Kirilovsky, D., and Kerfeld, C.A. (2010). Structural determinants underlying photoprotection in the photoactive orange carotenoid protein of cyanobacteria. J. Biol. Chem. *285*, 18364–18375.
Wilson, A., Gwizdala, M., Mezzetti, A., Alexandre, M., Kerfeld, C.A., and Kirilovsky, D. (2012). The essential role of the n-terminal domain of the orange carotenoid protein in cyanobacterial photoprotection: importance of a positive charge for phycobilisome binding. Plant Cell *24*, 1972–1983.
Woodson, J.D., Perez-Ruiz, J.M., and Chory, J. (2011). Heme synthesis by plastid ferrochelatase I regulates nuclear gene expression in plants. Curr. Biol. *21*, 897–903.
Xiang, T., Nelson, W., Rodriguez, J., Tolleter, D., and Grossman, A.R. (2015). Symbiodinium transcriptome and global responses of cells to immediate changes in light intensity when grown under autotrophic or mixotrophic conditions. Plant J. *82*, 67–80.
Xiao, Y., Savchenko, T., Baidoo, E.E., Chehab, W.E., Hayden, D.M., Tolstikov, V., Corwin, J.A., Kliebenstein, D.J., Keasling, J.D., and Dehesh, K. (2012). Retrograde signaling by the plastidial metabolite MEcPP regulates expression of nuclear stress-response genes. Cell *149*, 1525–1535.
Yokono, M., Takabayashi, A., Akimoto, S., and Tanaka, A. (2015). A megacomplex composed of both photosystem reaction centres in higher plants. Nat. Commun. *6*, 6675.
Zhang, H., Liu, H., Niedzwiedzki, D.M., Prado, M., Jiang, J., Gross, M.L., and Blankenship, R.E. (2014). Molecular mechanism of photoactivation and structural location of the cyanobacterial orange carotenoid protein. Biochemistry *53*, 13–19.
Zhang, P., Allahverdiyeva, Y., Eisenhut, M., and Aro, E.-M. (2009). Flavodiiron proteins in oxygenic photosynthetic organisms: photoprotection of photosystem II by Flv2

and Flv4 in *Synechocystis* sp PCC 6803. PLoS One *4*, e5331.

Zhang, P., Eisenhut, M., Brandt, A.-M., Carmel, D., Silen, H.M., Vass, I., Allahverdiyeva, Y., Salminen, T.A., and Aro, E.-M. (2012). Operon flv4-flv2 provides cyanobacterial photosystem II with flexibility of electron transfer. Plant Cell *24*, 1952–1971.

Zito, F., Finazzi, G., Delosme, R., Nitschke, W., Picot, D., and Wollman, F.A. (1999). The Q_o site of cytochrome b(6) f complexes controls the activation of the LHCII kinase. EMBO J. *18*, 2961–2969.

Photoinhibition and the Damage Repair Cycle of Photosystem II

7

Yasusi Yamamoto and Miho Yoshioka-Nishimura

Abstract

Excessive illumination of photosystem II in oxygenic photosynthetic organisms, such as cyanobacteria, algae and higher plants, causes photoinhibition and blocks electron transport in photosystem II. In this event, the reaction centre D1 protein is damaged primarily by reactive oxygen species or other endogenous radicals produced by the photochemical reaction, and degradation and/or irreversible aggregation of the damaged proteins occur subsequently. When the D1 protein is photodamaged under moderate high light or weak light, the damaged protein is proteolysed and replaced by a newly synthesized copy. The photoinhibition under these conditions is reversible and photosystem II activity recovers rapidly in the dark. In contrast, irreversible aggregation of the D1 protein, which is caused by excessive illumination, prevents proper D1 turnover. Once the aggregated products accumulate in photosystem II complexes their removal by proteases is hampered. The photoinhibition observed under these conditions is irreversible. In higher plant chloroplasts, illumination with high light leads to dynamic changes in the structure of the thylakoids at the molecular and membrane levels. It has become clear recently that these structural changes are necessary so that photosystem II can endure the effects of light stress effectively.

Introduction: response of photosystem II to excessive light

Light intensity from the sun is dependent on the time of day and weather conditions. Light conditions also vary over a longer period because of seasonal fluctuations. To capture light energy efficiently under fluctuating light conditions photosynthesizing cells that move freely in water change their location, whereas terrestrial plants reorient their leaves or grow in a light-dependent direction. Conversely, when these cells or leaves receive excessive light from the sun, they either avoid or tolerate the light stress using both short and long term strategies. These protective mechanisms work even at the level of the thylakoid membranes (Fig. 7.1). Under the conditions where light intensity is excessive but not extremely high photosystem II (PSII) complexes and the light-harvesting complexes of PSII (LHCII), which are usually associated with each other, dissociate and change their location and/or orientation in the thylakoid membranes to avoid light stress (Goral *et al.*, 2010; Herbstova *et al.*, 2012; Yamamoto *et al.*, 2014). During excessive illumination local membrane fluidity may increase in the grana where the PSII–LHCII supercomplex is abundant. Aggregation of LHCII is a well-known phenomenon and it is a reversible process related to efficient dissipation of excessive excitation energy captured in the PSII–LHCII supercomplexes as heat (Horton *et al.*, 1991; Niyogi, 1999). This mechanism has been conventionally monitored using a pulse amplitude modulation (PAM) fluorometer (Walz, Germany) (Schreiber *et al.*, 1986) as the energy-dependent quenching (qE) in the non-photochemical quenching of chlorophyll fluorescence (NPQ) (Horton *et al.*, 1996; Niyogi, 1999). In this step, the so-called xanthophyll cycle is activated through de-epoxidation of xanthophyll pigments (Demmig-Adams, 1990; Yamamoto *et al.*, 1972). A PsbS protein in land plants (Li *et al.*, 2000) and a LHCSR in green algae (Peers *et al.*, 2009), which are both members of the LHC gene family, sense the luminal pH and may induce specific aggregated forms of LHCII through protonation of the sensor proteins.

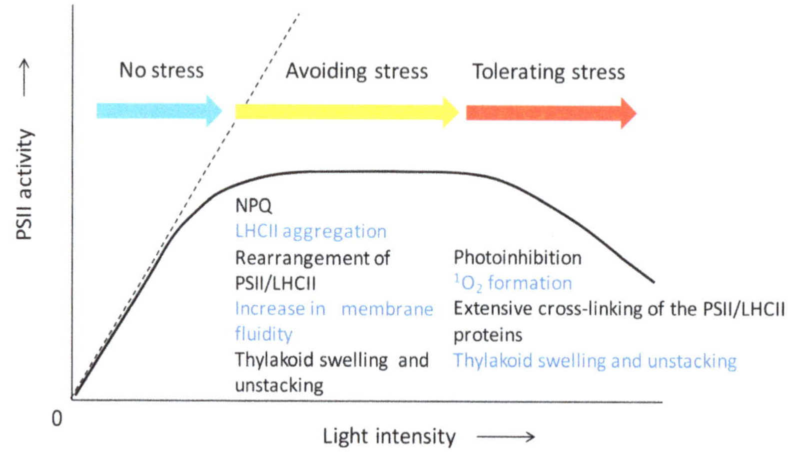

Figure 7.1 Light response curve for PSII activity and the events that occur when the thylakoids are not exposed to or avoid and tolerate light stress. Under low light, PSII activity increases with increasing light intensity. PSII activity gradually decreases and reaches a plateau when the light intensity increases further. Chloroplasts first try to avoid the light stress by dissipating the excessive light energy as heat (qE of NPQ). However, when the light intensity increases, photoinhibition of PSII is apparent. Irreversible aggregation of PSII proteins occurs under severe light stress conditions.

Photoinhibition of PSII takes place when the light intensity increases further and avoidance of the light stress becomes difficult (Kyle *et al.*, 1984; Powles, 1984), and then the tolerance mechanism activates. A typical way to tolerate the light stress is preferential damage to and specific degradation of the reaction centre D1 protein (Aro *et al.*, 1993; Barber and Andersson, 1992; Mattoo *et al.*, 1981; Ohad *et al.*, 1984). The PSII–LHCII supercomplexes are disorganized and the photodamaged D1 protein is quickly replaced by a newly synthesized copy in PSII. Thus, the photoinhibition is reversible and is referred to as 'reversible photoinhibition' herein. The mechanism of damage to the D1 protein under light stress has been studied extensively at the molecular level (Aro *et al.*, 1993; Barber and Andersson, 1992; Yamamoto *et al.*, 2008). Through reversible photoinhibition a dynamic change in the distribution and orientation of PSII–LHCII supercomplexes and of the components related to the repair of the photodamaged PSII occurs on the thylakoids (Goral *et al.*, 2010; Herbstova *et al.*, 2012; Hind *et al.*, 2014; Yoshioka-Nishimura *et al.*, 2014). Changes in the structure of the thylakoids at a larger scale, such as stacking/unstacking and shrinkage/swelling of the thylakoids, are also postulated (Herbstova *et al.*, 2012; Yamamoto *et al.*, 2013; Yoshioka-Nishimura *et al.*, 2014). Phosphorylation of PSII was shown to regulate the stacking and unstacking of the thylakoids in *Arabidopsis* (Fristedt *et al.*, 2009).

Extremely high light intensity under physiological temperatures causes irreversible aggregation of the damaged D1 protein and neighbouring proteins (Ishikawa *et al.*, 1999; Yamamoto *et al.*, 2008). The protein aggregates prevent smooth movement of the proteins and lipids on the thylakoid membrane. Since membrane fluidity and protein movement are essential for maintaining homeostasis of PSII, accumulation of these protein aggregates induces photoinhibition of PSII that cannot be repaired by the conventional repair system (Yamamoto *et al.*, 2013). Hereafter, this kind of photoinhibition is referred to as 'irreversible photoinhibition'. Irreversible photoinhibition may eventually cause dysfunction of chloroplasts and cell death. Here, we describe in detail the various tolerating responses of PSII and the thylakoid membranes against high light exposure.

Mechanism of photodamage to PSII

Acceptor-side photoinhibition

Photoreaction under excessive illumination of PSII induces several different events in PSII, which are dependent on the ambient temperature. Under physiologically normal temperatures, excessive illumination may induce over-reduction of the acceptor side of PSII and thereby two non-physiological electron transport pathways become prominent. One pathway is the consecutive one-electron reduction of oxygen by the over-reduced electron acceptors of PSII. This reaction leads to the formation of superoxide anion radicals ($O_2^{\bullet -}$). Many components at the acceptor side of PSII can donate an electron to molecular oxygen, including the reduced primary electron acceptor pheophytin ($Pheo^-$), the reduced plastoquinone acceptor Q_A^-, plastosemiquinone and plastoquinol, and even reduced cytochrome b_{559} (Pospisil, 2009). The other pathway is reversed electron flow from $Pheo^-$ to the oxidized primary electron donor $P680^+$ and through this charge recombination process ground-state chlorophylls are excited to semi-stable triplet state chlorophylls, which in turn react with molecular oxygen, forming singlet oxygen ($^1O_2$) molecules (Telfer, 2014). Extensive studies carried out over decades indicate that $^1O_2$ produced becomes the major cause of the acceptor-side photoinhibition (Barber and Andersson, 1992).

The $^1O_2$ molecules produced here have short lives, indicating that they react immediately at the site of production, namely with the reaction centre proteins. The D1 protein is very vulnerable to light stress. The D2 protein has been reported to be occasionally photodamaged as well, but it is more stable than D1 (Jansen *et al.*, 1999). The site of photodamage to the D1 protein is probably on the stroma-exposed DE-loop that connects the membrane spanning helices D and E (Aro *et al.*, 1993; Barber and Andersson, 1992). The acceptor side photoinhibition of PSII has been well studied *in vitro* using isolated PSII membranes and PSII complexes (Yamamoto *et al.*, 2008), but the contribution of this photoinhibition mechanism *in vivo* under natural light conditions is not completely resolved and requires further investigation.

Donor-side photoinhibition

Endogenous cationic radicals are produced at the donor-side of PSII by illumination at unusually high or low temperature conditions or under other anomalous conditions such as C_l depletion or bicarbonate depletion (Klimov *et al.*, 1997), and these radicals may also contribute to the photodamage of the D1 protein. Under these conditions, oxygen-evolving activity of PSII is impaired and electrons are not efficiently transferred from the Mn_4O_5Ca complex to the oxidized primary electron donor $P680^+$. This leads to stabilization of $P680^+$ and the oxidized form of the secondary electron donor Tyr Z, and they may oxidatively damage the donor side of PSII. For donor-side photoinhibition strong light is not necessary and weak illumination is sufficient because the quantum efficiency of photoinhibition is high. The site of the donor-side photoinhibition is proposed to be the lumen-exposed loops of the D1 protein connecting the membrane-spanning helices (Aro *et al.*, 1993; Barber and Andersson, 1992).

Malfunction of the oxygen-evolving process of PSII under various stress conditions also induces various ROS (Pospisil, 2009). These ROS may react immediately with the neighbouring proteins at the donor-side of PSII and inevitably damage them. Like the acceptor-side photoinhibition, the donor-side photoinhibition has been studied extensively *in vitro* (Blubaugh and Cheniae, 1990; Klimov *et al.*, 1990; Yamamoto *et al.*, 2008), but how much the donor-side photoinhibition mechanism contributes to photoinhibition under natural conditions remains unclear. Weak light-induced photoinhibition at higher temperature (35-40°C) in spinach leaves is one of the rare examples showing donor-side photoinhibition *in vivo* (Ohira *et al.*, 2005).

Other mechanisms of photoinhibition

Besides the two mechanisms of photoinhibition of PSII described above, other mechanisms of photoinhibition have been proposed. Absorption of visible light at the Mn cluster in the oxygen-evolving complex of PSII is suggested to inactivate PSII. Comparison between the action spectrum of photoinhibition and the absorption spectrum of PSII components lends support for this concept (Hakala *et al.*, 2005; Ohnishi *et al.*, 2005). As Mn compounds show absorbance in the visible and UV regions, Mn may contribute to photoinhibition under visible light. However, the details of the possible involvement of the Mn cluster in photoinhibition are still under debate. Another group suggested participation of singlet oxygen produced by cytochrome b_6f in photoinhibition (Suh *et al.*, 2000).

Disassembly of the PSII complexes in reversible photoinhibition

To remove the photodamaged D1 protein from the PSII complex, a coordinated and regulated disassembly of the PSII complex is essential (Fig. 7.2). Initially, PSII–LHCII supercomplexes disassemble, where the central PSII dimers and the surrounding LHCII timers and minor LHCs dissociate. The PSII dimers containing the photodamaged D1 protein then dissociate. Earlier biochemical analyses showed that CP43 is the first component to be released from the PSII complex containing the photodamaged D1 protein (Aro *et al.*, 2005; Barbato *et al.*, 1992). In the cyanobacterium *Synechocystis* sp. PCC6803, Psb27 assembly factor was shown to bind and stabilize unassembled CP43 during disassemble process of PSII complexes by high light. PsbK may also form a complex with CP43. CP43 is one of the several phosphoproteins in the core part

Sequence of events in the damage-repair cycle of PSII in higher plant chloroplasts (1)

Light stress

↓

Phosphorylation of the PSII core subunit
Photodamage to the D1 protein

↓

Disassembly of the PSII-LHCII super complex

↓

Monomerization of the PSII dimer

↓

Dislocation of the damaged D1 protein from the PSII complex

↓

Dephosphorylation of the D1 protein

↓

Degradation of the D1 protein by FtsH and Deg proteases

Sequence of events in the damage-repair cycle of PSII in higher plant chloroplasts (2)

Singlet oxygen-stimulated gene expression of *PsbA* gene

↓

De novo synthesis of the D1 protein

↓

Insertion of the new D1 copy into the PSII complex

↓

Reassembly of the PSII dimer

↓

Reassembly of the PSII-LHCII super complex

Figure 7.2 Flow of events occurring in the damage-repair cycle of PSII under light stress. Because the light stress on PSII is moderate, photoinhibition is reversible and the photodamaged D1 protein is removed from the PSII complex and replaced by a newly synthesized copy.

of PSII. Other phosphoproteins in the PSII core include the D1 protein, D2 protein, the PsbH subunit and the TSP9 protein (Bennett, 1980; Vener, 2007). PSII core proteins are strongly phosphorylated under high light conditions (Rintamaki *et al.*, 1996). The PSII core proteins are phosphorylated mainly by STN8 kinase (Bonardi *et al.*, 2005); although, the STN7 kinase also shows minor activity (Vainonen *et al.*, 2005). The use of *stn7* and *8* mutants has enabled a clear relationship between protein phosphorylation and disorganization of the PSII complex to be established (Tikkanen and Aro, 2012). Phosphorylation of the core part of the PSII complex under excessive illumination was shown to help disassembly of the damaged PSII complex (Fristedt and Vener, 2011; Tikkanen *et al.*, 2008). After CP43 is removed from the PSII complex, the damaged and phosphorylated D1 protein is dislocated and moved to the membrane area to be dephosphorylated by a phosphatase(s) and then digested by a protease(s). Phosphatases involved in these steps have been identified (Samol *et al.*, 2012; Sirpio *et al.*, 2008; Vener *et al.*, 1999).

Degradation of the D1 protein in reversible photoinhibition

Accumulation of photodamaged proteins in PSII complexes is a potential hazard to PSII, and the damaged D1 protein is removed promptly by degradation with the aid of specific proteases. However, before proteases can break down the damaged D1 protein, the photodamaged D1-containing PSII complexes must disassemble and through this process the damaged D1 proteins are released from the PSII complexes as described above. Several proteases are considered to target damaged D1 protein and currently the likely two candidates are the ATP-dependent zinc-metalloprotease FtsH (filamentation temperature-sensitive H) (Komenda *et al.*, 2006; Silva *et al.*, 2003) and an ATP-independent serine protease Deg (degradation of periplasmic proteins) (Itzhaki *et al.*, 1998; Kapri-Pardes *et al.*, 2007; Kato *et al.*, 2012; Sun *et al.*, 2007). FtsH is a thylakoid membrane-bound processive endopeptidase and Deg is a thylakoid lumen-localized endopeptidase (actually in the four Deg proteases present in *Arabidopsis* chloroplast, Deg1, 5 and 8 are located in the lumen, whereas Deg2 is attached to the stroma side of the thylakoids). The D1 proteins that are damaged by ROS or endogenous cationic radicals are recognized by either of the two proteases, with recognition likely to be defined by the location of the site, namely the donor side (luminal side) or the acceptor side (stromal side) of the D1 protein. The proteases cleave the damaged protein into primary fragments (Yoshioka-Nishimura and Yamamoto, 2014). These primary fragments are further digested to smaller fragments and these fragments are subsequently broken down to amino acids. The details of the actions of these proteases are currently under debate.

There remains disparity about where the degradation of the photodamaged D1 protein occurs in the thylakoid membranes. The degradation is believed to occur at the stroma thylakoids where the thylakoids are not stacked and membrane-bound oligomeric protease complexes can access the damaged proteins easily from the stroma-exposed thylakoids to the grana where PSII complexes are abundant. For this postulate, PSII complexes retaining the photodamaged D1 must move from the grana core to the stroma thylakoids even though the grana are crowded with PSII complexes and other proteins. In a recent study, however, it was shown that the grana margins and also the grana end membranes are the actual sites of D1 degradation (Yoshioka *et al.*, 2010). In this case, the PSII complexes containing the damaged D1 proteins do not need to move a long distance through the grana to reach the degradation sites. When degradation of the D1 protein occurs at the luminal side of the protein by lumen-localized Deg proteases, the grana stack will not be impeded in accessing the damaged PSII complexes; although, extrinsic proteins such as PsbO, P and Q shield the D1 protein from being directly accessible to the proteases and this may limit access of the lumen proteases to the damaged D1 protein. In such a case, release of these extrinsic proteins from PSII is necessary for dislocation of the damaged D1 protein and subsequent degradation of the protein (Henmi *et al.*, 2004). More recently, macromolecular crowding in the lumen of stacked grana has been specified with concrete evidence, and it is strongly suggested that light-induced increase in size of the lumen is necessary for the access of the luminal proteases to the damaged proteins.

De novo synthesis and insertion of the D1 protein to the D1-depleted PSII complex and reassembly of the PSII complex

After degradation of the D1 protein, a new copy of the D1 protein is synthesized on the thylakoid-bound ribosomes (Baena-Gonzalez and Aro, 2002). It is not clear what the signal is that induces the expression of the *psbA* gene encoding the D1 protein. The $^1O_2$ and the other ROS produced by strong light are good candidates for the signals (Fischer *et al.*, 2013; Krieger-Liszkay *et al.*, 2008; Laloi *et al.*, 2007). Upon initiation of translation of the D1 protein, the elongating polypeptide chain is directly inserted into the thylakoid membrane and associates with the PSII core. The elongation step of the D1 protein translation has been suggested to be susceptible to ROS and that this inhibition of the repair of the D1 protein is responsible for photoinhibition of PSII (Nishiyama *et al.*, 2004).

Formation of the multisubunit PSII complex usually starts with accumulation of the D2 protein, followed by incorporation of the D1 protein and then antenna chlorophyll-binding protein CP47 associates to form the core of the PSII complex (Baena-Gonzalez and Aro, 2002; Komenda *et al.*, 2004). The α-subunit of cytochrome b_{559} also appears at a relatively early stage of PSII assembly (Hashimoto A, 1993; Morais *et al.*, 1998). Considering all these facts, intactness of the D2 protein and other core subunits during photoinhibition is important for reassembly of PSII complexes. As already described, CP43 is removed first with some association factors when the PSII complex disassembles under light stress, and the converse may be true when the PSII complex is organized: CP43 is the last protein to complete the formation of the core part of PSII in the reassembly process. Extrinsic proteins of PSII are partly located in the pool in the thylakoid lumen as stable free forms (Hashimoto *et al.*, 1996) and they bind to the PSII core at specific times points in the assembly process (Hashimoto, 1993; Hashimoto *et al.*, 1997). This elaborate turnover process of the D1 protein ensures that PSII activity is quickly restored after its impairment by photoinhibition.

Aggregation of the D1 protein and neighbouring proteins in PSII irreversible photoinhibition

By contrast to the reversible process observed in PSII under moderate light stress described above, the aggregation of the D1 protein with nearby polypeptides in the PSII complexes under extreme high light creates a situation that is more prohibitive to PSII activity. The aggregation corresponds to covalent cross-linking between the photodamaged D1 protein and the neighbouring polypeptides, including the D2 protein, CP43 and the α-subunit of cytochrome b_{559} (Yamamoto *et al.*, 2008). This protein aggregate was identified by SDS/urea-PAGE and Western blot analysis with specific antibodies, but there is the possibility that aggregation is not limited to these polypeptides. A broader spectrum of irreversible protein aggregates may be involved in photoinhibition. Since the protein aggregates are not removed readily by proteases, the formation of aggregates is fatal to the activity of PSII.

Dynamic changes to the thylakoids under light stress

A photodamaged D1 protein must be removed immediately from the PSII complex, otherwise damaged proteins accumulate and inhibit the repair cycle of PSII. The proteins involved in the PSII repair, including specific proteases, should therefore move freely between the grana core and the stroma thylakoids when the thylakoids are irradiated with excessive light. In the degradation of the damaged D1 protein by FtsH protease, extrusion of the hydrophilic part of FtsH may limit its movement from the stroma thylakoids to the grana. FtsH protease is anchored to the thylakoid membrane and has a large hydrophilic domain at the C-terminus, which is 6.5 nm in height and is exposed to the stroma. Since the width of the stromal gap between the two adjacent grana membranes is only ~3.5 nm (Daum *et al.*, 2010; Kirchhoff *et al.*, 2011), the FtsH protease is thought to be excluded from the grana thylakoids. To answer the question how the FtsH protease gains access to the damaged D1 protein in the grana core, a relationship between structural changes to the thylakoid membranes and movement of the FtsH protease has been studied recently. Estimation of the contents of the grana by digitonin treatment of spinach thylakoids and subsequent centrifugation (Chow *et al.*, 1980) revealed that grana stacking decreases irreversibly under strong light (Khatoon *et al.*, 2009). In accordance with this observation, transmission electron microscopy (TEM) showed dynamic structural changes to the thylakoids following intense illumination of spinach leaves (Yoshioka-Nishimura *et al.*, 2014). Thylakoid membranes bent outward at both ends of the grana and the gap between the grana margins widened. Distribution of FtsH proteases in the thylakoid membranes was assayed with immunoelectron microscopy before and after strong illumination for 30 min, and the FtsH proteases labelled with gold particles were observed in the stroma thylakoids and the grana margins in the dark, whereas subsequent strong illumination increased immunogold in the grana stacks. Degradation of the D1 protein occurred in parallel. The conclusion drawn from these results was that under light stress the FtsH protease can migrate to the grana and degrade damaged D1 proteins in the grana with the aid of structural changes to the thylakoids, in particular in the grana margins. Recently, 3D tomography has revealed that when spinach thylakoid membranes are exposed to strong light for an hour, the stacked grana loosen not only at the grana margins but also throughout the entire thylakoid membranes (Yoshioka-Nishimura *et al.*, 2014). The unstacking of the thylakoids should facilitate free migration of proteins in the thylakoid membranes. Thus, the unstacking of the thylakoid membranes is initiated from the grana margins and the whole thylakoid membranes might unstack eventually.

Conclusions

Excessive illumination of photosystem II (PSII) in oxygenic photosynthetic organism causes photoinhibition and blocks electron transport in PSII. During these processes, the reaction centre D1 protein of PSII is photodamaged by reactive oxygen species or other radicals produced by the photochemical reaction of PSII. This photodamage leads to degradation and/or irreversible aggregation of the damaged proteins, and the extent and type of damage is dependent on the intensity of the light. Under moderate photoinhibitory conditions or weak light, the damaged protein is proteolysed and replaced by a newly synthesized copy. Through this turnover of the D1 protein, PSII activity recovers rapidly in the dark. In contrast, irreversible aggregation of the D1 protein, which is caused by strong excessive illumination, prevents proper D1 turnover. Aggregation of the D1 protein with neighbouring proteins in PSII creates protein aggregates that accumulate in PSII complexes and are not eliminated readily by proteases. The photoinhibition observed under such a condition becomes irreversible. When the ambient temperature is higher than physiological temperatures, this irreversible photoinhibition occurs easily even under weak illumination. Accompanying these reversible and irreversible photoinhibition processes are dynamic changes to the thylakoids. They occur both at the molecular and membrane levels. It has become clear recently that these structural changes to the thylakoids are necessary for PSII to endure light stress effectively. All these processes take place in the thylakoid membranes to maintain optimal activity of the PSII under light stress.

Future trends

Repair of photodamaged PSII is a dynamic process that includes dissociation and reorganization of PSII complexes, and is highly dependent on the mobility of many proteins and protein complexes as well as lipids in the thylakoid membranes. Future research in this field will focus on elucidating various changes in the molecular arrangement of PSII–LHCII protein complexes and movement of other components essential for the repair process under light stress. Visualization of the process is probably most important to realize the dynamism of the thylakoid membranes and many sophisticated methods and current technology will be applied for this purpose.

References

Aro, E.M., Suorsa, M., Rokka, A., Allahverdiyeva, Y., Paakkarinen, V., Saleem, A., Battchikova, N., and Rintamaki, E. (2005). Dynamics of photosystem II: a proteomic approach to thylakoid protein complexes. J. Exp. Bot. *56*, 347–356.

Aro, E.M., Virgin, I., and Andersson, B. (1993). Photoinhibition of Photosystem II. Inactivation, protein damage and turnover. Biochim. Biophys. Acta *1143*, 113–134.

Baena-Gonzalez, E., and Aro, E.M. (2002). Biogenesis, assembly and turnover of photosystem II units. Phil. Trans. R. Soc. Lond. B Biol. Sci *357*, 1451–1459; discussion 1459–1460.

Barbato, R., Friso, G., Rigoni, F., Dalla Vecchia, F., and Giacometti, G.M. (1992). Structural changes and lateral redistribution of photosystem II during donor side photoinhibition of thylakoids. J. Cell Biol. *119*, 325–335.

Barber, J., and Andersson, B. (1992). Too much of a good thing: light can be bad for photosynthesis. Trends Biochem. Sci. *17*, 61–66.

Bennett, J. (1980). Chloroplast phosphoproteins. Evidence for a thylakoid-bound phosphoprotein phosphatase. Eur. J. Biochem. *104*, 85–89.

Blubaugh, D.J., and Cheniae, G.M. (1990). Kinetics of photoinhibition in hydroxylamine-extracted photosystem II membranes: relevance to photoactivation and sites of electron donation. Biochemistry *29*, 5109–5118.

Bonardi, V., Pesaresi, P., Becker, T., Schleiff, E., Wagner, R., Pfannschmidt, T., Jahns, P., and Leister, D. (2005). Photosystem II core phosphorylation and photosynthetic acclimation require two different protein kinases. Nature *437*, 1179–1182.

Chow, W.S., Thorne, S.W., Duniec, J.T., Sculley, M.J., and Boardman, N.K. (1980). The stacking of chloroplast thylakoids. Effects of cation screening and binding, studied by the digitonin method. Arch. Biochem. Biophys. *201*, 347–355.

Daum, B., Nicastro, D., Il, J.A., McIntosh, J.R., and Kuhlbrandt, W. (2010). Arrangement of photosystem II and ATP synthase in chloroplast membranes of spinach and pea. Plant Cell 22, 1299–1312.

Demmig-Adams, B. (1990). Carotenoids and photoprotection in plants: a role for the xanthophyll zeaxanthin. Biochim. Biophys. Acta *1020*, 1–24.

Fischer, B.B., Hideg, E., and Krieger-Liszkay, A. (2013). Production, detection, and signaling of singlet oxygen in photosynthetic organisms. Antioxid Redox Signal *18*, 2145–2162.

Fristedt, R., and Vener, A.V. (2011). High light induced disassembly of photosystem II supercomplexes in *Arabidopsis* requires STN7-dependent phosphorylation of CP29. PLoS One *6*, e24565.

Fristedt, R., Willig, A., Granath, P., Crevecoeur, M., Rochaix, J.D., and Vener, A.V. (2009). Phosphorylation of photosystem II controls functional macroscopic folding of photosynthetic membranes in *Arabidopsis*. Plant Cell *21*, 3950–3964.

Goral, T.K., Johnson, M.P., Brain, A.P., Kirchhoff, H., Ruban, A.V., and Mullineaux, C.W. (2010). Visualizing the mobility and distribution of chlorophyll proteins in higher plant thylakoid membranes: effects of photoinhibition and protein phosphorylation. Plant J. *62*, 948–959.

Hakala, M., Tuominen, I., Keranen, M., Tyystjarvi, T., and Tyystjarvi, E. (2005). Evidence for the role of the oxygen-evolving manganese complex in photoinhibition of Photosystem II. Biochim. Biophys. Acta *1706*, 68–80.

Hashimoto, A., Akasaka, T., and Yamamoto, Y. (1993). Characteristics of the assembly of the 33 kDa oxygen–evolving complex protein in the etioplasts and the developing chloroplasts of barley seedlings. Biochim. Biophys. Acta *1183*, 397–407.

Hashimoto, A., Ettinger, W.F., Yamamoto, Y., and Theg, S.M. (1997). Assembly of newly imported oxygen-evolving complex subunits in isolated chloroplasts: sites of assembly and mechanism of binding. Plant Cell *9*, 441–452.

Hashimoto, A., Yamamoto, Y., and Theg, S.M. (1996). Unassembled subunits of the photosynthetic oxygen–evolving complex present in the thylakoid lumen are long-lived and assembly-competent. FEBS Lett. *391*, 29–34.

Henmi, T., Miyao, M., and Yamamoto, Y. (2004). Release and reactive-oxygen-mediated damage of the oxygen–evolving complex subunits of PSII during photoinhibition. Plant Cell Physiol. *45*, 243–250.

Herbstova, M., Tietz, S., Kinzel, C., Turkina, M.V., and Kirchhoff, H. (2012). Architectural switch in plant photosynthetic membranes induced by light stress. Proc. Natl. Acad. Sci. U.S.A. *109*, 20130–20135.

Hind, G., Wall, J.S., Varkonyi, Z., Istokovics, A., Lambrev, P.H., and Garab, G. (2014). Membrane crystals of plant light-harvesting complex II disassemble reversibly in light. Plant Cell Physiol 55, 1296–1303.

Horton, P., Ruban, A.V., Rees, D., Pascal, A.A., Noctor, G., and Young, A.J. (1991). Control of the light-harvesting function of chloroplast membranes by aggregation of the LHCII chlorophyll–protein complex. FEBS Lett. *292*, 1–4.

Horton, P., Ruban, A.V., and Walters, R.G. (1996). Regulation of Light Harvesting in Green Plants. Annu. Rev. Plant. Physiol. Plant Mol. Biol *47*, 655–684.

Ishikawa, Y., Nakatani, E., Henmi, T., Ferjani, A., Harada, Y., Tamura, N., and Yamamoto, Y. (1999). Turnover of the aggregates and cross-linked products of the D1 protein generated by acceptor-side photoinhibition of photosystem II. Biochim. Biophys. Acta *1413*, 147–158.

Itzhaki, H., Naveh, L., Lindahl, M., Cook, M., and Adam, Z. (1998). Identification and characterization of DegP, a serine protease associated with the luminal side of the thylakoid membrane. J. Biol. Chem. *273*, 7094–7098.

Jansen, M.A., Mattoo, A.K., and Edelman, M. (1999). D1-D2 protein degradation in the chloroplast. Complex light saturation kinetics. Eur. J. Biochem. *260*, 527–532.

Kapri-Pardes, E., Naveh, L., and Adam, Z. (2007). The thylakoid lumen protease Deg1 is involved in the repair of photosystem II from photoinhibition in *Arabidopsis*. Plant Cell *19*, 1039–1047.

Kato, Y., Sun, X., Zhang, L., and Sakamoto, W. (2012). Cooperative D1 degradation in the photosystem II repair mediated by chloroplastic proteases in *Arabidopsis*. Plant. Physiol. *159*, 1428–1439.

Khatoon, M., Inagawa, K., Pospisil, P., Yamashita, A., Yoshioka, M., Lundin, B., Horie, J., Morita, N., Jajoo, A., and Yamamoto, Y. (2009). Quality control of photosystem II: Thylakoid unstacking is necessary to avoid further damage to the D1 protein and to facilitate D1 degradation under light stress in spinach thylakoids. J. Biol. Chem. *284*, 25343–25352.

Kirchhoff, H., Hall, C., Wood, M., Herbstova, M., Tsabari, O., Nevo, R., Charuvi, D., Shimoni, E., and Reich, Z. (2011). Dynamic control of protein diffusion within the granal thylakoid lumen. Proc. Natl Acad Sci U.S.A. *108*, 20248–20253.

Klimov, V.V., Baranov, S.V., and Allakhverdiev, S.I. (1997). Bicarbonate protects the donor side of photosystem II against photoinhibition and thermoinactivation. FEBS Lett. *418*, 243–246.

Klimov, V.V., Shafiev, M.A., and Allakhverdiev, S.I. (1990). Photoinactivation of the reactivation capacity of photosystem II in pea subchloroplast particles after a complete removal of manganese. Photosynth. Re.s *23*, 59–65.

Komenda, J., Barker, M., Kuvikova, S., de Vries, R., Mullineaux, C.W., Tichy, M., and Nixon, P.J. (2006). The FtsH protease slr0228 is important for quality control of photosystem II in the thylakoid membrane of *Synechocystis* sp. PCC 6803. J. Biol. Chem. *281*, 1145–1151.

Komenda, J., Reisinger, V., Muller, B.C., Dobakova, M., Granvogl, B., and Eichacker, L.A. (2004). Accumulation of the D2 protein is a key regulatory step for assembly of the photosystem II reaction center complex in *Synechocystis* PCC 6803. J. Biol. Chem. *279*, 48620–48629.

Krieger-Liszkay, A., Fufezan, C., and Trebst, A. (2008). Singlet oxygen production in photosystem II and related protection mechanism. Photosynth. Res. *98*, 551–564.

Kyle, D.J., Ohad, I., and Arntzen, C.J. (1984). Membrane protein damage and repair: Selective loss of a quinone-protein function in chloroplast membranes. Proc. Natl. Acad. Sci. U.S.A. *81*, 4070–4074.

Laloi, C., Stachowiak, M., Pers-Kamczyc, E., Warzych, E., Murgia, I., and Apel, K. (2007). Cross-talk between singlet oxygen- and hydrogen peroxide-dependent signaling of stress responses in *Arabidopsis thaliana*. Proc. Natl. Acad. Sci. U.S.A. *104*, 672–677.

Li, X.P., Bjorkman, O., Shih, C., Grossman, A.R., Rosenquist, M., Jansson, S., and Niyogi, K.K. (2000). A pigment-binding protein essential for regulation of photosynthetic light harvesting. Nature *403*, 391–395.

Mattoo, A.K., Pick, U., Hoffman-Falk, H., and Edelman, M. (1981). The rapidly metabolized 32,000-dalton polypeptide of the chloroplast is the 'proteinaceous shield' regulating photosystem II electron transport and mediating diuron herbicide sensitivity. Proc. Natl. Acad. Sci. U.S.A. *78*, 1572–1576.

Morais, F., Barber, J., and Nixon, P.J. (1998). The chloroplast-encoded alpha subunit of cytochrome b-559 is required for assembly of the photosystem two complex in both the light and the dark in *Chlamydomonas reinhardtii*. J. Biol. Chem. *273*, 29315–29320.

Nishiyama, Y., Allakhverdiev, S.I., Yamamoto, H., Hayashi, H., and Murata, N. (2004). Singlet oxygen inhibits the repair of photosystem II by suppressing the translation elongation of the D1 protein in *Synechocystis* sp. PCC 6803. Biochemistry *43*, 11321–11330.

Niyogi, K.K. (1999). Photoprotection revised: Genetic and molecular approaches. Annu. Rev. Plant. Physiol. Plant Mol. Biol. *50*, 333–359.

Ohad, I., Kyle, D.J., and Arntzen, C.J. (1984). Membrane protein damage and repair: removal and replacement of inactivated 32-kilodalton polypeptides in chloroplast membranes. J. Cell Biol. *99*, 481–485.

Ohira, S., Morita, N., Suh, H.J., Jung, J., and Yamamoto, Y. (2005). Quality control of photosystem II under light stress – turnover of aggregates of the D1 protein in vivo. Photosynth. Res. *84*, 29–33.

Ohnishi, N., Allakhverdiev, S.I., Takahashi, S., Higashi, S., Watanabe, M., Nishiyama, Y., and Murata, N. (2005). Two-step mechanism of photodamage to photosystem II: step 1 occurs at the oxygen-evolving complex and step 2 occurs at the photochemical reaction center. Biochemistry *44*, 8494–8499.

Peers, G., Truong, T.B., Ostendorf, E., Busch, A., Elrad, D., Grossman, A.R., Hippler, M., and Niyogi, K.K. (2009). An ancient light-harvesting protein is critical for the regulation of algal photosynthesis. Nature *462*, 518–521.

Pospisil, P. (2009). Production of reactive oxygen species by photosystem II. Biochim. Biophys. Acta *1787*, 1151–1160.

Powles, S.B. (1984). Photoinhibition of photosynthesis induced by visible light. Annu. Rev. Plant. Physiol. *35*, 14–44.

Rintamaki, E., Kettunen, R., and Aro, E.M. (1996). Differential D1 dephosphorylation in functional and photodamaged photosystem II centers. Dephosphorylation is a prerequisite for degradation of damaged D1. J. Biol. Chem. *271*, 14870–14875.

Samol, I., Shapiguzov, A., Ingelsson, B., Fucile, G., Crevecoeur, M., Vener, A.V., Rochaix, J.D., and Goldschmidt-Clermont, M. (2012). Identification of a photosystem II phosphatase involved in light acclimation in *Arabidopsis*. Plant Cell *24*, 2596–2609.

Schreiber, U., Schliwa, U., and Bilger, W. (1986). Continuous recording of photochemical and non-photochemical chlorophyll fluorescence quenching with a new type of modulation fluorometer. Photosynth Res. *10*, 51–62.

Silva, P., Thompson, E., Bailey, S., Kruse, O., Mullineaux, C.W., Robinson, C., Mann, N.H., and Nixon, P.J. (2003). FtsH is involved in the early stages of repair of photosystem II in *Synechocystis* sp PCC 6803. Plant Cell *15*, 2152–2164.

Sirpio, S., Khrouchtchova, A., Allahverdiyeva, Y., Hansson, M., Fristedt, R., Vener, A.V., Scheller, H.V., Jensen, P.E., Haldrup, A., and Aro, E.M. (2008). AtCYP38 ensures early biogenesis, correct assembly and sustenance of photosystem II. Plant J. *55*, 639–651.

Suh, H.J., Kim, C.S., and Jung, J. (2000). Cytochrome b6/f complex as an indigenous photodynamic generator of singlet oxygen in thylakoid membranes. Photochem. Photobiol. *71*, 103–109.

Sun, X., Peng, L., Guo, J., Chi, W., Ma, J., Lu, C., and Zhang, L. (2007). Formation of DEG5 and DEG8 complexes and their involvement in the degradation of photodamaged photosystem II reaction center D1 protein in *Arabidopsis*. Plant Cell *19*, 1347–1361.

Telfer, A. (2014). Singlet oxygen production by PSII under light stress: mechanism, detection and the protective role of beta-carotene. Plant Cell Physiol. *55*, 1216–1223.

Tikkanen, M., and Aro, E.M. (2012). Thylakoid protein phosphorylation in dynamic regulation of photosystem II in higher plants. Biochim. Biophys. Acta *1817*, 232–238.

Tikkanen, M., Nurmi, M., Kangasjarvi, S., and Aro, E.M. (2008). Core protein phosphorylation facilitates the repair of photodamaged photosystem II at high light. Biochim. Biophys. Acta *1777*, 1432–1437.

Vainonen, J.P., Hansson, M., and Vener, A.V. (2005). STN8 protein kinase in *Arabidopsis thaliana* is specific in phosphorylation of photosystem II core proteins. J. Biol. Chem. *280*, 33679–33686.

Vener, A.V. (2007). Environmentally modulated phosphorylation and dynamics of proteins in photosynthetic membranes. Biochim Biophys Acta *1767*, 449–457.

Vener, A.V., Rokka, A., Fulgosi, H., Andersson, B., and Herrmann, R.G. (1999). A cyclophilin-regulated PP2A-like protein phosphatase in thylakoid membranes of plant chloroplasts. Biochemistry *38*, 14955–14965.

Yamamoto, H.Y., Kamite, L., and Wang, Y.Y. (1972). An ascorbate-induced absorbance change in chloroplasts from violaxanthin de-epoxidation. Plant. Physiol. *49*, 224–228.

Yamamoto, Y., Aminaka, R., Yoshioka, M., Khatoon, M., Komayama, K., Takenaka, D., Yamashita, A., Nijo, N., Inagawa, K., Morita, N., *et al.* (2008). Quality control of photosystem II: impact of light and heat stresses. Photosynth Res *98*, 589–608.

Yamamoto, Y., Hori, H., Kai, S., Ishikawa, T., Ohnishi, A., Tsumura, N., and Morita, N. (2013). Quality control of Photosystem II: reversible and irreversible protein aggregation decides the fate of Photosystem II under excessive illumination. Front Plant Sci *4*, 433.

Yamamoto, Y., Kai, S., Ohnishi, A., Tsumura, N., Ishikawa, T., Hori, H., Morita, N., and Ishikawa, Y. (2014). Quality control of PSII: behavior of PSII in the highly crowded grana thylakoids under excessive light. Plant Cell Physiol. *55*, 1206–1215.

Yoshioka-Nishimura, M., Nanba, D., Takaki, T., Ohba, C., Tsumura, N., Morita, N., Sakamoto, H., Murata, K., and Yamamoto, Y. (2014). Quality control of photosystem II: direct imaging of the changes in the thylakoid structure and distribution of FtsH proteases in spinach chloroplasts under light stress. Plant Cell Physiol. *55*, 1255–1265.

Yoshioka-Nishimura, M., and Yamamoto, Y. (2014). Quality control of Photosystem II: the molecular basis for the action of FtsH protease and the dynamics of the thylakoid membranes. J. Photochem. Photobiol. B *137*, 100–106.

Yoshioka, M., Nakayama, Y., Yoshida, M., Ohashi, K., Morita, N., Kobayashi, H., and Yamamoto, Y. (2010). Quality control of photosystem II: FtsH hexamers are localized near photosystem II at grana for the swift repair of damage. J. Biol. Chem. *285*, 41972–41981.

Plastoglobules: Lipid Droplets at the Thylakoid Membrane

8

Thibaut Pralon and Felix Kessler

Abstract

Plastoglobules (PG) are lipid droplets that are structurally and functionally associated with the thylakoid membranes of chloroplasts. PG thus effectively form a thylakoid membrane microdomain. The thylakoid membranes provide the environment for the photosynthetic light reactions. The thylakoid membranes in majority consist of bilayer forming galactolipids and also harbour neutral lipids, including carotenoids, chlorophylls, prenylquinones (plastoquinone, phylloquinone, tocopherols), and others. The neutral lipids participate directly in the photosynthetic light reactions, act as anti-oxidants or are metabolic products, respectively. The biosynthesis, abundance, redox state and other parameters of these compounds are tightly controlled. PG store these compounds but also actively participate in their metabolism. PG thereby contribute to continuous remodelling of the thylakoid membrane. This depends on the presence of an assortment of enzymes and regulatory kinases at PG that are involved in a wide range of processes affecting thylakoid membrane composition. Here, we review the current state of the PG field.

Discovery of PG

Ultrastructure of the chloroplast reveals osmiophilic bodies, the plastoglobules

In the middle of the twentieth century, electron microscopy gave scientists a new tool to visualize plant subcellular organization. Soon, scientists began to describe the details of chloroplast ultrastructure in different photosynthetic organisms (Cohen and Bowler, 1953; Wolken and Palade, 1952). Among the discoveries were spherical particles that are ubiquitous in plastids. Owing to the presence of unsaturated lipids they were intensely contrasted by osmium tetroxide and therefore initially termed osmiophilic globules (Greenwood *et al.*, 1963). Their diameter varied from 10 to 500 nm depending on the plastid origin and developmental state (Greenwood *et al.*, 1963; Lichtenthaler, 1968). The globules consist of a polar membrane lipid monolayer on the outside and neutral lipids on the inside (Hansmann and Sitte, 1982). In addition, proteins coat the surface of globules (Kessler *et al.*, 1999; Pozueta-Romero *et al.*, 1997). Therefore, structurally, the osmiophilic globules, now termed plastoglobules (PG), were categorized as lipoprotein particles or lipid droplets. Typical for lipid droplets, the lipid to protein ratio of PG is high resulting in a very low buoyant density. Due to their identification as lipid storing particles, the composition of the isolated PG became a topic of interest.

Plastoglobule lipid composition

In order to determine lipid composition, plastoglobules needed to be isolated. Taking advantage of their low density, plastoglobules could easily be isolated and separated from the denser envelope and thylakoid membranes by floatation centrifugation using sucrose density gradients. Lipid composition of isolated PG was then analysed by thin-layer chromatography. First studies of the lipid composition of plastoglobules revealed the presence of triacylglycerol (TAG), plastoquinone/-ol, α-tocopherol (vitamin E), galactolipids, but chlorophyll was notably absent (Bailey and Whyborn, 1963; Greenwood *et al.*, 1963). Follow up investigations of the PG lipid composition largely confirmed the earlier

Table 8.1 Main lipids found in plastoglobules of different plastids

Plastids	Lipid	References
Chloroplast	α-Tocopherol (vitamin E); fatty acid phytyl esters (FAPE); galactolipids; phylloquinone (vitamin K1); plastoquinone/ol (PQ/PQH2); triacylglycerol (TAG)	Bailey and Whyborn (1963), Gaude *et al.* (2007), Greenwood *et al.* (1963), Lichtenthaler and Peveling (1967), Lichtenthaler and Sprey (1966), Steinmüller and Tevini (1985), Zbierzak *et al.* (2010)
Chromoplast	α-Tocopherol (vitamin E); carotenoids; carotenoid esters; plastoquinone (PQ); triacylglycerol (TAG)	Deruère *et al.* (1994), Hansmann and Sitte (1982), Steinmüller and Tevini (1985)
Gerontoplast	Fatty acid phytyl esters (FAPE); triacylglycerol (TAG)	Gaude *et al.* (2007), Tevini and Steinmüller (1985)

results while adding some new compounds such as phylloquinone (vitamin K_1) (Lichtenthaler, 1966; Lichtenthaler and Peveling, 1967; Lichtenthaler and Sprey, 1966). Recently, fatty acid phytyl esters (FAPE) were added to the PG lipid list (Gaude *et al.*, 2007) (Table 8.1). Triacylglycerol was found to be the major lipid constituent of chloroplast PG, whereas carotenoids and their esters were the most abundant in chromoplast PG (Hansmann and Sitte, 1982; Lichtenthaler, 1969a; Steinmüller and Tevini, 1985; Tevini and Steinmüller, 1985). So far, a complete PG lipidome using state of the art mass spectrometry techniques has not been determined. Therefore, additional, presumably less abundant lipid species may still be discovered. Apart from lipids some early studies on PG composition also hinted at the presence of proteins (Hansmann and Sitte, 1982; Steinmüller and Tevini, 1985). In those days, however, the technology to identify and analyse the proteins was not yet sufficiently well developed.

Lipid trafficking between PG and the thylakoid membrane

Given the common nature of lipids in thylakoids and PG, an important question concerns the possibility and mechanism of lipid exchange between the two compartments. An electron tomographic study of plastoglobules revealed that PG are normally in contact with the margins of stromal thylakoids. High-resolution image analysis showed that the outer lipid leaflet of PG is contiguous with that of the thylakoid membrane (Austin *et al.*, 2006). The contact sites between PG and thylakoid membranes therefore result in a tight patch that connects the interior hydrophobic phase of PG with the thylakoid bilayer. This configuration could therefore allow the simultaneous transfer of polar membrane lipids and hydrophobic molecules between the two compartments thereby providing a physical basis for molecular trafficking between the two compartments and explanation for their molecular cooperation.

PG morphological changes in response to developmental transitions and stress adaptation

Using electron microscopy on various plant tissues, developmental states and stress conditions, plastid ultrastructural modifications implicating both PG and thylakoid membranes were observed. These changes provide striking visual evidence of physical membrane remodelling. For instance, the presence of numerous PG in proplastids and etioplasts that decrease in number during chloroplast formation while thylakoids emerge (Lichtenthaler, 1968; Sprey and Lichtenthaler, 1966) led to the hypothesis of an involvement of PG in thylakoid membrane formation. During chloroplast senescence PG become supersized while the thylakoids progressively disappear (Lichtenthaler, 1969b; Tuquet and Newman, 1980) suggesting that PG participate in thylakoid disassembly. Similarly, during chromoplast development in red pepper fruit maturation, the disappearance of the thylakoid membranes concomitantly with the proliferation and enlargement of PG was observed. Ultimately, pigment-filled fibrils, derived from globular PG emerged (Deruère *et al.*, 1994). The chloroplast to chromoplast transition in other species follows an analogous pattern but the PG themselves serve to store the pigments. In summary, the electron microscopic evidence implicates PG in several types of plastid developmental transitions: from etioplast to chloroplast, from chloroplast to gerontoplast and from chloroplast to chromoplast, respectively.

Electron microscopic observation of plastids after various stress treatments showed that ultrastructural changes in PG and the thylakoid membrane are not limited to developmental transitions. Various stress types such as high salt, drought, nitrogen deprivation, heat, high light as well as heavy metals increase the size and number of PG (Ben Salem-Fnayou *et al.*, 2011; Gaude *et al.*, 2007; Locy *et al.*, 1996; Rey *et al.*, 2000; Zhang *et al.*, 2010). The aforementioned stress types result in leaf bleaching and altered photosynthetic parameters, which in turn reflect chlorophyll degradation and thylakoid membrane perturbation and disassembly. Thus, the various stress types to some extent reiterate senescence related phenotypes. As for the plastid developmental transitions, ultrastructural changes during stress adaptation provide evidence for the implication of PG in thylakoid lipid remodelling. One particularly striking example is *Arabidopsis* under nitrogen-depleted conditions (Gaude *et al.*, 2007): by electron microscopy, thylakoids appear strongly diminished whereas PG become supersized. Here, the visible physical remodelling correlates with a reduction of chlorophyll and monogalactosyldiacylglyerol (MGDG) levels and an increase of triacylglycerol and fatty acid phytyl esters (Lippold *et al.*, 2012).

For a long time, however, the metabolic roles of PG (apart of course from lipid storage (Lichtenthaler, 1968; Lichtenthaler and Peveling, 1967; Sprey and Lichtenthaler, 1966)) in plastid developmental transitions and in stress adaptations were unknown. Only with advent of modern molecular biology techniques evidence started to emerge indicating that it is the protein complement of PG that enables the active participation of PG in thylakoid membrane remodelling (Fig. 8.1).

PG protein composition

In the 1980s, proteins were detected in PG (Hansmann and Sitte, 1982; Steinmüller and Tevini, 1985). In red pepper chromoplasts, carotenoids are stored in globulo-tubular structures termed fibrils that contain an abundant 30 kDa protein that was termed fibrillin (Deruère *et al.*, 1994; Knoth *et al.*, 1986). The amino acid sequence of fibrillin was determined by Deruère *et al.* 1994 making it the first known PG protein. Related proteins, called PAP (Plastid-lipid Associated Proteins) belonging to the emerging fibrillin family were also identified and found to be major PG proteins in chloroplasts and chromoplasts of several species (Kessler *et al.*, 1999; Pozueta-Romero *et al.*, 1997).

In the meantime, the PG proteomes from chloroplasts have been determined and refined revealing thirty consensus proteins (Table 8.2) (Lundquist *et al.*, 2012). In addition, the proteome of red pepper chromoplast PG has been determined (Ytterberg *et al.*, 2006). PG proteins can be classified in three main groups: (1) structural proteins (PAP/Fibrillins), (2) chloroplast metabolic enzymes and (3) unknown proteins (Vidi *et al.*, 2006). Genome-wide coexpression analysis of the genes encoding PG proteins resulted in a modular network consisting of four distinct modules each containing PG proteins. Specific functional enrichment in each of the modules was attributed to chlorophyll degradation/senescence, isoprenoid biosynthesis, plastid proteolysis, redox regulators and phosphoregulators of electron flow (Lundquist *et al.*, 2012).

The discovery of a surprising variety of associated proteins resulted in the new hypothesis proposing that PG do not only serve in lipid storage but also in lipid metabolism. Moreover, it appears that PG proteomes are functionally tailored to the plastid type. In chromoplast PG, for instance, enzymes involved in the synthesis and accumulation of coloured carotenoids are specifically recruited (Ytterberg *et al.*, 2006). In the following, we will review the identified and characterized PG proteins as well as their implication in lipid metabolism and remodelling of the thylakoid membrane during plastid developmental transitions and stress adaptation.

PG proteins and metabolic activities

Fibrillins

Fibrillin was originally discovered through its association with carotenoid-sequestering fibrils of red pepper (*Capsicum annuum*) chromoplasts (Deruère *et al.*, 1994). Despite their fibrillar appearance, fibrils are structurally closely related to plastoglobules having an outer polar lipid monolayer lined with proteins and enclosing a hydrophobic core

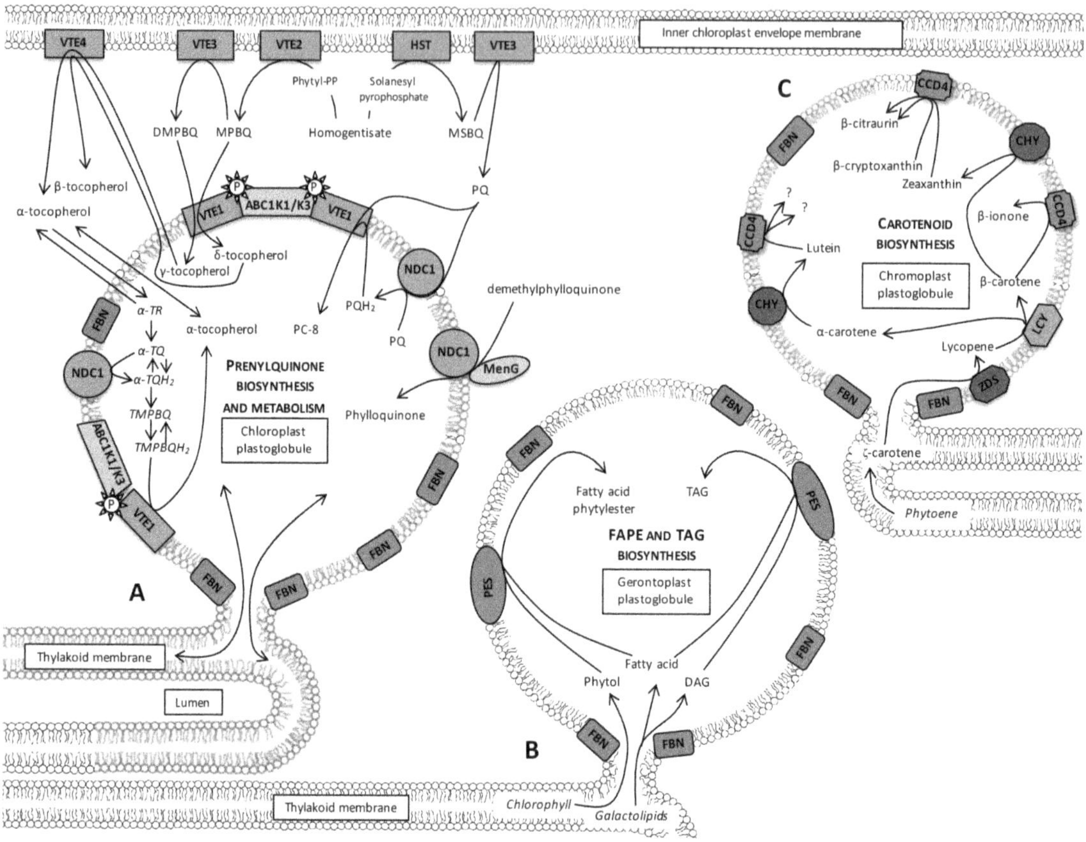

Figure 8.1 Lipid metabolism in plastoglobules. (A) Prenylquinone biosynthesis and metabolism in chloroplast plastoglobules. NDC1 is required for phylloquinone biosynthesis. NDC1 is also implicated in plastochromanol (PC-8) biosynthesis and in tocopherol recycling by reducing substrates, plastoquinol and TMPBQ, to their quinol forms. VTE1 uses reduced substrates for the biosynthesis of plastochromanol and tocopherol recycling. ABC1K1 kinase is required for increased tocopherol production under highlight. ABC1K3 kinase is required for normal PC-8 accumulation and tocopherol recycling. Both ABC1K1 and -K3 do so by acting on VTE1 activity, presumably by phosphorylation. (B) During senescence-induced thylakoid disassembly, fatty acid phytyl esters and triacylglycerols are synthetized by phytyl ester synthase 1 and 2 (PES1 and -2) in gerontoplast plastoglobules. (C) In chromoplast plastoglobules LCY, CHY, ZDS are recruited to promote carotenoid biosynthesis and accumulation. CCD4 participates in flower colour production and volatile emission. ABC1K, activity of bc1-like kinase; α-TQH$_2$, α-tocopherol-quinol; α-TQ, α-tocopherol-quinone; α-TR, α-tocopheroxyl-radical; CCD4, carotenoid cleavage dioxygenase 4; CHY, β-carotene β-hydroxylase; DMPBQ, dimethylphytylbenzoquinone; FAPE, fatty acid phytyl ester; FBN, fibrillins; HST, homogentisate prenyl transferase; LCY, lycopene β-cyclase; MenG, demethylmenaquinone methyl transferase; MPBQ, methylphytylbenzoquinone; MSBQ, methylsolanesylbenzoquinone; NDC1, NAD(P)H dehydrogenase C1; P, phosphorylation; PC-8, plastochromanol-8; PES, phytyl ester synthases; PQ, plastoquinone-9; PQH$_2$, plastoquinol-9; TAG, triacylglycerol; VTE1, tocopherol cyclase, vitamin E deficient 1; VTE2, homogentisate phytyl transferase, vitamin E deficient 2; VTE3, tocopherol methyltransferase, vitamin E deficient 3; VTE4, γ-tocopherol methyltransferase, vitamin E deficient 4; ZDS, ζ-carotene desaturase.

containing coloured carotenoids. Fibril carotenoids are mostly present as esters, capsanthin diester being the most abundant at more than 50% in red pepper (Deruère *et al.*, 1994). It is interesting to note that in tomato chromoplasts no fibrils are present and plastoglobules take over their function (Simkin *et al.*, 2007). The precise structure of carotenoid-containing structures has been attributed to the lipid to protein ratio. A high proportion of protein favours the formation of fibrillar rather than globular structures (Wuttke, 1976). Fibrillin is the most abundant protein of carotenoid fibrils. Purified fibrillin protein was able to reconstitute fibrils when it was mixed with polar lipids and

Table 8.2 Refined PG proteome according to Lundquist *et al.* (2012) and containing three carotenoid metabolic enzymes from chromoplast PG. Ytterberg *et al.* (2006)

Accession	Common name	Symbol	Function(s)
Known and characterized plastoglobule proteins			
At4g04020	Fibrillin 1a	FBN1a; PGL35	PG structure
At4g22240	Fibrillin 1b	FBN1b; PGL33	
At2g35490	Fibrillin 2	FBN2; PGL40	
At3g23400	Fibrillin 4	FBN4; PGL30.4	
At4g32770	Tocopherol cyclase	VTE1	Tocopherol biosynthesis and regeneration
At5g08740	NAD(P)H dehydrogenase C1	NDC1	Phylloquinone biosynthesis PC-8 biosynthesis PQ redox state regulation
	β-Carotene β-hydroxylase	CHY; CrtR-β	Carotenoid metabolism (identified in red pepper chromoplast PG)
	ζ-Carotene desaturase	ZDS	
	Lycopene β-cyclase	LCY; LYC-β	
At4g19170	Carotenoid cleavage dioxygenase 4	CCD4; NCED4	
At1g54570	Phytyl ester synthase 1	PES1; DGAT3	TAG and FAPE biosynthesis
At3g26840	Phytyl ester synthase 2	PES2; DGAT4	
At4g31390	Activity of bc1-like kinase 1	ABC1K1	Prenylquinone biosynthesis and metabolic regulation
At1g79600	Activity of bc1-like kinase 3	ABC1K3	
At3g07700	Activity of bc1-like kinase 7	ABC1K7	
Uncharacterized plastoglobule proteins			
At3g58010	Fibrillin 7a	FBN7a; PGL34	PG structure
At2g42130	Fibrillin 7b	FBN7b; PGL30	
At2g46910	Fibrillin 8	FBN8	
At1g71810	Activity of bc1-like kinase 5	ABC1K5	Regulation of metabolic process
At3g24190	Activity of bc1-like kinase 6	ABC1K6	
At5g05200	Activity of bc1-like kinase 9	ABC1K9	
At1g32220	Flavin reductase-related 1		Unknown
At2g34460	Flavin reductase-related 2		
At1g78140	UbiE methyltransferase-related 1		Prenyl lipid biosynthesis
At2g41040	UbiE methyltransferase-related 2		
At3g10130	SOUL domain protein		Tetrapyrrole metabolism
At1g06690	Aldo/keto reductase		Unknown
At4g39730	PLAT/LH2-1		Lipase, lipoxygenase
At3g27110	M48 protease		Protein homeostasis
At5g41120	Esterase 1		TAG/FAPE biosynthesis
At1g73750	Unknown SAG		Participation in senescence associated processes
At4g13200	Unknown 1		Unknown
At3g43540	Unknown 2	DUF1350	Unknown

carotenoids at their natural stoichiometric ratio. The highest yields were achieved with the esterified xanthophylls, zeaxanthin diester and capsanthin diester (Deruère *et al.*, 1994). Reconstitution did not occur in the absence of fibrillin. Based on the above findings, fibrillins were attributed a structural role in fibril formation. Overexpression of fibrillin in tobacco leaves led to an increase of plastoglobule numbers and clusters reinforcing the idea of structural role (Rey *et al.*, 2000). Homologues of fibrillin exist in many other species such as potato (plastid lipid associated proteins, PAP) (Langenkämper *et al.*, 2001) and pea (plastoglobulins, PGL) (Kessler *et al.*, 1999).

In *Arabidopsis*, quantitative proteomics indicated that fibrillins are the most abundant proteins of chloroplast PG, four homologues (FBN1a, -1b, -2 and -4) (At4g04020, At4g22240, At2g35490, At3g23400) making up almost 50% of the PG protein mass (Lundquist *et al.*, 2012). Interestingly also, the same study indicated that different fibrillin homologues partition to varying and predictable degrees between the plastoglobules and the thylakoid membrane. Low isoelectric points and a higher hydrophobicity index favoured PG over thylakoid association.

A study on the apple *fib4* (the orthologue of *Arabidopsis* FBN4) knock down mutant revealed reduced accumulation of plastoquinone in PG while the total cellular plastoquinone remained constant. Fib4 contains a conserved lipocalin motif (DLDKLQGKWRLLY) that could potentially facilitate binding and transport of lipids (Singh *et al.*, 2012; Singh *et al.*, 2010). These findings led to the hypothesis that fibrillins apart from a structural role may also have a role in metabolite trafficking between the thylakoid membrane and plastoglobules. Fibrillins such as Fib4 would therefore be expected to be present at the neck-like structure at the interface between PG and the thylakoid membrane.

The databases indicate that fibrillins are ancient, evolutionarily conserved proteins originating from cyanobacteria. Fibrillins may therefore be omnipresent in photosynthetic species (Cunningham *et al.*, 2010). While *Arabidopsis* has many fibrillin homologues, *Synechocystis* sp. PCC 6803 has only two. Therefore, this cyanobacterial strain is an interesting model to test fibrillin function. The deletion of the two homologues of *Synechocystis* sp. PCC 6803 fibrillin (PGL1 and -2) was viable and photosynthetically active. However, the carotenoid composition was altered suggesting compromised resistance to high light stress (Cunningham *et al.*, 2010). In cyanobacteria lipid globules are rather rare, so their association with the fibrillins PGL1 and -2 was not readily apparent. It was therefore suggested that fibrillins in cyanobacteria may function in neutral lipid sequestration outside the PG context.

It is not known how proteins are targeted to PG but they do not have apparent targeting sequences. It has been shown for *Arabidopsis* PGL34/FBN7a (At3g58010) (Vidi *et al.*, 2007) that the entire protein with the exception of a small C-terminal sequence is required for assembly on PG. As PG are frequently observed at the stromal margins of thylakoids it has been proposed that the membrane curvature together perhaps with 'lipid-associated proteins' such as the fibrillins may facilitate the budding of PG.

NDC1, a type II NAD(P)H dehydrogenase required for phylloquinone biosynthesis

NDC1 (At5g08740) was identified as a bona fide PG protein in three independent proteomics studies (Lundquist *et al.*, 2012; Vidi *et al.*, 2006; Ytterberg *et al.*, 2006). NDC1 belongs to the family of monomeric Type II NAD(P)H-dependent dehydrogenases (NDH-2). Type II NAD(P)H-dependent dehydrogenases serve as an alternative to the multi-subunit respiratory complex I (type I NADH dehydrogenase (NDH-1)) to catalyse electron transfer from NADH to ubiquinone in the mitochondrial respiratory chain (Feng *et al.*, 2012). In *Arabidopsis*, a family of seven NDH-2 homologues exists. Six of these are located in mitochondria (Michalecka *et al.*, 2003), but only one, NDC1, is present in chloroplasts.

NDC1 is conserved from cyanobacteria, which provides an evolutionary explanation for its presence in chloroplasts, although NDC1 was originally localized in mitochondria (Michalecka *et al.*, 2003). There is also evidence for dual mitochondrial and chloroplast localization of the protein (Carrie *et al.*, 2008; Eugeni-Piller *et al.*, 2011). In chloroplasts, not only proteomics but also fluorescence localization and subfractionation indicate that NDC1 sublocalizes to PG (Eugeni-Piller *et al.*, 2011). Ubiquinone, the usual substrate of NDH-2 enzymes, is not present in chloroplasts. But it was shown that NDC1 is a multifunctional enzyme accepting as substrates a variety of quinone derivatives such as decyl-ubiquinone and decyl-plastoquinone as well as isolated PG (Eugeni-Piller *et al.*, 2011). As NDC1 should well be able to use plastoquinone as substrate for NAD(P)H-dependent reduction, it is imaginable that NDC1 functions in cyclic electron flow around photosystem I (PSI). In *Arabidopsis*, however, mutant analysis has revealed that the contribution of NDC1 to cyclic electron flow around PSI is negligible and that the chloroplast NDH complex is responsible for this task. But, in *Chlamydomonas* the chloroplast NDH complex is lacking and an NDH-2 homologue (Nda-2) has been implicated in cyclic electron flow (Jans *et al.*, 2008).

Despite the lack of involvement in cyclic electron flow NDC1 affects the plastoquinone redox equilibrium: total plastoquinone in the *ndc1* mutant is significantly more oxidized than in the wild type (Eugeni-Piller *et al.*, 2011). Plastoquinone is distributed between the thylakoids and plastoglobules. The photochemically active plastoquinone pool is localized in the thylakoid membrane but a large proportion of the plastoquinone is present in plastoglobules. Presumably, the plastoglobule-localized plastoquinone is selectively reduced by NDC1 without interfering with the plastoquinone participating in photosynthetic electron transport. This suggests that the thylakoid plastoquinone pool and the PG plastoquinone reservoir are well separated and do not rapidly equilibrate. The redox state of the plastoquinone pool is known to control the level of phosphorylation of the chloroplast light-harvesting complex II (LHCII) (Allen *et al.*, 1981). It will therefore be interesting to test whether and if to what extent NDC1 together with PG-localized plastoquinone contributes to plastoquinone redox regulation.

In a surprising twist, phylloquinone was completely absent from the *ndc1* mutant which instead accumulated demethylphylloquinone, lacking the 3-methyl group (Eugeni-Piller *et al.*, 2011). It is unclear why demethylphylloquinone accumulates in the *ndc1* mutant because the methylation step is known to be carried out by the AtMenG enzyme in *Arabidopsis* (Lohmann *et al.*, 2006). AtMenG is normally expressed in the *ndc1* mutant background. Despite the lack of phylloquinone, the *ndc1* mutant has no visible phenotype. In *Synechocystis* (Sakuragi *et al.*, 2002) and *Arabidopsis* (Lohmann *et al.*, 2006) *menG* mutant's demethylphylloquinone effectively replaces phylloquinone. In *Arabidopsis*, however, the maximal photosynthetic efficiency was reduced in *atmenG* plants grown under high light. It therefore needs to be noted that despite the similarities the molecular phenotypes of *atmenG* and *ndc1* are not identical. Several hypothetical explanations for the role of NDC1 in phylloquinone synthesis may be proposed. One is that AtMenG-dependent methylation prefers reduced demethylphylloquinone as its substrate and the reduction of demethylphylloquinone depends on NDC1. As we will see later, substrate reduction by NDC1 is also of importance for reactions catalysed by tocopherol cyclase (VTE1, At4g32770). Another possibility is that NDC1 regulates AtMenG activity. For example, NDC1 could serve as platform to recruit AtMenG to the PHYLLO metabolon (Gross *et al.*, 2006). The PHYLLO metabolon consists of MenF, MenD, MenC, and MenH in a single polypeptide and catalyses the preceding steps of phylloquinone biosynthesis. In support of this possibility, like NDC1-YFP both PHYLLO-DsRed and AtMenG–GFP fusion proteins gave punctate fluorescence within chloroplasts that were reminiscent of PG. To test this, it will be interesting to see whether the punctate patterns of the latter two fluorescent fusion proteins is disrupted in the *ndc1* mutant background.

Implication of PG in tocochromanol biosynthesis and metabolism

Role of PG in tocopherol biosynthesis

Tocochromanols (including tocopherol, tocotrienols and plastochromanol-8 (PC-8)) are important lipid antioxidants in the chloroplast protecting thylakoid membrane lipids from oxidation (Dörmann, 2007; Mène-Saffrané and DellaPenna, 2010). All of the enzyme activities of the tocopherol biosynthesis pathway were originally localized to the inner chloroplast envelope membrane where biosynthesis is thought to occur (Schultz and Soll, 1980). The tocopherol cyclase VTE1 (At4g32770) catalyses the second to last step in *a*-tocopherol synthesis by introducing the chromanol ring. Surprisingly, however, proteomics studies indicated that the majority of VTE1 protein is localized in PG. This finding was confirmed by expression of a fluorescent VTE1-fusion protein, immunoelectron microscopy (Austin *et al.*, 2006) and membrane fractionation studies (Vidi *et al.*, 2006). In the *vte1* mutant, the substrate of VTE1 dimethylphytylbenzoquinone (DMPBQ) accumulated to a large extent in plastoglobules providing evidence for the implication of

PG as a reaction site in the tocopherol cyclase step (Eugeni-Piller *et al.*, 2012).

PG and VTE1 in the tocopherol redox cycle

The thylakoid membrane is exposed to the production of reactive oxygen species due to the presence of the photosynthetic apparatus. Tocopherols act as antioxidants by rescuing membrane lipids from oxidation. Tocopherols scavenge lipid peroxy radicals yielding a tocopheroxyl radical followed by chromanol ring opening that results in α-tocoquinone. In the tocopherol repair cycle, α-tocoquinone is dehydrated resulting in trimethylphytylbenzoquinone (TMPBQ) before the re-introduction of chromanol ring by VTE1. The tocopherol redox cycle is believed to occur at the thylakoid membrane and in plastoglobules. In favour of this scenario, α-tocoquinone can be detected under high light conditions (Kobayashi and DellaPenna, 2008) and accumulates to much higher levels in the *ndc1* mutant background (Eugeni-Piller *et al.*, 2014). This finding suggests that VTE1 preferentially acts on a reduced substrate in the tocopherol repair cycle.

PG and VTE1 in plastochromanol-8 synthesis

While tocopherols and tocotrienols have been well studied, knowledge on plastochromanol-8, that is derived from plastoquinol by tocopherol cyclase activity, is rather scarce despite its wide presence in crop species (Kruk *et al.*, 2014). Together with tocopherol, PC-8 protects *Arabidopsis* seed oil from oxidation and is essential for seed longevity (Mène-Saffrané *et al.*, 2010). PC-8 has a chromanol-containing headgroup identical to that of γ-tocopherol but the longer solanesyl side chain of plastoquinone. Surprisingly, overexpression of VTE1 led to the accumulation of large quantities of PC-8 rather than tocopherol. A large proportion of the plastochromanol accumulated in PG (Zbierzak *et al.*, 2010). In the VTE1 overexpressing line, plastoglobules were more numerous and present in larger clusters than in the wild type. Interestingly also, PC-8 levels were reduced in *ndc1* mutant background. This suggests that also in the case of PC-8, VTE1 prefers reduced plastoquinol as the substrate.

PG and metabolite trafficking within the chloroplast

Localization of VTE1 at PG while the other enzymes of the tocopherol pathway are localized at the inner envelope membrane suggests that metabolites are trafficked from the inner envelope membrane to PG and back to complete the pathway. This would particularly concern DMPBQ, the substrate of VTE1 and γ-tocopherol, the substrate of the VTE4 methylase that is localized at the inner envelope membrane (Zbierzak *et al.*, 2010). How likely is this to occur? Fascinating experimentation in *Arabidopsis*, fittingly described as transorganellar complementation, demonstrated that VTE1 when targeted to the endoplasmic reticulum (ER) is still able to complement the *vte1* mutant (Mehrshahi *et al.*, 2013). The work by Mehrshahi *et al.* (2013) thus indicates that metabolites are sufficiently mobile for DMPBQ to reach the ER and γ-tocopherol to return to the chloroplast. Presumably, this does not implicate specific transport proteins localized at the chloroplast envelope membrane but an interorganellar pathway relying on membrane connections between the ER and the chloroplast outer envelope membrane that allows to complete the tocopherol cyclase reaction outside the chloroplast (Mehrshahi *et al.*, 2013). These connections may implicate a shared outer lipid layer between the chloroplast and the ER (Mehrshahi *et al.*, 2013) not unlike the one observed between PG and the thylakoid membrane. If DMPBQ can reach the ER, it appears likely that DMPBQ can also reach PG to complete the tocopherol cyclase reaction. Still, the question remains why the tocopherol cyclase reaction would be located in PG and not at the inner envelope membrane. One reason may lie in the fact that VTE1 catalyses chromanol ring formation also in other molecules than DMPBQ. This concerns TMPBQ, that is recycled to α-tocopherol at the end of the tocopherol redox cycle (Eugeni-Piller *et al.*, 2014) and plastoquinone that is converted to plastochromanol-8 by VTE1 (Zbierzak *et al.*, 2010). It has been shown that chromanol ring formation preferentially occurs in the reduced state of the substrate in the case of tocopherol (Grütter *et al.*, 2006). NDC1 which is present in PG has the ability to reduce a variety of quinolic substrates. Owing to the presence of NDC1, PG may present a favourable, reductive reaction environment in which chromanol ring formation by VTE1 may readily occur.

Roles of PG in carotenoid metabolism

PG play a key role in chloroplast to chromoplast transition. During this transition chlorophyll is degraded, thylakoids are disassembled and replaced by large carotenoid-containing plastoglobules in tomato (Simkin *et al.*, 2007) or fibrils in red pepper fruit (Deruère *et al.*, 1994), to give two examples. In red pepper chromoplasts, four enzymes implicated in carotenoid biosynthesis are specifically recruited to PG: these are ζ-carotene desaturase (ZDS), lycopene β-cyclase (LCY) and two β-carotene β-hydroxylases (CHY) (Ytterberg *et al.*, 2006). ZDS introduces conjugated double bonds in ζ-carotene contributing to the synthesis of lycopene, a red pigment of tomato fruit (Ben-Shaul and Naftali, 1969). LCY catalyses the formation of ionone rings at either ends of lycopene resulting in α- or β-carotene. CHY catalyses the addition of hydroxyl groups to the ionone rings giving the xanthophyll derivatives lutein and zeaxanthin, respectively (Liu *et al.*, 2015; Nogueira *et al.*, 2013; Ruiz-Sola and Rodríguez-Concepción, 2012). Hypothetically, the PG location of these enzymes may facilitate substrate channelling and subsequent accumulation of the carotenoid products within PG.

Carotenoids are subject to oxidative cleavage by carotenoid cleavage dioxygenases (CCDs) giving rise to a large variety of so-called apocarotenoids (Auldridge *et al.*, 2006). Apocarotenoids take on a multitude of physiological functions including those of phytohormones (abscisic acid, strigolactones), pollinator attracting volatiles (β-ionone, geranial) and chloroplast to nucleus retrograde signalling (β-cyclocitral) (Gomez-Roldan *et al.*, 2008; Hirayama and Shinozaki, 2007; Jia *et al.*, 2013; Ramel *et al.*, 2012). Of the known CCDs only CCD4 (At4g19170) has been identified in PG (Lundquist *et al.*, 2012; Vidi *et al.*, 2006; Ytterberg *et al.*, 2006).

Homologues of CCD4 carry out a variety of cleavage reactions in different species and tissues. For instance the *Chrysanthemum morifolium* homologue CmCCD4 produces white colour in chrysanthemum flowers. It presumably does so by degrading lutein that is abundant in *cmccd4* mutant yellow flowers (Zhu *et al.*, 2010). Although the association of chrysanthemum CmCCD4 with PG has not been demonstrated it appears likely that its products accumulate in the chromoplast PG (Yoshioka *et al.*, 2012). Similarly, CitCCD4 in the Satsuma mandarin fruit is responsible for the accumulation of the orange-coloured pigment β-citraurin by specific cleavage of β-cryptoxanthin and zeaxanthin (Ma *et al.*, 2013). During *Crocus sativa* stigma development CsCCD4 cleaves β-carotene into β-ionone and β-cyclocitral (Rubio-Moraga *et al.*, 2014; Rubio *et al.*, 2008).

In an elegant genetic screen, *Arabidopsis* AtCCD4 was found to degrade carotenoids during seed maturation as well as leaf senescence (Gonzalez-Jorge *et al.*, 2013). *ccd4* mutant seeds accumulate β-carotene which consequently was identified as a probable substrate of AtCCD4. Lack of β-carotene (provitamin A) in food is the leading cause for vitamin A deficiency that may cause xerophthalmia and total blindness (Sommer, 1990). In the past, genetic engineering has been used to introduce the β-carotene biosynthetic pathway into the 'Golden rice' endosperm (Ye *et al.*, 2000). The research on *Arabidopsis* CCD4 opens up exciting new perspectives for provitamin A biofortification using smart breeding in seed crops (Gonzalez-Jorge *et al.*, 2013).

AtCCD4 is strongly induced during senescence and degrades β-carotene under dark-induced senescence conditions (Gonzalez-Jorge *et al.*, 2013). While many steps of chlorophyll degradation have been elucidated, this is an essential enzyme known to be implicated in carotenoid degradation. It will be interesting to see whether and which CCD4 homologues are implicated in the colour production in the fall foliage of perennials.

Phytol ester synthases: mediators of thylakoid remodelling during senescence and nitrogen deprivation

During senescence and under stress conditions such as nitrogen deprivation PG become supersized while the extent of thylakoid membranes is reduced (Besagni and Kessler, 2013). Typically chlorophyll and galactolipids are catabolized under these conditions. In the second step of chlorophyll degradation, free phytol is liberated from pheophytine by the pheophytine pheophorbide hydrolase (PPH) (Schelbert *et al.*, 2009; Zhang *et al.*, 2014). Free phytol is considered toxic and does not accumulate. In a salvage pathway, free phytol is converted to phytol diphosphate by phytol kinase (VTE5) and incorporated into tocopherols

(Ischebeck *et al.*, 2006). Alternatively free phytol may be incorporated into fatty acid phytyl esters (FAPE) by esterification with free fatty acids stemming from galactolipid hydrolysis (Gaude *et al.*, 2007). Free fatty acids may also be used to synthesize triacylglycerol. Together FAPE and TAG may account for PG supersizing during senescence and under nitrogen stress (Besagni and Kessler, 2013). Therefore, the two conditions present a striking case for thylakoid lipid remodelling, both at the biochemical and ultrastructural levels.

Interestingly, two enzymes, phytyl ester synthases 1 and 2 (PES1 and -2) (At1g54570 and At3g26840) largely responsible for the formation of FAPE and TAG were identified as components of the PG proteome. PES1 and -2 are both able to transfer fatty acids from galactolipids directly to free phytol and diacylglycerol. PES1 and -2 belong to the seven-member diacylglycerol acyltransferase (DGAT) family in *Arabidopsis* and are both strongly induced during senescence and nitrogen starvation. The *pes1pes2* double mutant had strongly reduced levels of phytyl esters under nitrogen starvation indicating that the two enzymes are predominant in FAPE synthesis (Lippold *et al.*, 2012). Complementation of the H1246 yeast strain lacking acyltransferase activity showed that both PES1 and -2 restored TAG and sterol ester synthesis. 14:0 Acyl-CoA was the preferred substrate in the yeast system, but monogalactosyldiacylglycerol also worked as an acyl donor (Pan *et al.*, 2013). Finally, chlorophyll degradation and thylakoid membrane disassembly were delayed in the *pes1pes2* double mutant. In summary the evidence makes a compelling case for the participation of PES1 and -2 in thylakoid membrane remodelling during senescence and under nitrogen starvation.

Homologues of phytyl ester synthase also play an important role in carotenoid ester biosynthesis and chromoplast biogenesis in flower organs. It has recently been demonstrated in tomato that PYP1 (Pale Yellow Petal 1) is required for the accumulation in petals and anthers of high levels of xanthophylls that were mostly neoxanthin and violaxanthin esterified with myristic and palmitic acids (Ariizumi *et al.*, 2014). Xanthophyll esters were absent from *pyp1* mutant alleles, which also had a reduced total carotenoid content and perturbed chromoplasts.

ABC1-like kinases, metabolic regulators in PG

The plant kinase family is very large, but surprisingly few kinases are known to reside in chloroplasts (Bayer *et al.*, 2012). Of the chloroplast kinases, the family of ABC1-like kinases constitutes the largest group: nine homologues are known or predicted to be in chloroplasts, six of which were localized to PG using proteomics. Next to the fibrillins the ABC1-like kinases are the most abundant group of PG proteins both in number and in contribution to total PG protein mass (Lundquist *et al.*, 2012).

The first ABC1-like kinase was discovered in yeast mitochondria and is required for the activity of the cytochrome BC1 complex (Bousquet *et al.*, 1991). It is also known as COQ8 (coenzyme Q biosynthesis). The *abc1/coq8* mutant is respiration deficient and was found to lack ubiquinone (Do *et al.*, 2001). An analogous phenotype was caused by mutation of the bacterial ubiB gene, a homologue of ABC1/COQ8 that also lacks ubiquinone and instead accumulates the octaprenylphenol biosynthetic intermediate (Poon *et al.*, 2000). In humans, the mutation of the ABC1/COQ8 homologues ADCK3 and ADCK4 (ABC1 domain-containing kinase) are linked to disease such as cerebellar ataxia (Lagier-Tourenne *et al.*, 2008) and steroid-resistant nephrotic syndrome (Ashraf *et al.*, 2013). The symptoms have been attributed to reduced ubiquinone levels and in some cases can be alleviated by ubiquinone supplementation. COQ8 probably functions as a regulatory kinase in the ubiquinone biosynthesis pathway. In favour of this hypothesis, COQ3, -5 and -7 of the ubiquinone biosynthesis pathway were phosphorylated in an ABC1/COQ8-dependent manner (Xie *et al.*, 2011).

Given the known role of ABC1-like kinases in mitochondrial ubiquinone biosynthesis, the strong presence of ABC1-like kinases in chloroplasts (termed ABC1Ks) seems surprising and begs the question of their function. By analogy to mitochondrial ABC1-like kinases an implication in chloroplast prenylquinone metabolism may be proposed. Reverse genetic analyses provided confirmatory evidence for this hypothesis: in the *abc1k1* mutant (At4g31390), α-, γ- and δ-tocopherol levels were lower after high light treatment than in the wild type (Martinis *et al.*, 2014). At the same time, α-TQ, the α-tocopherol oxidation product was increased suggesting a defect in the tocopherol repair cycle. In the *abc1k3* mutant (At1g79600), no effect was observed on α-tocopherol levels, but PC-8 was reduced and α-TQ increased (Martinis *et al.*, 2013) suggesting a defect in VTE1-dependant metabolic steps but excluding tocopherol biosynthesis. The common denominator of the *abc1k1* and -*k3* molecular phenotypes is their dependence on VTE1 activities. In fact, the observed mutant effects could be explained by deregulation of VTE1. Indeed, VTE1 behaves as an *in vitro* substrate of recombinant ABC1K1 as well as ABC1K3 (Martinis *et al.*, 2013). In addition, VTE1 is an experimentally confirmed phosphoprotein having a phosphorylation hotspot near its N-terminus (PhosphAt database, University of Hohenheim, http://phosphat.uni-hohenheim.de/).

However, the activities of the ABC1K family extend beyond plastid prenylquinone metabolism. In the *abc1k1* mutant, β-carotene levels were reduced even under moderate light conditions. After exposure to highlight and return to moderate light conditions, the *abc1k1* mutant failed to accumulate starch normally producing higher amounts of soluble sugars instead (Martinis *et al.*, 2014).

The ABC1K1 was initially identified as *pgr6* (Proton Gradient Regulator 6) in a genetic screen for *Arabidopsis* mutants with reduced NPQ and hence increase chlorophyll fluorescence. *pgr* mutants are defective in photosynthetic electron transport and in pH gradient generation across the thylakoid membrane (Shikanai *et al.*, 1999). Currently it is not clear why the *abc1k1* mutant has a *pgr* phenotype. Potentially, the *pgr* phenotype may be a consequence of the metabolic perturbation occurring in the *abc1k1* mutant. Alternatively, ABC1K1 may directly phosphorylate target(s) participating in the photosynthetic electron transport chain up to the level of the cytochrome b_6f complex.

The existence of additional targets of the ABC1-like kinases is supported by a study showing that ABC1K1 and -K3 form a complex (Lundquist *et al.*, 2013). The study consequently focused on the double mutant and found the double mutant to degreen under different types of stress and to senesce prematurely. This phenotype was accompanied by remodelling of the PG proteome and the recruitment of lipoxygenases 3 and 4 (At1g17420 and At1g72520) to PG. These two enzymes participate in the jasmonate biosynthetic pathway that is highly active during senescence and nitrogen stress conditions. This may reflect the increased galactolipid turnover that would be expected under the aforementioned conditions.

As a consequence of the reduced ability of *abc1k1/abc1k3* to adapt to stress, reactive oxygen species (ROS) production may be elevated. This notion is supported by the increased levels of β-cyclocitral that in turn mediate retrograde signalling from the chloroplast to the nucleus (Lundquist *et al.*, 2013; Ramel *et al.*, 2012).

ABC1K8/OSA1 is located at the chloroplast envelope membrane and is up-regulated upon cadmium exposure. *abc1k8* mutants (At5g64940) are more sensitive to a variety of stress types including cadmium and high light. The ABC1K7/SIA1 homologue has been implicated in salt stress tolerance (Yang *et al.*, 2012), overexpression resulting in higher tolerance. The *abc1k7* (At3g07700) and *abc1k8* mutants also synthesized lower levels of highly unsaturated DGDG and MGDG (Manara *et al.*, 2015). ABC1K7 and ABC1K8 were previously implicated in oxidative stress resistance as the respective mutants accumulated more ferritin, superoxide and had reduced tolerance to ROS. Interestingly, *abc1k8* and particularly the *abc1k7/abc1k8* double mutant accumulated increased levels of DGDG oxylipins that are product of ROS and a source for the synthesis of jasmonate. In this context it is interesting to note that the jasmonate pathway is recruited to PG in the *abc1k1/abc1k3* double mutant. It is therefore possible that the ABC1K1, -3, -7 and -8 cooperate in the regulation of thylakoid membrane disassembly and jasmonate production during senescence.

Uncharacterized PG proteins

Despite the advances made in recent years, PG still hold numerous secrets. A good number of the PG proteins have not been characterized so far and may hold clues to currently unidentified PG functions. Combinations of reverse genetics with –omics tools have proven to be very successful in the functional analysis of PG proteins and will probably continue to do so in the future.

Among the uncharacterized PG proteins are several fibrillins (FBN7a, FBN7b, FBN8) (At3g58010, At2g4213, At2g46910). The functions of fibrillins in the chloroplast remain poorly understood. Therefore the characterization of additional fibrillins may reveal precious new information on their specific roles role in development and under various stresses. In particular it will be interesting to see whether the classification of fibrillins as structural proteins holds up to experimental scrutiny and to test whether fibrillins play a larger role in metabolite trafficking between the PG and thylakoid compartments.

Recent research on the ABC1-like kinases provided evidence for the role of ABC1K1, -K3, -K7 and -K8 in regulation of range of chloroplast metabolic pathways. Three ABC1-like kinases (ABC1K5, -K6, -K9) (At1g71810, At3g24190, At1g05200) associated with PG remain to be explored. Their characterization may well reveal other aspects of chloroplast metabolism and organellar processes that are under the control of the ABC1-like kinase family. Phosphorylation is a reversible process and therefore phosphatases may be expected to play a role in ABC1-like kinase regulatory cycles. While no phosphatases were found in the PG proteome, they exist elsewhere in the chloroplast.

A number of uncharacterized and predicted metabolic enzymes also remain in the PG proteome. Among these are two UbiE-related methyltransferases (At1g78140, At2g41040). UbiE (known as CoQ5 in yeast) is a methyltransferase required for ubiquinone biosynthesis in bacteria and mitochondria. Based on the analogy between ubiquinone and the chloroplast prenylquinones, it is tempting to hypothesize, that the UbiE-related methyltransferases participate in an unrecognized methylation step in PG prenyl lipid metabolism.

A third predicted esterase/lipase/thioesterase (At5g41120) highly homologous to PES1 and PES2 is present in the PG proteome. This protein is an excellent candidate for a third phytyl ester synthase, responsible for the residual accumulation of FAPE in the *pes1pes2* double mutant (Lippold *et al.*, 2012).

Two uncharacterized PG proteins, an unknown SAG (senescence-associated gene) (At1g73750) and a predicted M48 protease (At3g27110) are highly up-regulated during senescence and may shed new light on PG functions during senescence (Lundquist *et al.*, 2012).

Yet other uncharacterized proteins, such as PLAT/LH2-1 (At4g39730), are likely to be involved in lipid metabolism in PG. Furthermore, additional undiscovered PG proteins may exist. Some of these proteins may associate with PG only under specific conditions. Others may only loosely associate with PG and dissociate during the purification procedure. The advances in mass spectrometric technology may also allow the detection of very low abundance PG proteins in the near future.

Future applications

Recombinant protein production

Plants have the potential for almost unlimited production of recombinant proteins (Ma *et al.*, 2005). In particular, the transformation of the chloroplast genome and protein expression inside chloroplasts of so-called transplastomic plants results in high protein yields (Bock, 2014). Plastoglobules have also been explored for their potential as a destination for recombinant proteins (Vidi *et al.*, 2007). The proposed advantage of PG is their low density that would allow enriching a recombinant protein in PG by floatation centrifugation before proceeding to protein purification. Transplastomic plants have been engineered to produce chimeric proteins of the mature form of the fibrillin PGL35/FBN1a fused to HIV p24 (PGL35–HIVp24) as well as hepatitis C virus core protein (PGL35-HCV) (Shanmugabalaji *et al.*, 2013). The PGL35–HIVp24 fusion protein partitioned between PG and thylakoid membranes. Targeting proteins to PG and thylakoids by the means of transplastomic fibrillin fusion proteins may be optimized in the future to become an effective tool in plant biotechnology.

Metabolic engineering in fruit and leaves

Several PG lipid components are vitamins (tocopherols and phylloquinone) or commercially important pigment crop traits such as lycopene and other carotenoids. Enzymes implicated in their biosynthesis and metabolism reside at the PG surface. It may therefore be possible to modify the nature and abundance of PG lipid components in leaves, flowers and fruit. An example of metabolic engineering is CCD4 in chrysanthemum flowers. The chrysanthemum flowers are normally white. The chrysanthemum *ccd4* mutant of has yellow instead of white flowers due to the accumulation of lutein. The *Arabidopsis ccd4* mutant has enhanced carotenoid levels. Engineering or selecting crop varieties with reduced CCD4 activity represents a potential short cut to new 'golden' crop varieties that hold promise to combat xerophthalmia in developing countries. Systematic tilling of PG genes crop species will very likely turn up new alleles of potential commercial interest. For instance, mutants in the regulatory ABC1-like kinases are known to affect prenyl lipid levels as well as senescence. It may therefore be possible to identify ABC1K-like alleles that are enhanced for tocopherol and carotenoid content or feature delayed senescence and therefore accumulate more biomass.

Acknowledgements

This work was supported by the University of Neuchâtel and the Swiss National Science Foundation grants 31003A_127380, 31003A_144156, and IZEBZO-143169/1.

References

Allen, J.F., Bennett, J., Steinback, K.E., and Arntzen, C.J. (1981). Chloroplast protein-phosphorylation couples plastoquinone redox state to distribution of excitation-energy between photosystems. Nature *291*, 25–29.

Ariizumi, T., Kishimoto, S., Kakami, R., Maoka, T., Hirakawa, H., Suzuki, Y., Ozeki, Y., Shirasawa, K., Bernillon, S., Okabe, Y., *et al.* (2014). Identification of the carotenoid modifying gene *PALE YELLOW PETAL 1* as an essential factor in xanthophyll esterification and yellow flower pigmentation in tomato (*Solanum lycopersicum*). Plant J. *79*, 453–465.

Ashraf, S., Gee, H.Y., Woerner, S., Xie, L.X., Vega-Warner, V., Lovric, S., Fang, H., Song, X., Cattran, D.C., Avila-Casado, C., *et al.* (2013). *ADCK4* mutations promote steroid-resistant nephrotic syndrome through CoQ_{10} biosynthesis disruption. J. Clin. Invest. *123*, 5179–5189.

Auldridge, M.E., McCarty, D.R., and Klee, H.J. (2006). Plant carotenoid cleavage oxygenases and their apocarotenoid products. Curr. Opin. Plant. Biol. *9*, 315–321.

Austin, J.R., Frost, E., Vidi, P.A., Kessler, F., and Staehelin, L.A. (2006). Plastoglobules are lipoprotein subcompartments of the chloroplast that are permanently coupled to thylakoid membranes and contain biosynthetic enzymes. Plant Cell *18*, 1693–1703.

Bailey, J.L., and Whyborn, A.G. (1963). Osmiophilic globules of chloroplasts. II. Globules of spinach-beet chloroplast. Biochim. Biophys. Acta *78*, 163–174.

Bayer, R.G., Stael, S., Rocha, A.G., Mair, A., Vothknecht, U.C., and Teige, M. (2012). Chloroplast-localized protein kinases: a step forward towards a complete inventory. J. Exp. Bot. *63*, 1713–1723.

Ben-Shaul, Y., and Naftali, Y. (1969). The development and ultrastructure of lycopene bodies in chromoplasts of *Lycopersicum esculentum*. Protoplasma *67*, 333–344.

Ben Salem-Fnayou, A., Bouamama, B., Ghorbel, A., and Mliki, A. (2011). Investigations on the leaf anatomy and ultrastructure of grapevine (*Vitis vinifera*) under heat stress. Microsc. Res. Tech. *74*, 756–762.

Besagni, C., and Kessler, F. (2013). A mechanism implicating plastoglobules in thylakoid disassembly during senescence and nitrogen starvation. Planta *237*, 463–470.

Bock, R. (2014). Genetic engineering of the chloroplast: novel tools and new applications. Curr. Opin. Biotechnol. *26*, 7–13.

Bousquet, I., Dujardin, G., and Slonimski, P.P. (1991). *ABC1*, a novel yeast nuclear gene has a dual function in mitochondria: it suppresses a cytochrome b mRNA translation defect and is essential for the electron transfer in the bc 1 complex. EMBO J. *10*, 2023–2031.

Carrie, C., Murcha, M.W., Kuehn, K., Duncan, O., Barthet, M., Smith, P.M., Eubel, H., Meyer, E., Day, D.A., Millar, A.H., *et al.* (2008). Type II NAD(P)H dehydrogenases are targeted to mitochondria and chloroplasts or peroxisomes in *Arabidopsis thaliana*. FEBS Lett. *582*, 3073–3079.

Cohen, M., and Bowler, E. (1953). Lamellar structure of the tobacco chloroplast. Protoplasma *42*, 414–416.

Cunningham, F.X., Jr., Tice, A.B., Pham, C., and Gantt, E. (2010). Inactivation of genes encoding plastoglobuli-like proteins in *Synechocystis* sp. PCC 6803 leads to a light-sensitive phenotype. J. Bacteriol. *192*, 1700–1709.

Deruère, J., Römer, S., Dharlingue, A., Backhaus, R.A., Kuntz, M., and Camara, B. (1994). Fibril assembly and carotenoid overaccumulation in chromoplasts – a model for supramolecular lipoprotein structures. Plant Cell *6*, 119–133.

Do, T.Q., Hsu, A.Y., Jonassen, T., Lee, P.T., and Clarke, C.F. (2001). A defect in coenzyme Q biosynthesis is responsible for the respiratory deficiency in *Saccharomyces cerevisiae abc1* mutants. J. Biol. Chem. *276*, 18161–18168.

Dörmann, P. (2007). Functional diversity of tocochromanols in plants. Planta *225*, 269–276.

Eugeni-Piller, L., Besagni, C., Ksas, B., Rumeau, D., Bréhélin, C., Glauser, G., Kessler, F., and Havaux, M. (2011). Chloroplast lipid droplet type II NAD(P)H quinone oxidoreductase is essential for prenylquinone metabolism and vitamin K_1 accumulation. Proc. Natl. Acad. Sci. U.S.A. *108*, 14354–14359.

Eugeni-Piller, L., Glauser, G., Kessler, F., and Besagni, C. (2014). Role of plastoglobules in metabolite repair in the tocopherol redox cycle. Front. Plant Sci. *5*.

Eugeni-Piller, L.E., Abraham, M., Dörmann, P., Kessler, F., and Besagni, C. (2012). Plastid lipid droplets at the crossroads of prenylquinone metabolism. J. Exp. Bot. *63*, 1609–1618.

Feng, Y., Li, W.F., Li, J., Wang, J.W., Ge, J.P., Xu, D., Liu, Y.J., Wu, K.Q., Zeng, Q.Y., Wu, J.W., *et al.* (2012). Structural insight into the type-II mitochondrial NADH dehydrogenases. Nature *491*, 478–482.

Gaude, N., Bréhélin, C., Tischendorf, G., Kessler, F., and Dörmann, P. (2007). Nitrogen deficiency in *Arabidopsis* affects galactolipid composition and gene expression and results in accumulation of fatty acid phytyl esters. Plant J. *49*, 729–739.

Gomez-Roldan, V., Fermas, S., Brewer, P.B., Puech-Pagès, V., Dun, E.A., Pillot, J.P., Letisse, F., Matusova, R., Danoun,

S., Portais, J.C., *et al.* (2008). Strigolactone inhibition of shoot branching. Nature *455*, 189–194.

Gonzalez-Jorge, S., Ha, S.H., Magallanes-Lundback, M., Gilliland, L.U., Zhou, A.L., Lipka, A.E., Nguyen, Y.N., Angelovici, R., Lin, H.N., Cepela, J., *et al.* (2013). Carotenoid cleavage dioxygenase4 is a negative regulator of beta-carotene content in *Arabidopsis* seeds. Plant Cell *25*, 4812–4826.

Greenwood, A.D., Leech, R.M., and Williams, J.P. (1963). Osmiophilic globules of chloroplasts: I. Osmiophilic globules as a normal component of chloroplasts and their isolation and composition in *Vicia Faba* L. Biochim. Biophys. Acta *78*, 148–162.

Gross, J., Cho, W.K., Lezhneva, L., Falk, J., Krupinska, K., Shinozaki, K., Seki, M., Herrmann, R.G., and Meurer, J. (2006). A plant locus essential for phylloquinone (vitamin K_1) biosynthesis originated from a fusion of four eubacterial genes. J. Biol. Chem. *281*, 17189–17196.

Grütter, C., Alonso, E., Chougnet, A., and Woggon, W.D. (2006). A biomimetic chromanol cyclization leading to alpha-tocopherol. Angew. Chem. Int. Edit. *45*, 1126–1130.

Hansmann, P., and Sitte, P. (1982). Composition and molecular structure of chromoplast globules of *Viola tricolor*. Plant Cell Rep. *1*, 111–114.

Hirayama, T., and Shinozaki, K. (2007). Perception and transduction of abscisic acid signals: keys to the function of the versatile plant hormone ABA. Trends Plant Sci. *12*, 343–351.

Ischebeck, T., Zbierzak, A.M., Kanwischer, M., and Dörmann, P. (2006). A salvage pathway for phytol metabolism in *Arabidopsis*. J. Biol. Chem. *281*, 2470–2477.

Jans, F., Mignolet, E., Houyoux, P.A., Cardol, P., Ghysels, B., Cuine, S., Cournac, L., Peltier, G., Remacle, C., and Franck, F. (2008). A type II NAD(P) H dehydrogenase mediates light-independent plastoquinone reduction in the chloroplast of *Chlamydomonas*. Proc. Natl. Acad. Sci. U.S.A. *105*, 20546–20551.

Jia, C.G., Zhang, L.P., Liu, L.H., Wang, J.S., Li, C.Y., and Wang, Q.M. (2013). Multiple phytohormone signalling pathways modulate susceptibility of tomato plants to *Alternaria alternata* f. sp. *lycopersici*. J. Exp. Bot. *64*, 637–650.

Kessler, F., Schnell, D., and Blobel, G. (1999). Identification of proteins associated with plastoglobules isolated from pea (*Pisum sativum* L.) chloroplasts. Planta *208*, 107–113.

Knoth, R., Hansmann, P., and Sitte, P. (1986). Chromoplasts of *Palisota barteri*, and the molecular structure of chromoplast tubules. Planta *168*, 167–174.

Kobayashi, N., and DellaPenna, D. (2008). Tocopherol metabolism, oxidation and recycling under high light stress in *Arabidopsis*. Plant J. *55*, 607–618.

Kruk, J., Szymanska, R., Cela, J., and Munné-Bosch, S. (2014). Plastochromanol-8: fifty years of research. Phytochemistry *108*, 9–16.

Lagier-Tourenne, C., Tazir, M., López, L.C., Quinzii, C.M., Assoum, M., Drouot, N., Busso, C., Makri, S., Ali-Pacha, L., Benhassine, T., *et al.* (2008). ADCK3, an ancestral kinase, is mutated in a form of recessive ataxia associated with coenzyme Q_{10} deficiency. Am. J. Hum. Genet. *82*, 661–672.

Langenkämper, G., Manac'h, N., Broin, M., Cuiné, S., Becuwe, N., Kuntz, M., and Rey, P. (2001). Accumulation of plastid lipid-associated proteins (fibrillin/CDSP34) upon oxidative stress, ageing and biotic stress in Solanaceae and in response to drought in other species. J. Exp. Bot. *52*, 1545–1554.

Lichtenthaler, H.K. (1966). Plastoglobuli und Plastidenstruktur. Ber. Deut. Bot. Ges. *79*, 82–88.

Lichtenthaler, H.K. (1968). Plastoglobuli and fine structure of plastids. Endeavour *27*, 144–149.

Lichtenthaler, H.K. (1969a). Plastoglobuli and lipoquinone content of chloroplasts from *Cereus Peruvianus* (L.) Mill. Planta *87*, 304–310.

Lichtenthaler, H.K. (1969b). Plastoglobuli of spinach – Their size and composition during chloroplast degeneration. Protoplasma *68*, 315–326.

Lichtenthaler, H.K., and Peveling, E. (1967). Plastoglobuli in verschiedenen Differenzierungsstadien der Plastiden bei *Allium Cepa* L. Planta *72*, 1–13.

Lichtenthaler, H.K., and Sprey, B. (1966). Über die osmiophilen globularen Lipideinschlusse der Chloroplasten. Z. Naturforsch. Pt. B. *B 21*, 690–697.

Lippold, F., vom Dorp, K., Abraham, M., Holzl, G., Wewer, V., Yilmaz, J.L., Lager, I., Montandon, C., Besagni, C., Kessler, F., *et al.* (2012). Fatty acid phytyl ester synthesis in chloroplasts of *Arabidopsis*. Plant Cell *24*, 2001–2014.

Liu, L.H., Shao, Z.Y., Zhang, M., and Wang, Q.M. (2015). Regulation of carotenoid metabolism in tomato. Molecular Plant *8*, 28–39.

Locy, R.D., Chang, C.C., Nielsen, B.L., and Singh, N.K. (1996). Photosynthesis in salt-adapted heterotrophic tobacco cells and regenerated plants. Plant Physiol. *110*, 321–328.

Lohmann, A., Schottler, M.A., Bréhélin, C., Kessler, F., Bock, R., Cahoon, E.B., and Dörmann, P. (2006). Deficiency in phylloquinone (vitamin K_1) methylation affects prenyl quinone distribution, photosystem I abundance, and anthocyanin accumulation in the *Arabidopsis AtmenG* mutant. J. Biol. Chem. *281*, 40461–40472.

Lundquist, P.K., Poliakov, A., Bhuiyan, N.H., Zybailov, B., Sun, Q., and van Wijk, K.J. (2012). The functional network of the *Arabidopsis* plastoglobule proteome based on quantitative proteomics and genome-wide coexpression analysis. Plant Physiol. *158*, 1172–1192.

Lundquist, P.K., Poliakov, A., Giacomelli, L., Friso, G., Appel, M., McQuinn, R.P., Krasnoff, S.B., Rowland, E., Ponnala, L., Sun, Q., *et al.* (2013). Loss of plastoglobule kinases ABC1K1 and ABC1K3 causes conditional degreening, modified prenyl-lipids, and recruitment of the jasmonic acid pathway. Plant Cell *25*, 1818–1839.

Ma, G., Zhang, L.C., Matsuta, A., Matsutani, K., Yamawaki, K., Yahata, M., Wahyudi, A., Motohashi, R., and Kato, M. (2013). Enzymatic formation of beta-citraurin from beta-cryptoxanthin and zeaxanthin by carotenoid cleavage dioxygenase4 in the flavedo of citrus fruit. Plant Physiol. *163*, 682–695.

Ma, J.K., Drake, P.M., Chargelegue, D., Obregon, P., and Prada, A. (2005). Antibody processing and engineering in plants, and new strategies for vaccine production. Vaccine *23*, 1814–1818.

Manara, A., DalCorso, G., Guzzo, F., and Furini, A. (2015). Loss of the atypical kinases ABC1K7 and ABC1K8 changes the lipid composition of the chloroplast membrane. Plant Cell Physiol. *56*, 1193–1204.

Martinis, J., Glauser, G., Valimareanu, S., and Kessler, F. (2013). A chloroplast ABC1-like kinase regulates vitamin E metabolism in *Arabidopsis*. Plant Physiol. *162*, 652–662.

Martinis, J., Glauser, G., Valimareanu, S., Stettler, M., Zeeman, S.C., Yamamoto, H., Shikanai, T., and Kessler, F. (2014). ABC1K1/PGR6 kinase: a regulatory link between photosynthetic activity and chloroplast metabolism. Plant J. *77*, 269–283.

Mehrshahi, P., Stefano, G., Andaloro, J.M., Brandizzi, F., Froehlich, J.E., and DellaPenna, D. (2013). Transorganellar complementation redefines the biochemical continuity of endoplasmic reticulum and chloroplasts. Proc. Natl. Acad. Sci. U.S.A. *110*, 12126–12131.

Mène-Saffrané, L., and DellaPenna, D. (2010). Biosynthesis, regulation and functions of tocochromanols in plants. Plant Physiol. Bioch. *48*, 301–309.

Mène-Saffrané, L., Jones, A.D., and DellaPenna, D. (2010). Plastochromanol-8 and tocopherols are essential lipid-soluble antioxidants during seed desiccation and quiescence in *Arabidopsis*. Proc. Natl. Acad. Sci. U.S.A. *107*, 17815–17820.

Michalecka, A.M., Svensson, A.S., Johansson, F.I., Agius, S.C., Johanson, U., Brennicke, A., Binder, S., and Rasmusson, A.G. (2003). *Arabidopsis* genes encoding mitochondrial type II NAD(P)H dehydrogenases have different evolutionary orgin and show distinct responses to light. Plant Physiol. *133*, 642–652.

Nogueira, M., Mora, L., Enfissi, E.M.A., Bramley, P.M., and Fraser, P.D. (2013). Subchromoplast sequestration of carotenoids affects regulatory mechanisms in tomato lines expressing different carotenoid gene combinations. Plant Cell 25, 4560–4579.

Pan, X., Siloto, R.M., Wickramarathna, A.D., Mietkiewska, E., and Weselake, R.J. (2013). Identification of a pair of phospholipid: diacylglycerol acyltransferases from developing flax (*Linum usitatissimum* L.) seed catalyzing the selective production of trilinolenin. J. Biol. Chem. *288*, 24173–24188.

Poon, W.W., Davis, D.E., Ha, H.T., Jonassen, T., Rather, P.N., and Clarke, C.F. (2000). Identification of *Escherichia coli ubiB*, a gene required for the first monooxygenase step in ubiquinone biosynthesis. J. Bacteriol. *182*, 5139–5146.

Pozueta-Romero, J., Rafia, F., Houlné, G., Cheniclet, C., Carde, J.P., Schantz, M.L., and Schantz, R. (1997). A ubiquitous plant housekeeping gene, *PAP*, encodes a major protein component of bell pepper chromoplasts. Plant Physiol. *115*, 1185–1194.

Ramel, F., Birtic, S., Ginies, C., Soubigou-Taconnat, L., Triantaphylidès, C., and Havaux, M. (2012). Carotenoid oxidation products are stress signals that mediate gene responses to singlet oxygen in plants. Proc. Natl. Acad. Sci. U.S.A. *109*, 5535–5540.

Rey, P., Gillet, B., Römer, S., Eymery, F., Massimino, J., Peltier, G., and Kuntz, M. (2000). Over-expression of a pepper plastid lipid-associated protein in tobacco leads to changes in plastid ultrastructure and plant development upon stress. Plant J. *21*, 483–494.

Rubio-Moraga, A., Rambla, J.L., Fernández-de-Carmen, A., Trapero-Mozos, A., Ahrazem, O., Orzáez, D., Granell, A., and Gómez-Gómez, L. (2014). New target carotenoids for CCD4 enzymes are revealed with the characterization of a novel stress-induced carotenoid cleavage dioxygenase gene from *Crocus sativus*. Plant Mol. Biol. *86*, 555–569.

Rubio, A., Rambla, J.L., Santaella, M., Gómez, M.D., Orzaez, D., Granell, A., and Gómez-Gómez, L. (2008). Cytosolic and plastoglobule-targeted carotenoid dioxygenases from *Crocus sativus* are both involved in beta-ionone release. J. Biol. Chem. *283*, 24816–24825.

Ruiz-Sola, M.A., and Rodríguez-Concepción, M. (2012). Carotenoid biosynthesis in *Arabidopsis*: a colorful pathway. In The *Arabidopsis* book (Rockville, USA: The American Society of Plant Biologists), p. e0158.

Sakuragi, Y., Zybailov, B., Shen, G.Z., Jones, A.D., Chitnis, P.R., van der Est, A., Bittl, R., Zech, S., Stehlik, D., Golbeck, J.H., *et al.* (2002). Insertional inactivation of the *menG* gene, encoding 2-phytyl-1,4-naphthoquinone methyltransferase of *Synechocystis* sp. PCC 6803, results in the incorporation of 2-phytyl-1,4-naphthoquinone into the A_1 site and alteration of the equilibrium constant between A_1 and Fx in photosystem I. Biochemistry-Us *41*, 394–405.

Schelbert, S., Aubry, S., Burla, B., Agne, B., Kessler, F., Krupinska, K., and Hörtensteiner, S. (2009). Pheophytin pheophorbide hydrolase (pheophytinase) is involved in chlorophyll breakdown during leaf senescence in *Arabidopsis*. Plant Cell *21*, 767–785.

Schultz, G., and Soll, J. (1980). Biosynthesis of alpha-tocopherol (vitamin E), phylloquinone (2-methyl-3-phytylnaphtoquinone, vitamin K1) and other prenylquinones in plants. On the problem of inability of the biosynthesis in animals – a survey (author's translation). Dtsch. Tierarztl. Wochenschr. *87*, 410–412.

Shanmugabalaji, V., Besagni, C., Piller, L.E., Douet, V., Ruf, S., Bock, R., and Kessler, F. (2013). Dual targeting of a mature plastoglobulin/fibrillin fusion protein to chloroplast plastoglobules and thylakoids in transplastomic tobacco plants. Plant Mol. Biol. *81*, 13–25.

Shikanai, T., Munekage, Y., Shimizu, K., Endo, T., and Hashimoto, T. (1999). Identification and characterization of *Arabidopsis* mutants with reduced quenching of chlorophyll fluorescence. Plant Cell Physiol. *40*, 1134–1142.

Simkin, A.J., Gaffé, J., Alcaraz, J.P., Carde, J.P., Bramley, P.M., Fraser, P.D., and Kuntz, M. (2007). Fibrillin influence on plastid ultrastructure and pigment content in tomato fruit. Phytochemistry *68*, 1545–1556.

Singh, D.K., Laremore, T.N., Smith, P.B., Maximova, S.N., and McNellis, T.W. (2012). Knockdown of *FIBRILLIN4* gene expression in apple decreases plastoglobule plastoquinone content. PloS One 7, e47547.

Singh, D.K., Maximova, S.N., Jensen, P.J., Lehman, B.L., Ngugi, H.K., and McNellis, T.W. (2010). *FIBRILLIN4* is required for plastoglobule development and stress resistance in apple and *Arabidopsis*. Plant Physiol. *154*, 1281–1293.

Sommer, A. (1990). Xerophthalmia, keratomalacia and nutritional blindness. Int. Ophthalmol. *14*, 195–199.

Sprey, B., and Lichtenthaler, H.K. (1966). Zur Frage der Beziehungen zwischen Plastoglobuli und

Thylakoidgenese in Gerstenkeimlingen. Z. Naturforsch. Pt. B. *B 21*, 697–699.

Steinmüller, D., and Tevini, M. (1985). Composition and function of plastoglobuli. I. Isolation and purification from chloroplasts and chromoplasts. Planta *163*, 201–207.

Tevini, M., and Steinmüller, D. (1985). Composition and function of plastoglobuli. II. Lipid-composition of leaves and plastoglobuli during beech leaf senescence. Planta *163*, 91–96.

Tuquet, C., and Newman, D.W. (1980). Aging and regreening in soybean cotyledons. 1. Ultrastructural-changes in plastids and plastoglobuli. Cytobios *29*, 43–59.

Vidi, P.A., Kanwischer, M., Baginsky, S., Austin, J.R., Csucs, G., Dörmann, P., Kessler, F., and Bréhélin, C. (2006). Tocopherol cyclase (VTE1) localization and vitamin E accumulation in chloroplast plastoglobule lipoprotein particles. J. Biol. Chem. *281*, 11225–11234.

Vidi, P.A., Kessler, F., and Bréhélin, C. (2007). Plastoglobules: a new address for targeting recombinant proteins in the chloroplast. BMC Biotechnol. 7.

Wolken, J.J., and Palade, G.E. (1952). Fine structure of chloroplasts in two flagellates. Nature *170*, 114–115.

Wuttke, H.G. (1976). Chromoplasts in *Rosa rugosa*: Development and chemical characterization of tubular elements. Z. Naturforsch. C. *31*, 456–460.

Xie, L.T.X., Hsieh, E.J., Watanabe, S., Allan, C.M., Chen, J.Y., Tran, U.C., and Clarke, C.F. (2011). Expression of the human atypical kinase ADCK3 rescues coenzyme Q biosynthesis and phosphorylation of Coq polypeptides in yeast *coq8* mutants. BBA-Mol. Cell. Biol. L. *1811*, 348–360.

Yang, S., Zhang, Q., Li, T., Du, J., Yang, S., and Yang, C. (2012). *AtSIA1*, an *ABC1*-like kinase, regulates salt response in *Arabidopsis*. Biologia *67*, 1107–1111.

Ye, X.D., Al-Babili, S., Klöti, A., Zhang, J., Lucca, P., Beyer, P., and Potrykus, I. (2000). Engineering the provitamin A (beta-carotene) biosynthetic pathway into (carotenoid-free) rice endosperm. Science *287*, 303–305.

Yoshioka, S., Aida, R., Yamamizo, C., Shibata, M., and Ohmiya, A. (2012). The *carotenoid cleavage dioxygenase 4 (CmCCD4a)* gene family encodes a key regulator of petal color mutation in chrysanthemum. Euphytica *184*, 377–387.

Ytterberg, A.J., Peltier, J.B., and van Wijk, K.J. (2006). Protein profiling of plastoglobules in chloroplasts and chromoplasts. A surprising site for differential accumulation of metabolic enzymes. Plant Physiol. *140*, 984–997.

Zbierzak, A.M., Kanwischer, M., Wille, C., Vidi, P.A., Giavalisco, P., Lohmann, A., Briesen, I., Porfirova, S., Bréhélin, C., Kessler, F., *et al.* (2010). Intersection of the tocopherol and plastoquinol metabolic pathways at the plastoglobule. Biochem. J. *425*, 389–399.

Zhang, R., Wise, R.R., Struck, K.R., and Sharkey, T.D. (2010). Moderate heat stress of *Arabidopsis thaliana* leaves causes chloroplast swelling and plastoglobule formation. Photosynth. Res. *105*, 123–134.

Zhang, W., Liu, T.Q., Ren, G.D., Hörtensteiner, S., Zhou, Y.M., Cahoon, E.B., and Zhang, C.Y. (2014). Chlorophyll degradation: the tocopherol biosynthesis-related phytol hydrolase in *Arabidopsis* seeds is still missing. Plant Physiol. *166*, 70–79.

Zhu, C., Bai, C., Sanahuja, G., Yuan, D., Farré, G., Naqvi, S., Shi, L., Capell, T., and Christou, P. (2010). The regulation of carotenoid pigmentation in flowers. Arch. Biochem. Biophys. *504*, 132–141.

Redox Regulation in Chloroplasts

9

Monica Balsera, Peter Schürmann and Bob B. Buchanan

Abstract

The reversible oxidation of cysteine (Cys) residues is commonly used to regulate a range of cellular processes throughout biology. Modification of regulatory Cys alters protein solubility and activity and, under control conditions, ultimately results in the fine-tuning of processes in response to metabolic demands. In oxyphotosynthetic organisms, the reversible modification of protein thiols is a major mechanism for direct regulation of chloroplast enzymes in a light-dependent manner and in response to different types of stress. The main redox regulatory system in chloroplasts is the ferredoxin-dependent thioredoxin system (FTS) which coordinates metabolic pathways of oxygenic photosynthesis via thioredoxin-mediated dithiol/disulfide (SH/S-S) transitions. NADP-linked thioredoxin reductase C and the glutathione/glutaredoxin systems complement the FTS in maintaining redox balance under certain environmental conditions.

Introduction

In chloroplasts, the coordinated activities of two specialized protein–pigment complexes within the thylakoid membrane, termed the photosystems, catalyse light-driven electron transfer from water to ferredoxin (Fdx) – a small electronegative iron–sulfur protein in the stroma (Nelson and Ben-Shem, 2004; Nevo *et al.*, 2012). Reduced Fdx distributes reducing power to a number of metabolic pathways via specific interactions with enzymes functional in $NADP^+$ reduction, and nitrogen and sulfur assimilation (Fig. 9.1) (Hanke and Mulo, 2013). During the light-driven electron transfer reactions, an electrochemical potential difference of protons generated across the thylakoid membrane drives ATP synthesis in the stroma (Fig. 9.1). Photosynthetically generated NADPH and ATP are used to produce sugar-phosphates, amino acids, and numerous other metabolites needed by the plant cell. During photosynthesis, a fraction of the fixed CO_2 accumulates transiently as starch in the stroma (Weise *et al.*, 2011). At night, chloroplasts operate as heterotrophs, using the oxidative pentose phosphate pathway (OPP) to generate reducing power (in the form of NADPH) and mobilize starch as carbon source to sustain the plant until light is restored (Kruger and von Schaewen, 2003). Chloroplasts are also the site of other metabolic processes that include the synthesis of amino acids, lipids, and secondary metabolites needed for growth and survival of the plant (Lunn, 2006; Rolland *et al.*, 2012). To coordinate their diverse functions, chloroplasts have developed sophisticated regulatory mechanisms linked to the diurnal activity of the plant that bring about reversible changes in the redox state of cysteine (Cys) thiol groups of selected proteins (Balsera *et al.*, 2014). In particular, thiol-disulfide exchange in proteins linked to light via the Fdx-thioredoxin (Trx) system (FTS) (Fig. 9.2) acts as a signalling element, affecting carbohydrate metabolism, nitrogen assimilation, tetrapyrrole synthesis, and lipid metabolism among others (Schürmann and Buchanan, 2008).

Photosynthesis not only provides the energy and reducing power – in the form of ATP and NADPH, respectively –to fix carbon and perform other metabolic reactions, but also produces molecular oxygen as a by-product. As a consequence of the spatial and temporal concurrence of electron and energy transfer reactions with photosynthetic oxygen evolution within the thylakoid membrane significant amounts

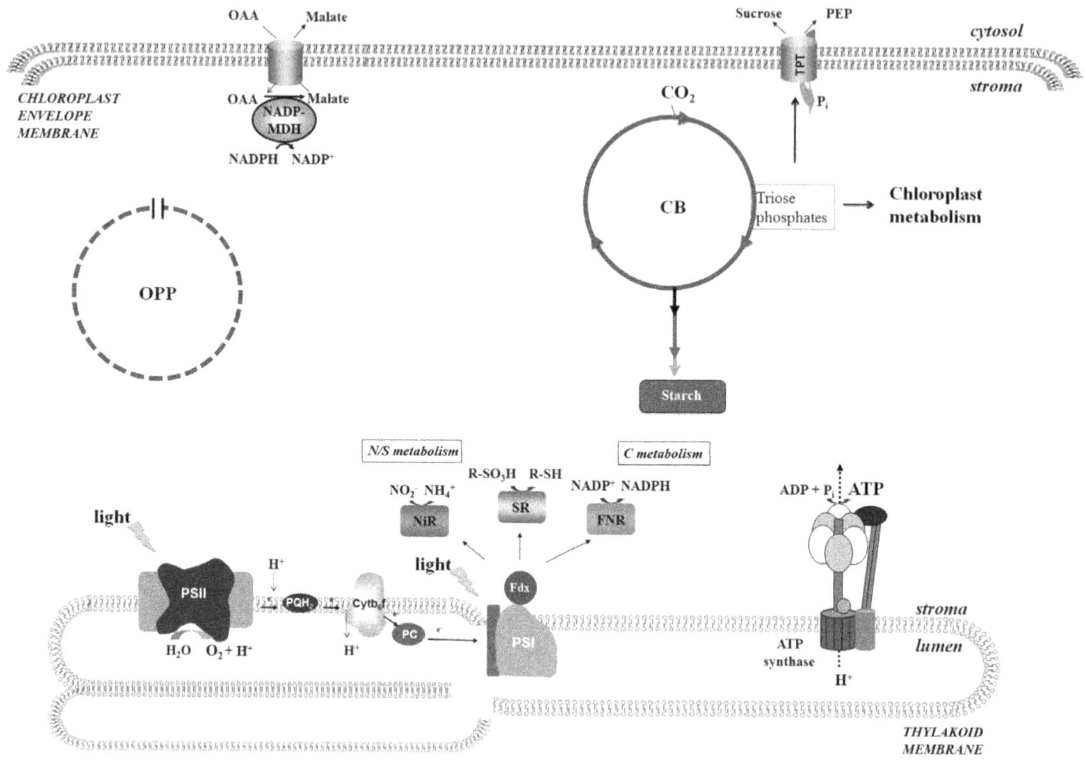

Figure 9.1 Oxygenic photosynthesis and associated metabolic processes in chloroplasts. Chloroplasts have the capacity to perform oxygenic photosynthesis, that is, to convert light energy into reducing power (NADPH) and ATP with the concomitant evolution of molecular oxygen. The photosynthetic machinery is made up of four protein complexes, namely photosystem II (PSII), cytochrome b_6f, photosystem I (PSI) and ATP synthase, whose coordinated activities couple electron transport to the chemiosomotic process. Fdx acts as the central distributor of photosynthetically derived-electrons to different partner proteins, particularly ferredoxin–NADP reductase (FNR), nitrite reductase (NiR) and sulfite reductase (SR). Most of the NADPH and ATP are used in the CB cycle. The triose phosphates produced are either exported to the cytosol by the Pi-triose phosphate translocator or used within the chloroplast. A fraction of the fixed CO_2 accumulates as transitory starch. During the day, the OPP is deactivated. The malate valve (NADP-MDH) controls the export of reducing equivalents via metabolic exchange during the day.

of reactive oxygen species (ROS) are produced (Galvez-Valdivieso and Mullineaux, 2010). Light-dependent ROS generation in chloroplasts includes superoxide (O_2^-), hydrogen peroxide (H_2O_2), and singlet oxygen (O_2^1) molecules (Apel, 2004) (Fig. 9.3). Because ROS participate in key signalling events on one hand, but may cause irreversible injury to proteins and membranes on the other, chloroplasts have developed an antioxidant protection system to control intracellular redox changes and ROS accumulation in response to developmental and environmental changes. The system consists of soluble hydrophilic antioxidants – such as ascorbate, carotenoids and glutathione (GSH; g-L-glutamyl-L-cysteinylglycine) – and enzymatic scavenging mechanisms that include superoxide dismutase, ascorbate peroxidase, and catalase (Apel, 2004). Most amino acids in proteins are susceptible to ROS, but the ease of oxidation of the sulfur-containing amino acids, Cys and methionine (Met), makes them major sites for change. It has been found that the FTS, together with the GSH/glutaredoxin (Grx) system, can restore the activity of certain enzymes following oxidative deactivation by ROS (Meyer *et al.*, 2012; Schürmann and Buchanan, 2008).

The regulation of biochemical processes by light and dark is a distinct feature of oxygenic photosynthesis of both prokaryotes and eukaryotes. Initially identified in the 1960s in studies on photosynthetic

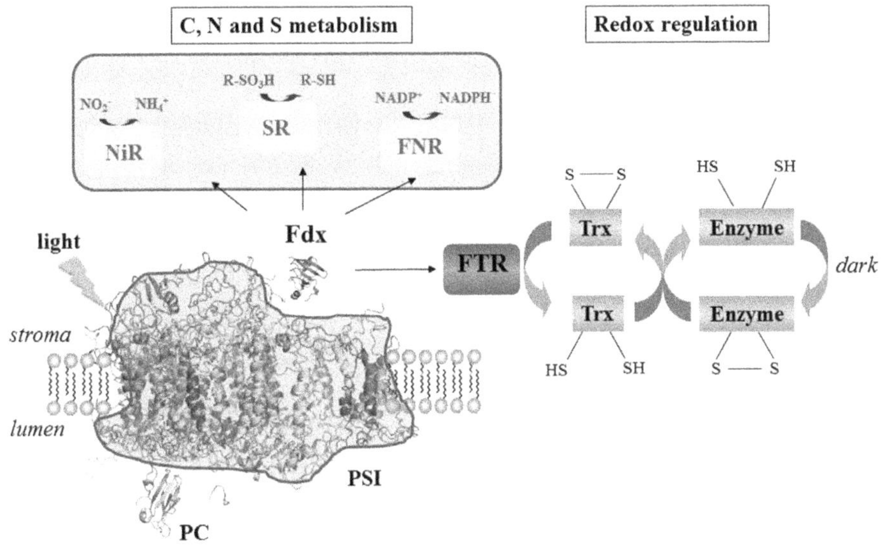

Figure 9.2 Ferredoxin plays a central role in enzyme regulation by light-dependent redox transitions. In addition to its metabolic function above (also Fig. 9.1), Fdx is important in regulating numerous enzymes of chloroplasts. On illumination, photosynthetic electrons from Fdx are transferred to Trx via FTR. Trx, in turn, reduces target enzymes thereby modifying their activity.

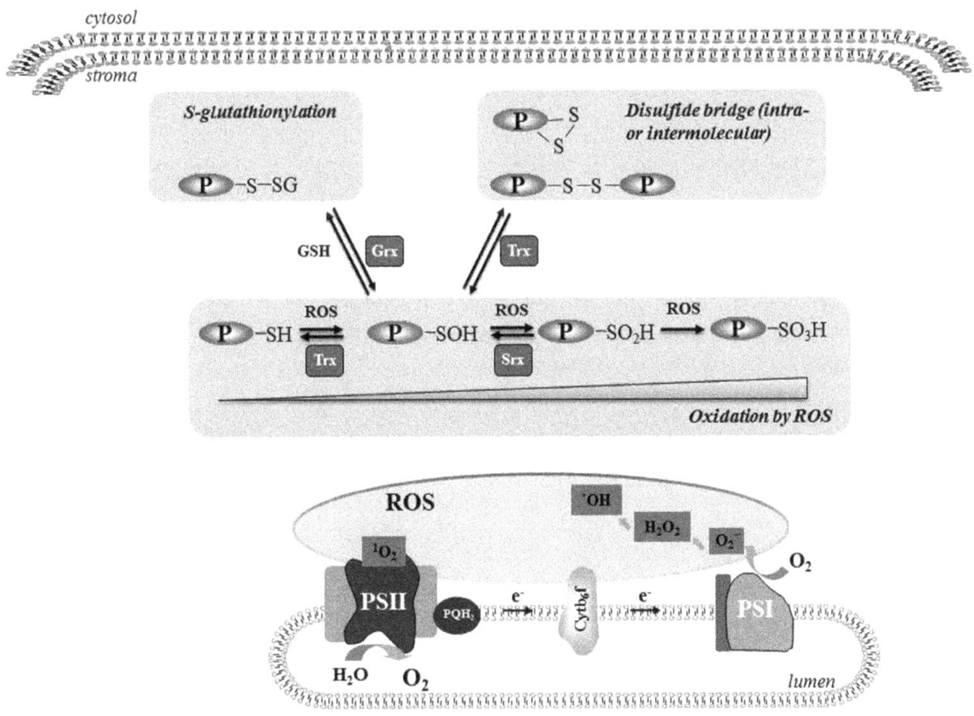

Figure 9.3 Regulatory cysteines of chloroplast enzymes are modified by different types of oxidation/reduction reactions. Since molecular oxygen production and electron transport are linked in thylakoids, photosynthetic cells have to cope continuously with the inevitable formation of ROS. The nucleophilic side chain of Cys is susceptible to several types of oxidative modifications linked to ROS in addition to thiol/disulfide changes. The Trx and Grx systems maintain the redox state of thiols in proteins.

carbon fixation, redox-based regulation in chloroplasts has been extended to form a sophisticated network of processes based on light and ROS (Buchanan, 1980; Buchanan and Balmer, 2005; Buchanan *et al.*, 2012).

Oxidative Cys modifications in chloroplast proteins

It is well known that cells apply post-translational modification reactions (PTMs) in proteins to a variety of amino acid residues and thereby dynamically regulate cellular processes and physiological states. Among the effective amino acid residues, Cys is by far the most reactive since its sulfur atom is nucleophilic and thus prone to oxidation, leading to a number of reversible and irreversible PTMs (Astier *et al.*, 2011; Dalle-Donne *et al.*, 2009; Klomsiri *et al.*, 2011; Poole *et al.*, 2004). In addition to serving as structural stabilizers, participating in catalysis and coordinating metal ions, Cys can function as a redox switch, thereby affecting the activity of enzymes and other proteins (Fomenko *et al.*, 2008). However, not all Cys residues of a protein are equally sensitive to chemical modification (Marino and Gladyshev, 2012). The reactivity of a particular Cys is largely dependent on ionization to the thiolate anion (RS^-) that is determined by a number of factors including its pK_a, itself affected by the microenvironment of the thiol group in a protein, and solvent accessibility (Roos *et al.*, 2013). In chloroplasts in particular, the reversible oxidation of selected Cys residues in redox sensing protein targets has been evolutionarily exploited to alter function and, with that, regulate metabolic processes in coordination with environmental conditions and needs of the cell (Balsera *et al.*, 2014; Lindahl *et al.*, 2011; Zaffagnini *et al.*, 2012b). In other cases, enzymes use the reversible formation and breaking of disulfide bonds of Cys as part of their catalytic cycle.

The possibilities for the chemical modification of reactive Cys residues in proteins are diverse (Fig. 9.3) (Couturier *et al.*, 2013; Kim *et al.*, 2015). In chloroplasts, the activities of specific enzymes are subject to regulation by reversible thiol/disulfide exchange reactions via the Trx system in a light-dependent manner (Fig. 9.2) (Schürmann and Buchanan, 2008). Classical examples of light/dark regulation through redox modulation include enzymes of the Calvin–Benson cycle (CB) and associated processes, such as the OPP, NADP-dependent malate dehydrogenase (NADP-MDH), and the gamma-subunit of ATP-synthase. Additionally, the thiol side chain of redox sensitive Cys (Cys-SH) is a major target of ROS (particularly H_2O_2 and ROOH) in biological systems. The primary oxidation product of Cys by ROS is sulfenic acid (Cys-SOH), that can be further oxidized to sulfinic acid (Cys-SO_2H), or react with GSH or other Cys residues resulting in glutathione-protein mixed disulfides (termed glutathionylation; Cys-SSG) or intra/intermolecular disulfides (Dalle-Donne *et al.*, 2009; Poole *et al.*, 2004) (Fig. 9.3). Protein glutathionylation is reversible primarily through the action of the GSH/Grx system (see below). In chloroplasts, sulfenic acids have been identified in the catalytic cycle of enzymes, such as 2-Cys peroxiredoxin (Prx) (Hall *et al.*, 2009). Sulfinic acid can undergo further oxidation to sulfonic acid (Cys-SO_3H), or be reduced back to the sulfhydryl form by sulfiredoxins (Iglesias-Baena *et al.*, 2010) (Fig. 9.3). The formation of sulfonic acid in biological systems is, however, irreversible and leads to protein degradation (Fig. 9.3). In undergoing redox change, thiols can also covalently attach a nitrogen monoxide group in a reaction termed S-nitrosylation (Cys-SNO) (Lindermayr *et al.*, 2005; Lindermayr, 2009) (not shown in Fig. 9.3). It has been proposed that S-nitrosylation of redox sensitive Cys residues promotes conformational changes that prevent irreversible protein carbonylation triggered by ROS.

Regulation of proteins in chloroplasts by thiol redox transitions linked to thioredoxin and glutaredoxin

Two types of oxidoreductases, Trx and Grx, are largely responsible for controlling the redox state of Cys groups in proteins of most eukaryotes and many prokaryotes (Meyer *et al.*, 2009, 2012). Although the two proteins are structurally similar, amino acid sequence and active site differences result in dissimilar catalytic reactions and typically substrate specificity (Carvalho *et al.*, 2006).

Trxs participate in dithiol/disulfide exchange reactions in a manner that may result in different changes in individual target proteins (Meyer *et al.*,

2005; Schürmann and Buchanan, 2008). Canonical members of the Trx family contain a common motif in the active site, WC(G/P)PC, with two strictly conserved Cys involved in a disulfide bridge that is maintained in the reduced state by a Trx reductase (TR) (Jacquot *et al.*, 2009). In chloroplasts, canonical Trxs link the activity of selected enzymes to diurnal changes. The redox-active site of Trxs is positioned in a structural framework that confers special properties to the most N-terminal Cys, known as catalytic Cys (Cys_{cat} or Cys_N). Under physiological conditions, Cys_{cat} is deprotonated and carries out a nucleophilic attack on the disulfide bond of a target protein. The ensuing mixed disulfide is resolved by a second Cys in Trx, called resolving Cys (Cys_{res} or Cys_C), resulting in oxidized Trx and reduced target (Fig. 9.4A). Chloroplasts have five major types of canonical Trxs, namely Trx f, m, x, y and z (Chibani *et al.*, 2009; Collin *et al.*, 2003; Gelhaye *et al.*, 2005) (Fig. 9.5). Additionally, a number of so-called atypical Trxs are present in chloroplasts – a group consisting of redox-active proteins with a Trx-fold, but with variations in the canonical active site (CxxC/S) (Cain *et al.*, 2009; Chibani *et al.*, 2009). Atypical chloroplast Trxs include CDSP32 (chloroplast drought-induced protein of 32 kDa), Trx, like 2.1 (or WCRKC protein) and lilium Trx [or ACHT (atypical Cys and histidine rich Trx) protein], is implicated in the oxidative stress response.

In chloroplasts, reduction of canonical Trxs relies on Fdx-dependent TR (FTR), which catalyses the transfer of photosynthetic electrons from Fdx to Trx (Fig. 9.2) (Schürmann and Buchanan, 2008). Composed of a catalytic (FTRc) and a variable subunit (FTRv), FTR is a unique enzyme that utilizes a 4Fe–4S centre and a redox-active disulfide to mediate electron transfer from a one-electron donor (Fdx) to a two-electron acceptor (Trx) (Dai *et al.*, 2007). In eukaryotic and most prokaryotic cell types, as well as the cytosol and mitochondria of plants, Trx is reduced not by FTR, but by a flavoenzyme, an NADP-dependent TR (NTR) (Fig. 9.6) (Jacquot *et al.*, 2009). In chloroplasts, FTR coexists with a third type of TR (NTRC) – a unique NTR present exclusively in oxyphototrophs (Serrato *et al.*, 2004) (Fig. 9.5). The unique feature of NTRC is the presence of two distinct functional domains, namely an NTR and a Trx domain, fused in a single polypeptide chain. Because the activity of NTRC depends on NADPH, the enzyme can function in both light and dark by receiving electrons, respectively, from photosynthetic electron flow or the OPP. Although NTRC has enzymatic activity characteristic of each of its two domains, the enzyme reduces its own Trx and has little interaction with other chloroplast Trxs, thus ensuring that the plant is able to distinguish and differentially regulate diurnal and nocturnal processes (Bohrer *et al.*, 2012; Serrato *et al.*, 2004). Thus, while the Fdx- and NADP-dependent systems can act in a complementary manner (as with the regulation of ADP-glucose pyrophoshorylase), they are basically independent of one another as seen with the CB cycle where the FTS serves as the primary regulator (Lepistö *et al.*, 2009). NTRC plays an important role in antioxidant metabolism such that chloroplasts must coordinate the activities of the complementary FTS and NTRC redox systems (Cejudo *et al.*, 2012; Perez-Ruiz *et al.*, 2006). It has been proposed that chloroplast proteins named CBSX1 and CBSX2, with a cystathionine b-synthase (CBS) domain, can modify Trx activities (Yoo *et al.*, 2011). However, details of the mechanism responsible and the effect on Trx redox state are scanty.

The Grx system confers additional reducing capacity in chloroplasts, namely, in deglutathionylation and in the assembly of iron–sulfur clusters in appropriate proteins (Couturier *et al.*, 2011; Rouhier *et al.*, 2007, 2008). Grxs contain a specific catalytic motif CxxC/S and are grouped in different classes based on amino acid sequences (Couturier *et al.*, 2009a). Chloroplasts contain representatives of class I Grxs (GrxS12 and GrxC5) and class II (GrxS14 and GrxS16). Class I Grxs are dependent on GSH and typically follow a monothiol mechanism for target deglutathionylation (Fig. 9.4B). The monothiol mechanism involves the glutathionylation of the most N-terminal Cys in Grxs (Cys_{cat}) upon reduction of a GSH-target protein adduct. Glutathionylated Cys is resolved by a second GSH that results in oxidized glutathione (GSSG) and reduced Grx; GSSG is then reduced by NADP-dependent GSH reductase (GR) (Bandyopadhyay *et al.*, 2008; Couturier *et al.*, 2009b). Class II Grxs employ a dithiol mechanism – that is, a dithiol/disulfide exchange – that involves a second Cys (Cys_{res}) that competes with GSH to regenerate the glutathionylated Grx (Zaffagnini *et al.*, 2008)

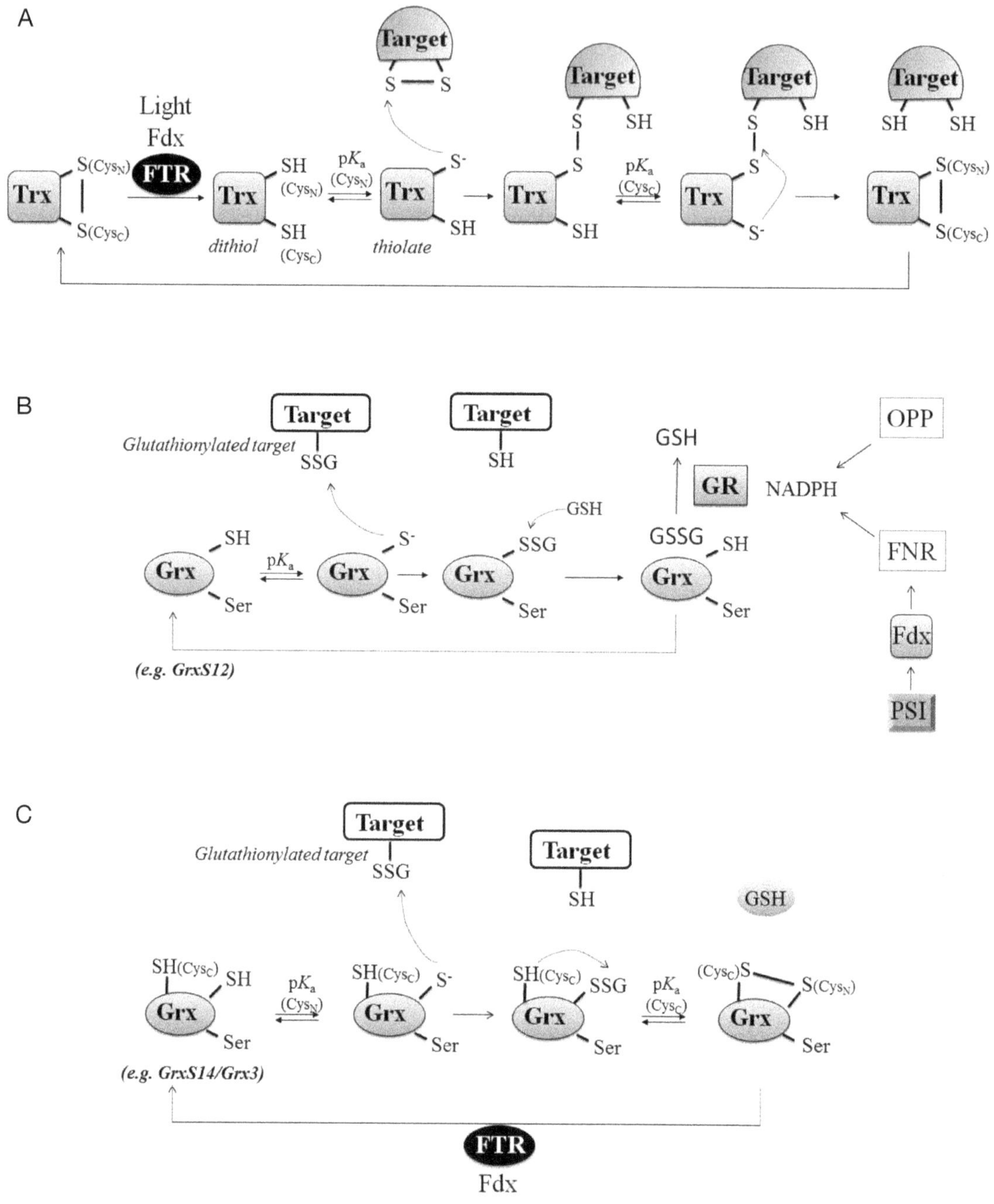

Figure 9.4 Mechanisms for reaction of thioredoxin and glutaredoxin with target proteins. Due to different physico-chemical redox properties, Trxs are mainly involved in protein disulfide reduction, whereas Grxs are primarily active in deglutathionylation reactions and Fe–S cluster assembly. (A) In illuminated chloroplasts, Trx is reduced by FTR. Under physiological conditions, the thiolate form of the catalytic Cys (Cys_N) of the CxxC active site initiates a nucleophilic attack on the disulfide bond in a target protein. Due to local conformational perturbations, the transient intermolecular disulfide formed is resolved by the resolving Cys in Trx (Cys_C), resulting in the formation of an intramolecular disulfide in Trx and the release of reduced target. (B) In Grx, the nucleophilic Cys of Grx forms a heterodisulfide with GSH upon reaction with the mixed disulfide of a glutathionylated target. Typically, a second GSH resolves the enzyme-glutathione mixed disulfide bond to generate the reduced Grx (e.g., GrxS12 and GrC5). GSH is kept reduced by the action of GR. (C) Some Grxs, such as GrxS14 and Grx3 – with the catalytic motif CxxC and CxxS, respectively – can also follow a mechanism in which the heterodisulfide in Grx is resolved by a second Cys (Cys_{res}), resulting in reduced target and oxidized Grx. In chloroplasts, GrxS14 is believed to be reduced via FTR.

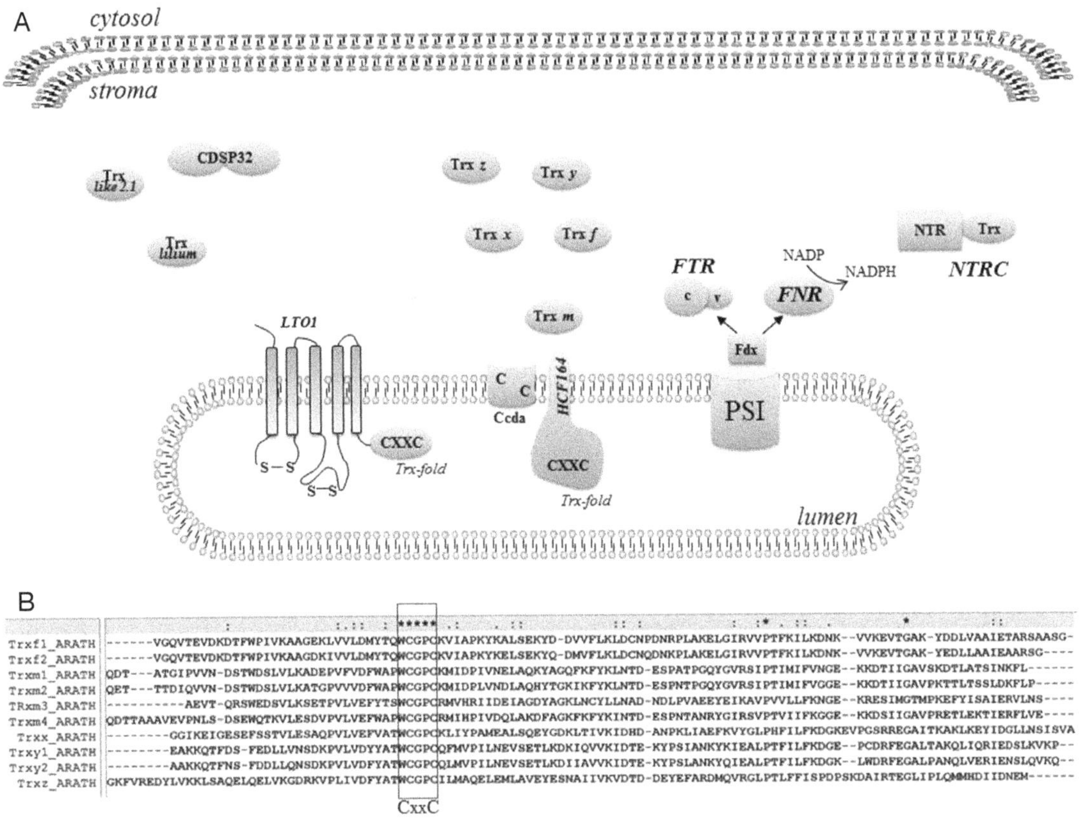

Figure 9.5 Thioredoxin redox systems of chloroplasts. (A) Inventory of canonical (f, m, x, y and z) and non-canonical (CDSP32, like 2.1, lilium) Trxs in chloroplasts. FTR, composed of catalytic (c) and variable (v) subunits, transfers reducing equivalents from photoreduced Fdx to chloroplast Trxs (Fig. 9.2). The bifunctional NADPH-thioredoxin reductase C (NTRC) receives reducing equivalents from NADPH generated either via FNR in the light (or by OPP in the dark; not shown). The evidence suggests that the ferredoxin- and NADP-dependent systems function independently of one another. Recent studies have demonstrated the existence of proteins in thylakoids that may participate in the transport of reducing equivalents from the stroma to the lumen. HCF164 contains a luminal Trx-like domain that seems to accept reducing equivalents from stromal Trx m via the polytopic membrane protein Ccda (cytochrome c defective A). LTO1 (lumen thiol oxidoreductase 1) – composed of a transmembrane domain with two pair of Cys fused to a soluble Trx-like domain – has been localized in thylakoids, adding a new participant in regulating thiol/disulfide formation in proteins within the lumen. (B) Multiple protein sequence alignment of canonical chloroplast Trxs from Arabidopsis.

(Fig. 9.4C). The disulfide formed between Cys_{cat} and Cys_{res} is reduced back by FTR.

The ferredoxin/thioredoxin system: a light-dependent redox mechanism regulating Calvin–Benson cycle and associated enzymes in chloroplasts

The light-dependent regulatory role of Trx in chloroplasts was originally uncovered by measuring the activity of key CB cycle enzymes of chloroplasts. Aside from fructose 1,6-bisphosphatase (FBPase), the first enzyme found to be activated by Trx (Buchanan *et al.*, 1967, 1971), target enzymes were selected by virtue of their catalysing irreversible reactions. These enzymes were found to be reductively activated by light via the FTS, whereas another enzyme, glucose 6-phosphate dehydrogenase (G6PDH), an enzyme of the OPP, was reductively deactivated under illumination and oxidatively activated in the dark (Buchanan, 1980; Schürmann and Buchanan, 2008). Other chloroplast enzymes, such as NADP-MDH, and ATP-synthase were also light activated. It was found that most target enzymes in the stroma are activated and modulated according

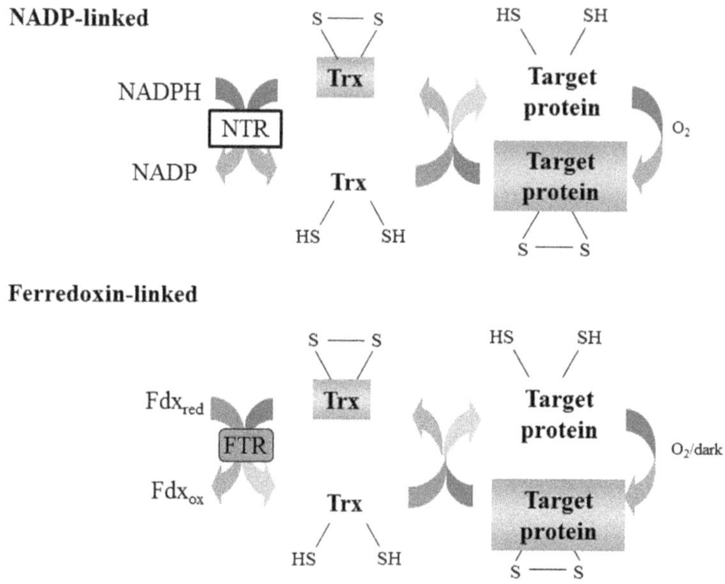

Figure 9.6 Fdx- and NADP-dependent thioredoxin reductases. In most organisms, the intramolecular disulfide bond of Trx receives reducing equivalents from NADPH via a specific Trx reductase (NTR). During evolution, oxyphotosynthetic organisms linked the reduction of Trx to light via FTR (see Fig. 9.2). In chloroplasts, FTR coexists with a special type of NTR (NTRC) which has the NTR and Trx domains fused in a single polypeptide chain.

to light intensity, whereas participants of the OPP and proteins in the lumen were active when oxidized (Buchanan and Luan, 2005).

Redox regulation of Trx-linked enzymes is possible due to the presence of a pair of redox-active Cys residues in target proteins (Balsera *et al.*, 2014; Schürmann and Buchanan, 2008). No common principle has been found for the molecular mechanisms by which redox modulates the activity of particular target enzymes. The reversible modification of regulatory Cys can either alter the catalytic properties of the interacting enzyme, modify the binding of activators and inhibitors, or lead to a conformational change with subsequent modulation of activity or protein–protein/protein–ligand interactions. In addition to modulating protein function, the reversible oxidation of Cys under physiological conditions serves to protect proteins from further oxidation. In most cases, the enzymes are regulated at multiple levels – change in light-induced pH or binding of substrates or allosteric effectors in response to metabolic changes.

Enzymes of the CB cycle turned out to be highly regulated by Trx to coordinate carbon assimilation with the availability of CO_2 and light. In chloroplasts, five central members of the cycle – Rubisco via Rubisco activase (RCA), glyceraldehyde-3-phosphate dehydrogenase (GAPDH), phosphoribulokinase (PRK), FBPase and sedoheptulose-1,7-bisphosphatase (SBPase) – are diurnally regulated by Trx. In RCA and GAPDH (Fig. 9.7A), C-terminal extensions harbouring two conserved redox-active cysteine residues act as an auto-inhibitory domain in the oxidized state (Fermani *et al.*, 2007; Portis *et al.*, 2008; Sparla *et al.*, 2002; Zhang and Portis, 1999). On the other hand, FBPase (Fig. 9.7C), SBPase and PRK have internal insertions in their amino acid sequences that contain the regulatory conserved cysteines (Balsera *et al.*, 2013; Chiadmi *et al.*, 1999; Dunford *et al.*, 1998; Miziorko, 2000). With these enzymes, reduction induces conformational changes with effects on their activity. In other cases, reversible changes in the redox and activation states are accompanied by the formation of transient complexes. The small protein CP12 turned out to be a central regulatory element in oxygenic photosynthesis (Balsera *et al.*, 2014; Graciet *et al.*, 2003; Marri *et al.*, 2014). CP12 is a disordered protein in solution that contains one or two regulatory disulfides (Pohlmeyer *et al.*, 1996; Wedel *et al.*, 1997). The formation of a disulfide promotes a conformational change in the protein that

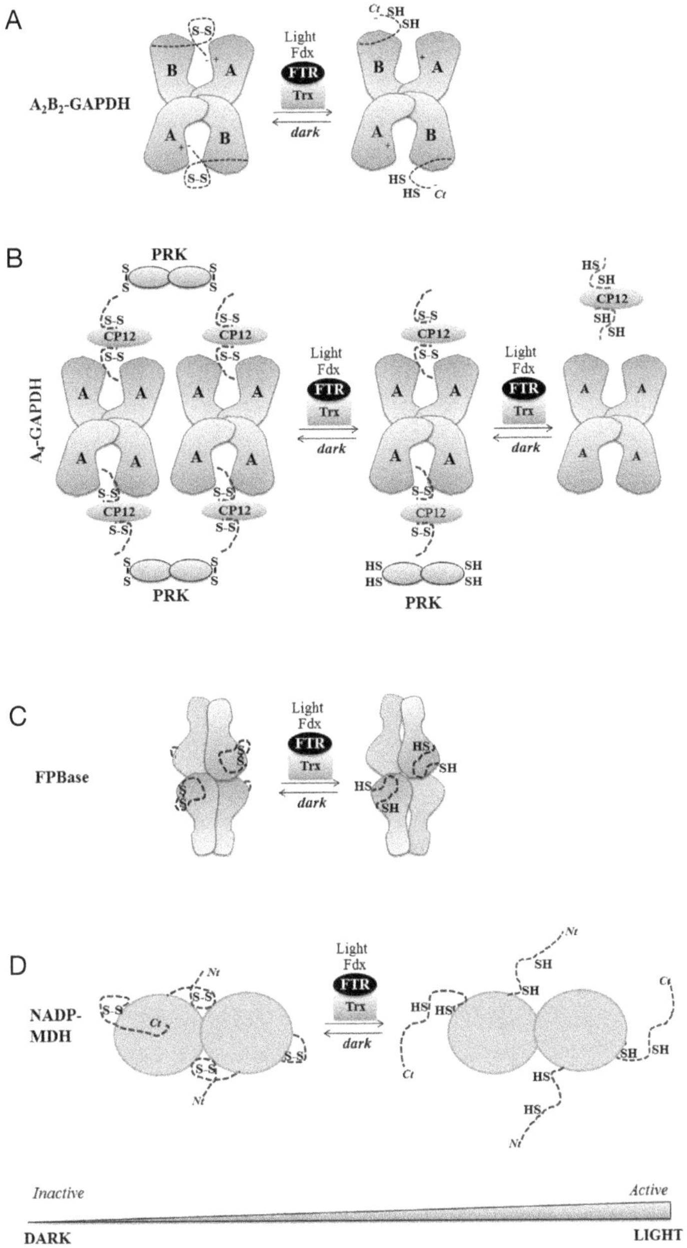

Figure 9.7 Models for activation of selected regulatory enzymes by FTS. (A) Chloroplasts contain two types of GAPDH subunits, A and B, that differ by the presence of a C-terminal tail in the B subunit harbouring a regulatory Cys that undergoes reversible diurnal redox changes in response to Trx. In chloroplasts, the homotetramer A_4–GAPDH and the heterotetramer A_2B_2–GAPDH coexist. Structural studies have shown that in the A_2B_2–GAPDH complex, the oxidized C-terminal tail of the B subunit packs into a cleft between a pair of A/B subunits and occupies the catalytic site, deactivating the enzyme (left). In this conformation, a negatively charged residue at the C-terminus of the B subunit (–) interacts with a conserved positively charged residue of the A-subunit (+). On reduction by Trx, the C-terminal tail unfolds, allowing access to the catalytic site and activating the enzyme. (B) The oxidation of the regulatory peptide CP12 deactivates PRK and A_4-GAPDH by forming a ternary complex (left). Complex dissociation depends mainly on the reduction of CP12 by Trx. Following dissociation, PRK must be further activated by reduced Trx. (C) In FBPase, the oxidation of a redox-sensitive loop involves a disorder-order transition that stabilizes its inactive conformation (left); following reduction, the enzyme assumes a conformation in which the enzyme is active. (D) In NADP-MDH, the oxidized N-terminal disulfide is inserted in a cleft at the subunit interface of the dimer, distorting the catalytic site; the oxidized C-terminal disulfide loops the extension back into the active site, inhibiting access of the substrate. On reduction the enzyme unfolds, allowing the substrate to access the active site.

increases its affinity initially for GAPDH and subsequently for PRK (Fermani *et al.*, 2012; Howard *et al.*, 2008; Matsumura *et al.*, 2011; Trost *et al.*, 2006) (Fig. 9.7B). The ternary complex that oxidized CP12 forms with PRK and GAPDH in the dark results in the reversible inhibition of these enzymes. Complex dissociation is mainly dependent on the reduction of CP12 by the FTS (Marri *et al.*, 2009). Trx is also linked to the activation of NADP-MDH, a central metabolic enzyme that functions as a malate valve (Lemaire *et al.*, 2005; Ocheretina *et al.*, 2000). The enzyme contains N- and C-terminal extensions that each harbour a regulatory disulfide (Fig. 9.7D). The N-terminal disulfide inserts into a cleft at the subunit interface in the dimer, distorting the catalytic site. As a consequence of its disulfide, the C-terminal end of the enzyme is trapped inside the active site, acting as an internal inhibitor by obstructing substrate access (Johansson *et al.*, 1999; Krimm *et al.*, 1999). Chloroplast Trxs also influence ATP synthesis via modification of regulatory Cys in an insertion loop of the gamma subunit of ATP synthase. It seems that formation of the disulfide promotes a conformational change in the regulatory loop that directly influences rotation of the enzyme during catalysis (Hisabori *et al.*, 2013; Samra *et al.*, 2006). It has been proposed that the catalytic activity of CF1 is tightly redox-regulated to prevent futile ATP hydrolysis in the dark, when ATP synthesis is inactive. In a unique mechanism the formation of the regulatory disulfide of G6PDH activates the enzyme due to increased ability of substrate to access the active site and ability to bind NADP cofactor (Nee *et al.*, 2014).

Owing to the sessile nature of plants, chloroplasts must adjust to environmental changes. To obtain insight into chloroplast processes regulated by Trx under different light conditions, Yoshida *et al* (2014) carried out a comparative study of the *in vivo* redox dynamics of several Trx-linked chloroplast enzymes in *Arabidopsis*. They found that the enzymes responded to light differently. ATP synthase was rapidly activated, via the reduction of its gamma subunit, even under low-light conditions, likely to enable the enzyme to function early on during the induction of photosynthesis (Konno *et al.*, 2012; Yoshida *et al.*, 2014). Other enzymes such as FBPase and SBPase were gradually reduced in accord with increase in light intensity (Yoshida *et al.*, 2014). The authors hypothesized that this distinct redox behaviour might result from the different localization of the enzymes within the organelle, thus allowing efficient adaptation to the light environment.

Trx-redox regulation beyond the Calvin–Benson and oxidative pentose phosphate pathways

Light-dependent Trx-regulation in chloroplasts is not restricted to carbon fixation. It is now evident that this mechanism applies to the regulation of enzymes and proteins functional in many chloroplast processes (Balmer *et al.*, 2003; Montrichard *et al.*, 2009; Motohashi *et al.*, 2001). These include the regulation of starch metabolism, fatty acid and chlorophyll synthesis, plastid gene expression and photosystem II (PSII) assembly and state-transitions among others.

As discussed above, chloroplasts are effective in coordinating electron transfer reactions in thylakoids with stromal enzymes functional in CO_2 fixation by the CB cycle. The phosphorylated intermediates formed are either exported to the cytosol as triose phosphates or are used within the chloroplast for synthesizing transient starch from fructose-6-phosphate formed by FBPase of the CB cycle. A body of evidence has shown that starch synthesis is controlled by light and coordinated with CO_2 fixation (Glaring *et al.*, 2012). The committed step of starch biosynthesis is catalysed by ADP-glucose pyrophosphorylase (AGPase), a heterotetramer composed of two large and two smaller subunits. The enzyme is activated by reduction of an intermolecular disulfide bridge between the two smaller subunits via FTS (Ballicora *et al.*, 2000; Hendriks *et al.*, 2003; Thormählen *et al.*, 2013). The regulation of starch metabolism is, however, complex as the enzyme is also subject to allosteric regulation, being activated by the first product of the Calvin–Benson cycle, 3-phosphoglyceric acid (3-PGA), and inhibited by inorganic phosphate (Pi) (Hwang *et al.*, 2005). Additionally, AGPase is also regulated by NTRC, thereby complementing the FTS in the dark when the OPP, in effect, replaces the CB cycle for certain reactions (Michalska *et al.*, 2009). There is evidence that allosteric and redox regulation act synergistically on AGPase to ensure that starch synthesis is efficiently controlled in the light and dark. In line with this behaviour, it has recently been reported that the metabolism of trehalose, which influences starch accumulation in chloroplasts, particularly under stress conditions, is linked to Trx and the FTS via trehalose 6-phosphate phosphatase isoform D (TPPD) (Krasensky *et al.*, 2014). The activity of the enzyme is controlled by formation of an intramolecular disulfide bridge that leads to deactivation. TPPD is reactivated on reduction by Trx.

Plastids are the site of *de novo* fatty acid synthesis in plants. In chloroplasts, these reactions consume intermediate CB cycle products as well as photosynthetically generated ATP and NADPH (Ohlrogge and Browse, 1995). The activity of acetyl-CoA carboxylase (ACCase), which catalyses the first committed step of fatty acid synthesis, is stimulated by light via FTS (Kozaki and Sasaki, 1999; Sasaki *et al.*, 1997). In chloroplasts ACCase is a multi-enzyme complex composed of four subunits: the biotin carboxylase, biotin carboxyl carrier protein and carboxyltransferase which is made up of the nuclear-encoded a- and the chloroplast-encoded b-subunits. The reversible formation of an intermolecular disulfide bond between the two subunits of the carboxyltransferase seems to be responsible for the regulatory properties of the ACCase complex (Kozaki *et al.*, 2001).

The biosynthesis of chlorophyll takes place in the chloroplast. Due to the phototoxicity of intermediates of the pathway, chlorophyll synthesis is strictly regulated in photosynthetic cells. The Mg–chelatase complex is a heterotrimeric enzyme, composed of CHLI, CHLD and CHLH subunits, that catalyse the insertion of Mg^{2+} into protoporphyrin IX. The FTS was found to reductively activate the ATPase activity of the CHLI subunit (Balmer *et al.*, 2003; Ikegami *et al.*, 2007; Luo *et al.*, 2012). In related studies, NTRC has also been linked to enzymes of chlorophyll synthesis (Pérez-Ruiz *et al.*, 2014; Richter *et al.*, 2013). It is envisaged that NTRC regulates enzymatic activity during light/dark growth periods or temporally under light-limiting conditions.

In addition to regulation at the protein level, chloroplasts use light and redox transitions to control the expression of specific genes linked to changes in development and environment (Arsova *et al.*, 2010; Rochaix, 2013; Wang *et al.*, 2014). The FTS has been found to participate in the plastid-encoded polymerase (PEP)-dependent expression of chloroplast genes through the regulation of two components of a transcriptionally active chromosome (TAC) complex in plastids – namely fructokinase-like proteins 1 and 2 (FLN1 and FLN2) (Arsova *et al.*, 2010). In another experimental approach, AtECB1 (*Arabidopsis* early chloroplast biogenesis 1) protein was identified as a regulator of PEP-dependent expression (Yua *et al.*, 2014). Interestingly, AtECB1 has an SxxC motif in a predicted Trx-like fold; *in vitro* experiments suggested that the protein has disulfide reductase activity. A putative interaction with Trx z led the authors to propose a role for Trx in signalling (Yua *et al.*, 2014). However, further experiments are needed for assigning this function.

There is also evidence for a role for Trx in translation. Several nuclear-encoded factors have been found to act as redox-sensor components that modulate chloroplast translation via FTS (Barnes and Mayfield, 2003; Steiner *et al.*, 2009). Light-dependent reduction of regulatory disulfides in mRNA-binding proteins, such as RB60, RB47 and TBA1, has been reported to activate the translation of *psbA* and *psbD* mRNAs in plastids in *Chlamydomonas* (Danon and Mayfield, 1994; Fong *et al.*, 2000; Kim and Mayfield, 1997; Somanchi *et al.*, 2005; Trebitsh *et al.*, 2000). It has been proposed that NTRC also enables the regulation of plastid gene expression by redox transitions during the night (Schwarz *et al.*, 2012).

A role for the FTS at yet another level is emerging – namely, the biogenesis of the photosynthetic apparatus. It has been shown that Trx plays a role in assisting the proper assembly of CP47 into the PSII core complex by modulating the redox state of intermolecular disulfide bridges in PSII subunits. The mechanism of activation is largely unknown and future experiments are needed to identify the thiol acceptors in PSII (Wang *et al.*, 2013). Further, there is evidence that Trx functions in regulating state transition – that is, the reversible phosphorylation of proteins of light harvesting complex II (LHCII) that enables LHCII to migrate from PSII to photosystem I (PSI). It was found that LHCII kinase, STN7, is inactivated on reduction of a regulatory disulfide bridge by high light in a process mediated by FTS (Rintamäki *et al.*, 2000; Wunder *et al.*, 2013). The mechanism of regulation likely involves a redox-dependent conformational change that interferes with ATP binding by the kinase.

During the last decade, redox-active proteins have also been detected in the thylakoid lumen, opening an area previously unexplored (Buchanan and Luan, 2005). Unlike their stromal counterparts, the luminal redox-active proteins examined so far are active in the oxidized state and less active when reduced – a change compatible with a role for light in regulation. Thus, while enzymes of the stroma recognize light by sensing reduced Fdx, those of the lumen appear to respond to photochemically generated oxygen. So far, the immunophilin FKBP13, the oxygen-evolving complex extrinsic subunit OE33 (or OEE1/PsbO), violaxanthin de-epoxidase (VDE), and plastidic type I signal peptidase have been identified as candidate Trx (or Trx-like) targets of the lumen (Gopalan *et al.*, 2004; Hall *et al.*, 2010; Midorikawa *et al.*, 2014; Roberts *et al.*, 2012). Whereas the lumen does not appear to contain free Trx (or Grx), recent work has revealed the presence of a thiol–disulfide exchange system in the compartment (Karamoko *et al.*, 2013; Motohashi and Hisabori, 2006). HCF164, a thylakoid protein containing a soluble Trx-like domain with redox-active Cys exposed to the lumen, has been proposed to act as transmembrane transducer of reducing equivalents from stromal Trx m via the polytopic membrane protein Ccda (cytochrome c defective A) (Motohashi and Hisabori, 2006) (Fig. 9.5A). Luminal targets of HCF164 are also surfacing, for example PsaN, a PSI protein facing the lumen. Further, a protein containing a vitamin K epoxide oxidoreductase (VKOR) domain fused to a lumen-exposed Trx-like domain containing redox-active Cys (lumen thiol oxidoreductase 1 or LTO1) has been localized in thylakoids (Fig. 9.5A) (Feng *et al.*, 2011; Furt *et al.*, 2010; Karamoko *et al.*, 2011). The latter is proposed to introduce a disulfide bond into PsbO in the biogenesis of PSII.

Light/dark versus oxidative regulation

The chloroplast thioredoxin system is not only associated with diurnal regulation, it also functions in an oxidative type of regulation displayed by chloroplast proteins that respond to ROS (Balmer *et al.*, 2003; Buchanan and Balmer, 2005). A number of Trx-linked chloroplast enzymes and other proteins are sensitive to low ROS concentrations. For example, H_2O_2 is recognized as a potent inhibitor of photosynthesis, because even at low concentrations CO_2 fixation is inhibited by more than 50% due to the oxidation of thiol-modulated enzymes of the CB cycle (Muthuramalingam *et al.*, 2013). It has been found that Trx may reactivate oxidized targets after an oxidative event as found recently in a methanogenic archaeon (Susanti *et al.*, 2014). Rubisco has been detected as a primary oxidation target in chloroplasts (Moreno *et al.*, 2008; Muthuramalingam *et al.*, 2013). The enzyme was identified as a recurrent target in Trx-affinity chromatography approaches, although it lacks regulatory Cys. An oxidative type of regulation is envisaged (Balsera *et al.*, 2014). Current evidence suggests that the activity of several subunits of the protein import apparatus in chloroplasts is linked to redox by way of Trx in response to oxidative events (Balsera *et al.*, 2009, 2010).

Specificity for thioredoxin

Based on their mode of interaction, Trx target proteins can be categorized into two groups (Hara and Hisabori, 2013): (i) regulatory proteins, whose activity is modulated by the reversible oxidation of Cys in response to perturbations of the cellular redox state; (ii) catalytic targets, in which proteins undergo thiol/disulfide exchange/disulfide bond formation during a catalytic cycle, such as antioxidant enzymes Prx and Met sulfoxide reductase (MSR). Because of the diversity of types of Trx in chloroplasts, the specific role of each isoform has been under investigation. Except for Trx f, evolutionary studies indicate that chloroplasts inherited their suite of canonical Trxs from cyanobacteria, the putative ancestors of plant chloroplasts (Balsera *et al.*, 2014; Hartman *et al.*, 1990; Lemaire *et al.*, 2007; Sahrawy *et al.*, 1996). The Trx f gene was likely acquired from the eukaryotic host after the endosymbiotic event. The reductant for canonical plastid Trxs seems to be FTR, although the reduction of Trx z remains controversial (Bohrer *et al.*, 2012; Schürmann and Buchanan, 2008). Whereas some experiments link Trx z directly to the reduction of FTR, others suggest that different plastid Trxs are able to reduce Trx z, at least *in vitro* (Bohrer *et al.*, 2012; Chibani *et al.*, 2011).

The high similarity in their protein sequences (Fig. 9.5B) suggests some degree of functional redundancy among the different chloroplast Trx isoforms. Over the years, the specific role of each Trx has been addressed. The conclusions from a number of studies indicate that Trxs f and m are involved mainly in the light-dependent regulation of chloroplast enzymes, whereas Trxs x and y serve as a reductant for enzymes performing antioxidant functions. Trx z participates in the redox regulation of plastid gene expression (Arsova *et al.*, 2010; Schröter *et al.*, 2010) while NTRC plays a key role in antioxidant metabolism as well as in chloroplast processes still being explored. The mechanism by which Trxs recognize individual target proteins is still largely unknown. A combination of specific structural recognition motifs such as surface charge plus subcellular localization and expression patterns seem to be key in this regard (Collin *et al.*, 2003; Michelet *et al.*, 2005; Schürmann and Buchanan, 2008; Yoshida *et al.*, 2015).

Several reports suggest that Trxs f and m have distinct non-overlapping functions in plants. Likely, Trx f became more specific for a regulatory role as chloroplasts developed after the endosymbiotic event – i.e. when the regulation of FBPase, SBPase, Rubisco activase and AGPase was linked to light (Balsera *et al.*, 2013; Schürmann and Buchanan, 2008). Whereas most plants contain a single *Trx f* gene, some species such as *Arabidopsis* have two Trx f isoforms, Trx f1 and f2, likely as the result of a recent gene duplication event (Meyer *et al.*, 2005). Trx f seems to be most effective in regulating enzymes of CO_2 fixation via the CB cycle in chloroplasts, a function that is more restricted in cyanobacteria. Due to its eukaryotic origin, cyanobacteria lack Trx f, dictating that Trx m and possibly other Trxs fills a major regulatory role (Balsera *et al.*, 2014). Loss of function of *Trx f1* in *Arabidopsis* was found to impair light activation of ADP-glucose pyrophosphorylase (Thormählen *et al.*, 2013), while overexpression of the enzyme in tobacco chloroplasts resulted in excessive starch accumulation (Sanz-Barrio *et al.*, 2013). Surprisingly, the overexpression of Trx f in tobacco had no significant effect on photosynthesis or growth (Rey *et al.*, 2013). Similar findings have been reported in *Arabidopsis* and pea (Luo *et al.*, 2012; Thormählen *et al.*, 2013).

By contrast, changes in photosynthetic activity and oxidative stress tolerance have been detected in tobacco plants overexpressing Trx m (Rey *et al.*, 2013). These plants, however, failed to show significant change in starch content. Recent evidence suggests that Trx m acts in a broader range of physiological functions than Trx f (Rey *et al.*, 2013). Trx m is the only type common to all oxyphotosynthetic organisms. During evolution, the number of Trx m isoforms was expanded in plants: *Arabidopsis* contains four isoforms of Trx m (Trx m1, m2, m3 and m4) and tomato two (Trx m1 and m2/m4). Trx m3 – not common to all plants – has unique properties (Benitez-Alfonso and Jackson, 2009; Collin *et al.*, 2003). In *Arabidopsis*, the predominant expression of Trx m3 in non-green plastids has been related to a putative role in the redox regulation of symplastic protein trafficking and meristem development (Benitez-Alfonso and Jackson, 2009). Structurally similar, Trx m1, m2 and m4 have been found to be peripherally associated with the thylakoid membrane at the stromal side in chloroplasts (Friso *et al.*, 2004; Peltier *et al.*, 2002). No obvious phenotype was observed in *Arabidopsis* plants lacking Trx m1 or m4, and functional compensation for the three isoforms has been suggested (Collin *et al.*, 2003; Laugier *et al.*, 2013; Wang *et al.*, 2013). There is, however, evidence that Trx m4 functions in cyclic electron transport (Courteille *et al.*, 2013) and that the knock-down of Trx m in rice resulted in abnormal chloroplast development and impaired growth (Chi *et al.*, 2008). It has been argued that this discrepancy might originate from the fact that Trx m isoforms fulfil redundant functions in *Arabidopsis*, but not in rice.

Trxs x and y mainly serve as electron donors to catalytic targets. In particular, Trx x is an effective reductant for 2 Cys-Prx (Collin *et al.*, 2003), and Trx y reduces Q-type Prxs (Prx-Q) and type II Prxs (Prx II) (Collin *et al.*, 2004; Navrot *et al.*, 2006). Prxs are thiol-based peroxidases that detoxify hydrogen and organic peroxides (Dietz, 2011; Lamkemeyer *et al.*, 2006). Their catalytic mechanism involves the oxidation of a Cys_{cat} (or peroxidatic Cys) by hydroperoxides to sulfenic acid. Depending on the type of Prx, the enzyme is regenerated to its reduced state by the Trx or GSH/Grx systems (Zaffagnini *et al.*, 2008). For the 2-Cys Prx class, sulfenic acid is reduced via the formation of an intermolecular disulfide bridge using a resolving Cys found on another subunit, whereas in Prx-Q the resolving Cys is located on the same subunit and forms an intramolecular disulfide bridge

(Rouhier and Jacquot, 2002). Trx-y also functions as electron donor to plastid MSR with two redox active Cys, like MSRB2 and MSRA4 (Laugier *et al.*, 2013).

Trx z has unique structural properties (Bohrer *et al.*, 2012; Chibani *et al.*, 2011). Although Trx z has been implicated in the redox regulation of PEP-dependent gene expression (Arsova *et al.*, 2010), recent studies suggest that the redox activity of thioredoxin z and FLN1 are dispensable for the autotrophic growth of *Arabidopsis* (Wimmelbacher and Börnke, 2014). It has been proposed that the redox activity might only play a role in the fine tuning of PEP function (Wimmelbacher and Börnke, 2014).

Certain plastid Trxs have more specialized functions such as CDSP32, an atypical Trx involved in the oxidative stress response during environmental constraints (Broin *et al.*, 2002; Rey *et al.*, 2005). CDSP32 is unusual in being able to directly reduce sulfenic acid, a property not shared with any other chloroplast Trx. Proteins targeted by CDSP32 include two Prxs (Q and type II), and 1-Cys MSR type B. Trx like2.1, which shows Grx-like activity serves as a reductant for plastid thiol peroxidases and MSRs, employing a single cysteine residue for catalysis in an unusual mechanism involving glutathionylation of Cys_{cat} and subsequent regeneration of the sulfhydryl bond by the NADPH/GR/GSH reducing system (Chibani *et al.*, 2012). Lilium Trx is mainly associated with thylakoids, where it seems to function as an oxidative sensor (Dangoor *et al.*, 2009, 2012). Together with 2-Cys Prx, lilium Trx regulates photosynthetic electron transport by oxidative feedback in response to illumination of dark-adapted plants at low to moderate light intensity.

A unique type of Trx is found in NTRC. Here, reducing equivalents from NADPH are transferred to NTR and then directly to a fused Trx that, in turn, reduces selected target enzymes (Bohrer *et al.*, 2012; Lee *et al.*, 2012; Perez-Ruiz *et al.*, 2006). NTRC turns out to be a highly efficient electron donor to 2Cys-Prx (Kirchsteiger *et al.*, 2009; Moon *et al.*, 2006). In addition to its role in oxidative stress, NTRC has been implicated in the redox control of chloroplast processes, including starch synthesis, chlorophyll metabolism, and aromatic amino acid biosynthesis (Cejudo *et al.*, 2012; Michalska *et al.*, 2009; Pérez-Ruiz *et al.*, 2014; Richter *et al.*, 2013). Loss of NTRC function in *Arabidopsis* resulted in a pale phenotype, impaired photosynthesis and increased sensitivity to oxidative stress, high temperature and prolonged darkness (Chae *et al.*, 2013; Lepistö *et al.*, 2009; Perez-Ruiz *et al.*, 2006). On the other hand, overexpression of wild-type NTRC promoted growth by increasing leaf size and biomass yield of the rosettes (Toivola *et al.*, 2013).

Regulation by glutathionylation

The reversible modification of proteinaceous Cys by GSH (glutathionylation) has emerged as a complementary mechanism of redox regulation in chloroplasts (Rouhier *et al.*, 2008; Zaffagnini *et al.*, 2012b). Glutathionylation reactions have been implicated in protection of proteins against Cys over-oxidation, regulation and redox signalling (Dalle-Donne *et al.*, 2009; Zaffagnini *et al.*, 2012c).

GSH is the predominant non-protein thiol in chloroplasts (Noctor *et al.*, 2012). Either prompted by ROS or as a mandatory step of the catalytic cycle, GSH can react with susceptible Cys in proteins, resulting in the formation of a mixed disulfide bridge. Subsequent deglutathionylation involves the participation of Grxs (Fig. 9.4B and C). One of the targets modulated by glutathionylation is Trx f (Michelet *et al.*, 2005). This feature is made possible by the presence of a surface-accessible 'third' Cys, outside the CxxC active site. It was found that glutathionylated Trx f is not efficiently reduced by FTR, and, consequently, its capacity to activate CB enzymes by light was decreased (Michelet *et al.*, 2005).

Other glutathionylated chloroplast enzymes targeted by Grx include PRK and GAPDH isoform A. It has been proposed that glutathionylation accompanying the oxidation of PRK is a major mode for regulating the enzyme under oxidative stress (Rouhier *et al.*, 2005; Zaffagnini *et al.*, 2012a). GAPDH isoform A, a Trx-independent isoform of GAPDH, was found to be inhibited by glutathionylation (Zaffagnini *et al.*, 2007, 2008). Several studies support the possibility that the activity of CB cycle enzymes decreases due to glutathionylation under oxidative stress, thus constituting a mechanism for down-regulating carbohydrate metabolism in plants. Recently, the putative regulation of chloroplast fatty acid synthesis has been linked to glutathionylation. Proteomic studies have provided evidence for glutathionylation of biotin carboxyl carrier and biotin carboxylase – the Trx-independent subunits of ACCase (Dixon *et al.*, 2005; Michelet *et al.*, 2008). In short, glutathionylation reactions seem to play an active role in the redox regulatory network of chloroplasts by modulating of the activity of signalling proteins, protecting metabolic enzymes and regenerating the activity of stress-specific enzymes, particularly during an oxidative event.

Conclusions

Plants have evolved regulatory mechanisms to coordinate chloroplast activities with the diurnal cycle. One way this mode of regulation is achieved is with redox-active Cys that enable the modification of selected enzymes in direct response to light and dark via the FTS. The reversible oxidation of Cys is used to stabilize proteins in their active (or inactive) conformation, to modify the catalytic properties of enzymes or promote protein–protein associations. The Trx regulation of chloroplast enzymes is further refined by fine-tuning mechanisms through intermediary metabolites or ATP/ADP and NADPH/NADP ratios, enabling cellular reactions to proceed efficiently in a highly controlled fashion. Moreover, because plants have to cope with environmental fluctuations that can compromise metabolism as well as growth and development, redox-based regulation was extended to dynamically change proteins in response to oxidative stress. ROS-mediated Cys oxidation activates antioxidant enzymes, such as Prxs, or decreases the activity of enzymes linked to carbon fixation or ATP utilization. The capability to coordinate biochemical processes with the diurnal cycle and in response to oxidative stress was likely critical to the evolutionary development of oxygenic photosynthesis.

Future trends

Our understanding of the redox signalling pathways functional in chloroplasts has greatly increased in past few years, thereby opening new opportunities for manipulating photosynthesis, particularly under variable environmental conditions. However, the function of the Trx system has been studied mainly in *in vitro* experiments, most recently in *Arabidopsis.* More in-depth *in vivo* studies should be undertaken and include plants of agronomic interest. A comparative study of the proteins showing oxidative cysteine modification, including specificity of the type and sites of modification remains to be undertaken. It would be of interest to elucidate, at the structural level, the redox-based modifications relevant for protective and signalling purposes under normal conditions and under different types of stress, including oxidative stress. This would include describing the extent of thiol oxidation and how the different oxidative modifications are coordinated in response to various stimuli. Finally, recent evidence that manipulation of the expression level of selected components of the Trx system in chloroplasts enhances biomass yield should be extended with obvious implications for agriculture and bioenergy production. In particular, understanding the mechanisms by which modified plants accumulate increased levels of starch deserves further investigation.

References

Apel, K. (2004). Reactive oxygen species: metabolism, oxidative stress, and signal transduction. Annu. Rev. Plant Biol. *55*, 373–399.

Arsova, B., Hoja, U., Wimmelbacher, M., Greiner, E., Üstün, Ş., Melzer, M., Petersen, K., Lein, W., and Börnke, F. (2010). Plastidial thioredoxin z interacts with two fructokinase-like proteins in a thiol-dependent manner: Evidence for an essential role in chloroplast development in *Arabidopsis* and *Nicotiana benthamiana*. Plant Cell *22*, 1498–1515.

Astier, J., Rasul, S., Koen, E., Manzoor, H., Besson-Bard, A., Lamotte, O., Jeandroz, S., Durner, J., Lindermayr, C., and Wendehenne, D. (2011). S-nitrosylation: An emerging post-translational protein modification in plants. Plant Sci. *181*, 527–533.

Ballicora, M.A., Frueauf, J.B., Fu, Y., Schürmann, P., and Preiss, J. (2000). Activation of the potato tuber ADP-glucose pyrophosphorylase by thioredoxin. J. Biol. Chem. *275*, 1315–1320.

Balmer, Y., Koller, A., del Val, G., Manieri, W., Schürmann, P., and Buchanan, B.B. (2003). Proteomics gives insight into the regulatory function of chloroplast thioredoxins. Proc. Natl. Acad. Sci. U.S.A. *100*, 370–375.

Balsera, M., Soll, J., and Buchanan, B. (2009). Protein Import in chloroplasts. Adv. Bot. Res. *52*, 277–332.

Balsera, M., Soll, J., and Buchanan, B. (2010). Redox extends its regulatory reach to chloroplast protein import. Trends Plant Sci. *15*, 515–521.

Balsera, M., Uberegui, E., Susanti, D., Schmitz, R., Mukhopadhyay, B., Schürmann, P., and Buchanan, B. (2013). Ferredoxin:thioredoxin reductase (FTR) links the regulation of oxygenic photosynthesis to deeply rooted bacteria. Planta *237*, 619–635.

Balsera, M., Uberegui, E., Schürmann, P., and Buchanan, B.B. (2014). Evolutionary development of redox regulation in chloroplasts Antioxid Redox Signal *21*, 1327–1355.

Bandyopadhyay, S., Gama, F., Molina-Navarro, M.M., Gualberto, J.M., Claxton, R., Naik, S.G., Huynh, B.H., Herrero, E., Jacquot, J.P., Johnson, M.K., *et al.* (2008). Chloroplast monothiol glutaredoxins as scaffold proteins for the assembly and delivery of [2Fe–2S] clusters. EMBO J. *27*, 1122–1133.

Barnes, D., and Mayfield, S. (2003). Redox control of posttranscriptional processes in the chloroplast. Antioxid. Redox Signal 5, 89–94.

Benitez-Alfonso, Y., and Jackson, D. (2009). Redox homeostasis regulates plasmodesmal communication in *Arabidopsis* meristems. Plant Signal Behav. *4*, 655–659.

Bohrer, A.-S., Massot, V., Innocenti, G., Reichheld, J.-P., Issakidis-Bourguet, E., and Vanacker, H. (2012). New insights into the reduction systems of plastidial thioredoxins point out the unique properties of thioredoxin z from *Arabidopsis*. J. Exp. Bot. *63*, 6315–6323.

Broin, M., Cuiné, S., Eymery, F., and Rey, P. (2002). The plastidic 2-cysteine peroxiredoxin is a target for a thioredoxin involved in the protection of the photosynthetic apparatus against oxidative damage. Plant Cell *14*, 1417–1432.

Buchanan, B.B., Kalberer, P.P., and Arnon, D.I. (1967). Ferredoxin-activated fructose diphosphatase in isolated chloroplasts. Biochem. Biophys. Res. Commun. *29*, 74–79.

Buchanan, B.B., Schürmann, P., and Kalberer, P.P. (1971). Ferredoxin-activated fructose diphosphatase of spinach chloroplasts: Resolution of the system, properties of the alkaline fructose diphosphatase component, and physiological significance of the ferredoxin-linked activation. J. Biol. Chem. *246*, 5952–5959.

Buchanan, B.B. (1980). Role of light in the regulation of chloroplast enzymes. Annu. Rev. Plant Physiol. *31*, 341–374.

Buchanan, B.B., and Balmer, Y. (2005). Redox regulation: A broadening horizon. Annu. Rev. Plant Biol. *56*, 187–220.

Buchanan, B.B., and Luan, S. (2005). Redox regulation in the chloroplast thylakoid lumen: a new frontier in Photosynth. Res. J. Exp. Bot. *56*, 1439–1447.

Buchanan, B.B., Holmgren, A., Jacquot, J.P., and Scheibe, R. (2012). Fifty years in the thioredoxin field and a bountiful harvest. Biochim. Biophys. Acta *1820*, 1822–1829.

Cain, P., Hall, M., Schroder, W., Kieselbach, T., Robinson, C., and Schröder, W. (2009). A novel extended family of stromal thioredoxins. Plant Mol. Biol. *70*, 273–281.

Carvalho, A.P., Fernandes, P.A., and Ramos, M.J. (2006). Similarities and differences in the thioredoxin superfamily. Prog. Biophys. Mol. Biol. *91*, 229–248.

Cejudo, F.J., Ferrandez, J., Cano, B., Puerto-Galan, L., and Guinea, M. (2012). The function of the NADPH thioredoxin reductase C-2-Cys peroxiredoxin system in plastid redox regulation and signalling. FEBS Lett. *586*, 2974–2980.

Chae, H., Moon, J., Shin, M., Chi, Y., Jung, Y., Nawkar, G., Jung, H., Hyun, J., Kim, W., Kang, C., *et al.* (2013). Thioredoxin Reductase Type C (NTRC) orchestrates enhanced thermotolerance to *Arabidopsis* by its redox-dependent holdase chaperone function. Mol. Plant *6*, 323–336.

Chi, Y.H., Moon, J.C., Park, J.H., Kim, H.-S., Zulfugarov, I.S., Fanata, W.I., Jang, H.H., Lee, J.R., Lee, Y.M., Kim, S.T., *et al.* (2008). Abnormal chloroplast development and growth inhibition in rice thioredoxin m knock-down plants. Plant Physiol. *148*, 808–817.

Chiadmi, M., Navaza, A., Miginiac-Maslow, M., Jacquot, J.-P., and Cherfils, J. (1999). Redox signalling in the chloroplast: structure of oxidized pea fructose-1,6-bisphosphate phosphatase. EMBO J. *18*, 6809–6815.

Chibani, K., Wingsle, G., Jacquot, J.-P., Gelhaye, E., and Rouhier, N. (2009). Comparative genomic study of the thioredoxin family in photosynthetic organisms with emphasis on *Populus trichocarpa*. Mol. Plant 2, 308–322.

Chibani, K., Tarrago, L., Schürmann, P., Jacquot, J.P., and Rouhier, N. (2011). Biochemical properties of poplar thioredoxin z. FEBS Lett. *585*, 1077–1081.

Chibani, K., Tarrago, L., Gualberto, J.M., Wingsle, G., Rey, P., Jacquot, J.-P., and Rouhier, N. (2012). Atypical thioredoxins in poplar: the glutathione-dependent thioredoxin-like 2.1 supports the activity of target enzymes possessing a single redox active cysteine. Plant. Physiol. *159*, 592–605.

Collin, V., Issakidis-Bourguet, E., Marchand, C., Hirasawa, M., Lancelin, J.-M., Knaff, D.B., and Miginiac-Maslow, M. (2003). The *Arabidopsis* plastidial thioredoxins: new functions and new insights into specificity. J. Biol. Chem. *278*, 23747–23752.

Collin, V., Lamkemeyer, P., Miginiac-Maslow, M., Hirasawa, M., Knaff, D.B., Dietz, K.J., and Issakidis-Bourguet, E. (2004). Characterization of plastidial thioredoxins from *Arabidopsis* belonging to the new y-type. Plant. Physiol. *136*, 4088–4095.

Courteille, A., Vesa, S., Sanz-Barrio, R., Cazale, A.-C., Becuwe-Linka, N., Farran, I., Havaux, M., Rey, P., and Rumeau, D. (2013). Thioredoxin m4 controls photosynthetic alternative electron pathways in *Arabidopsis*. Plant. Physiol. *161*, 508–520.

Couturier, J., Jacquot, J.-P., and Rouhier, N. (2009a). Evolution and diversity of glutaredoxins in photosynthetic organisms. Cell Mol Life Sci. 66, 2539–2557.

Couturier, J., Koh, C.S., Zaffagnini, M., Winger, A.M., Gualberto, J.M., Corbier, C., Decottignies, P., Jacquot, J.-P., Lemaire, S.D., Didierjean, C., *et al.* (2009b). Structure–function relationship of the chloroplastic glutaredoxin S12 with an atypical WCSYS active site. J. Biol. Chem. *284*, 9299–9310.

Couturier, J., Stroeher, E., Albetel, A.-N., Roret, T., Muthuramalingam, M., Tarrago, L., Seidel, T., Tsan, P., Jacquot, J.-P., Johnson, M.K., *et al.* (2011). *Arabidopsis* chloroplastic glutaredoxin C5 as a model to explore molecular determinants for iron–sulfur cluster binding into glutaredoxins. J. Biol. Chem. *286*, 27515–27527.

Couturier, J., Chibani, K., Jacquot, J.-P., and Rouhier, N. (2013). Cysteine-based redox regulation and signalling in plants. Front. Plant. Sci. *4*, 259.

Dai, S., Friemann, R., Glauser, D.A., Bourquin, F., Manieri, W., Schürmann, P., and Eklund, H. (2007). Structural snapshots along the reaction pathway of ferredoxin-thioredoxin reductase. Nature *448*, 92–96.

Dalle-Donne, I., Rossi, R., Colombo, G., Giustarini, D., and Milzani, A. (2009). Protein S-glutathionylation:

a regulatory device from bacteria to humans. Trends Biochem. Sci. *34*, 85–96.

Dangoor, I., Peled-Zehavi, H., Levitan, A., Pasand, O., and Danon, A. (2009). A small family of chloroplast atypical thioredoxins. Plant. Physiol. *149*, 1240–1250.

Dangoor, I., Peled-Zehavi, H., Wittenberg, G., and Danon, A. (2012). A chloroplast light-regulated oxidative sensor for moderate light intensity in *Arabidopsis*. Plant Cell *24*, 1894–1906.

Danon, A., and Mayfield, S. (1994). Light-regulated translation of chloroplast messenger RNAs through redox potential. Science *266*, 1717–1719.

Dietz, K.-J. (2011). Peroxiredoxins in plants and cyanobacteria. Antioxid. Redox Signal. *15*, 1129–1159.

Dixon, D.P., Skipsey, M., Grundy, N.M., and Edwards, R. (2005). Stress-induced protein S-glutathionylation in *Arabidopsis*. Plant. Physiol. *138*, 2233–2244.

Dunford, R., Durrant, M., Catley, M., and Dyer, T. (1998). Location of the redox-active cysteines in chloroplast sedoheptulose-1,7-bisphosphatase indicates that its allosteric regulation is similar but not identical to that of fructose-1,6-bisphosphatase. Photosynth. Res. *58*, 221–230.

Feng, W.-K., Wang, L., Lu, Y., and Wang, X.-Y. (2011). A protein oxidase catalysing disulfide bond formation is localized to the chloroplast thylakoids. FEBS J. *278*, 3419–3430.

Fermani, S., Sparla, F., Falini, G., Martelli, P.L., Casadio, R., Pupillo, P., Ripamonti, A., and Trost, P. (2007). Molecular mechanism of thioredoxin regulation in photosynthetic A2B2-glyceraldehyde-3-phosphate dehydrogenase. Proc. Natl. Acad. Sci. U.S.A. *104*, 11109–11114.

Fermani, S., Trivelli, X., Sparla, F., Thumiger, A., Calvaresi, M., Marri, L., Falini, G., Zerbetto, F., and Trost, P. (2012). Conformational selection and folding-upon-binding of intrinsically disordered protein CP12 regulate photosynthetic enzymes assembly. J. Biol. Chem. *287*, 21372–21383.

Fomenko, D., Marino, S., and Gladyshev, V. (2008). Functional diversity of cysteine residues in proteins and unique features of catalytic redox-active cysteines in thiol oxidoreductases. Mol. Cells *26*, 228–235.

Fong, C.L., Lentz, A., and Mayfield, S.P. (2000). Disulfide bond formation between RNA binding domains is used to regulate mRNA binding activity of the chloroplast poly(A)-binding protein. J. Biol. Chem. *275*, 8275–8278.

Friso, G., Giacomelli, L., Ytterberg, A.J., Peltier, J.-B., Rudella, A., Sun, Q., and Wijk, K.J.v. (2004). In-depth analysis of the thylakoid membrane proteome of *Arabidopsis thaliana* chloroplasts: New Proteins, new functions, and a plastid proteome database. Plant Cell *16*, 478–499.

Furt, F., Oostende, C.v., Widhalm, J.R., Dale, M.A., Wertz, J., and Basset, G.J.C. (2010). A bimodular oxidoreductase mediates the specific reduction of phylloquinone (vitamin K1) in chloroplasts. Plant J,=. *64*, 38–46.

Galvez-Valdivieso, G., and Mullineaux, P.M. (2010). The role of reactive oxygen species in signalling from chloroplasts to the nucleus. Physiol. Plant 138, 430–439.

Gelhaye, E., Rouhier, N., Navrot, N., and Jacquot, J.P. (2005). The plant thioredoxin system. Cell. Mol. Life Sci. 62, 24–35.

Glaring, M.A., Skryhan, K., Kötting, O., Zeeman, S.C., and Blennow, A. (2012). Comprehensive survey of redox sensitive starch metabolising enzymes in *Arabidopsis thaliana*. Plant. Physiol. Biochem. *58*, 89–97.

Gopalan, G., He, Z., Balmer, Y., Romano, P., Gupta, R., Héroux, A., Buchanan, B.B., Swaminathan, K., and Luan, S. (2004). Structural analysis uncovers a role for redox in regulating FKBP13, an immunophilin of the chloroplast thylakoid lumen. Proc. Natl. Acad. Sci. U.S.A. *101*, 13945–13950.

Graciet, E., Gans, P., Wedel, N., Lebreton, S., Camadro, J.-M., and Gontero, B. (2003). The small protein CP12: A protein linker for supramolecular complex assembly. Biochemistry *42*, 8163–8170.

Hall, A., Karplus, P.A., and Poole, L.B. (2009). Typical 2-Cys peroxiredoxins – structures, mechanisms and functions. FEBS J. *276*, 2469–2477.

Hall, M., Mata-Cabana, A., Akerlund, H.E., Florencio, F.J., Schroder, W.P., Lindahl, M., and Kieselbach, T. (2010). Thioredoxin targets of the plant chloroplast lumen and their implications for plastid function. Proteomics *10*, 987–1001.

Hanke, G.U.Y., and Mulo, P. (2013). Plant type ferredoxins and ferredoxin-dependent metabolism. Plant Cell Environ. *36*, 1071–1084.

Hara, S., and Hisabori, T. (2013). Kinetic analysis of the interactions between plant thioredoxin and target proteins. Front Plant Sci. *4*, 508.

Hartman, H., Syvanen, M., and Buchanan, B.B. (1990). Contrasting evolutionary histories of chloroplast thioredoxins f and m. Mol. Biol. Evol. *7*, 247–254.

Hendriks, J.H.M., Kolbe, A., Gibon, Y., Stitt, M., and Geigenberger, P. (2003). ADP-glucose pyrophosphorylase is activated by posttranslational redox-modification in response to light and to sugars in leaves of *Arabidopsis* and other plant species. Plant. Physiol. *133*, 838–849.

Hisabori, T., Sunamura, E.-I., Kim, Y., and Konno, H. (2013). The chloroplast ATP synthase features the characteristic redox regulation machinery. Antioxid. Redox Signal *19*, 1846–1854.

Howard, T.P., Metodiev, M., Lloyd, J.C., and Raines, C.A. (2008). Thioredoxin-mediated reversible dissociation of a stromal multiprotein complex in response to changes in light availability. Proc. Natl. Acad. Sci. U.S.A. *105*, 4056–4061.

Hwang, S.-K., Salamone, P., and Okita, T. (2005). Allosteric regulation of the higher plant ADP-glucose pyrophosphorylase is a product of synergy between the two subunits. FEBS Lett. *579*, 983–990.

Iglesias-Baena, I., Barranco-Medina, S., Lázaro-Payo, A., López-Jaramillo, F.J., Sevilla, F., and Lázaro, J.-J. (2010). Characterization of plant sulfiredoxin and role of sulphinic form of 2-Cys peroxiredoxin. J. Exp. Bot. *61*, 1509–1521.

Ikegami, A., Yoshimura, N., Motohashi, K., Takahashi, S., Romano, P.G.N., Yoshimura, N., Hisabori, T., Takamiya, K.-I., and Masuda, T. (2007). The CHLI1 subunit of *Arabidopsis thaliana* magnesium chelatase is a target protein of the chloroplast thioredoxin. J. Biol. Chem. *282*, 19282–19291.

Jacquot, J.-P., Eklund, H., Rouhier, N., and Schürmann, P. (2009). Structural and evolutionary aspects of thioredoxin reductases in photosynthetic organisms. Trends Plant Sci. *14*, 336–343.

Johansson, K., Ramaswamy, S., Saarinen, M., Lemaire-Chamley, M., Issakidis-Bourguet, E., Miginiac-Maslow, M., and Eklund, H. (1999). Structural basis for light activation of a chloroplast enzyme: The structure of sorghum NADP-malate dehydrogenase in its oxidized form. Biochemistry 38, 4319–4326.

Karamoko, M., Cline, S., Redding, K., Ruiz, N., and Hamel, P.P. (2011). Lumen Thiol Oxidoreductase1, a disulfide bond-forming catalyst, is required for the assembly of Photosystem II in *Arabidopsis*. Plant Cell 23, 4462–4475.

Karamoko, M., Gabilly, S., and Hamel, P.P. (2013). Operation of trans-thylakoid thiol-metabolizing pathways in photosynthesis. Front. Plant Sci. *4*, 476.

Kim, H.-J., Ha, S., Lee, H.Y., and Lee, K.-J. (2015). ROSics: Chemistry and proteomics of cysteine modifications in redox biology. Mass Spectrom. Rev. *34*, 184–208.

Kim, J., and Mayfield, S.P. (1997). Protein disulfide isomerase as a regulator of chloroplast translational activation. Science *278*, 1954–1957.

Kirchsteiger, K., Pulido, P., González, M., and Cejudo, F.J. (2009). NADPH thioredoxin reductase C controls the redox status of chloroplast 2-Cys peroxiredoxins in *Arabidopsis thaliana*. Mol. Plant 2, 298–307.

Klomsiri, C., Karplus, P.A., and Poole, L. (2011). Cysteine-based redox switches in enzymes. Antioxid. Redox Signal 14, 1065–1077.

Konno, H., Nakane, T., Yoshida, M., Ueoka-Nakanishi, H., Hara, S., and Hisabori, T. (2012). Thiol modulation of the chloroplast ATP synthase is dependent on the energization of thylakoid membranes. Plant Cell Physiol. *53*, 626–634.

Kozaki, A., and Sasaki, Y. (1999). Light-dependent changes in redox status of the plastidic acetyl-CoA carboxylase and its regulatory component. Biochem. J. *339*, 541–546.

Kozaki, A.K., Mayumi, K., and Sasaki, Y. (2001). Thiol-disulfide exchange between nuclear-encoded and chloroplast-encoded subunits of pea acetyl-CoA carboxylase. J. Biol. Chem. *276*, 39919–39925.

Krasensky, J., Broyart, C., Rabanal, F., and Jonak, C. (2014). The redox-sensitive chloroplast trehalose-6-phosphate phosphatase AtTPPD regulates salt stress tolerance. Antioxid. Redox Signal *21*, 1289–1304.

Krimm, I., Goyer, A., Issakidis-Bourguet, E., Miginiac-Maslow, M., and Lancelin, J.-M. (1999). Direct NMR observation of the thioredoxin-mediated reduction of the chloroplast NADP-malate dehydrogenase provides a structural basis for the relief of autoinhibition. J. Biol. Chem. *274*, 34539–34542.

Kruger, N.J., and von Schaewen, A. (2003). The oxidative pentose phosphate pathway: structure and organisation. Curr. Opin. Plant Biol. 6, 236–246.

Lamkemeyer, P., Laxa, M., Collin, V., Li, W., Finkemeier, I., Schöttler, M.A., Holtkamp, V., Tognetti, V.B., Issakidis-Bourguet, E., Kandlbinder, A., *et al.* (2006). Peroxiredoxin Q of *Arabidopsis thaliana* is attached to the thylakoids and functions in context of photosynthesis. Plant J. 45, 968–981.

Laugier, E., Tarrago, L., Courteille, A., Innocenti, G., Eymery, F., Rumeau, D., Issakidis-Bourguet, E., and Rey, P. (2013). Involvement of thioredoxin y2 in the preservation of leaf methionine sulfoxide reductase capacity and growth under high light. Plant Cell Environ. *36*, 670–682.

Lee, Y., Kim, S., Lazar, P., Moon, J.C., Hwang, S., Thangapandian, S., Shon, Y., Lee, K.O., Lee, S.Y., and Lee, K.W. (2012). Comparative molecular modeling sudy of *Arabidopsis* NADPH-dependent thioredoxin reductase and its hybrid protein. PLoS One 7, e46279.

Lemaire, S.D., Quesada, A., Merchan, F., Corral, J.M., Igeno, M.I., Keryer, E., Issakidis-Bourguet, E., Hirasawa, M., Knaff, D.B., and Miginiac-Maslow, M. (2005). NADP-Malate dehydrogenase from unicellular green alga *Chlamydomonas reinhardtii*. A first step toward redox regulation? Plant. Physiol. *137*, 514–521.

Lemaire, S.D., Michelet, L., Zaffagnini, M., Massot, V., and Issakidis-Bourguet, E. (2007). Thioredoxins in chloroplasts. Curr. Genet. *51*, 343–365.

Lepistö, A., Kangasjärvi, S., Luomala, E.-M., Brader, G., Sipari, N., Keränen, M., Keinänen, M., and Rintamäki, E. (2009). Chloroplast NADPH-thioredoxin reductase interacts with photoperiodic development in *Arabidopsis*. Plant. Physiol. *149*, 1261–1276.

Lindahl, M., Mata-Cabana, A., and Kieselbach, T. (2011). The disulfide proteome and other reactive cysteine proteomes: analysis and functional significance. Antioxid Redox Signal. *14*, 2581–2641.

Lindermayr, C., Saalbach, G., and Durner, J. (2005). Proteomic identification of S-nitrosylated proteins in *Arabidopsis*. Plant. Physiol. *137*, 921–930.

Lindermayr, C. (2009). S-Nitrosylation in plants: pattern and function. J. Proteomics 73, 1–9.

Lunn, J.E. (2006). Compartmentation in plant metabolism. J. Exp. Bot. 58, 35–47.

Luo, T., Fan, T., Liu, Y., Rothbart, M., Yu, J., Zhou, S., Grimm, B., and Luo, M. (2012). Thioredoxin redox regulates ATPase activity of magnesium chelatase CHLI subunit and modulates redox-mediated signaling in tetrapyrrole biosynthesis and homeostasis of reactive oxygen species in pea plants. Plant. Physiol. *159*, 118–130.

Marino, S.M., and Gladyshev, V.N. (2012). Analysis and functional prediction of reactive cysteine residues. J. Biol. Chem. *287*, 4419–4425.

Marri, L., Zaffagnini, M., Collin, V., Issakidis-Bourguet, E., Lemaire, S.D., Pupillo, P., Sparla, F., Miginiac-Maslow, M., and Trost, P. (2009). Prompt and easy activation by specific thioredoxins of Calvin cycle enzymes of *Arabidopsis thaliana* associated in the GAPDH/CP12/PRK supramolecular complex. Mol. Plant 2, 259–269.

Marri, L., Thieulin-Pardo, G., Lebrun, R., Puppo, R., Zaffagnini, M., Trost, P., Gontero, B., and Sparla, F. (2014). CP12-mediated protection of Calvin–Benson cycle enzymes from oxidative stress. Biochimie 97, 228–237.

Matsumura, H., Kai, A., Maeda, T., Tamoi, M., Satoh, A., Tamura, H., Hirose, M., Ogawa, T., Kizu, N., Wadano, A., *et al.* (2011). Structure basis for the regulation of glyceraldehyde-3-phosphate dehydrogenase activity via the intrinsically disordered protein CP12. Structure *19*, 1846–1854.

Meyer, Y., Reichheld, J., and Vignols, F. (2005). Thioredoxins in *Arabidopsis* and other plants. Photosynth. Res. *86*, 419–433.

Meyer, Y., Buchanan, B.B., Vignols, F., and Reichheld, J.-P. (2009). Thioredoxins and glutaredoxins: unifying elements in redox biology. Annu. Rev. Genet. *43*, 335–367.

Meyer, Y., Belin, C., Delorme-Hinoux, V., Reichheld, J.-P., and Riondet, C. (2012). Thioredoxin and glutaredoxin systems in plants: Molecular mechanisms, crosstalks, and functional significance. Antioxid. Redox Signal. *17*, 1124–1160.

Michalska, J., Zauber, H., Buchanan, B.B., Cejudo, F.J., and Geigenberger, P. (2009). NTRC links built-in thioredoxin to light and sucrose in regulating starch synthesis in chloroplasts and amyloplasts. Proc. Natl. Acad. Sci. U.S.A. *106*, 9908–9913.

Michelet, L., Zaffagnini, M., Marchand, C., Collin, V., Decottignies, P., Tsan, P., Lancelin, J.-M., Trost, P., Miginiac-Maslow, M., Noctor, G., *et al.* (2005). Glutathionylation of chloroplast thioredoxin f is a redox signaling mechanism in plants. Proc. Natl. Acad. Sci. U.S.A. *102*, 16478–16483.

Michelet, L., Zaffagnini, M., Vanacker, H., Le Maréchal, P., Marchand, C., Schroda, M., Lemaire, S.D., and Decottignies, P. (2008). *In vivo* targets of S-thiolation in *Chlamydomonas reinhardtii*. J. Biol. Chem. *283*, 21571–21578.

Midorikawa, T., Endow, J.K., Dufour, J., Zhu, J., and Inoue, K. (2014). Plastidic type I signal peptidase 1 is a redox-dependent thylakoidal processing peptidase. Plant J. *80*, 592–603.

Miziorko, H.M. (2000). Phosphoribulokinase: current perspectives on the structure/function basis for regulation and catalysis. Adv Enzymol. Relat. Areas Mol. Biol. 74, 95–127.

Montrichard, F., Alkhalfioui, F., Yano, H., Vensel, W.H., Hurkman, W.J., and Buchanan, B.B. (2009). Thioredoxin targets in plants: The first 30 years. J. Proteomics 72, 452–474.

Moon, J.C., Jang, H.H., Chae, H.B., Lee, J.R., Lee, S.Y., Jung, Y.J., Shin, M.R., Lim, H.S., Chung, W.S., Yun, D.-J., *et al.* (2006). The C-type *Arabidopsis* thioredoxin reductase ANTR-C acts as an electron donor to 2-Cys peroxiredoxins in chloroplasts. Biochem. Biophys. Res. Commun. *348*, 478–484.

Moreno, J., García-Murria, M.J., and Marín-Navarro, J. (2008). Redox modulation of Rubisco conformation and activity through its cysteine residues. J. Exp. Bot. *59*, 1605–1614.

Motohashi, K., Kondoh, A., Stumpp, M.T., and Hisabori, T. (2001). Comprehensive survey of proteins targeted by chloroplast thioredoxin. Proc. Natl. Acad. Sci. U.S.A. *98*, 11224–11229.

Motohashi, K., and Hisabori, T. (2006). HCF164 receives reducing equivalents from stromal thioredoxin across the thylakoid membrane and mediates reduction of target proteins in the thylakoid lumen. J. Biol. Chem. *281*, 35039–35047.

Muthuramalingam, M., Matros, A., Scheibe, R., Mock, H.-P., and Dietz, K.-J. (2013). The hydrogen peroxide-sensitive proteome of the chloroplast *in vitro* and in vivo. Front Plant Sci. *4*, 54.

Navrot, N., Collin, V., Gualberto, J., Gelhaye, E., Hirasawa, M., Rey, P., Knaff, D.B., Issakidis, E., Jacquot, J.-P., and Rouhier, N. (2006). Plant glutathione peroxidases are functional peroxiredoxins distributed in several subcellular compartments and regulated during biotic and abiotic stresses. Plant. Physiol. *142*, 1364–1379.

Nee, G., Aumont-Nicaise, M., Zaffagnini, M., Nessler, S., Valerio-Lepiniec, M., and Issakidis-Bourguet, E. (2014). Redox regulation of chloroplastic G6PDH activity by thioredoxin occurs through structural changes modifying substrate accessibility and cofactor binding. Biochem. J. *457*, 117–125.

Nelson, N., and Ben-Shem, A. (2004). The complex architecture of oxygenic photosynthesis. Nat. Rev. Mol. Cell Biol. *5*, 971–982.

Nevo, R., Charuvi, D., Tsabari, O., and Reich, Z. (2012). Composition, architecture and dynamics of the photosynthetic apparatus in higher plants. Plant J. *70*, 157–176.

Noctor, G., Mhamdi, A., Chaouch, S., Han, Y.I., Neukermans, J., Marquez-Garcia, B., Queval, G., and Foyer, C.H. (2012). Glutathione in plants: an integrated overview. Plant Cell Environ. *35*, 454–484.

Ocheretina, O., Haferkamp, I., Tellioglu, H., and Scheibe, R. (2000). Light-modulated NADP-malate dehydrogenases from mossfern and green algae: insights into evolution of the enzyme's regulation. Gene *258*, 147–154.

Ohlrogge, J., and Browse, J. (1995). Lipid biosynthesis. Plant Cell *7*, 957–970.

Peltier, J.-B., Emanuelsson, O., Kalume, D.E., Ytterberg, J., Friso, G., Rudella, A., Liberles, D.A., Söderberg, L., Roepstorff, P., von Heijne, G., *et al.* (2002). Central functions of the luminal and peripheral thylakoid proteome of *Arabidopsis* determined by experimentation and genome-wide prediction. Plant Cell *14*, 211–236.

Perez-Ruiz, J.M., Spinola, M.C., Kirchsteiger, K., Moreno, J., Sahrawy, M., and Cejudo, F.J. (2006). Rice NTRC is a high-efficiency redox system for chloroplast protection against oxidative damage. Plant Cell *18*, 2356–2368.

Pérez-Ruiz, J.M., Guinea, M., Puerto Galan, L., Javier Cejudo, F., Pérez Ruiz, J., Puerto Galán, L., and Cejudo, F. (2014). NADPH thioredoxin reductase C is involved in redox regulation of the Mg-Chelatase I subunit in *Arabidopsis thaliana* chloroplasts. Mol. Plant *7*, 1252–1255.

Pohlmeyer, K., Paap, B.K., Soll, J., and Wedel, N. (1996). CP12: a small nuclear-encoded chloroplast protein provides novel insights into higher-plant GAPDH evolution. Plant Mol Biol. *32*, 969–978.

Poole, L.B., Karplus, P.A., and Claiborne, A. (2004). Protein sulfenic acids in redox signaling. Annu. Rev. Pharmacol. Toxicol. *44*, 325–347.

Portis, A.R., Li, C., Wang, D., and Salvucci, M.E. (2008). Regulation of Rubisco activase and its interaction with Rubisco. J. Exp. Bot. *59*, 1597–1604.

Rey, P., Cuiné, S., Eymery, F., Garin, J., Court, M., Jacquot, J.-P., Rouhier, N., and Broin, M. (2005). Analysis of the proteins targeted by CDSP32, a plastidic thioredoxin participating in oxidative stress responses. Plant J. *41*, 31–42.

Rey, P., Sanz-Barrio, R., Innocenti, G., Ksas, B., Courteille, A., Rumeau, D., Issakidis-Bourguet, E., and Farran, I. (2013). Overexpression of plastidial thioredoxins f and m differentially alters photosynthetic activity and response to oxidative stress in tobacco plants. Front. Plant Sci. *4*, 390.

Richter, A.S., Peter, E., Rothbart, M., Schlicke, H., Toivola, J., Rintamäki, E., and Grimm, B. (2013). Posttranslational influence of NADPH-dependent thioredoxin reductase C on enzymes in tetrapyrrole synthesis. Plant. Physiol. *162*, 63–73.

Rintamäki, E., Martinsuo, P., Pursiheimo, S., and Aro, E.-M. (2000). Cooperative regulation of light-harvesting complex II phosphorylation via the plastoquinol and ferredoxin-thioredoxin system in chloroplasts. Proc. Natl. Acad. Sci. U.S.A. *97*, 11644–11649.

Roberts, I.N., Lam, X.T., Miranda, H., Kieselbach, T., and Funk, C. (2012). Degradation of PsbO by the Deg protease HhoA is thioredoxin dependent. PLoS One *7*, e45713.

Rochaix, J.-D. (2013). Redox regulation of thylakoid protein kinases and photosynthetic gene expression. Antioxid. Redox Signal. *18*, 2184–2201.

Rolland, N., Curien, G., Finazzi, G., Kuntz, M., Maréchal, E., Matringe, M., Ravanel, S., and Seigneurin-Berny, D. (2012). The biosynthetic capacities of the plastids and integration between cytoplasmic and chloroplast processes. Annu. Rev. Genet. *46*, 233–264.

Roos, G., Foloppe, N., and Messens, J. (2013). Understanding the pK_a of redox cysteines: the key role of hydrogen bonding. Antioxid. Redox Signal *18*, 94–127.

Rouhier, N., and Jacquot, J.-P. (2002). Plant peroxiredoxins: alternative hydroperoxide scavenging enzymes. Photosynth Res. 74, 259–268.

Rouhier, N., Villarejo, A., Srivastava, M., Gelhaye, E., Keech, O., Droux, M., Finkemeier, I., Samuelsson, G., Dietz, K., Jacquot, J.-P., *et al.* (2005). Identification of plant glutaredoxin targets. Antioxid. Redox Signal. *7*, 919–929.

Rouhier, N., Unno, H., Bandyopadhyay, S., Masip, L., Kim, S.-K., Hirasawa, M., Gualberto, J.M., Lattard, V., Kusunoki, M., Knaff, D.B., *et al.* (2007). Functional, structural, and spectroscopic characterization of a glutathione-ligated [2Fe–2S] cluster in poplar glutaredoxin C1. Proc. Natl. Acad. Sci. U.S.A. *104*, 7379–7384.

Rouhier, N., Lemaire, S.D., and Jacquot, J.-P. (2008). The role of glutathione in photosynthetic organisms: Emerging functions for glutaredoxins and glutathionylation. Annu. Rev. Plant Biol. *59*, 143–166.

Sahrawy, M., Hecht, V., Lopez-Jaramillo, J., Chueca, A., Chartier, Y., and Meyer, Y. (1996). Intron position as an evolutionary marker of thioredoxins and thioredoxin domains. J. Mol. Evol. *42*, 422–431.

Samra, H.S., Gao, F., He, F., Hoang, E., Chen, Z., Gegenheimer, P.A., Berrie, C.L., and Richter, M.L. (2006). Structural analysis of the regulatory dithiol-containing domain of the chloroplast ATP synthase γ subunit. J. Biol. Chem. *281*, 31041–31049.

Sanz-Barrio, R., Corral-Martinez, P., Ancin, M., Segui-Simarro, J.M., and Farran, I. (2013). Overexpression of plastidial thioredoxin f leads to enhanced starch accumulation in tobacco leaves. Plant Biotechnol. J. *11*, 618–627.

Sasaki, Y., Kozaki, A., and Hatano, M. (1997). Link between light and fatty acid synthesis: Thioredoxin-linked reductive activation of plastidic acetyl-CoA carboxylase. Proc. Natl. Acad. Sci. U.S.A. *94*, 11096–11101.

Schröter, Y., Steiner, S., Matthäi, K., and Pfannschmidt, T. (2010). Analysis of oligomeric protein complexes in the chloroplast sub-proteome of nucleic acid-binding proteins from mustard reveals potential redox regulators of plastid gene expression. Proteomics *10*, 2191–2204.

Schürmann, P., and Buchanan, B.B. (2008). The ferredoxin/thioredoxin system of oxygenic photosynthesis. Antioxid. Redox Signal *10*, 1235–1273.

Schwarz, C., Bohne, A.-V., Wang, F., Cejudo, F.J., and Nickelsen, J. (2012). An intermolecular disulfide-based light switch for chloroplast psbD gene expression in *Chlamydomonas reinhardtii*. Plant J. *72*, 378–389.

Serrato, A.J., Perez-Ruiz, J.M., Spinola, M.C., and Cejudo, F.J. (2004). A novel NADPH thioredoxin reductase, localized in the chloroplast, which deficiency causes hypersensitivity to abiotic stress in *Arabidopsis thaliana*. J. Biol. Chem. *279*, 43821–43827.

Somanchi, A., Barnes, D., and Mayfield, S.P. (2005). A nuclear gene of *Chlamydomonas reinhardtii*, Tba1, encodes a putative oxidoreductase required for translation of the chloroplast psbA mRNA. Plant J. *42*, 341–352.

Sparla, F., Pupillo, P., and Trost, P. (2002). The C-terminal extension of glyceraldehyde-3-phosphate dehydrogenase subunit b acts as an autoinhibitory domain regulated by thioredoxins and nicotinamide adenine dinucleotide. J. Biol. Chem. *277*, 44946–44952.

Steiner, S., Dietzel, L., Schroeter, Y., Fey, V., Wagner, R., Schröter, Y., and Pfannschmidt, T. (2009). The role of phosphorylation in redox regulation of photosynthesis genes psaA and psbA during photosynthetic acclimation of mustard. Mol. Plant *2*, 416–429.

Susanti, D., Wong, J.H., Vensel, W.H., Loganathan, U., DeSantis, R., Schmitz, R.A., Balsera, M., Buchanan, B.B., and Mukhopadhyay, B. (2014). Thioredoxin targets fundamental processes in a methane-producing archaeon, *Methanocaldococcus jannaschii*. Proc. Natl. Acad. Sci. U.S.A. *111*, 2608–2613.

Thormählen, I., Ruber, J., Von Roepenack-Lahaye, E., Ehrlich, S.-M., Massot, V., Huemmer, C., Tezycka, J., Issakidis-Bourguet, E., and Geigenberger, P. (2013). Inactivation of thioredoxin *f*1 leads to decreased light activation of ADP-glucose pyrophosphorylase and altered diurnal starch turnover in leaves of *Arabidopsis* plants. Plant Cell Environ. *36*, 16–29.

Toivola, J., Nikkanen, L., Dahlström, K.M., Salminen, T.A., Lepistö, A., Vignols, F., and Rintamäki, E. (2013). Overexpression of chloroplast NADPH-dependent thioredoxin reductase in *Arabidopsis* enhances leaf growth and elucidates in-vivo function of reductase and thioredoxin domains. Front Plant Sci. *4*, 389.

Trebitsh, T., Levitan, A., Sofer, A., and Danon, A. (2000). Translation of chloroplast psbA mRNA is modulated in the light by counteracting oxidizing and reducing activities. Mol. Cell Biol. *20*, 1116–1123.

Trost, P., Fermani, S., Marri, L., Zaffagnini, M., Falini, G., Scagliarini, S., Pupillo, P., and Sparla, F. (2006). Thioredoxin-dependent regulation of photosynthetic glyceraldehyde-3-phosphate dehydrogenase: autonomous vs. CP12-dependent mechanisms. Photosynth. Res. 89, 263–275.

Wang, P., Liu, J., Liu, B., Feng, D., Da, Q., Shu, S., Su, J., Zhang, Y., Wang, J., and Wang, H.-B. (2013). Evidence for a role of chloroplastic m-type thioredoxins in the biogenesis of Photosystem II in *Arabidopsis*. Plant. Physiol. *163*, 1710–1728.

Wang, P., Liu, J., Liu, B., Da, Q., Feng, D., Su, J., Zhang, Y., Wang, J., and Wang, H.-B. (2014). Ferredoxin:Thioredoxin Reductase is required for proper chloroplast development and is involved in the regulation of plastid gene expression in *Arabidopsis thaliana*. Mol. Plant 7, 1586–1590.

Wedel, N., Soll, J., and Paap, B.K. (1997). CP12 provides a new mode of light regulation of Calvin cycle activity in higher-plants. Proc. Natl. Acad. Sci. U.S.A. *94*, 10479–10484.

Weise, S.E., van Wijk, K.J., and Sharkey, T.D. (2011). The role of transitory starch in C3, CAM, and C4 metabolism and opportunities for engineering leaf starch accumulation. J. Exp. Bot. *62*, 3109–3118.

Wimmelbacher, M., and Börnke, F. (2014). Redox activity of thioredoxin z and fructokinase-like protein 1 is dispensable for autotrophic growth of *Arabidopsis thaliana*. J. Exp. Bo.t 65, 2405–2413.

Wunder, T., Liu, Q., Aseeva, E., Bonardi, V., Leister, D., and Pribil, M. (2013). Control of STN7 transcript abundance and transient STN7 dimerisation are involved in the regulation of STN7 activity. Planta *237*, 541–558.

Yoo, K.S., Ok, S.H., Jeong, B.-C., Jung, K.W., Cui, M.H., Hyoung, S., Lee, M.-R., Song, H.K., and Shin, J.S. (2011). Single Cystathionine β-synthase domain-containing proteins modulate development by regulating the thioredoxin system in *Arabidopsis*. Plant Cell *23*, 3577–3594.

Yoshida, K., Matsuoka, Y., Hara, S., Konno, H., and Hisabori, T. (2014). Distinct redox behaviors of chloroplast thiol enzymes and their relationships with photosynthetic electron transport in *Arabidopsis thaliana*. Plant Cell Physiol. *55*, 1415–1425.

Yoshida, K., Hara, S., and Hisabori, T. (2015). Thioredoxin selectivity for thiol-based redox regulation of target proteins in chloroplasts. J. Biol. Chem. *290*, 14278–14288.

Yua, Q.-B., Ma, Q., Kong, M.-M., Zhao, T.-T., Zhang, X.-L., Zhou, Q., Huang, C., Chong, K., and Yang, Z.-N. (2014). AtECB1/MRL7, a thioredoxin-like fold protein with disulfide reductase activity, regulates chloroplast gene expression and chloroplast biogenesis in *Arabidopsis thaliana*. Mol. Plant 7, 206–217.

Zaffagnini, M., Michelet, L., Marchand, C., Sparla, F., Decottignies, P., Le Maréchal, P., Miginiac-Maslow, M., Noctor, G., Trost, P., and Lemaire, S.D. (2007). The thioredoxin-independent isoform of chloroplastic glyceraldehyde-3-phosphate dehydrogenase is selectively regulated by glutathionylation. FEBS J. *274*, 212–226.

Zaffagnini, M., Michelet, L., Massot, V., Trost, P., and Lemaire, S.D. (2008). Biochemical characterization of glutaredoxins from *Chlamydomonas reinhardtii* reveals the unique properties of a chloroplastic CGFS-type glutaredoxin. J. Biol. Chem. *283*, 8868–8876.

Zaffagnini, M., Bedhomme, M., Groni, H., Marchand, C.H., Puppo, C., Gontero, B., Cassier-Chauvat, C., Decottignies, P., and Lemaire, S.D. (2012a). Glutathionylation in the photosynthetic model organism *Chlamydomonas reinhardtii:* A proteomic survey. Mol Cell Proteomics 11, M111.

Zaffagnini, M., Bedhomme, M., Lemaire, S.D., and Trost, P. (2012b). The emerging roles of protein glutathionylation in chloroplasts. Plant Sci. *185–186*, 86–96.

Zaffagnini, M., Bedhomme, M., Marchand, C., Morisse, S., Trost, P., and Lemaire, S. (2012c). Redox regulation in photosynthetic organisms: focus on glutathionylation. Antioxid. Redox Signal *16*, 567–586.

Zhang, N., and Portis, A.R. (1999). Mechanism of light regulation of Rubisco: A specific role for the larger Rubisco activase isoform involving reductive activation by thioredoxin-f. Proc. Natl. Acad. Sci. U.S.A. *96*, 9438–9443.

The Transporters of Plastids – New Insights into an Old Field

10

Karsten Fischer, Andreas P.M. Weber and Hans-Henning Kunz

Abstract

Plastids are the hallmark organelles of all photosynthetic eukaryotes and are the site of major anabolic pathways. Metabolic pathways within chloroplasts are highly connected with pathways occurring in the surrounding cytosol and in other organelles, which mandates a massive flux of organic solutes across the chloroplast envelope membranes. In addition, inorganic ions, taken up by the roots and transported to the other plant organs, have to be distributed within the cell to the different organelles, including plastids. Hence, a plethora of metabolite and ion transport systems that mediate these fluxes are found in plastids. During the evolution of chloroplasts from a cyanobacterial ancestor, the majority of the chloroplast envelope membrane transporters were recruited from the pre-existing repertoire of host transporters whilst only few are of cyanobacterial provenance. We will review the current knowledge on chloroplast metabolite and ion transporters from an evolutionary perspective and for some discuss their physiological functions.

Introduction

A specific feature of all photosynthetic eukaryotic cells, in contrast to primary non-photosynthetic eukaryotes, is that they contain plastids, which are semi-autonomous organelles of endosymbiotic origin (Reyes-Prieto *et al.*, 2007). Plastids play essential roles in plants since they represent the sites of photosynthesis and a multitude of other essential metabolic routes. They provide, for example, the building blocks for a wide variety of primary and secondary metabolites as well as phytohormones. Moreover, plastids play an essential role in plant nitrogen and sulfur assimilation. The compounds synthesized in plastids can be either downstream products of photoassimilates, e.g. phosphorylated sugars, fatty acids, amino acids and nucleotides, or are derived from chloroplast components, e.g. membrane lipids or pigments that are backbones for vitamins and plant hormones such as oxylipins and abscisic acid. Fixed carbon, nitrogen and sulfur in the form of sugar phosphates, amino acids, fatty acids, nucleotides and phytohormones are essential for other organelles, and thus have to be exported. At the same extent as the chloroplast provides substrates to the rest of the cell, it also relies on import of metabolites to fuel several plastid biosynthesis pathways which are dependent on precursors from the cytosol. Pyruvate uptake for instance is crucial for fatty acid, terpenoid, and branched-chain amino acid biosynthesis (Furumoto *et al.*, 2011), while glucose-6-phosphate imported into non-green plastids is the substrate for starch biosynthesis and the oxidative pentose phosphate pathway (Niwiadomski *et al.*, 2005). In addition, several biochemical pathways involve coordinated catalytic steps in different organelles. One prominent example is the photorespiration pathway, which is split between chloroplasts, peroxisomes, and mitochondria (see below). For this integration of plastids into the cell, plastid metabolism is intertwined with that of the surrounding cytosol and the other cellular compartments, which causes considerable traffic of metabolites across the plastid envelope membranes (Weber, 2004; Weber and Fischer, 2007). Moreover, plastids have to take up several inorganic cations and anions, which are important for photosynthesis and other processes within plastids. Since most of these small inorganic and organic molecules are charged or polar and hence cannot pass the envelope membrane by passive diffusion, various transporters are required to enable their directional transport across the plastid membranes. Understanding metabolite transport between plastid and cytosol is thus of crucial importance for understanding plant metabolism (Weber, 2004; Weber and Fischer, 2007).

Plastids of higher plants are bounded by two concentric membranes, the inner and outer plastid envelope membranes. Transport across the outer envelope membrane is enabled by a set of β-barrel proteins (e.g. OEP16, 21, 23, 24, and 37) that form pores that are cation or anion selective and in some cases are regulated by metabolites (Bölter and Soll, 2001; Goetze *et al.*, 2015). The role the outer envelope membrane and its pores play in the regulation of transport into and out of plastids has been intensively discussed and the interested reader is referred to Flügge (2000) and Soll *et al.* (2000) for an overview. The transport across the inner envelope, in contrast, is catalysed by a large set of α-helical membrane transporters, which are highly specific for their substrates (Fischer, 2011). Based on *in silico* analyses and data from proteomics projects the 'permeome' of the inner envelope membrane of higher plants has been predicted to comprise about 150 different transporters (Weber *et al.*, 2005). Up

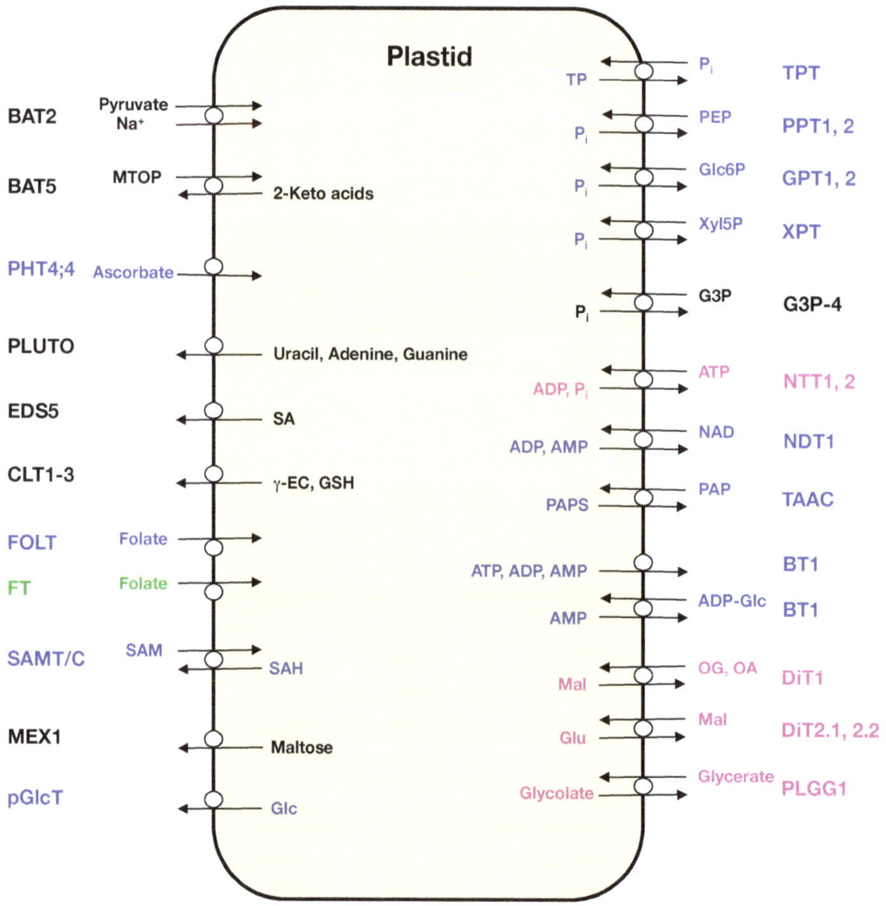

Figure 10.1 Transport of organic substances across the inner envelope membrane of plastids from higher plants. Only transporters which were characterized at the molecular level are shown. All names of transporters refer to the proteins in *Arabidopsis thaliana* except for the BT1 protein from maize. The origin of the transporters is indicated by a colour code. Blue, Host cell; green, cyanobacterium; pink, horizontal gene transfer; black, unknown. G3P, glycerol-3-P; Glc, glucose; Glc6P, glucose-6-phosphate; Glu, glutamate; GSH, glutathione; Mal, malate; MTOP, 4-methylthio-2-oxobutanoate; NAD, nicotinamide adenine dinucleotide; OA, oxaloacetate; OG, 2-oxoglutarate; P_i, inorganic phosphate; PAP, adenosine phosphosulfate; PAPS, 3′-phosphoadenosine 5 phosphosulfate; PEP, phosphoenolpyruvate; SA, salicylic acid; SAH, S-adenosylhomocysteine; SAM, S-adenosylmethionine; TP, triose phosphates; Xyl5P, xyluloxe-5-phosphate; γ-EC, γ-glutamylcysteine.

to now, about 40 of these transporters have been analysed at the molecular level and have an ascribed physiological function (Figs. 10.1 and 10.2 and Table 10.1).

Because a number of excellent reviews about various aspects of plastid transport were published in the last years (e.g. Weber and Fischer, 2007; 2009; Facchinelli and Weber, 2011; Fischer, 2011; Rolland *et al.*, 2012) here we will discuss only the most recent progress in understanding the metabolic connections between plastid and cytosol. In particular, we will focus on three different topics:

1. the endosymbiotic origin and evolution of plastids and the role plastid transporters have played in these processes;
2. the structure and function of some recently identified and characterized organic metabolite transporters;
3. the importance of different transport systems for inorganic ions.

We will mainly discuss transporters from flowering plants, especially *Arabidopsis thaliana*, and focus on their physiological and biochemical functions, i.e. their functions in the context of the cellular metabolism.

Establishment of endosymbiosis and the role of transporters

Plastids evolved from cyanobacteria

Since the ground-breaking publication by Mereschkowsky (1905; for an English translation see: Martin and Kowallik, 1999), who suggested that the chloroplasts of plant cells evolved from cyanobacteria, the 'endosymbiont' theory is now generally accepted (Margulis, 1971; Bhattacharya *et al.*, 2004). The endosymbiont theory states that a primitive eukaryotic cell that already contained mitochondria and other cellular compartments took up a photosynthetic cyanobacterium that eventually evolved into a novel organelle, the plastid, bounded by two envelope membranes, a process now called primary endosymbiosis (Reyes-Prieto *et al.*, 2007). This endosymbiont-containing 'protoalga' gave rise to the Archaeplastida, i.e. the three photosynthetic eukaryotic lineages containing primary plastids: red and green algae (including land plants), and the glaucophytes (Bhattacharya *et*

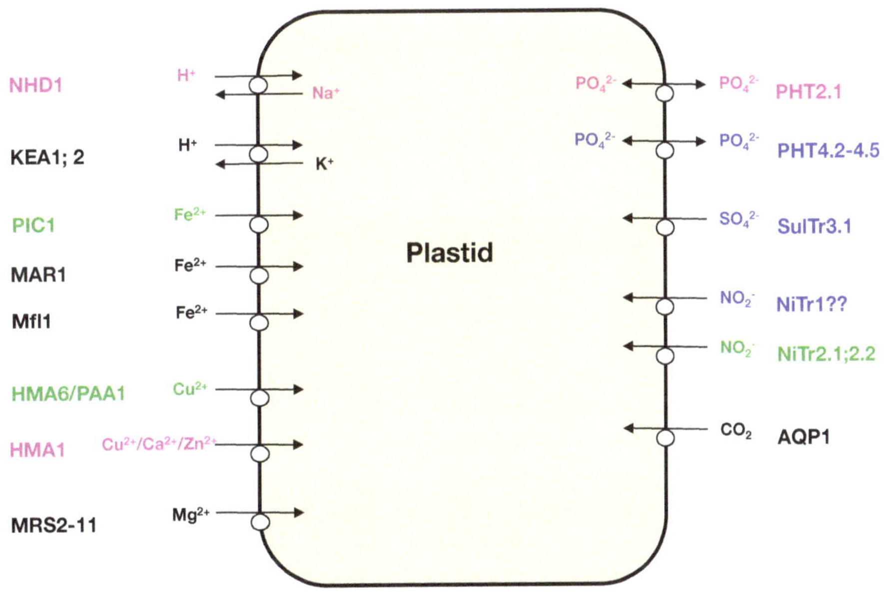

Figure 10.2 Transport of inorganic substances across the inner envelope membrane of plastids from higher plants. Only transporters which were characterized at the molecular level are shown. All names of transporters refer to the proteins in *Arabidopsis thaliana*. The origin of the transporters is indicated by a colour code. Blue, Host cell; green, cyanobacterium; pink, horizontal gene transfer; black, unknown.

Table 10.1 Transporters of the inner envelope membrane of plastids from higher plants. Only transporters discussed in the text are included. HGT, Horizontal gene transfer

Name	Activity	Substrates	Gene ID	Ocurrence	Origin	Identification
PLGG1	Plastid glycollate/ glycerate transporter 1	Glycollate, Glycerate	At1g32080	Plants, Bacteria	HGT	Co-expression analysis
PHT4.2	Phosphate transporter	(Na^+)-PO_4^{2-}	At2g38060	Eukaryotes	Host cell	Sequence homology
PHT4.3	Phosphate transporter	(Na^+)-PO_4^{2-}	At3g46980	Eukaryotes	Host cell	Sequence homology
PHT4.5	Phosphate transporter	(Na^+)-PO_4^{2-}	At5g20380	Eukaryotes	Host cell	Sequence homology
PHT4.4	Phosphate transporter	(Na^+)-PO_4^{2-} Ascorbate	At4g00370	Eukaryotes	Host cell	Sequence homology
BT1	Brittle 1 protein (*Zea mays*)	ADP-Glucose, AMP	BT016796	Eukaryotes	Host cell	Forward genetics
BT1	Brittle 1 protein	ATP, ADP, AMP	At4g32400	Eukaryotes	Host cell	Sequence homology
PLUTO	Plastid nucleobase transporter	Adenine, Guanine, Uracil	At5g03555	Eukaryotes Prokaryotes	Unknown	Sequence homology
EDS5	Enhanced disease susceptibility 5	Salicylic acid	At4g39030	Eukaryotes Prokaryotes	Unknown	Forward genetic screen
G3Pp4	G3P permease 4	Glycerol-3-P	At4g17550	Plants, Bacteria	Unknown	Sequence homology
KEA1	K^+ efflux antiporter 1	K^+/H^+	At1g01790	Plants, Bacteria	Unknown	Sequence homology
KEA2	K^+ efflux antiporter 2	K^+/H^+	At4g00630	Plants, Bacteria	Unknown	Sequence homology
NHD1	Na^+/H^+ antiporter D	Na^+/H^+	At3g19490	Plants, Bacteria	HGT	Sequence homology
PIC1	Permease in chloroplasts 1	Fe^{2+}/H^+	At2g15290	Plants	Cyanobacterium	Sequence homology
MAR1	Multiple antibiotic resistance 1	Fe^{2+}, antibiotics	At5g26820	Plants	Unknown	Forward genetic screen
Mfl1	MitoFerrinlike1	Fe^{2+}	At5g42130	Eukaryotes	Unknown	Sequence homology
HMA1	Heavy metal ATPase 1	Cu^{2+}, Zn^{2+}, Co^{2+}, Ca^{2+}, Cd^{2+}	At4g37270	Plants Chlamydia	HGT	Proteomics
PAA1	Copper P-type ATPase	Cu^{2+}	At4g33520	Plants, Bacteria	Cyanobacterium	Forward genetics

al., 2004). Additional photosynthetic eukaryotes evolved through higher order endosymbiosis in which either a red or green alga, or protists containing plastids of secondary endosymbiotic origin, were engulfed by other eukaryotic cells. These new endosymbionts were subsequently reduced to secondary or higher order plastids, which, in contrast to primary plastids, are enveloped by three to four membranes (Gibbs, 1978; Cavalier-Smith, 2000; Yoon *et al.*, 2002; Weber and Fischer, 2009). The plastid envelope membranes are thus relics of the evolutionary origin of plastids: the first, innermost envelope membrane has evolved from the cyanobacterial plasma membrane while the second envelope membrane (i.e. the outer membrane of primary plastids) has a chimeric origin, being remnant of both the cyanobacterial outer membrane and a host membrane derived from the endomembrane system (Cavalier-Smith 2000, 2003). In secondary plastids, the third membrane is derived from the plasma membrane of the eukaryotic endosymbiont while the fourth membrane (if existing), is (in many cases) continuous with the endomembrane system of the host (Stork *et al.*, 2013).

Extensive phylogenetic and phylogenomic work in the last decade showed that the plastids of the Archaeplastida are of monophyletic origin, that is, they can be traced back to a single endosymbiotic event (Reyes-Prieto *et al.*, 2007; Price *et al.*, 2012). This work also revealed a molecular timeline for the origin of the photosynthetic eukaryotes, with the primary endosymbiotic event having taken place about 1.6 billion years ago (BYA), followed by the split into red algae, green algae and glaucophytes between 1.6 and 1.5 BYA. The secondary endosymbiosis involving a red alga occurred about 1.3 BYA (Yoon *et al.*, 2004). Thus, these data support an ancient origin of photosynthetic eukaryotes, i.e. next to the origin of eukaryotes. The transformation of the bacterial endosymbiont to a *bona fide* cellular organelle was driven by several processes, namely:

a The loss of the bacterial cell wall in red and green algae and their descendants. Only the glaucophytes maintained a peptidoglycan layer between the plastid envelope membranes (Pfanzagl *et al.*, 1996).
b The reduction of the former bacterial genome by about 95%, achieved by outright gene losses and a massive transfer of genes into the nuclear genome of the host cell, a process called endosymbiotic gene transfer (EGT) (Leister, 2003; Qiu *et al.*, 2013).
c The establishment of a sophisticated protein import apparatus in both envelope membranes that permits the specific and efficient uptake of now nuclear encoded proteins, which are targeted to plastids as their final location. In fact, the majority of plastid proteins is encoded on the nuclear genome and post-translationally imported into plastids (Soll and Schleiff, 2004; Gutensohn *et al.*, 2006).
d The rebuilding of the membranes surrounding the endosymbiont, i.e. the transformation of bacterial membranes facing the environment into organellar membranes facing the cytosol, achieved by the loss of many bacterial transporters and the integration of new transport systems.

Metabolite transporters played an important role in the establishment of the endosymbiosis

The importance of EGT and protein import for the evolution of the endosymbiotic relationship is extensively discussed in the literature (e.g. Bhattacharya *et al.*, 2007; Archibald, 2009). Both processes enabled the host cell to get the endosymbiont under control, i.e. to adapt division of the endosymbiont to the host cell cycle and to change and control the metabolism of the endosymbiont. Finally, the former bacterium completely lost the ability to survive without its host preventing the 'escape' of the endosymbiont. However, the role of transporters in this process has been ignored for a long time but came into focus in the last two decades. It has been proposed that the integration of new transporters was a critical and very early step in plastid evolution (Cavalier-Smith, 2000) based on the following arguments:

1 The merger of two free-living organisms required the integration and coordination of two previously independent metabolic entities, in this case a photoautotrophic primary producer and a heterotrophic organism. The stabilization of this relationship would require an 'immediate' benefit for (at least) one or both of the partners, otherwise the endosymbiont would have been lost again.
2 The host cell derives benefits from the complementation or extension of its limited metabolic capabilities by the biochemical versatility of the endosymbiont (Nowack and Melkonian, 2010). In the case of plastids, the endosymbiont introduced the capacity for oxygenic photosynthesis into eukaryotic cells, a process that converted a heterotrophic eukaryotic cell into a photoautotrophic organism.
3 To benefit from photosynthesis, the host cell somehow must have been able to tap into the cyanobacterial carbon pool, i.e. into the products of photosynthetic carbon fixation. This had to be achieved in a controlled manner to guarantee survival of the newly obtained organelle.
4 Since contemporary cyanobacterial genomes (and most likely also those of the cyanobacterial ancestor of chloroplasts) do not encode

for transporters that permit the export of Calvin cycle intermediates across their plasma membranes, it is likely that transporters exporting metabolites from the endosymbiont were provided by the host cell (or a third organism; see below).

Assuming that the presented arguments are correct, it can be expected that the host already owned a set of metabolite transporters that was (re)targeted to the membrane surrounding the cyanobacterium. What kind of transporters would that be? Insights gained from research on metabolite transport in flowering plants are very helpful in this respect. It is known for decades that triose phosphates represent the most important metabolites of the Calvin–Benson cycle that are exported to the cytosol (Flügge, 1999; Heldt, 2002). In land plants, the triose phosphate export is mediated by a triose phosphate/phosphate antiporter (TPT) (Flügge *et al.*, 1989). The TPT is a member of the plastid phosphate translocator family (pPTs), which can be further divided into four subfamilies (TPT, PPT, GPT, XPT). The names are according to their main transport substrates, which are, besides triose phosphates, phosphoenolpyruvate (Fischer *et al.*, 1997), glucose 6-phosphate (Kammerer *et al.*, 1998) and xylulose 5-phosphate (Eicks *et al.*, 2002). The pPTs connect the primary (and secondary) metabolism of plastids with that of the cytosol (for an overview see Flügge *et al.*, 2003). Recent progress in genomics showed that pPT proteins are not only found in green land plants, but in almost all photosynthetic eukaryotes for which genome or comprehensive EST information is available, including photosynthetic organisms containing secondary plastids, such as diatoms, or non-photosynthetic parasites like the *Apicomplexa* which evolved from algae (with the notable exception of glaucophytes (see below)) (Weber *et al.*, 2006; Brooks *et al.*, 2010; Price *et al.*, 2012). Extensive analysis of genomic and EST-sequence data from a broad range of organisms showed that the pPTs evolved from transport proteins of the eukaryotic endomembrane system, specifically from nucleotide sugar transporters (NSTs) of the ER and Golgi membranes (Knappe *et al.*, 2003; Weber *et al.*, 2006). Genes encoding NSTs are ubiquitously present in eukaryotic genomes but absent from prokaryotes. This means, a metabolite transporter originally residing in the host endomembrane system was routed to the bacterial plasma membrane (i.e. the plastid inner envelope membrane) and thus enabled the host to connect to the photosynthetic carbon pool of the endosymbiont.

The model described here proposes that nucleotide sugars were the first substrates exported from the endosymbiont to feed the carbohydrate metabolism of the host cell. By analysing the starch and sugar metabolism of contemporary plants and algae Ball and co-workers (Ball *et al.*, 2011) reconstructed the metabolism of the common ancestor of Archaeplastida showing that starch synthesis originally occurred in the cytosol. They suggested that ADP-glucose export defined the original metabolic flux by which the endosymbiont provided photosynthate to its host. Recently, however, the sequencing of the genome of the first glaucophyte alga, *Cyanophora paradoxa*, revealed that this species lacks pPT proteins but possesses instead two bacterial type UhpC hexose phosphate translocators which are closely related to proteins in chlamydiae (Price *et al.*, 2012). This led to the hypothesis that a hexose phosphate transporter represents the first carbon exporter of the endosymbiont. Although the nature of the first transporters integrated into the endosymbiont membrane is still an open question, comprehensive analysis of the phylogeny of plastid envelope-localized transporters of plants and algae showed that most of the envelope 'permeome' evolved from pre-existing host proteins whereas only a small portion (< 15%) was contributed by the endosymbiont (Tyra *et al.*, 2007; see Fig. 10.1). Most of the cyanobacterial transporters are involved in the transport of inorganic ions or metabolites of secondary metabolism while all of the plastid transporters involved in primary metabolism were derived from the host cell (Figs. 10.1 and 10.2). This clearly shows that integration of plastid and host metabolism was predominantly a host-driven process, a fact that corroborates the theory that the relationship between the cyanobacterium/plastid and the host cell was and is not mutualism, but helotism (slavery) (Cavalier-Smith, 2000).

A surprisingly large share of plastid envelope membrane transporters (Fig. 10.1), such as the adenine nucleotide transporters (NTTs; Kampfenkel *et al.*, 1995), the dicarboxylate translocators (DiTs; Weber *et al.*, 1995), metal-transporting ATPases (Seigneurin *et al.*, 2006), the Na^+/H^+ antiporter

(Müller *et al.*, 2014) and the above mentioned UhpC transporters have their evolutionary origin in prokaryotic intracellular parasites (i.e. *Chlamydia* and/or *Rickettsia*). Indeed, a comprehensive study of genomes from 17 distinct plant species found 55 genes, which have been introduced into the genome of the protoalga by horizontal transfer (HGT) from *Chlamydiae* (Moustafa *et al.*, 2008). Most of these genes encode plastid-targeted proteins, including transporters. The surprisingly high number of *Chlamydia*-derived genes in all plant genomes let to the hypothesis that a *Chlamydia*-type bacterium was directly involved in the establishment of the cyanobacterium–host relationship by providing important genes/proteins for this process (Huang and Gogarten, 2007; Ball *et al.*, 2013; Facchinelli *et al.*, 2013). *Chlamydiae* were originally identified as a group of obligate intracellular parasites of vertebrates, but the phylum now also comprises several intracellular endosymbionts of arthropods and protists. Although there are no contemporary plant-connected *Chlamydiae* known, it has been speculated that primordial – now extinct – parasites of plants existed once (Subtil *et al.*, 2014). This essential contribution to plastid evolution of a third partner in an 'ménage à trois' would also explain why the successful transformation of a phototrophic bacterium to an organelle was such a rare event, considering the fact that the uptake of (phototrophic) bacteria by eukaryotic protists was and still is a very common process.

How was the early integration of transporters achieved?

If the theory about the early integration of one or more transporters into the membranes of the endosymbiont is correct then the question remains how this has been achieved without the existence of the canonical TOC-TIC protein import pathway found in all extant plastids (with the exception of the chromatophores of *Paulinella*) (Soll and Schleiff, 2004; Gutensohn *et al.*, 2006)? A solution to this problem came with the identification of an increasing number of plastid proteins, which are imported via a second route, i.e. the import into the Endoplasmic Reticulum (ER) and transport within vesicles via the Golgi apparatus to the plastids (Kleffmann *et al.*, 2004; Villarejo *et al.*, 2005). Although there is an ongoing debate which of the two distinct protein import pathways is more ancient (Bhattacharya *et al.*, 2007; Gagat *et al.*, 2013), it is unlikely that one of these sophisticated processes was responsible for the early integration of transporters. However, there is mounting evidence for a direct physical contact of the ER membranes with the envelope membranes of plastids:

1. Physical contact, i.e. continuity, between plastids and the ER has been shown frequently in light- and electron-microscopical investigations in plants and algae (e.g. Wooding and Northcote, 1965; Whatley *et al.*, 1991; Kaneko and Keegstra, 1996).
2. A strong force (>400 pN) of attraction between chloroplasts and the ER has been directly determined by using laser scalpels and optical tweezers (Andersson *et al.*, 2007). The attraction is at least partially mediated through protein–protein interactions.
3. In an elegant experimental strategy named 'transorganellar complementation' DellaPenna and co-workers also demonstrated a metabolic continuity of ER and plastids (Mehrshahi *et al.*, 2013). Here, the knock down of a plastid-localized step in the synthesis of an essential hydrophobic compound, e.g. tocopherol, was complemented by the expression of the corresponding enzyme in the ER. The complementation of the mutant phenotype shows that the ER localized enzyme had access to substrates of the plastid, which could be explained by the transport of the hydrophobic metabolites to the ER, most likely via the contact sites. Interestingly, similar contact sites between mitochondria and the ER have been found and intensively studied. These are involved in a number of processes, e.g. mitochondrial division and distribution but also biochemical processes like Ca^{2+} and phospholipid transport (Klecker *et al.*, 2014).

Based on these data, a model of the structure of these contact sites has been proposed which is based on membrane hemifusion (Mehrshahi *et al.*, 2014). According to this model, the outer leaflets of two opposing membranes would merge into one chimeric leaflet, while the two inner leaflets would form a hemifused bilayer at the contact site. These structures would enable the transport of lipids but also other hydrophobic compounds between the two organelles. It is tempting to speculate that such contact sites not only exist in extant plants but also existed in the ancestor of all eukaryotic phototrophs and that this contact sites were the port of entry for the first transporter proteins. Such an early targeting of host proteins has been demonstrated in several endosymbiotic relationships between bacteria and plants or insects, with the goal to enable the host to control and manipulate the endosymbiont. In all these cases, the endomembrane system of the host was involved in targeting (Nowack, 2014).

Newly discovered transporters and their functions

The first transporter involved in photorespiration, the glycolate/glycerate transporter

Photorespiration is the light-dependent consumption of oxygen and release of CO_2 in plants, algae and cyanobacteria (Hagemann *et al.*, 2013). In the 1970s it was shown by Ogren and co-workers that the cause of the oxygen consumption is the oxygenase activity of the CO_2-fixing enzyme present in all photosynthetic organisms, whereupon the enzyme was renamed ribulose 1,5-bisphosphate carboxylase/oxygenase (RubisCO) (Bowes *et al.*, 1971; Somerville, 2001). Obviously, CO_2 and O_2 compete for the same binding site of the enzyme. Thus, under the conditions of the present atmosphere with a 500-fold excess of oxygen (21% versus 0.04%) RubisCO cannot efficiently discriminate between both substrates. Under these conditions, carboxylation exceeds oxygenation by about three fold, i.e. for every three CO_2 one O_2 is assimilated. However, at higher CO_2 concentrations (0.4–1%) or lower oxygen concentrations (2%) the oxygenase activity of RubisCO is almost completely suppressed.

The transfer of O_2 onto the acceptor RuBP leads to the production of one molecule 3-phosphoglycerate (3-PGA) and one molecule 2-phosphoglycolate (2-PG). While 3-PGA is used in the Calvin–Benson cycle, 2-PG has to be removed to avoid both toxic effects of this metabolite and the depletion of the Calvin–Benson cycle. This is achieved in a series of eight reactions, the C2 cycle, that converts two 2-PG into one 3-PGA and one CO_2 (Bauwe *et al.*, 2010). In short, 2-PG is de-phosphorylated in chloroplasts to glycolate, which is oxidized to glyoxylate and further transaminated to glycine, both reactions occurring in peroxisomes. After the uptake of glycine into mitochondria it is converted to serine, a reaction, which leads to the release of CO_2 and ammonium. Both compounds are re-assimilated in chloroplasts by the Calvin–Benson cycle and the glutamine synthetase/glutamate synthase system, respectively. Serine is transported back to peroxisomes where it is deaminated to hydroxypyruvate, which is then reduced to glycerate. Glycerate is imported into chloroplasts and finally phosphorylated to 3-PGA. Taken together, the C2 cycle detoxifies 2-PG while recovering 75% of the carbon that is contained in 2-PG.

Photorespiration and photosynthesis are essentially linked, i.e. phototrophic organisms cannot perform photosynthesis without functional photorespiration under ambient concentrations of CO_2 and O_2. This has been clearly shown by the analysis of mutants that display a photorespiratory phenotype, i.e. stunted growth and chlorotic leaves (and eventually plant death) at low CO_2 concentrations but normal growth at high CO_2 concentrations (Somerville, 2001). Forward genetics screens exploited this phenotype, which successfully led to the identification of most of the enzymes of the C2 cycle and their corresponding genes (Somerville, 2001).

Because the C2 cycle is distributed to three different organelles (chloroplasts, peroxisomes, mitochondria) metabolite transporters in the organellar membranes are crucial to connect the compartments and ensure an efficient metabolite flux through photorespiration. A comprehensive analysis of the compartmentalization of the C2 cycle and of tightly associated pathways, namely those for recycling of ammonium, for the distribution of redox equivalents between the organelles, and the transport of essential coenzymes, revealed nine transport processes in the core cycle and up to 16 in the associated pathways (Eisenhut *et al.*, 2013). The first and only transporters identified so far that are involved both in ammonium fixation and redox shuttle are the two plastid dicarboxylate transporters DiT1 and DiT2 which transport the metabolites needed for both processes (2-oxoglutarate, glutamate, malate, oxaloacetate) across the plastid inner envelope membrane (reviewed in, for example, Weber and Fischer, 2007; Eisenhut *et al.*, 2014). Only recently, the first transporter that is part of the core C2 cycle has been identified and characterized, namely the plastid glycolate/glycerate transporter PLGG1 (Fig. 10.1) (Pick *et al.*, 2013). Plastid glycolate and glycerate transport was initially analysed by Howitz and McCarty employing isolated chloroplasts and purified and reconstituted proteins to measure transport activities (Howitz and McCarty, 1985a,b, 1991). The obtained data revealed that:

1. Both substrates are transported by the same protein.
2. The protein has two different transport modes, a glycolate/glycerate antiport and a substrate/H^+ symport, which explains how the transporter manages the export of two molecules of glycolate and the import of one molecule of glycerate.

However, the molecular identity of the protein and the corresponding gene remained elusive because it was not identified in the above-mentioned genetic screens. Instead, Pick and colleagues used a different strategy based on co-expression analysis. This 'guilty-by-association' approach is based on the concept that genes involved in the same process often display similar transcription profiles. This phenomenon makes it feasible to identify candidate genes in expression databases based on their co-expression with genes of known function (Eisen *et al.*, 1998). The *Arabidopsis thaliana* gene *At1g32080* (*PLGG1*) was identified due to its co-expression with almost all known photorespiratory genes. The encoded protein has 12 predicted membrane spanning helices, thus most likely represents a transporter protein. The actual function of the protein was elucidated by the extensive analysis of a *plgg1* T-DNA insertion mutant (Pick *et al.*, 2013). First, the *plgg1* mutant displayed a *bona fide* 'photorespiratory' phenotype, i.e. yellow and bleached leaves at ambient CO_2 concentrations but green leaves and normal growth at 0.3% CO_2. Second, detailed comparative metabolic analyses of mutant and WT plants revealed an accumulation of C2 cycle metabolites, especially glycolate and glycerate, but not glyoxylate at low CO_2 concentrations, clearly indicating a disturbance in photorespiration. The fact that metabolites downstream of glycolate (export) accumulate in the *plgg1* mutant can be explained by a residual passive diffusion of glycolate and glycerate through the lipid bilayer. Third, both the glycerate-dependent oxygen evolution of isolated chloroplasts and the flux through the C2 cycle were diminished in the mutant plants. Based on these data, the authors concluded that PLGG1 represents the long sought glycolate/glycerate transporter of chloroplasts.

Interestingly, the *PLLG1* gene had been identified before by two other groups based on its homology to the bacterial *LrgB* gene (Yang *et al.*, 2012) and in a screen for *A. thaliana* mutants with albino or pale green leaves (Yamaguchi *et al.*, 2012). It was therefore named *AtLrgB*. Thorough analyses of the mutant leaf phenotype and the chloroplasts ultrastructure led to the conclusion that this gene is involved in chloroplast development, the regulation of cell death and leaf senescence. However, as described above, the mutant phenotype can now be explained by the accumulation of toxic metabolites due to severe reduction of photorespiration (Pick *et al.*, 2013). Nonetheless is the homology to the bacterial *LrgB* gene very interesting. *LrgB* and *LrgA* are encoded on the same operon in *Staphylococcus aureus*. This operon is part of a holin/antiholin system controlling cell death and lysis in many bacteria (Rice and Bales, 2008). Holins are membrane proteins believed to form pores in the bacterial plasma membranes that enable export of murein hydrolases. The activity of the murein hydrolases destabilizes the bacterial cell wall and induces cell lysis (Rice and Bales, 2008). Antiholins are also membrane proteins and somehow inhibit holin activity. However, the actual function at the molecular level still remains unknown. Holin/antiholin systems are involved in processes such as the release of bacteriophages from infected cells, control of cell death and sporulation, lysis in response to oxygen availability and glucose concentration, and biofilm establishment (Rice and Bales, 2008).

In *S. aureus*, holins are encoded by the *CidA* and *CidB* genes, while *LrgA* and *LrgB* encode antiholins. Strikingly, the *CidA*/*LrgA* and *CidB*/*LrgB* genes share a considerable degree of homology and might have evolved from single genes (Saier and Reddy, 2015). *LrgB* genes have been found in a large number of bacterial species but surprisingly not in cyanobacteria, which means that *LrgB* sequences found in plants and algae were not derived by EGT from the endosymbiont but by HGT from a yet unknown bacterial species or from a eukaryotic source (Tyra *et al.*, 2007; Yang *et al.*, 2012). It has been proposed that plant PLLG1 evolved by fusion of the bacterial *LrgA* and *LrgB* genes, i.e. *LrgA* and *LrgB* encode the N-terminal and C-terminal part of PLLG1, respectively (Yang *et al.*, 2012). To summarize, the glycolate/glycerate transporter in chloroplasts have evolved from a protein export system in bacteria after HGT and adoption of a new function, namely metabolite transport.

Vitamin C and nucleobase transporters

Vitamin C – synthesis and transport

Ascorbate (vitamin C) is an essential compound in a wide range of organisms including plants where it has multiple functions, e.g. it is involved in the control of cell division and plant growth, cell wall synthesis, regulation of gene expression and hormone synthesis, and pathogen defence (Toth *et al.*, 2013). In chloroplasts, ascorbate is, together with glutathione, the major redox buffer. In the stroma, it participates in the enzymatic scavenging of Reactive Oxygen Species (ROS) that are produced in the 'Mehler reaction' by photosystem I. ROS are in a first step detoxified by superoxide dismutase to hydrogen peroxide, which is further reduced to water by ascorbate peroxidase (Foyer and Noctor, 2011). The electron donor ascorbate is thereby oxidized to monodehydroascorbate (MDHA) that disproportionates into ascorbate and dehydroascorbate (DHA). In the thylakoid lumen ascorbate is the electron donor for violaxanthin de-epoxidase (VDE), which converts violaxanthin to zeaxanthin. The latter is an important factor in non-photochemical quenching (NPQ) of chlorophyll fluorescence, a process by which excess excitation energy is dissipated as heat to prevent photoinhibition of photosynthesis (Demmig-Adams and Adams, 2006). Because of its numerous functions in plants, ascorbate is present in all cell compartments, accumulating to concentrations of 20 mM or higher in chloroplasts and the cytosol (Zechmann *et al.*, 2011). In plants, ascorbate is synthesized *de novo* via three different pathways: the L-galactose pathway (Wheeler *et al.*, 1998), the L-gulose pathway (Wolucka and Van Montagu, 2003) and the myo-inositol pathway (Lorence *et al.*, 2004). In addition, in ripening fruits a fourth salvage pathway has been detected that starts with D-galacturonic acid, a product of pectin degradation (Agius *et al.*, 2003). However, analysis of deletion mutants determined the Smirnoff–Wheeler pathway as the physiological most relevant one (Dowdle *et al.*, 2007). All pathways are located in the cytosol, with the exception of the last, common step, which is located in the mitochondrial inner membrane (Millar *et al.*, 2003). This implies the need for ascorbate transport systems across all internal cellular membranes but also across the plasma membrane for long-distance transport via the phloem (Franceschi and Tarlyn, 2002).

Transport activities for ascorbate have been measured in the plasma membrane (reviewed in Horemans *et al.*, 2000) and in isolated chloroplasts (Beck *et al.*, 1983; Foyer and Lelandais, 1996) from different plants. Across the plasma membrane ascorbate and DHA transport occurs via different transport systems, namely facilitated diffusion, proton driven co-transport and an ascorbate/DHA antiport (Horemans *et al.*, 2000). Uptake of ascorbate into chloroplasts is achieved by a low affinity carrier ($K_M \approx 20$ mM), which most likely functions as an ascorbate/DHA antiporter (Beck *et al.*, 1983; Foyer and Lelandais, 1996). These data indicated that the ascorbate/DHA transport in plants differ from that in animals. In the latter, DHA is taken up by a plasma membrane localized facilitative glucose transporter (GLUT), that accepts both glucose and DHA as substrates, but not ascorbate (Vera *et al.*, 1993). In contrast, the plant glucose transporters analysed so far do not transport DHA (or ascorbate) (Beck *et al.*, 1983; Foyer and Lelandais, 1996). Ascorbate on the other hand is imported into mammalian cells by two homologous sodium-driven ascorbate carriers (SVCT1, 2; now renamed SCL23A1, A2) (Daruwala *et al.*, 1999; Tsukaguchi *et al.*, 1999; May, 2011). Both proteins belong to the family of nucleobase-ascorbate transporters (NAT), also known as the nucleobase:cation symporter-2 family (NCS2) or, in the newest nomenclature, the SLC23 family. The NAT family is found in all domains of life, i.e. prokaryotes, protists, fungi, plants and animals (de Koning and Diallinas, 2000). In contrast to the mammalian NAT proteins, those in bacteria and fungi, e.g. *Aspergillus nidulans*, transport nucleobases like xanthine, uric acid and, with lower capacities, other purines, but not ascorbate (de Koning and Diallinas, 2000).

Families of 11–12 NAT proteins have been identified in *Arabidopsis*, rice and tomato (Maurino *et al.*, 2006; Cai *et al.*, 2014). These proteins show significant homology to NATs from both fungi and animals. All proteins analysed so far are located in the plasma membrane (Maurino *et al.*, 2006; Niopek-Witz *et al.*, 2014). A first hint at the function of plant NAT proteins came from the analysis of a homologue from maize, the LPE1 protein, which turned out to transport the purine nucleobases xanthine and uric acid, but not ascorbate (Argyrou *et al.*, 2001). A detailed analysis of NAT3 and NAT12 from *Arabidopsis* also revealed their functions as nucleobase transporters, in this case for adenine, guanine and uracil (Niopek-Witz *et al.*, 2014).

Thus, the molecular identity of the ascorbate transporters in plants has remained elusive until very recently when Miyaji *et al.* (2015) characterized the first ascorbate transporter in *Arabidopsis*, which belongs to the PHT4 family of phosphate transporters. So far, four families of inorganic phosphate transporters have been identified (PHT1–4). While PHT1 and PHT3 represent phosphate carriers of plasma membranes and mitochondria, PHT2 is located in the inner envelope membrane of plastids (see Fig. 10.2; Versaw and Harrisson, 2002). The PHT4 family, first described in plants by Roth and co-workers (Roth *et al.*, 2004), consists of six members (PHT4;1 to PHT4;6) in *Arabidopsis*, with five of them being located in plastids, while PHT4;6 is located in the Golgi (Guo *et al.*, 2008). All members of the family catalyse the transport of phosphate and are thus involved in phosphate homeostasis in chloroplasts, amyloplasts and the Golgi (see Fig. 10.2; Guo *et al.*, 2008; Hassler *et al.*, 2012; Miyaji *et al.*, 2015). In amyloplasts, for example, PHT4;2 mediates the export of excess phosphate that otherwise would inhibit starch synthesis (Irigoyen *et al.*, 2011). Plant PHT4 proteins possess substantial homology to members of the solute carrier family 17 (SLC17) found in animals and other eukaryotes. These proteins mainly reside in the plasma membrane, but also in endomembranes, especially Golgi derived vesicles (Reimer, 2013). In animals, SLC17 represents a functionally diverse family of proteins that mediate the transport of a wide spectrum of different organic anions (besides the transport of phosphate and chloride), e.g. glutamate (Omote *et al.*, 2011), urate (Miyaji *et al.*, 2013), sialic acid, and nucleotides (Reimer, 2013). Substrate transport is Cl^--dependent and in most cases driven by the membrane potential $\Delta\psi$. Based on the broad substrate specificities of the animal transporters, Miyaji *et al.* (2015) hypothesized that members of the plant PHT4 family might also function as organic anion transporters. To prove their hypothesis, one member of each of the four subgroups of the PHT4 family from *Arabidopsis* (*PHT4;3, 4;4, 4;5 and 4;6*) was expressed in *Escherichia coli* and the transport activities of the purified proteins were measured after incorporation into liposomes. First, the authors showed that all four transporters indeed catalyse a Na^+-dependent phosphate transport. Thus, these proteins represent a second sodium-driven transport system in the inner envelope membranes of plastids, the Na^+-pyruvate carrier being the other one (Furumoto *et al.*, 2011). One of the proteins, PHT4;4, exhibited a significant ascorbate transport activity while the others did not. Ascorbate transport activity required millimolar concentrations of Cl^- and was triggered by a membrane potential but not by a proton or sodium gradient. Thus, the transport kinetics resembled those of the homologous mammalian transporters. Ascorbate transport was not inhibited by other organic anions like glutamate or ATP or by DHA, indicating a high specificity for ascorbate. To elucidate its physiological function the authors analysed *pht4;4* deletion mutants of *Arabidopsis*. The mutant plants did not show any visible phenotype, neither under conditions of low light nor under high light when the antioxidant properties of ascorbate would be most needed. However, *pht4;4* plants had a decreased amount of ascorbate in leaves and showed a lower capacity for NPQ. The lack of significant physiological differences between the deletion mutants and wild-type plants may be due to a redundancy in antioxidant systems in chloroplasts and plants. First, PHT4;4 might represent only the high affinity uptake system (K_M = 1.2 mM) since measurements on isolated chloroplasts point to the existence of an additional system with lower affinity (Beck *et al.*, 1983) which has not been identified at the molecular level. Second, other antioxidant compounds like glutathione might compensate for a reduction in the ascorbate system.

In summary, animals and plants possess different transport systems for the transport of ascorbate and DHA, although surprisingly the different transporter families involved are found in both kingdoms. In animals, ascorbate and DHA are transported by proteins of the NAT and glucose transporter families, respectively, while in plants ascorbate transport is catalysed by a member of the PHT4 family.

Nucleobase transporters are essential for the distribution of purines and pyrimidines

Purine and pyrimidine nucleobases are the defining part of nucleotides which themselves constitute a central part of the metabolism of every living cell as building blocks of DNA and RNA and as the 'energy currency' of the cell. Because of their importance for life almost all organisms (with the exception of some parasitic species) are able to synthesize nucleobases and nucleotides *de novo*. In addition, more energy efficient salvage pathways exist in which nucleobases are converted to their corresponding nucleosides and nucleotides. The enormous importance of these compounds is also reflected by the large number of transporters responsible for their distribution within and between cells. The *Arabidopsis* genome, for example, encodes for 49 proteins that are capable of transporting nucleobases (and nucleosides). These belong to six different transporter families (Girke *et al.*, 2014).

In purine *de novo* synthesis, inosine monophosphate (IMP) is the first product with a complete purine ring. IMP is then converted to AMP, GMP and other nucleotides or nucleobases (Zrenner *et al.*, 2006). In plants, purine biosynthesis is exclusively located in plastids (Smith and Atkins, 2002), which implies that the resulting nucleotides have to be exported from plastids for use in other compartments. Although plastids possess an ATP/ADP transporter, this protein functions as an antiporter, i.e. its activity does not lead to a net transport of nucleotides (Reiser *et al.*, 2004). Ten years ago, however, a first uniporter providing the cytosol and other compartments with adenine nucleotides (AMP, ADP, ATP) synthesized in plastids was identified (Leroch *et al.*, 2005; see Fig. 10.1). This transporter belongs to the Brittle1 (BT1) proteins which are named after the *Brittle-1* maize mutant that has kernels with a collapsed angular appearance (Wentz, 1926), a phenotype that is caused by a severely reduced starch content in the endosperm. It has been shown that maize BT1 transports ADP-glucose (ADP-Glc), the substrate for plastid starch synthesis, in strict counter exchange with ADP (or AMP) (Kirchberger *et al.*, 2007). This transport system is necessary in the endosperm of maize and other cereals because ADP-Glc synthesis takes place in the cytosol in the seed storage tissue of these plants. In all other mono- and dicotyledonous plants, including the non-endosperm tissue of maize and other cereals, ADP-Glc synthesis is located in the plastid stroma. Here, the BT1 proteins have a different physiological function, the above-mentioned export of adenine nucleotides. This is a striking finding because it means that these closely related transporters not only have different substrate specificities but also show different transport mechanisms (uniport versus antiport).

In contrast to purine biosynthesis, the pyrimidine synthesis in plants is distributed to three different compartments. While the first two steps occur in plastids, the remaining reactions are located in mitochondria and the cytosol (Witz *et al.*, 2012). The last step forming uridine monophosphate (UMP) occurs in the cytosol. In contrast, the salvage pathway for pyrimidine nucleotides, i.e. the conversion of the pyrimidine nucleobases to their corresponding nucleotides, is located in plastids (Mainguet *et al.*, 2009). Based on the compartmentation of these pathways, Witz *et al.* (2012) proposed a model where UMP in the cytosol is hydrolysed to uracil, which is then taken up by plastids where it serves as substrate for pyrimidine nucleotide synthesis. To confirm their hypothesis the authors first showed that isolated plastids from Cauliflower indeed possess an uptake system for uracil. The uracil transporter protein in *Arabidopsis*, encoded by the gene *At5g03555*, was then identified by its homology to the uracil transporter FUR4 from yeast (Jund *et al.*, 1988). These genes belong to the nucleobase:cation symporter family 1 (NCS1), also known as purine-related transporters. The NCS1 family consists of more than 1000 proteins from bacteria, archaea, fungi and plants (Girke *et al.*, 2014). However, *At5g03555* and its homologues found in maize, grapevine, and poplar seemed to be the sole members of the NCS1 family in plants. Because the protein encoded by *At5g03555* contains a predicted N-terminal plastid targeting sequence Witz and co-workers determined the intracellular localization by transient expression of a protein–GFP fusion. The NCS1 protein was indeed targeted to chloroplasts in *Arabidopsis* and was therefore named PLUTO (for plastidic nucleobase transporter). The substrate specificity of PLUTO was analysed by expression of the protein in an *E. coli* mutant lacking the endogenous uracil permease uraA. It turned out that PLUTO not only transports uracil, but also the purine nucleobases adenine and guanine (Witz *et al.*, 2012), confirming results achieved by expression of the transporter in yeast (Mourad *et al.*, 2012). Recently, a homology model of the tertiary structure of PLUTO based on the crystal structure of the MPH1 transporter from *Microbacterium liquefaciens* was proposed (Witz *et al.*, 2014). According to that model PLUTO consists of 12 transmembrane helices, ten of them forming a compact core. In addition, four amino acid residues, which form the substrate-binding site, were identified and their function was verified by mutational analysis (Witz *et al.*, 2014).

To date, PLUTO represents the only plant nucleobase transporter clearly located in plastids (Girke *et al.*, 2014). Two other transporters for adenine and guanine (Azg1 and 2) with homology to fungal Azg nucleobase transporters have been identified in *Arabidopsis* (Mansfield *et al.*, 2009). Although both are predicted to be targeted to plastids no experimental evidence for their plastid localization has been presented so far.

Plastid transport and pathogen defence – a new connection

Plants have developed diverse defence mechanisms against an enormous number of different viral, bacterial, fungal and animal pathogens. Systemic acquired resistance (SAR) is a defence mechanism in which infection in one part of the plant activates resistance not only at the infection site but throughout the plant which results in a broad and durable resistance to subsequent pathogen infections (Durrant and Dong, 2004). This is achieved by the generation of mobile signals at the infection site, which are translocated to distal tissues and activate defence programs there. A number of mobile compounds involved in SAR have been identified including methyl salicylic acid (MeSA), the diterpenoid dehydroabietadiene (DA), the lysine catabolite pipecolic acid, azelaic acid (AzA), glycerol-3-phosphate (G3P) and the DIR1 protein that shows homology to lipid transfer proteins (Shah *et al.*, 2014 and references therein). Some of these metabolites are synthesized in plastids, i.e. this organelle and its transporters are obviously involved in plant defence. Here we focus on the two metabolites salicylic acid (SA) and G3P because plastid transporters for both compounds have been identified and characterized recently. In addition, we will highlight the transport of two other hormones across the envelope membranes, namely jasmonic acid and abscisic acid.

Export of salicylic acid from plastids

SA is a key regulator of plant defence against biotrophic pathogens. It achieves its function by binding to NPR proteins in the cytosol, which, after import into the nucleus, induce a large set of defence genes (Shah *et al.*, 2014). Despite its enormous importance the synthesis of SA in plants is still not fully resolved. However, efforts to elucidate the SA biosynthetic pathway uncovered evidence for two distinct pathways, the isochorismate (IC) route and the phenylalanine ammonia lyase (PAL) route (Dempsey *et al.*, 2011). The latter one starts with the conversion of phenylalanine to cinnamic acid that is further converted to SA by up to three different, not completely resolved routes. In the IC pathway, chorismate, the end product of the shikimate pathway, is converted to isochorismate, from which SA is directly generated. The first step is catalysed by isochorismate synthase (ICS), which in *Arabidopsis* is encoded by two genes (*ICS1, 2*) (Wildermuth *et al.*, 2001; Garcion *et al.*, 2008) while the second step is proposed to be catalysed by a so far unidentified enzyme. Analysis of several mutants with disturbed SA metabolism showed that the ISC1 pathway is the main contributor to SA production both in infected and non-infected plants. Both ICS isoforms are located in plastids while the location of the second step of the pathway is unknown (Strawn *et al.*, 2007; Garcion *et al.*, 2008). Interestingly, plants expressing a plastid-localized SA hydroxylase from *Pseudomonas putida* are unable to accumulate SA in response to biotic stress, indicating that SA synthesis occurs exclusively in plastids (Fragniere *et al.*, 2011). Thus, the different localization of biosynthesis and physiological activity requires an export of SA from plastids into the cytosol.

Indeed, a plastid SA transporter has been characterized recently (Serrano *et al.*, 2013). The gene encoding this transporter was initially discovered in a forward-genetic screen for *Arabidopsis* plants with enhanced disease susceptibility. The 12 independent loci detected were thus named *EDS1–12* (Glazebrook *et al.*, 1996). The *EDS5* gene was isolated by positional cloning and shown to encode a MATE transporter (which stands for Multidrug and Toxin Extrusion) (Nawrath *et al.*, 2002). The expression of EDS5 is induced by UV-C light, SA treatment and after pathogen inoculation of WT plants, corroborating its involvement in the response to abiotic and biotic stress. Recently, Yamasaki *et al.* (2013) showed that the EDS5 protein is localized to the inner envelope membrane of plastids, a result that was verified by Serrano *et al.* (2013). Importantly, these authors demonstrated the function of EDS5 as SA transporter by different methods. First, isolated chloroplasts from WT *Arabidopsis* plants and from transgenic plants overexpressing EDS5 accumulated radiolabelled SA, while *esd5* mutants did not. Second, yeast expressing EDS5 showed an import activity for SA. Third, the authors chose a genetic strategy to verify the transport function of EDS5. Here, transgenic plants in WT or *eds5* mutant background were constructed with a constitutively active IC pathway, both resulting in high levels of SA in the absence of stimuli but with a different subcellular distribution. In the WT background the chloroplasts contained only low levels of SA due to the export activity of EDS5, whereas in the *eds5* mutant background chloroplast accumulated high levels of SA. These data clearly indicated that EDS5 represents the long sought SA exporter of plastids.

MATE transporters form a very large protein family, with members found in all phylogenetic groups, i.e. bacteria, Archaea, animals, fungi and plants. The first MATE transporters were identified in *Vibrio parahaemolyticus* and *E. coli* (Morita *et al.*, 1998) and shown to transport a number of organic cations (reviewed in Omote *et al.*, 2006). They were named according to their function in bacteria and animals, the extrusion of drugs and toxic compounds. All MATE transporters characterized so far catalyse an antiport of an organic cation against either a proton or sodium ion, depending on the type of protein. In most cases, they have broad substrate specificities accepting a wide range of different organic molecules. While bacteria and animals only possess a few MATE proteins, plants usually have large gene families, e.g. the *Arabidopsis* genome contains 57 MATE-like genes (see: aramemnon.botanik.uni-koeln.de). The first plant MATE transporters analysed were ALF5 and DTX1 from *Arabidopsis* that are located in the plasma membrane and catalyse the efflux of xenobiotics, thus conferring resistance to toxins (Diener *et al.*, 2001; Li *et al.*, 2002). Other MATE transporters are located in the tonoplast membrane and mediate the transport of flavonoids (e.g. Marinova *et al.*, 2007) and nicotine (Morita *et al.*, 2009), i.e. these transporters are involved in the sequestration of secondary metabolites into the vacuole. Thus, the existence of these large number of plant MATE transporters might reflect the necessity of transporting the enormous variety of secondary metabolites synthesized by plants.

To summarize, EDS5 is the first MATE transporter found in plastids and the second family member involved in phytohormone transport. Previously, abscisic acid efflux across the plasma membrane was shown to be facilitated by *AtDTX/MATE50* (Zhang *et al.*, 2014).

Transport of jasmonic acid and abscisic acid

Jasmonic acid (JA) and abscisic acid (ABA) represent important phytohormones involved in responses to biotic and abiotic stress. Interestingly, although their signals are perceived in the cytosol or nucleus, respectively, both compounds are derived from nonpolar precursors synthesized in chloroplasts. Already 30 min after physical wounding of a plant leaf or upon herbivore attack JA accumulates in green tissues. In the JA biosynthesis pathway α-linolenic acid is converted via three enzymatic steps in the stroma to oxo-phytodienoic acid (OPDA), while the final biochemical steps of this pathway occur exclusively in peroxisomes. Therefore, OPDA has to be released from chloroplasts and shuttled via the cytosol into peroxisomes. It has been suggested that OPDA can be stored while esterified to the glycerol backbone and/or the galactose headgroup of chloroplast membrane galactolipids. This buffer would help to guarantee the prompt JA biosynthesis upon perception of a stress trigger. However, it still remains elusive what gene(s) encode the chloroplasts OPDA exporter. In the peroxisomal membrane the full-size ABC-transporter PXA1/CTS facilitates fatty acid import and most likely also OPDA uptake into the organelle. PXA1 possesses a homologue designated as PMP1 that was found in multiple independent proteomic studies in plastid membranes (Zybailov *et al.*, 2008; Takabayashi *et al.*, 2013). Thus PMP1 may play a role in OPDA release from chloroplasts. However, most likely other OPDA transport mechanisms exist in both organelles since a complete loss of JA biosynthesis results in male sterile plants. *pxa1/cts* mutant plants are not compromised in fertility and show only reduced JA accumulation after wounding. If only one locus would encode a chloroplast and peroxisomal OPDA carrier, respectively, forward genetics screens that were successfully in identifying other major players of the JA pathway, e.g. OPR3, should have also led to the transporter genes. Therefore it is possible, that gene families may encode multiple homologues with partially overlapping substrates or that an independent transport mechanism may exist which might be related to fatty acid transport systems. Using unicellular prokaryotic and eukaryotic systems it was shown that free fatty acids cannot be imported into mutant cells that lack acyl-activating enzymes (Pulsifer *et al.*, 2012; von Berlepsch *et al.*, 2012). In *A. thaliana*, up to 63 loci encode acyl-activating enzymes with different subcellular localizations (Shockey *et al.*, 2003). A study with *A. thaliana* mutants showed that two long-chain acyl-CoA synthetases, LACS4 in the endoplasmic reticulum and LACS9 in the outer chloroplast envelope membrane, are involved in lipid trafficking from ER to plastids (Jessen *et al.*, 2015). Because LACS9 facilitates fatty acid import, the fatty acid export mechanism of chloroplasts remained enigmatic for a long time. Recently, a gene designated as FAX1 (At3g57280) that potentially encodes the long-sought exporter was discovered in *A. thaliana* (Li *et al.*, 2015). The *fax1* mutants are compromised in male fertility, growth rate and accumulate plastid-produced lipid species. Loss of function *fax1* mutants might still be partially rescued by a putative second homologue *fax2* (At2g38550) that was identified many times in the chloroplast proteome. Given the fact that fatty acids and the JA precursor OPDA are imported into peroxisomes via the same mechanism, FAX1 and its homologue may also contribute to some extend to the plastid OPDA release. It is therefore tempting to test JA levels in plastid *fax* single or higher-order mutants. It is possible that the reduced fertility of *fax1* is partially due to decrease in JA production.

The situation for the other aforementioned phytohormone ABA is similarly elusive. ABA is produced in chloroplasts from carotenoids (Finkelstein, 2013). In the later steps of ABA biosynthesis, zeaxanthin is produced by two β-carotene hydroxylase homologues (BCH1 and BCH2). Zeaxanthin is further metabolized to violaxanthin by zeaxanthin epoxidase (ZEP). This represents a very interesting step because zeaxanthin to violaxanthin conversion is also an essential component of NPQ via the xanthophyll cycle to adapt to high-light conditions (see above). In other words, phytohormone production and an independent physiological feature, i.e. photosynthetic acclimation, is partly interrelated. After two further enzymatic reactions the last plastid product of ABA biosynthesis, xanthoxin, is formed. The last two remaining enzymatic steps occur in the cytosol (Ma *et al.*, 2009; Park *et al.*, 2009). Thus, it has been postulated for years that a xanthoxin export mechanism in the chloroplast envelope membrane exists, but no genes encoding this transporter(s) have been identified so far (Floss and Walter, 2009).

So far three transporter families were shown to be involved in either ABA import or release at the plasma membrane. These are ABC transporters (ABC-G22, 25, 40), a nitrate/peptide transporter (NRT1.2/AIT1) and a MATE carrier (DTX/MATE-type 50) (Jarzyniak and Jasinski, 2014). However, none of these carriers has close homologues with strong *in silico* predictions for chloroplast targeting and it is uncertain if any of these members could potentially accept xanthoxin as a transport substrate at all. Again genetic functional redundancy could be involved here. Due to the ease of detecting ABA signalling-related mutants, e.g. germination mutants, most ABA deficient mutants in the last thirty years were identified using extensive forward genetic screening. However, although all ABA-deficiency mutants isolated so far lack proteins downstream of zeaxanthin synthesis, a chloroplast transporter was never amongst them (Floss and Walter, 2009; Finkelstein, 2013). Therefore, it is possible that a protein family of unknown function in the chloroplast envelope membrane awaits discovery for the plastid xanthoxin export.

Glycerol-3-phosphate, its transport and physiological function

Glycerolipids consists of two (polar lipids) or three (triacylglycerol, TAG) fatty acids bound to a glycerol backbone. While polar lipids constitute the lipid bilayers of membranes, TAGs serve as storage reserves in many organisms. In most plant species, including *Arabidopsis*, glycerolipids are derived from two different pathways, the plastid-localized (prokaryotic) and the ER-localized (eukaryotic) pathway (Ohlrogge and Browse, 1995). In both, glycerolipids are synthesized by the transfer of fatty acids onto glycerol-3-phosphate (G3P) as acceptor molecule. G3P can be synthesized in plants by two reactions. First, glycerol is phosphorylated to G3P by a cytosolic glycerol kinase (Eastmond, 2004). Second, G3P is produced by the reduction of dihydroxyacetone phosphate (DHAP), a metabolite of glycolysis and the Calvin cycle. This reaction is catalysed by DHAP reductases, which are found in plastids, mitochondria and the cytosol of plant cells.

However, G3P is not only involved in lipid biosynthesis but is also a major player in plant pathogen defence. Initial evidence for this role of G3P came from analysis of an *Arabidopsis* mutant with increased susceptibility to a variety of bacterial and fungal pathogens that normally do not infect wild-type *Arabidopsis* plants (Kang *et al.*, 2003). This mutant is obviously impaired in 'non-host disease resistance', which refers to the phenomenon that most plant species are resistant to the pathogens of other plant species. The *nonhost1* (*nho1*) mutant has a single nucleotide substitution in the *NHO1* gene (At1g80460) leading to a premature stop codon in the second exon of the gene. The single copy *NHO1* gene encodes the cytosolic glycerol kinase indicating that G3P or metabolites derived from it are necessary for non-host disease resistance (Kang *et al.*, 2003). Shortly thereafter, Nandi and coworkers (Nandi *et al.*, 2004) demonstrated that the *Arabidopsis SFD1* gene, also known as *GLY1*, is required for the induction of SAR. *SFD1/GLY1* codes for a plastid-localized DHAP reductase. The loss of this enzyme activity not only leads to an altered lipid composition of plastids but also to a compromised SAR-conferred pathogen resistance (Nandi *et al.*, 2004; Lorenc-Kukula *et al.*, 2012). A detailed analysis of *nho1* and *sfd1/gly1* single and double mutants revealed that G3P acts as an important inducer of SAR in plants (Chanda *et al.*, 2011). This conclusion was drawn from data showing that the synthesis of G3P at the infection site and the transport to distal tissues of a compound derived from G3P is necessary for SAR. The transport depends on the cooperative interaction of G3P and DIR1, i.e. G3P is necessary for the transport of DIR1. However, it should be emphasized that it is not G3P itself that is transported together with DIR1 but a metabolite derived from it, most likely a lipid (derived) compound. Interestingly, feeding of exogenous G3P restores SAR in the G3P synthesis mutants indicating the existence of a G3P transport system in the plasma membrane.

In plants, G3P transport has first been shown in isolated spinach chloroplasts (Fliege *et al.*, 1978), although at that time the molecular identity of the transporter remained elusive. In bacteria, in contrast, the G3P transporter (GlpT) from *E. coli* (Larson *et al.*, 1982) and later from several other bacteria, e.g. *Haemophilus influenzae* (Song *et al.*, 1998), has been cloned and extensively studied. GlpT, a close homologue of the bacterial glucose-6-P transporter UhpT, catalyses the exchange of G3P against inorganic phosphate and thus enables the bacteria to grow on G3P as sole carbon source (reviewed in Lemieux *et al.*, 2004). The GlpT protein has been crystallized and its three-dimensional structure has been determined (Huang *et al.*, 2003). The protein consists of twelve transmembrane helices, which form two N- and C-terminal domains. Between both domains, a substrate-translocation pore exists. GlpT operates via a single binding site with alternating access of substrates from both sides of the membrane. Plant P3G transporters have been identified based on their homology to bacterial and animal transporters. *Arabidopsis* possesses five G3P permeases (AtG3Pp1–5) while the genomes of other plants contain four to six genes encoding potential G3P transporters (Ramaiah *et al.*, 2011; see also aramemnon.botanik.uni-koeln.de). The G3Pp genes in *Arabidopsis* exhibited differences in tissue-specific expression and also differed in their response to phosphate starvation. A loss-of-function mutant of G3Pp4 showed an enhanced growth of lateral roots and a lower seed lipid content suggesting a role both in root development and lipid synthesis (Ramaiah *et al.*, 2011; Kawai *et al.*, 2014). This relatively moderate mutant phenotype might indicate some degree of redundancy within the family or compensation by other biochemical mechanisms. The G3Pp4 transporter is localized to plastids while the localization of the other family members have not been experimentally verified (Kawai *et al.*, 2014). The transport activity of G3Pp4 was analysed by expression in an *E. coli* Δ*glpT* strain lacking the internal G3P permeases. G3Pp4 indeed catalysed an import of G3P into the *E. coli* cells, thus this protein represent the first characterized plastid G3P transporter (Kawai *et al.*, 2014). The effect of the absence of plastid G3P transport on plant defence has not been analysed thus far. Therefore, the significance of plastid G3P transport for plant defence remains to be determined.

Ion transport processes, recent discoveries and future directions

In the last years a lot of research activity has been dedicated to chloroplast ion transport. The field covers ion flux across the envelope membranes and ion distribution within the plastid sub-compartments, i.e. stroma and thylakoid lumen, respectively. Chloroplast ion transport has a high relevance for proper organelle function since the plastid ion homeostasis is linked to pH homeostasis and osmoregulation. Ion and proton gradients across the envelope and thylakoid membranes are used to store energy, to fuel ATP production and to drive transport processes. Thus it is obvious that perturbation of ion homeostasis has a major impact on the biochemical processes in the chloroplast and specifically on photosynthesis. Recently, the first reviews on the topic have summarized findings from the last years either by focusing on the macro and micronutrient function of ions in the organelle (Hanikenne *et al.*, 2014) or on the link between ion transport and photosynthesis (Checchetto *et al.*, 2013; Finazzi *et al.*, 2014; Pfeil *et al.*, 2014). When comparing the reports it becomes obvious that open questions still remain regarding involved ion transporter genes and the transport directions of characterized carriers. With a few exceptions we will focus here on the ion traffic across the chloroplast envelope membrane (Fig. 10.2).

Chloroplasts have a high metabolic activity and many of the enzymatic reactions require ions as cofactors or for full activation. Because of the high demand for various ions several different carriers in the inner envelope membrane are expected to function in the ion import and export business of the chloroplast. Their role is to prevent ion starvation on one hand and toxic over-accumulation of specific ions on the other hand. The plastid ion homeostasis is a carefully fine-tuned process balancing the free levels of the individual metals (Cu^{2+}, Fe^{2+}, Mn^{2+}, Zn^{2+}, K^+, Mg^{2+}, Ca^{2+}, Na^+) and inorganic ions (Cl^-, PO_4^{2-}, SO_4^{2-}, NO_3^-) that vary significantly from nano- to millimolar range. In addition, the ionic composition found in the thylakoid lumen is significantly different from the plastid stroma.

Metal ion transport I – the monovalent ions K^+ and Na^+

K^+ is the most abundant ion in plant tissues and with a concentration of 100–200 mM also represents the dominating ion in chloroplasts (Robinson *et al.*, 1983; Schroppelmeier and Kaiser, 1988). Unfortunately, information on sub-organellar K^+ concentrations, i.e. stroma vs thylakoid lumen is not available thus far but would be highly desirable. K^+ functions in the organelle in osmoregulation, plays an essential role for enzyme activation, as a pH buffer and as a counter ion. The significance of proper K^+ supply for plants was recognized a long time ago. K^+ starvation results in chlorophyll loss, leaf necrosis and poor photosynthesis (Marschner, 1995). During salt stress toxic Na^+ ions also build up in chloroplasts and outcompete K^+, which results in similar phenotypes (Schroppelmeier and Kaiser, 1988; Stepien and Johnson, 2009). Because of the relevance of K^+ for chloroplasts, scientists early on performed physiological experiments on isolated intact chloroplasts or on reconstituted chloroplast membranes to measure K^+ transport and to determine the transport characteristics. The data revealed a K^+ channel activity in both the envelope (Mi *et al.*, 1994; Heibert *et al.*, 1995) and thylakoid membranes (Tester and Blatt, 1989) and a K^+/H^+ antiport activity in the envelope membranes (Huber and Maury, 1980; Demmig and Gimmler, 1983; Wu and Berkowitz, 1992; Wang *et al.*, 1993). However, although these data were published decades ago, no genes encoding these channels and transporters could be identified. First evidence for the molecular identity of the thylakoid K^+ channel came from the discovery of the *SynK* locus in cyanobacteria (Zanetti *et al.*, 2010). In *A. thaliana* TPK3 and TPK5 represent the closest homologues of SynK and both carry an *in silico* predicted N-terminal transit peptides for trafficking to the chloroplast. Indeed, an antibody raised against a TPK3/TPK5 epitope detected the protein(s) in the *A. thaliana* thylakoid membranes. The signal was verified to stem from TPK3, since no significant signal reduction was detected in *tpk5* T-DNA insertion lines (Zanetti *et al.*, 2010). To overcome the lack of T-DNA insertion lines in the *TPK3* locus, a *tpk3*-RNAi knock-down approach was carried out. The isolated transgenic lines revealed light sensitivity only under moderate light conditions (90 μmol photons/m^2/s) and modified proton motive force (pmf) portioning (Carraretto *et al.*, 2013). The pmf which drives ATP production is thought to be equally split in higher plants into the chemical potential ΔpH (about Δ3) and the transmembrane electric potential ΔΨ. The thylakoid channel TPK3 is therefore supposed to release K^+ from the lumen into the stroma and to act as a modifier of the pmf composition via ion counterbalancing.

Although a K^+ influx transport activity was detected many times in chloroplast envelope membranes, no gene encoding the corresponding channel(s) has been identified thus far. The most recent study on this subject, performed on intact pea chloroplasts, suggested a rather unspecific ion channel, designated as fast-activating channel (FAAC) to be responsible for the K^+ loading into chloroplasts (Pottosin *et al.*, 2005). Interestingly, the channel also showed permeability for Na^+ ions and might therefore be a main contributor of Na^+ accumulation in chloroplasts during salt stress. In addition, also the Na^+-dependent transporters of the BASS (Furumoto *et al.*, 2011) and PHT families (Miyaji *et al.*, 2015) might play a role in this process (see above). To balance the K^+ influx into plastids and to avoid a dramatic increase in osmotic potential electro neutral K^+/H^+ exchangers in chloroplast envelope membranes were proposed years ago (Bernardi, 1999). The author compared the situation at the inner envelope membrane with the mitochondrial inner membrane and concluded that similar systems should be in place in both organelles to prevent osmotic bursting. Indeed, K^+/H^+ exchange activities were detected by independent groups (Huber and Maury, 1980; Demmig and Gimmler, 1983; Wang *et al.*, 1993) but the actual proteins responsible for this process remained elusive.

The electro neutral cation/proton antiporters (CPA) in *A. thaliana* represent a large superfamily with 44 predicted genes (Mäser *et al.*, 2001). The group can be further divided into the CPA1 and CPA2 families. CPA1 consists of the $Na^+(K^+)(Li^+)/H^+$ exchangers (NHX, eight members) that play an important role in salt stress tolerance. The CPA2 family covers two subfamilies, which include Cation/H^+ exchangers (CHX, 28 members) and putative K^+-efflux antiporters (KEA, six members). Interestingly, no NHX or CHX subfamily members have ever been detected in chloroplast or mitochondrial proteomic studies. However, CHX23 was thought to be a chloroplast $Na^+(K^+)/H^+$ antiporter based on the analysis of a *chx23*RNAi line and an EMS TILLING line (Song *et al.*, 2004). Although both lines revealed pale green phenotypes which may indicate a loss of chloroplast function, later two independent studies carried out on viable homozygous T-DNA insertion lines did not confirm these earlier observations (Lu *et al.*, 2011; Evans *et al.*, 2012). In both mutant lines no leaf paleness or compromised chloroplast function was detected. Moreover, the CHX23 carrier was found to localize to the endoplasmic reticulum and is preferentially expressed in pollen.

Three KEA members (KEA1–3) carry chloroplast transit peptides and were identified in proteomics studies. The KEA transporters originate from a bacterial transport system known as KefC/B. In *E. coli* their transport activities were studied extensively. KefC/B were shown to release K^+ for proton uptake to lower the pH of the cytosol upon electrophilic stress (Munro *et al.*, 1991; Ferguson *et al.*, 1995; Fujisawa *et al.*, 2007; Roosild *et al.*, 2009). Using a KEA2 fragment from *A. thaliana* Aranda-Sicilia and colleagues were able to confirm a K^+/H^+ activity for the plant homologue (Aranda-Sicilia *et al.*, 2012). Interestingly, the KEA2 fragment was also able to exchange Na^+ ions for protons, albeit with less affinity. Briefly thereafter, T-DNA insertion mutants of *KEA* genes were isolated and described for the first time (Kunz *et al.*, 2014). Higher-order *kea1kea2* mutant plants, but not single mutants, displayed dramatic growth phenotypes along with poor photosynthesis. Chloroplasts in these mutants were found to be swollen and often disrupted pointing to the proposed role of these carriers in balancing the massive K^+ influx of the unknown K^+ channel (Aranda-Sicilia *et al.*, 2012; Kunz *et al.*, 2014). In the light, a pH gradient of one unit was detected across the envelope membrane (Werdan *et al.*, 1975). This gradient should ensure K^+ efflux form the chloroplast via KEA1 and KEA2 carriers. Since KEA transport activity is H^+ dependent, loss-of-function *kea1kea2* double mutant also revealed perturbed plastid pH homeostasis.

Interestingly, a third plastid member of the KEA-family, KEA3, resides in the thylakoid membrane (Armbruster *et al.*, 2014; Kunz *et al.*, 2014; Tomizioli *et al.*, 2014). The K^+ efflux channel TPK3 calls for a luminal K^+ loading system. KEA3 driven by the steep ΔpH in the light may facilitate this task. However, the transport activity would need strict regulation to avoid an uncoupling of the pmf. Recently, Armbruster and colleagues discovered that KEA3 functions to adjust photosynthesis under fluctuating light conditions (Armbruster *et al.*, 2014). Upon rapid shifts from high-light to low-light *kea3* mutants revealed prolonged phases of non-photochemical quenching. This effect was attributed to a less efficient H^+ release from the lumen, which is needed to quickly adapt to changing light conditions that plants experience every day in nature. It still remains to be shown what mechanisms are in place to ensure precise transport regulation via KEA3. In the case of TPK3 it was demonstrated that Ca^{2+} and H^+ activate the channel (Carraretto *et al.*, 2013). The evolutionary origin of chloroplast KEAs is not entirely clear. A symbiont origin for all three plastid KEA members found in *A. thaliana* was suggested, although only KEA homologues with low similarity to plant proteins exist in some cyanobacteria, e.g. *Gloeobacter violaceus* (Chanroj *et al.*, 2012; Pfeil *et al.*, 2014).

The next chloroplast ion carrier discussed in the context of plastid ion transport is NHD1. Since the activity of Na^+-dependent carriers of the BASS and PHT families results in Na^+ import, a plastid Na^+ efflux carrier was postulated (Furumoto *et al.*, 2011; Miyaji *et al.*, 2015). Indeed, it was shown that NHD1 from *A. thaliana* works in concert with the pyruvate carrier BASS2 if expressed simultaneously in a heterologous *E. coli* system (Furumoto *et al.*, 2011). Recently, in an independent work the Na^+/H^+ antiport of NHD1 was confirmed *in vitro* and *in vivo* in *A. thaliana nhd1* T-DNA insertion lines (Müller *et al.*, 2014). Interestingly, mutant seedlings revealed increased sensitivity towards salt-stress or salt-shock treatments. However, under normal growth conditions mutant plants grew similar to WT plants and did not respond in a similar fashion to salt treatments. NHD1–VENUS fusion proteins were found to localize to the chloroplast envelope membrane. KEA1 and KEA2 also reside in the envelope membrane and the KEA2 fragment was found to also facilitate Na^+/H^+ exchange (Aranda-Sicilia *et al.*, 2012; Kunz *et al.*, 2014). This motivated a further study in which *kea1kea2nhd1* triple mutants were isolated (Kunz *et al.*, 2014). Indeed, a strong increase in phenotype intensity was observed if KEA-mediated function across the inner envelope membrane was lost in the *nhd1* mutant background. Given the fact that all three transporters are energized by the light-dependent proton gradient across the envelope membrane it can be hypothesized that they all share the same transport direction, i.e. K^+/Na^+ efflux from the chloroplast in exchange for H^+. However, it would be interesting to investigate if the transport direction switches in the absence of light or in plastids from heterotrophic tissues in which the pH of the stroma is similar to the cytosol (Heldt *et al.*, 1973; Werdan *et al.*, 1975). KEA1 and KEA2 were found in the envelope proteome of cauliflower proplastids and thus might also have important functions in non-green plastids, e.g. in plastid development (Bräutigam and Weber, 2009). Lastly, it should be noted that an in-depth thylakoid proteomic study found NHD1 proteins also in the thylakoid stroma-lamella fraction indicating a dual localization of NHD1 in plastids (Tomizioli *et al.*, 2014). Interestingly, a dual-targeting within chloroplasts was suggested for another plastid transporter (TAAC/PAPST1; At5g01500). Initially the protein was characterized as an ADP/ATP exchanger in thylakoids (Thuswaldner *et al.*, 2007). However, later an independent study showed that TAAC/PAPST1 also functions as an antiporter for 3′-phosphoadenosine 5′-phosphosulfate (PAPS) in exchange with 5′-phosphoadenosine 3′-phosphate (PAP) or ATP across the inner envelope membrane (Gigolashvili *et al.*, 2012). To confirm a similar dual-targeting phenomenon for NHD1 more detailed studies are needed.

Metal ion transport II – the divalent ions Cu^{2+}, Fe^{2+} and Ca^{2+}

Iron represents an essential nutrient in plants with major roles for chloroplast function and photosynthesis. In fact, the two photosystems, the cytochrome b6–f complex, and ferredoxin belong to the most iron-enriched systems found in plant cells (Duy *et al.*, 2007). Depending on the plant species and its nutrient status 80–90% of the iron in leaves is located in chloroplasts where it is found in a variety of haem and Fe–S cluster proteins. Because of their capacity as transient electron acceptors these proteins participate in several vital reactions, such as electron transport in photosynthesis (Rieske proteins), as redox partners in nitrogen and sulfur assimilation but also in non-redox reactions, gene expression regulation and in cellular O_2 and iron status sensing (Abdel-Ghany *et al.*, 2005b; Pilon *et al.*, 2006). Under conditions of iron starvation, predominantly younger leaves show iron deficiency symptoms that involve chlorosis due to loss of chlorophyll pigments (mainly Chlorophyll a), poor photosynthetic efficiency and decreased CO_2 fixation rates. This underpins the immense importance of iron for chloroplasts. Several different Fe^{2+} uptake systems in chloroplasts have been proposed or identified over the last years. The first biochemical studies regarding iron uptake revealed that illuminated barley chloroplasts accumulate more Fe^{2+} than those kept in darkness, which indicated a light-depended transport mechanism (Bughio *et al.*, 1997). Later Shingles and colleagues demonstrated a proton dependent symport (most likely light-depended) of Fe^{2+} into isolated pea chloroplasts (Shingles *et al.*, 2002).

In *A. thaliana*, permease in chloroplasts 1 (PIC1, At2g15290) was the first transporter discovered to facilitate the majority of plastid import of this transient metal (Duy *et al.*, 2007). The *PIC1* gene was identified in comparative genome studies of *Arabidopsis* with cyanobacteria and was grouped within the family of high-affinity Fe^{2+}/Pb^{2+} permeases. Heterologous expression of *PIC1* in the yeast *fet3fet4* double mutant that lacks both low- and high-affinity iron uptake systems was able to rescue the mutant growth phenotype on low-iron media. In addition, *pic1* loss-of-function mutants (Duy *et al.*, 2007) and *PIC1* overexpressing lines (Duy *et al.*, 2011) were found to display typical iron related phenotypes, i.e. iron-deficiency symptoms or plastid Fe^{2+} and ROS accumulation, respectively. However, the actual transport mechanism of PIC1 is still unknown and thus PIC1 remains to be confirmed as the major H^+-dependent Fe^{2+} mechanism that has been described for pea chloroplasts (Shingles *et al.*, 2002). It should also be mentioned that PIC1 (also designated as TIC21) might be part of the TIC/TOC protein import machinery (Teng *et al.*, 2006) and that it interacts with the putative nickel/cobalt transporter NiCo/AtZN1 (At2g16800) (Duy *et al.*, 2011). It was therefore suggested that PIC1 and NiCo might work in close cooperation. However, although AtZN1 and its close homologue AtZN2 (At4g35080) both carry an *in silico* predicted transit peptide, the rice homologue zebranecrosis1 (ZN1) was found to reside in the thylakoid membrane fraction (Li *et al.*, 2010). Thus it is obvious that the model for a PIC1 and NiCo/ZN1 interaction needs further verification.

Three other transporters were also suggested to be involved in iron uptake into chloroplasts. However, their involvement in the process is less well understood. These are MAR1 (Conte *et al.*, 2009), Mfl1 (Tarantino *et al.*, 2011) and NAP14 (Shimoni-Shor *et al.*, 2010). MAR1 was first described as an unspecific influx carrier for antibiotics into chloroplasts. Because chlorotic *mar1* mutants could be rescued by exogenous iron, an involvement in chloroplast iron homeostasis was also discussed (Conte *et al.*, 2009). Mfl1 was found in independent chloroplast proteomics studies and has homology to Mitoferrin2 (MFRN2), a mitochondrial iron importer from zebrafish. Two *atmfl1* loss-of-function mutants had significantly lower iron levels in leaf tissues than WT plants after exogenous iron application. However, no other mutant phenotypes were observed (Tarantino *et al.*, 2011). The authors suggest that Mfl1 could play a role in chloroplast Fe^{2+} influx or efflux, i.e. adjusts chloroplast iron concentrations depending on the H^+-gradient across the envelope membranes. NAP14 is a non-intrinsic ABC protein. It was shown to reside in the chloroplast envelope membranes both in proteomics studies and by expression of a NAP14–GFP fusion protein. The loss-off-function *nap14* mutant displayed a severe albino-like phenotype and over-accumulated Fe^{2+} in shoot tissues which led to the conclusion that it plays a role in plastid iron transport (Shimoni-Shor *et al.*, 2010). Thus it is hard to pinpoint the actual main plastid Fe^{2+} transporter. For PIC1 at least a robust complementation of an iron transporter deficient yeast mutant could be shown. To better understand the exact contribution of the different proteins to plastid Fe^{2+} import direct transport studies in an *in vitro* system, e.g. purified protein reconstituted in liposomes, would be highly desirable.

Copper represents another redox active transition metal with essential function for the chloroplast. Similarly to iron, many plastid proteins require Cu^{2+} as a cofactor, such as enzymes with major function in photosynthesis (plastocyanin) and ROS protection (superoxide dismutase). Around 50% of the more than 100 different Cu-containing proteins are localized in the chloroplast (Marschner, 1995). This calls for efficient plastid Cu^{2+} uptake carriers and for further distribution to the plastid sub-compartments. In *A. thaliana* mainly three carriers were identified to be responsible for the copper transport. HMA1 (At4g37270) and HMA6/PAA1 (At4g33520) both reside in the inner chloroplast envelope membrane (Shikanai *et al.*, 2003; Seigneurin-Berny *et al.*, 2006) and HMA8/PAA2 (At5g21930) in the thylakoid membranes (Abdel-Ghany *et al.*, 2005a; Tapken *et al.*, 2012; Tomizioli *et al.*, 2014). The HMA proteins belong to the family of P1B-type ATPases. These primary active transporters can possess rather broad substrate specificity for divalent heavy metal cations. By employing yeast mutants perturbed in Ca^{2+} homeostasis or lacking heavy metal carriers, HMA1 was shown to transport Cu^{2+} but also Zn^{2+}, Ca^{2+}, Cd^{2+}, and Co^{2+} (Moreno *et al.*, 2008). HMA6/PAA1 seems to be more specific and accepts only copper and silver ions (Catty *et al.*, 2011). Plant mutants that lack HMA6/PAA1 remain pale if grown without the addition of exogenous Cu^{2+}. However, this is not the case for *hma1* mutants which behave similarly to wild-type plants under normal growth conditions (Kim *et al.*, 2009). HMA1 and HMA6/PAA1 have distinct but partially overlapping functions in Cu^{2+} uptake into chloroplasts but, most likely a third alternative route for copper import across the inner envelope may exist and awaits discovery (Boutigny *et al.*, 2014). However, both *hma6/paa1* and *hma8/paa2* loss-off-function mutants reveal high chlorophyll fluorescence and defects in photosynthetic electron transport. A complete loss of both carriers in *hma6hma8* double mutants results in early seedling lethality underpinning the importance of copper for the chloroplast and for plant development.

The important role of calcium signalling for chloroplasts has been recently highlighted and reviewed by others (Stael *et al.*, 2012; Nomura and Shiina, 2014). The stroma has to maintain a very low concentration of free calcium (below 150 nM) and thus the majority of plastid Ca^{2+} is stored in the thylakoid lumen where it is bound to the negatively charged thylakoid membranes or to calcium-binding proteins (~15 mM) (Nomura and Shiina, 2014). Early biochemical studies suggested a light-dependent Ca^{2+} uptake mechanism across the inner envelope membrane (Roh *et al.*, 1998). However, not much is known about Ca^{2+} import into plastids or into the thylakoid lumen. Two candidates for envelope membrane transporters cited in the literature are HMA1 (which is discussed above) and ACA1 (Huang *et al.*, 1993). HMA1 has broad substrate specificity for divalent cations but most likely is rather needed for heavy metal trafficking and is thus unlikely to play a major role for Ca^{2+} import (Moreno *et al.*, 2008; Kim *et al.*, 2009). Chloroplast localization of ACA1 was never confirmed experimentally. In fact, ACA1 localization to the ER, vacuole or the plasma membrane is much more likely according to proteomics data (Tanz *et al.*, 2013). For the Ca^{2+} import into the thylakoid lumen a light- or ATP-dependent Ca^{2+}/H^+ antiport was detected in isolated thylakoid membranes but no transporters have been identified to facilitate the transport thus far (Ettinger *et al.*, 1999). Since understanding plastid Ca^{2+} import is key to unravel the many roles of calcium (signalling) not only for biochemical processes within the organelle but also for the entire plant cell, more research activity should be focusing on this important biological question in the future.

Conclusions

The central role of plastids in plant cells calls for a high degree of connectivity between the different organelles under all physiological conditions, as well as in response to abiotic and biotic stresses. The envelope membranes represent the interface controlling metabolite and ion fluxes into and out of plastids. Since both membranes are impermeable for ions and polar compounds, a variety of specialized membrane transporters and channels facilitate the import and export of compounds from plastids. In classic physiological studies on isolated intact plastids or reconstituted plastid membranes transport activity across the envelope membrane was demonstrated for a wide range of substrates. To date, approximately 40 of these transport processes have been analysed in detail (Figs. 10.1 and 10.2), i.e. the responsible transport proteins have been identified and characterized biochemically and the corresponding genes have been cloned and sequenced. This has been achieved by different approaches, namely biochemical techniques of protein purification, forward and reverse genetics, 'omics' techniques (proteomics, transcriptomics, genomics), and bioinformatics (reviewed in Fischer, 2011).

However, the functions of the majority of genes coding for plastid transporters are still unknown. The percentage of well-investigated or described membrane transporters of plastids is still fairly low and estimated to be only approximately 25% of the total permeome (Weber *et al.*, 2005; Fischer, 2011). This number only reflects the situation in the model organism *A. thaliana*. Given the considerable differences in metabolism between different plant species, in particular the variety of secondary compounds produced by plants, it is expected that many more specialized chloroplast transporters exist in other plant species that are barely investigated due to the lack of genome information or genetic tools. Another aspect that adds complexity to the issue is variation in substrate specificity for paralogous proteins from different plant species, for example the Brittle1 carrier (see above). These differences in substrate specificities have obviously a major impact on the physiological relevance of these carriers, which is reflected by the diverse phenotypes in knock-down mutants in different plant species. In a recent review Pick and Weber (2014) discussed this still unknown plastid permeome, e.g. the still missing carbamoyl aspartate transporter (pyrimidine *de novo* synthesis), the unknown amino acid transporters and several others.

Phylogenomics approaches have revealed the evolution of the plastid transporters. Chloroplast metabolite transporters have multiple evolutionary origins, having evolved from transporter proteins of the host endomembrane system, from members of the mitochondrial carrier family, or cyanobacterial ancestors. Some have been introduced by horizontal gene transfer. Thus the transporter proteome of the plastid envelope membrane has a mosaic origin, i.e. genes encoding these proteins have been recruited from multiple sources. In addition, the potential role of plastid transporters in the establishment of the symbiotic relationship between endosymbiont and host cell has become clear. The integration of transporters connecting endosymbiont to cytosolic metabolism, a process primarily driven by the host cell, was an early and crucial event in chloroplast evolution.

Future trends

Although considerable progress has been made in unravelling the molecular nature of components of the plastid permeome over the past two decades, the topic is still far from being fully understood. Plastid (and plant) metabolism can only be mechanistically understood once all of the involved transporters and channels have been identified and characterized. However, despite the high importance of metabolites and ions for chloroplast function the molecular identity of several of the essential transport mechanisms remains unknown. One reason for this limited knowledge is that many plastid transporters are encoded by gene families. As illustrated by the envelope K^+/H^+ exchangers, genetic functional redundancy does occur in chloroplast transport systems, which limits the possibilities of gene discoveries by employing forward genetic-screens (Kunz *et al.*, 2014). To overcome these pitfalls combinations of old and new tools for the identification and molecular characterization of plastid transporters are needed to pave way for future discoveries.

Next generation sequencing technologies will increase throughput and decrease cost of genomic sequencing, thus allowing for genomes to be sequenced not just from model organisms but also from any source from which DNA can be isolated. This massive increase in DNA sequence information will not only enable the identification of most of the plastid transporters throughout the plant kingdom but will also help to elucidate the physiological function of genes and proteins and thus will generate a new understanding in fundamental aspects of biology and evolution. In combination with other data sources, such as transcriptome, metabolome, and protein interaction data, multiple associations can be generated that provide valuable clues about protein function. A higher-order goal would be to deduce the metabolic network of a recently sequenced organism or of its cellular organelles from its genome sequence, i.e. *ab initio* metabolic reconstruction. Fundamental to the function of proteins is their localization which can be determined by computer algorithms and proteomic analysis of cellular compartments. However, the discovery of non-canonical protein targeting to the plastid through the endomembrane system and proteomics data showing dual-targeting of proteins to multiple cellular destinations require a re-assessment of the plastid transporter proteome, primarily by deep proteomic analysis of envelope membranes from various plastid subtypes and from all phyla of the photosynthetic eukaryotes.

Targeting gene families by artificial micro RNAs (Hauser *et al.*, 2013) or multiple target CRISPR/Cas9 systems (Xing *et al.*, 2014) are new methods that are expected to unleash such a positive impulse. Artificial micro RNAs have been used for the alteration of gene expression for some time and have proven many times as valuable tools for genetic studies in a broad variety of plant species and algae (Schwab *et al.*, 2006; Tiwari *et al.*, 2014). High expectations for plant science are connected to the recently discovered CRISPR/Cas9 technique (Feng *et al.* 2013; Li *et al.* 2013). The advantage of CRISPR/Cas9 is that it results *in situ* genome editing events, i.e. the alteration of the genome directly in the nucleus of the cell. It either allows for introducing early-stop/miss-sense mutations or the introduction of external DNA. This overcomes unwanted expression silencing effects that are known to occur over generations when employing post-transcriptional gene silencing systems, such as artificial micro RNAs (Alvarez *et al.* 2006; Hauser *et al.* 2013). On the other hand, CRISPR/Cas9-induced mutations have to be established in the germline, which makes detailed genotyping, backcrossing and propagation of seedlings for several generations necessary (Gao and Zhao 2014). This obviously becomes more of an issue if multiple gene family members are targeted. Depending on the application, gene family targeting by artificial micro RNAs that allow for mutant screening already in the T1 generation might be a better choice (Hauser *et al.* 2013).

Acknowledgements

APMW acknowledges continued support by the Deutsche Forschungsgemeinschaft (grants WE 2231/8–2, EXC 1028).

References

Abdel-Ghany, S.E., Muller-Moule, P., Niyogi, K.K., Pilon, M., and Shikanai, T. (2005a). Two P-type ATPases are required for copper delivery in *Arabidopsis thaliana* chloroplasts. Plant Cell *17*, 1233–1251.

Abdel-Ghany, S.E., Ye, H., Garifullina, G.F., Zhang, L.H., Pilon-Smits, E.A.H., and Pilon, M. (2005b). Iron-sulfur cluster biogenesis in chloroplasts. Involvement of the scaffold protein CpIscA. Plant Physiol. *138*, 161–172.

Agius, F., Gonzalez-Lamothe, R., Caballero, J.L., Munoz-Blanco, J., Botella, M.A., and Valpuesta, V. (2003). Engineering increased vitamin C levels in plants by overexpression of a D-galacturonic acid reductase. Nat. Biotech. *21*, 177–181.

Alvarez, J.P., Pekker, I., Goldschmidt, A., Blum, E., Amsellem, Z., and Eshed, Y. (2006). Endogenous and synthetic microRNAs stimulate simultaneous, efficient, and localized regulation of multiple targets in diverse species. Plant Cell *18*, 1134–1151.

Andersson, M.X., Goksör, M., and Sandelius, A.S. (2007). Optical manipulation reveals strong attracting forces at membrane contact sites between Endoplasmic Reticulum and chloroplasts. J. Biol. Chem. *282*, 1170–1174.

Aranda-Sicilia, M.N., Cagnac, O., Chanroj, S., Sze, H., Rodríguez-Rosales, M.P., and Venema, K. (2012). *Arabidopsis* KEA2, a homolog of bacterial KefC, encodes a K^+/H^+ antiporter with a chloroplast transit peptide. Biochim. Biophys. Acta *1818*, 2362–2371.

Argyrou, E., Sophianopoulou, V., Schultes, N., and Diallinas, G. (2001). Functional characterization of a maize purine transporter by expression in *Aspergillus nidulans*. Plant Cell *13*, 953–964.

Armbruster, U., Carrillo, L.R., Venema, K., Pavlovic, L., Schmidtmann, E., Kornfeld, A., Jahns, P., Berry, J.A., Kramer, D.M., and Jonikas, M.C. (2014). Ion antiport accelerates photosynthetic acclimation in fluctuating light environments. Nat. Commun. *5*, Art. 5439.

Archibald, J.M. (2009). The puzzle of plastid evolution. Curr. Biol. *19*, R81-R88.

Ball, S.G., Colleoni, C., Cenci, U., Raj, J.N., and Tirtiaux, C. (2011). The evolution of glycogen and starch metabolism in eukaryotes gives molecular clues to understand the establishment of plastid endosymbiosis. J. Exp. Bot. *62*, 1775–1801.

Ball, S.G., Subtil, A., Bhattacharya, D., Moustafa, A., Weber, A.P.M., Gehre, L., Colleoni, C., Arias, M.C., Cenci, U., and Dauvillée, D. (2013). Metabolic effectors secreted by bacterial pathogens: essential facilitators of plastid endosymbiosis. Plant Cell *25*, 7–21.

Bauwe, H., Hagemann, M., and Fernie, A.R. (2010). Photorespiration: players, partners, and origin. Trends Plant Sci. *15*, 330–336.

Beck, E., Burkert, A., and Hofmann, M. (1983). Uptake of L-ascorbate by intact spinach chloroplasts. Plant Physiol. *73*, 41–45.

Bernardi, P. (1999). Mitochondrial transport of cations: Channels, exchangers, and permeability transition. Physiol. Rev. *79*, 1127–1155.

Bhattacharya, D., Yoon, H.S., and Hackett, J.D. (2004). Photosynthetic eukaryotes unite: endosymbiosis connects the dots. BioEssays *26*, 50–60.

Bhattacharya, D., Archibald, J.M., Weber, A.P.M., and Reyes-Prieto, A. (2007). How do endosymbionts become organelles? Understanding early events in plastid evolution. BioEssays *29*, 1239–1246.

Bölter, B., and Soll, J. (2001). Ion channels in the outer membranes of chloroplasts and mitochondria: open doors or regulated gates? EMBO J. *20*, 935–940.

Boutigny, S., Sautron, E., Finazzi, G., Rivasseau, C., Frelet-Barrand, A., Pilon, M., Rolland, N., and Seigneurin-Berny, D. (2014). HMA1 and PAA1, two chloroplast-envelope P-IB-ATPases, play distinct roles in chloroplast copper homeostasis. J. Exp. Bot. *65*, 1529–1540.

Bowes, G., Ogren, W.L., and Hageman, R.H. (1971). Phosphoglycolate production by ribulose diphosphate carboxylase. Biochem. Biophys. Res. Comm. *45*, 716–722.

Bräutigam, A., and Weber, A.P.M. (2009). Proteomic analysis of the proplastid envelope membrane provides novel insights into small molecule and protein transport across proplastid membranes. Mol. Plant 2, 1247–1261.

Brooks, C.F., Johnsen, H., van Dooren, G.G., Muthalagi, M., Lin, S.S., Bohne, W., Fischer, K., and Striepen, B. (2010). The Toxoplasma apicoplast phosphate translocator links cytosolic and apicoplast metabolism and is essential for parasite survival. Cell Host Microbe 7, 62–73.

Bughio, N., Takahashi, M., Yoshimura, E., Nishizawa, N.K., and Mori, S. (1997). Light-dependent iron transport into isolated barley chloroplasts. Plant Cell Physiol. *38*, 101–105.

Cai, X., Ye, J., Hu, T., Zhang, Y., Ye, Z., and Li, H. (2014). Genome-wide classification and expression analysis of nucleobase-ascorbate transporter (NAT) gene family in tomato. Plant Growth Regul. 73, 19–30.

Carraretto, L., Formentin, E., Teardo, E., Checchetto, V., Tomizioli, M., Morosinotto, T., Giacometti, G.M., Finazzi, G., and Szabó, I. (2013). A thylakoid-located two-pore K^+ channel controls photosynthetic light utilization in plants. Science 342, 114–118.

Catty, P., Boutigny, S., Miras, R., Joyard, J., Rolland, N., and Seigneurin-Berny, D. (2011). Biochemical characterization of AtHMA6/PAA1, a chloroplast envelope Cu(I)-ATPase. J. Biol. Chem. *286*, 36188–36197.

Cavalier-Smith, T. (2000). Membrane heredity and early chloroplast evolution. Trends Plant Sci. 5, 174–182.

Cavalier-Smith, T. (2003). Genomic reduction and evolution of novel genetic membranes and protein-targeting machinery in eukaryote-eukaryote chimaeras (meta-algae). Phil. Trans. R. Soc. Lond. *358*, 109–134.

Chanda, B., Xia, Y., Mandal, M.K., Yu, K., Sekine, K.T., Gao, Q., Selote, D., Hu, Y., Stromberg, A., Navarre, D., Kachroo, A., and Kachroo, P. (2001). Glycerol-3-phosphate is a critical mobile inducer of systemic immunity in plants. Nat. Genet. 43, 421–427.

Chanroj, S., Wang, G., Venema, K., Zhang, M.W., Delwiche, C.F., and Sze, H. (2012). Conserved and diversified gene families of monovalent cation/H^+ antiporters from algae to flowering plants. Front. Plant Sci. 3, Art. 25.

Checchetto, V., Teardo, E., Carraretto, L., Formentin, E., Bergantino, E., Giacometti, G.M., and Szabo, I. (2013). Regulation of photosynthesis by ion channels in cyanobacteria and higher plants. Biophys. Chem. *182*, 51–57.

Conte, S., Stevenson, D., Furner, I., and Lloyd, A. (2009). Multiple antibiotic resistance in *Arabidopsis* is conferred by mutations in a chloroplast-localized transport protein. Plant Physiol. *151*, 559–573.

Daruwala, R., Song, J., Koh, W.S., Rumsey, S.C., and Levine, M. (1999). Cloning and functional characterization of the human sodium-dependent vitamin C transporters hSVCT1 and hSVCT2. FEBS Lett. *460*, 480–484.

de Koning, H., and Diallinas, G. (2000). Nucleobase transporters. Mol. Membr. Biol. 75, 75–94.

Demmig, B., and Gimmler, H. (1983). Properties of the isolated intact chloroplast at cytoplasmic K^+ concentrations: I. Light-induced cation uptake into intact chloroplasts is driven by an electrical potential difference. Plant Physiol. 73, 169–174.

Demmig-Adams, B., and Adams, W.W. (2006). Photoprotection in an ecological context: the remarkable complexity of thermal energy dissipation. New Phytol. *172*, 11–21.

Dempsey, D.A., Vlot, A.C., Wildermuth, M.C., and Klessig, D.F. (2011). Salicylic acid biosynthesis and metabolism. *Arabidopsis* Book 9, e0156.

Diener, A.C., Gaxiola, R.A., and Fink, G.R. (2001). *Arabidopsis* ALF5, a multidrug efflux transporter gene family member, confers resistance to toxins. Plant Cell *13*, 1625–1637.

Dowdle, J., Ishikawa, T., Gatzek, S., Rolinski, S., and Smirnoff, N. (2007). Two genes in *Arabidopsis thaliana* encoding GDP-L-galactose-phosphorylase are required for ascorbate biosynthesis and seedling viability. Plant J. *52*, 673–689.

Durrant, W.E., and Dong, X. (2004). Systemic acquired resistance. Annu. Rev. Phytopathol. *42*, 185–209.

Duy, D., Stuebe, R., Wanner, G., and Philippar, K. (2011). The chloroplast permease PIC1 regulates plant growth and development by directing homeostasis and transport of iron. Plant Physiol. *155*, 1709–1722.

Duy, D., Wanner, G., Meda, A.R., von Wiren, N., Soll, J., and Philippar, K. (2007). PIC1, an ancient permease in *Arabidopsis* chloroplasts, mediates iron transport. Plant Cell *19*, 986–1006.

Eastmond, P.J. (2004). Glycerol-insensitive *Arabidopsis* mutants: *gli1* seedlings lack glycerol kinase, accumulate glycerol and are more resistant to abiotic stress. Plant J. *37*, 617–625.

Eicks, M., Maurino, V., Knappe, S., Flügge, U.I., and Fischer, K. (2002). The plastidic pentose phosphate translocator represents a link between the cytosolic and the plastidic pentose phosphate pathways in plants. Plant Physiol. *128*, 512–522.

Eisen, M.B., Spellmann, P.T., Brown, P.O., and Botstein, D. (1998). Cluster analysis and display of genome-wide expression patterns. Proc. Natl. Acad. Sci. U.S.A. *95*, 14863–14868.

Eisenhut, M., Hocken, N., and Weber, A.P.M. (2014). Plastidial metabolite transporters integrate photorespiration with carbon, nitrogen, and sulfur metabolism. Cell Calcium *58*, 98–104.

Eisenhut, M., Pick, T.R., Bordych, C., and Weber, A.P.M. (2013). Towards closing the remaining gaps in photorespiration – the essential but unexplored role of transport proteins. Plant Biol. *15*, 676–685.

Ettinger, W.F., Clear, A.M., Fanning, K.J., and Peck, M.L. (1999). Identification of a Ca^{2+}/H^{+} antiport in the plant chloroplast thylakoid membrane. Plant Physiol. *119*, 1379–1385.

Evans, A.R., Hall, D., Pritchard, J., and Newbury, H.J. (2012). The roles of the cation transporters CHX21 and CHX23 in the development of *Arabidopsis thaliana*. J. Exp. Bot. *63*, 59–67.

Facchinelli, F., and Weber, A.P.M. (2011). The metabolite transporters of the plastid envelope: an update. Front. Plant Sci. *2*, Article 50.

Facchinelli, F., Colleoni, C., Ball, S.G., and Weber, A.P.M. (2013). Chlamydia, cyanobiont, or host: who was on top in the ménage a trois? Trends Plant Sci. *18*, 673–678.

Feng, Z., Zhang, B., Ding, W., Liu, X., Yang, D., Wei, P., Cao, F., Zhu, S., Zhang, F., Mao, Y., and Zhu, J. (2013). Efficient genome editing in plants using a CRISPR/Cas system. Cell Res. *23*, 1229–1232.

Ferguson, G.P., McLaggan, D., and Booth, I.R. (1995). Potassium channel activation by glutathione-S-conjugates in *Escherichia coli*: protection against methylglyoxal is mediated by cytoplasmic acidification. Mol. Microbiol. *17*, 1025–1033.

Finazzi, G., Petroutsos, D., Tomizioli, M., Flori, S., Sautron, E., Villanova, V., Rolland, N., and Seigneurin-Berny, D. (2014). Ions channels/transporters and chloroplast regulation. Cell Calcium *58*, 86–97.

Finkelstein, R. (2013). Abscisic acid synthesis and response. *Arabidopsis* Book *11*, e0166.

Fischer, K. (2011). The im- and export business in plastids: Transport processes across the inner envelope membrane. Plant Physiol. *155*, 1511–1519.

Fischer, K., Kammerer, B., Gutensohn, M., Arbinger, B., Weber, A., Häusler, R.E., and Flügge, U.I. (1997). A new class of plastidic phosphate translocators: A putative link between primary and secondary metabolism by the phosphoenolpyruvate/phosphate antiporter. Plant Cell *9*, 453–462.

Fliege, R., Flügge, U.I., Werdan, K., and Heldt, H.W. (1978). Specific transport of inorganic phosphate, 3-phosphoglycerate and triose phosphates across the inner membrane of the envelope in spinach chloroplasts. Biochim. Biophys. Acta *502*, 232–247.

Floss, D.S., and Walter, M.H. (2009). Role of carotenoid cleavage dioxygenase 1 (CCD1) in apo-carotenoid biogenesis revisited. Plant Signal. Behav. *4*, 172–175.

Flügge, U.I. (1999). Phosphate translocators in plastids. Annu. Rev. Plant Physiol. Plant Mol. Biol. *50*, 27–45.

Flügge, U.I. (2000). Transport in and out of plastids: does the outer envelope membrane control the flow? Trends Plant Sci. *5*, 135–137.

Flügge, U.I., Fischer, K., Gross, A., Sebald, W., Lottspeich, F., and Eckerskorn, C. (1989). The triose phosphate-3-phosphoglycerate-phosphate translocator from spinach chloroplasts: nucleotide sequence of a full-length cDNA clone and import of the *in vitro* synthesized precursor protein into chloroplasts. EMBO J. *8*, 39–46.

Flügge, U.I., Häusler, R.E., Ludewig, F., and Fischer, K. (2003). Functional genomics of phosphate antiport systems of plastids. Physiol. Plant. *118*, 475–482.

Foyer, C.H., and Lelandais, M. (1996). A comparison of the relative rates of transport of ascorbate and glucose across the thylakoid, chloroplast and plasma membranes of pea leaf mesophyll cells. J. Plant Physiol. *148*, 391–398.

Foyer, C.H., and Noctor, G. (2011). Ascorbate and glutathione: the heart of the redox hub. Plant Physiol. *155*, 2–18.

Fragniere, C., Serrano, M., Abou-Mansour, E., Metraux, J.P., and L'Harindon, F. (2011). Salicylic acid and its location in response to biotic and abiotic stress. FEBS Lett. *585*, 1847–1852.

Franceschi, V.R., and Tarlyn, N.N. (2002). L-Ascorbic acid is accumulated in source leaf phloem and transported to sink tissues in plants. Plant Physiol. *130*, 649–656.

Fujisawa, M., Ito, M., and Krulwich, T.A. (2007). Three two-component transporters with channel-like properties have monovalent cation/proton antiport activity. Proc. Natl. Acad. Sci. U.S.A. *104*, 13289–13294.

Furumoto, T., Yamaguchi, T., Ohshima-Ichie, Y., Nakamura, M., Tsuchida-Iwata, Y., Shimamura, M., Ohnishi, J., Hata, S., Gowik, U., Westhoff, P., Bräutigam, A., Weber, A.P.M., and Izui, K. (2011). A plastidial sodium-dependent pyruvate transporter. Nature *476*, 472–476.

Gagat, P., Bodyl, A., and Mackiewicz, P. (2013). How protein targeting to primary plastids via the endomembrane system could have evolved? A new hypothesis based on phylogenetic studies. Biol. Direct *8*, Article 18, DOI: 10.1186/1745-6150-8-18.

Gao, Y., and Zhao, Y. (2014). Specific and heritable gene editing in *Arabidopsis*. Proc. Natl. Acad. Sci. U.S.A. *111*, 4357–4358.

Garcion, C., Lohmann, A., Lamodiere, E., Catinot, J., Buchala, A., Doermann, P., and Metraux, J.P. (2008). Characterization and biological function of the *isochorismate synthase 2* gene of *Arabidopsis*. Plant Physiol. *147*, 1279–1287.

Gibbs, S.P. (1978). The chloroplasts of Euglena may have evolved from symbiotic green algae. Can. J. Bot. *56*, 2883–2889.

Gigolashvili, T., Geier, M., Ashykhmina, N., Frerigmann, H., Wulfert, S., Krueger, S., Mugford, S.G., Kopriva, S., Haferkamp, I., and Fluegge, U.I. (2012). The *Arabidopsis* thylakoid ADP/ATP carrier TAAC has an additional role in supplying plastidic phosphoadenosine 5 '-phosphosulfate to the cytosol. Plant Cell *24*, 4187–4204.

Girke, C., Daumann, M., Niopek-Witz, S., and Möhlmann, T. (2014). Nucleobase and nucleoside transport and integration into plant metabolism. Front. Plant Sci. *5*, Article 443.

Glazebrook, J., Rogers, E.E., and Ausubel, F.M. (1996). Isolation of *Arabidopsis* mutants with enhanced disease susceptibility by direct screening. Genetics *143*, 973–982.

Goetze, T.A., Patil, M., Jeshen, I., Bölter, B., Grahl, S., and Soll, J. (2015). Oep23 forms an ion channel in the chloroplast outer envelope. BMC Plant Biol. 15, Article 1.

Guo, B., Jin, Y., Wussler, C., Blancaflor, E.B., Motes, C.M., and Versaw, W.K. (2008). Functional analysis of the

Arabidopsis PHT4 family of intracellular phosphate transporters. New Phytol. *177*, 889–898.

Gutensohn, M., Fan, E., Frielingsdorf, S., Hanner, P., Hou, B., Hust, B., and Klösgen, R.B. (2006). Toc, Tic, Tat *et al.*: structure and function of protein transport machineries in chloroplasts. J. Plant Physiol. *163*, 333–347.

Hagemann, M., Fernie, A.R., Espie, G.S., Kern, R., Eisenhut, M., Reumann, S., Bauwe, H., and Weber, A.P.M. (2013). Evolution of the biochemistry of the photorespiratory C2 cycle. Plant Biol. *15*, 639–647.

Hanikenne, M., Bernal, M., and Urzica, E.I. (2014). Ion homeostasis in the chloroplast. In Plastid Biology, S.M. Theg and, F.A. Wollman, eds. (New York, USA: Springer), pp. 465–514.

Hassler, S., Lemke, L., Jung, B., Möhlmann, T., Krüger, F., Schumacher, K., Espen, L., Martinoia, E., and Neuhaus, H.E. (2012). Lack of Golgi phosphate transporter PHT4;6 causes strong developmental defects, constitutively activated disease resistance mechanisms and altered intracellular phosphate compartmentation in *Arabidopsis*. Plant J. *72*, 732–744.

Hauser, F., Chen, W., Deinlein, U., Chang, K., Ossowski, S., Fitz, J., Hannon, G.J., and Schroeder, J.I. (2013). A genomic-scale artificial microRNA library as a tool to investigate the functionally redundant gene space in *Arabidopsis*. Plant Cell 25, 2848–2863.

Heibert, T., Steinkamp, T., Hinnah, S., Schwarz, M., Flügge, U.I., Weber, A., and Wagner, R. (1995). Ion channels in the chloroplast envelope membrane. Biochemistry *34*, 15906–15917.

Heldt, H.W. (2002). Three decades in transport business: studies of metabolite transport in chloroplasts – a personal perspective. Photosynth. Res. 73, 265–272.

Heldt, H.W., Werdan, K., Milovancev, M., and Geller, G. (1973). Alkalization of the chloroplast stroma caused by light-dependent proton flux into the thylakoid space. Biochim. Biophys. Acta *314*, 224–241.

Horemans, N., Foyer, C.H., and Asada, H. (2000). Transport and action of ascorbate at the plant plasma membrane. Trends Plant Sci. 5, 263–267.

Howitz, K.T., and McCarty, R.E. (1985a). Kinetic characteristics of the chloroplast envelope glycolate transporter. Biochemistry 24, 2645–2652.

Howitz, K.T., and McCarty, R.E. (1985b). Substrate specificity of the pea chloroplast glycolate transporter. Biochemistry *24*, 3645–3650.

Howitz, K.T., and McCarty, R.E. (1991). Solubilization, partial purification, and reconstitution of the glycolate/glycerate transporter from chloroplast inner envelope membranes. Plant Physiol. *96*, 1060–1069.

Huang, J.L., and Gogarten, J.P. (2007). Did an ancient chlamydial endosymbiosis facilitate the establishment of primary plastids? Genome Biol. *8*, R99.

Huang, L.Q., Berkelman, T., Franklin, A.E., and Hoffman, N.E. (1993). Characterization of a gene encoding a Ca^{2+}-ATPase-like protein in the plastid envelope. Proc. Natl. Acad. Sci. U.S.A. *90*, 10066–10070.

Huang, Y., Lemieux, M.J., Song, J., Auer, M., and Wang, D.N. (2003). Structure and mechanism of the glycerol-3-phosphate transporter from *Escherichia coli*. Science *301*, 616–620.

Huber, S.C., and Maury, W. (1980). Effects of magnesium on intact chloroplasts: I. Evidence for activation of (sodium) potassium/proton exchange across the chloroplast envelope. Plant Physiol. *65*, 350–354.

Irigoyen, S., Karlsson, P.M., Kuruvilla, J., Spetea, C., and Versaw, W.K. (2011). The sink-specific plastidic phosphate transporter PHT4;2 influences starch accumulation and leaf size in *Arabidopsis*. Plant Physiol. *157*, 1765–1777.

Jarzyniak, K.M., and Jasinski, M. (2014). Membrane transporters and drought resistance – a complex issue. Front. Plant Sci. *5*, Article 687.

Jessen, D., Roth, C., Wiermer, M., and Fulda, M. (2015). Two activities of long-chain acyl-coenzyme A synthetase are involved in lipid trafficking between the endoplasmic reticulum and the plastid in *Arabidopsis*. Plant Physiol. *167*, 351–366.

Jund, R., Weber, E., and Chevallier, M.R. (1988). Primary structure of the uracil transport protein of *Saccharomyces cerevisiae*. Eur. J. Biochem. *171*, 417–424.

Kammerer, B., Fischer, K., Hilpert, B., Schubert, S., Gutensohn, M., Weber, A., and Flügge, U.I. (1998). Molecular characterization of a carbon transporter in plastids from heterotrophic tissues: The glucose6-phosphate/phosphate antiporter. Plant Cell *10*, 105–117.

Kampfenkel, K.H., Möhlmann, T., Batz, O., van Montague, M., Inzé, D., and Neuhaus, H.E. (1995). Molecular characterization of an *Arabidopsis thaliana* cDNA encoding a novel putative adenylate translocator of higher plants. FEBS Lett. *374*, 351–355.

Kaneko, Y., and Keegstra, K. (1996). Plastid biogenesis in embryonic pea leaf cells during early germination. Protoplasma *195*, 59–67.

Kang, L., Li, J., Zhao, T., Xiao, F., Tang, X., Thilmony, R., He, S.Y., and Zhou, J.M. (2003). Interplay of the *Arabidopsis* nonhost resistance gene *NHO1* with bacterial virulence. Proc. Natl. Acad. Sci. U.S.A. *100*, 3519–3524.

Kawai, H., Ishikawa, T., Mitsui, T., Kore-eda, S., Yamada-Kawai, M., and Ohnishi, J. (2014). *Arabidopsis* glycerol-3-phosphate permease 4 is localized in the plastids and involved in the accumulation of seed oil. Plant Biotech. *31*, 159–165.

Kim, Y.Y., Choi, H., Segami, S., Cho, H.T., Martinoia, E., Maeshima, M., and Lee, Y. (2009). AtHMA1 contributes to the detoxification of excess Zn(II) in *Arabidopsis*. Plant J. *58*, 737–753.

Kirchberger, S., Leroch, M., Huynen, M.A., Wahl, M., Neuhaus, H.E., and Tjaden, J. (2007). Molecular and biochemical analysis of the plastidic ADP-glucose transporter (ZmBT1) from *Zea mays*. J. Biol. Chem. *282*, 22481–22491.

Klecker, T., Böckler, S., and Westermann, B. (2014). Making connections: interorganelle contacts orchestrate mitochondrial behavior. Trends Cell Biol. *24*, 537–545.

Kleffmann, T., Russenberger, D., von Zychlinski, A., Christopher, W., Sjolander, K., Gruissem, W., and Baginsky, S. (2004). The *Arabidopsis thaliana* chloroplast proteome reveals pathway abundance and novel protein functions. Curr. Biol. *14*, 354–362.

Knappe, S., Flügge, U.I., and Fischer, K. (2003). Analysis of the plastidic phosphate translocator gene family in *Arabidopsis* and identification of new phosphate translocator-homologous transporters, classified by their putative substrate-binding site. Plant Physiol. *131*, 1178–1190.

Kunz, H.H., Gierth, M., Herdean, A., Satoh-Cruz, M., Kramer, D.M., Spetea, C., and Schroeder, J.I. (2014). Plastidial transporters KEA1, -2, and -3 are essential for chloroplast osmoregulation, integrity, and pH regulation in *Arabidopsis*. Proc. Natl. Acad. Sci. U.S.A. *111*, 7480–7485.

Larson, T.J., Schumacher, W., and Boos, W. (1982). Identification of the glpT-encoded sn-glycerol-3-phosphate permease of *Escherichia coli*, an oligomeric integral membrane protein. J. Bacteriol. *152*, 1008–1021.

Leister, D. (2003). Chloroplast research in the genomic age. Trends Plant Sci. *19*, 47–55.

Lemieux, M.J., Huang, Y., and Wang, D.N. (2004). Glycerol-3-phosphate transporter of *Escherichia coli*: Structure, function and regulation. Res. Microbiol. *155*, 623–629.

Leroch, M., Kirchberger, S., Haferkamp, I., Wahl, M., Neuhaus, H.E., and Tjaden, J. (2005). Identification and characterization of a novel plastidic adenine nucleotide uniporter from *Solanum tuberosum*. J. Biol. Chem. *280*, 17992–18000.

Li, J., Norville, J., Aach, J., McCormack, M., Zhang, D., Bush, J., Church, G., and Sheen, J. (2013). Multiplex and homologous recombination-mediated genome editing in *Arabidopsis* and *Nicotiana benthamiana* using guide RNA and Cas9. Nat. Biotech. *31*, 688–691.

Li, J., Pandeya, D., Nath, K., Zulfugarov, I.S., Yoo, S.C., Zhang, H., Yoo, J.H., Cho, S.H., Koh, H.J., Kim, D.S., Seo, H.S., Kang, B.C., Lee, C.H., and Paek, N.C. (2010). ZEBRA-NECROSIS, a thylakoid-bound protein, is critical for the photoprotection of developing chloroplasts during early leaf development. Plant J. *62*, 713–725.

Li, L.G., He, Z.Y., Pandey, G.K., Tsuchiya, T., and Luan, S. (2002). Functional cloning and characterization of a plant efflux carrier for multidrug and heavy metal detoxification. J. Biol. Chem. *277*, 5360–5368.

Li, N., Gugel, I.L., Giavalisco, P., Zeisler, V., Schreiber, L., Soll, J., and Philippar, K. (2015). FAX1, a novel membrane protein mediating plastid fatty acid export. PLOS Biol. *13*, e1002053.

Lorence, A., Chevone, B.I., Mendes, P., and Nessler, C.L. (2004). *Myo*-inositol oxygenase offers a possible entry point into plant ascorbate biosynthesis. Plant Physiol. *134*, 1200–1205.

Lorenc-Kukula, K., Chaturvedi, R., Roth, M., Welti, R., and Shah, J. (2012). Biochemical and molecular characterization of SFD1`s involvement in lipid metabolism and defense signaling. Front. Plant Sci. 3, Article 26.

Lu, Y., Chanroj, S., Zulkifli, L., Johnson, M.A., Uozumi, N., Cheung, A., and Sze, H. (2011). Pollen tubes lacking a pair of K^+ transporters fail to target ovules in *Arabidopsis*. Plant Cell *23*, 81–93.

Ma, Y., Szostkiewicz, I., Korte, A., Moes, D., Yang, Y., Christmann, A., and Grill, E. (2009). Regulators of PP2C phosphatase activity function as abscisic acid sensors. Science *324*, 1064–1068.

Mainguet, S.E., Gakiere, G., Majira, A., Pelletier, S., Bringel, F., Guerard, F., Caboche, M., Berthome, R., and Renou, J.P. (2009). Uracil salvage is necessary for early *Arabidopsis* development. Plant J. *60*, 280–291.

Mansfield, T.A., Schultes, N.P., and Mourad, G.S. (2009). AtAzg1 and AtAzg2 comprise a novel family of purine transporters in *Arabidopsis*. FEBS Lett. *583*, 481–486.

Margulis, L. (1971). Symbiosis and evolution. Sci. Am. *225*, 48–57.

Marinova, K., Pourcel, L., Weder, B., Schwarz, M., Barron, D., Routaboul, J.M., Debeaujon, I., and Klein, M. (2007). The *Arabidopsis* MATE transporter TT12 acts as a vacuolar flavonoid/H^+-antiporter active in proanthocyanidin-accumulating cells of the seed coat. Plant Cell *19*, 2023–2038.

Marschner, H. (1995). Mineral nutrition of higher plants. (London, UK: Academic Press).

Martin, W., and Kowallik, K. (1999). Annotated English translation of Mereschkowsky`s 1905 paper 'Über Natur und Ursprung der Chromatophoren im Pflanzenreiche'. Eur. J. Phycol. *34*, 287–295.

Mäser, P., Thomine, S., Schroeder, J.I., Ward, J.M., Hirschi, K., Sze, H., Talke, I.N., Amtmann, A., Maathuis, F.J.M., Sanders, D., Harper, J.F., Tchieu, J., Gribskov, M., Persans, M.W., Salt, D.E., Kim, S.A., and Guerinot, M.L. (2001). Phylogenetic relationships within cation transporter families of *Arabidopsis*. Plant Physiol. *126*, 1646–1667.

Maurino, V.G., Grube, E., Zielinski, J., Schild, A., Fischer, K., and Flügge, U.I. (2006). Identification and expression analysis of twelve members of the nucleobase-ascorbate transporter (NAT) gene family in *Arabidopsis thaliana*. Plant Cell Physiol. *47*, 1381–1393.

May, J.M. (2011). The SLC23 family of ascorbate transporters: ensuring that you get and keep your daily dose of vitamin C. British J. Pharm. *164*, 1793–1801.

Mehrshahi, P., Johnny, C., and DellaPenna, D. (2014). Redefining the metabolic continuity of chloroplasts and ER. Trends Plant Sci. *19*, 501–507.

Mehrshahi, P., Stefano, G., Andaloro, J.M., Brandizzi, F., Froehlich, J.E., and DellaPenna, D. (2013). Transorganellar complementation redefines the biochemical continuity of endoplasmic reticulum and chloroplasts. Proc. Natl. Acad. Sci. U.S.A. *110*, 12126–12131.

Mereschkowsky, C. (1905). Über Natur und Ursprung der Chromatophoren im Pflanzenreiche. Biol. Centralbl. *25*, 593–604.

Mi, F., Peters, J.S., and Berkowitz, G.A. (1994). Characterization of a chloroplast inner envelope K^+ channel. Plant Physiol. *105*, 955–964.

Millar, A.H., Mittova, V., Kiddle, G., Heazlewood, J.L., Bartoli, C.G., Theodoulou, F.L., and Foyer, C.H. (2003). Control of ascorbate synthesis by respiration and its implications for stress response. Plant Physiol. *133*, 443–447.

Miyaji, T., Kawasaki, T., Togawa, N., Omote, H., and Moriyama, Y. (2013). Type I sodium dependent phosphate transporter acts as a membrane potential driven urate exporter. Curr. Mol. Pharmacol. *6*, 88–94.

Miyaji, T., Kuromori, T., Takeuchi, Y., Yamaji, N., Yokosho, K., Shimazawa, A., Sugimoto, E., Omote, H., Ma, J.F., Shinozaki, K., and Moriyama, Y. (2015). AtPHT4;4 is a chloroplast-localized ascorbate transporter in *Arabidopsis*. Nature Comm. *6*, Article 5928.

Moreno, I., Norambuena, L., Maturana, D., Toro, M., Vergara, C., Orellana, A., Zurita-Silva, A., and Ordenes,

V.R. (2008). AtHMA1 is a thapsigargin-sensitive Ca^{2+}/heavy metal pump. J. Biol. Chem. *283*, 9633–9641.

Morita, M., Shitan, N., Sawada, K., van Montagu, M.C.E., Inze, D., Rischer, H., Goossens, A., Oksman-Caldentey, K.M., Moriyama, Y., and Yasaki, K. (2009). Vacuolar transport of nicotine is mediated by a multidrug and toxic compound extrusion (MATE) transporter in *Nicotiana tabacum*. Proc. Natl. Acad. Sci. U.S.A. *106*, 2447–2452.

Morita, Y., Kodama, K., Shiota, S., Mine, T., Kataoka, A., Mizushima, T., and Tsuchiya, T. (1998). NorM, a putative multidrug efflux protein of *Vibrio parahaemolyticus* and its homolog in *Escherichia coli*. Antimicrob. Agents Chemother. *42*, 1778–1782.

Mourad, G.S., Tippmann-Crosby, J., Hunt, K.A., Gicheru, Y., Bade, K., Mansfield, T.A., and Schultes, N.P. (2012). Genetic and molecular characterization reveals a unique nucleobase cation symporter 1 in *Arabidopsis*. FEBS Lett. *586*, 1370–1378.

Moustafa, A., Reyes-Prieto, A., and Bhattacharya, D. (2008). Chlamydiae has contributed at least 55 genes to Plantae with predominantly plastid functions. PLOS One 3, e2005.

Müller, M., Kunz, H.H., Schroeder, J.I., Kemp, G., Young, H.S., and Neuhaus, H.E. (2014). Decreased capacity for sodium export out of *Arabidopsis* chloroplasts impairs salt tolerance, photosynthesis and plant performance. Plant J. *78*, 646–658.

Munro, A.W., Ritchie, G.Y., Lamb, A.J., Douglas, R.M., and Booth, I.R. (1991). The cloning and DNA sequence of the gene for the glutathione-regulated potassium-efflux system KefC of *Escherichia coli*. Mol. Microbiol. *5*, 607–616.

Nandi, A., Welti, R., and Shah, J. (2004). The *Arabidopsis thaliana* dihydroxyacetone phosphate reductase gene SUPPRESSOR OF FATTY ACID DESATURASE DEFICIENCY1 is required for glycerolipid metabolism and for the activation of systemic acquired resistance. Plant Cell *16*, 465–477.

Nawrath, C., Heck, S., Parinthawong, N., and Metraux, J.P. (2002). EDS5, an essential component of salicylic acid-dependent signaling for disease resistance in *Arabidopsis*, is a member of the MATE transporter family. Plant Cell *14*, 275–286.

Niopek-Witz, S., Deppe, J., Lemieux, M.J., and Möhlmann, T. (2014). Biochemical characterization and structure–function relationship of two plant NCS2 proteins, the nucleobase transporters NAT3 and NAT12 from *Arabidopsis thaliana*. Biochim. Biophys. Acta *1838*, 3025–3035.

Niewiadomski, P., Knappe, S., Geimer, S., Fischer, K., Schulz, B., Unte, U.S., Rosso, M.G., Ache, P., Flügge, U.I., and Schneider, A. (2005). The *Arabidopsis* plastidic glucose 6-phosphate/phosphate translocator GPT1 is essential for pollen maturation and embryo sac development. Plant Cell *17*, 760–775.

Nomura, H., and Shiina, T. (2014). Calcium signaling in plant endosymbiotic organelles: mechanism and role in physiology. Mol. Plant 7, 1094–1104.

Nowack, E.C.M. (2014). *Paulinella chromatophora* – rethinking the transition from endosymbiont to organelle. Acta Soc. Bot. Pol. *83*, 387–397.

Nowack, E.C.M., and Melkonian, M. (2010). Endosymbiotic associations within protists. Phil. Trans. R. Soc. Lond. *365*, 699–712.

Omote, H., Hiasa, M., Matsumoto, T., Otsuka, M., and Moriyama, Y. (2006). The MATE proteins as fundamental transporters of metabolic and xenobiotic organic cations. Trends Pharmacol. Sci. *27*, 587–593.

Omote, H., Miyaji, T., Juge, N., and Moriyama, Y. (2011). Vesicular neurotransmitter transporter: bioenergetics and regulation of glutamate transport. Biochemistry *50*, 5558–5565.

Ohlrogge, J., and Browse, J. (1995). Lipid biosynthesis. Plant Cell 7, 957–970.

Park, S.Y., Fung, P., Nishimura, N., Jensen, D.R., Fujii, H., Zhao, Y., Lumba, S., Santiago, J., Rodrigues, A., Chow, T.F., *et al.* (2009). Abscisic acid inhibits type 2C protein phosphatases via the PYR/PYL family of START proteins. Science *324*, 1068–1071.

Pfanzagl, B., Zenker, A., Pittenhauer, E., Allmaier, G., Martinez-Torrecuadrada, J., Schmid, E.R., de Pedro, M.A., and Löffelhardt, W. (1996). Primary structure of cyanelle peptidoglycan of *Cyanophora paradoxa*: a prokaryotic cell wall as part of an organelle envelope. J. Bacteriol. *178*, 332–339.

Pfeil, B.E., Schoefs, B., and Spetea, C. (2014). Function and evolution of channels and transporters in photosynthetic membranes. Cell. Mol. Life Sci. *71*, 979–998.

Pick, T.R., Bräutigam, A., Schulz, M.A., Obata, T., Fernie, A.R., and Weber, A.P.M. (2013). *PLGG1*, a plastidic glycolate glycerate transporter, is required for photorespiration and defines a unique class of metabolite transporters. Proc. Natl. Acad. Sci. U.S.A. *110*, 3185–3190.

Pick, T.R., and Weber, A.P.M. (2014). Unknown components of the plastidial permeome. Front. Plant Sci. *5*, Article 410.

Pilon, M., Abdel-Ghany, S.E., Van Hoewyk, D., Ye, H., and Pilon-Smits, E.A. (2006). Biogenesis of iron-sulfur cluster proteins in plastids. Genet. Engin. *27*, 101–117.

Pottosin, I.I., Muniz, J., and Shabala, S. (2005). Fast-activating channel controls cation fluxes across the native chloroplast envelope. J. Membr. Biol. *204*, 145–156.

Price, D.C., Chan, C.X., Yoon, H.S., *et al.* (2012). *Cyanophora paradoxa* genome elucidates origin of photosynthesis in algae and plants. Science 335, 843–847.

Pulsifer, I.P., Kluge, S., and Rowland, O. (2012). *Arabidopsis* LONG-CHAIN ACYL-COA SYNTHETASE 1 (LACS1), LACS2, and LACS3 facilitate fatty acid uptake in yeast. Plant Physiol. Biochem. *51*, 31–39.

Qiu, H., Price, D.C., Weber, A.P.M., Facchinelli, F., Yoon, H.S., and Bhattacharya, D. (2013). Assessing the bacterial contribution to the plastid proteome. Trends Plant Sci. *18*, 680–687.

Ramaiah, M., Jain, A., Baldwin, J.C., Kartikeyan, A.S., and Raghothama, K.G. (2011). Characterization of the phosphate-starvation induced *Glycerol-3-phosphate permease* gene family in *Arabidopsis*. Plant Physiol. *157*, 279–291.

Reimer, R.J. (2013). SLC17: A functionally diverse family of organic anion transporters. Mol. Aspects Med. *34*, 350–359.

Reiser, J., Linka, N., Lemke, L., Jeblick, W., and Neuhaus, H.E. (2004). Molecular physiological analysis of the two plastidic ATP/ADP transporters from *Arabidopsis*. Plant Physiol. *136*, 3524–3536.

Reyes-Prieto, A., Weber, A.P.M., and Bhattacharya, D. (2007). The origin and establishment of the plastid in algae and plants. Annu. Rev. Genet. *41*, 147–168.

Rice, K.C., and Bayles, K.W. (2008). Molecular control of bacterial death and lysis. Microbiol. Mol. Biol. Rev. *72*, 85–109.

Robinson, S.P., Downton, W.J.S., and Millhouse, J.A. (1983). Photosynthesis and ion content of leaves and isolated chloroplasts of salt-stressed spinach. Plant Physiol. *73*, 238–242.

Roh, M.H., Shingles, R., Cleveland, M.J., and McCarty, R.E. (1998). Direct measurement of calcium transport across chloroplast inner-envelope vesicles. Plant Physiol. *118*, 1447–1454.

Rolland, N., Curien, G., Finazzi, G., Kuntz, M., Marechal, E., Matringe, M., Ravanel, S., and Seigneurin-Berny, D. (2012). The biosynthetic capacities of the plastids and integration between cytoplasmic and chloroplast processes. Annu. Rev. Genet. *46*, 233–264.

Roosild, T.P., Castronovo, S., Miller, S., Li, C., Rasmussen, T., Bartlett, W., Gunasekera, B., Choe, S., and Booth, I.R. (2009). KTN (RCK) domains regulate K^+ channels and transporters by controlling the dimer-hinge conformation. Structure *17*, 893–903.

Roth, C., Menzel, G., Petetot, J.M., Rochat-Hacker, S., and Poirier, Y. (2004). Characterization of a protein of the plastid inner envelope having homology to animal inorganic phosphate, chloride and organic-anion transporters. Planta *218*, 406–416.

Saier, M.H., and Reddy, B.L. (2015). Holins in bacteria, eukaryotes, and archaea: Multifunctional xenologues with potential biotechnological and biomedical applications. J. Bacteriol. *197*, 7–17.

Schroppelmeier, G., and Kaiser, W.M. (1988). Ion homeostasis in chloroplasts under salinity and mineral deficiency: I. Solute concentrations in leaves and chloroplasts from spinach plants under NaCl or $NaNO_3$ salinity. Plant Physiol. *87*, 822–827.

Schwab, R., Ossowski, S., Riester, M., Warthmann, N., and Weigel, D. (2006). Highly specific gene silencing by artificial microRNAs in *Arabidopsis*. Plant Cell *18*, 1121–1133.

Seigneurin-Berny, D., Gravot, A., Auroy, P., Mazard, C., Kraut, A., Finazzi, G., Grunwald, D., Rappaport, F., Vavasseur, A., Joyard, J., Richaud, P., and Rolland, N. (2006). HMA1, a new Cu-ATPase of the chloroplast envelope, is essential for growth under adverse light conditions. J. Biol. Chem. *281*, 2882–2892.

Serrano, M., Wang, B., Aryal, B., Garcion, C., Abou-Monsour, E., Heck, S., Geisler, M., Mauch, F., Nawrath, C., and Metraux, J.P. (2013). Export of salicylic acid from the chloroplast requires the multidrug and toxin extrusion-like transporter EDS5. Plant Physiol. *162*, 1815–1821.

Sha, J., Chaturvedi, R., Chowdhury, Z., Venables, B., and Petros, R.A. (2014). Signalling by small metabolites in systemic acquired resistance. Plant J. *79*, 645–658.

Shikanai, T., Muller-Moule, P., Munekage, Y., Niyogi, K.K., and Pilon, M. (2003). PAA1, a P-type ATPase of *Arabidopsis*, functions in copper transport in chloroplasts. Plant Cell *15*, 1333–1346.

Shimoni-Shor, E., Hassidim, M., Yuval-Naeh, N., and Keren, N. (2010). Disruption of Nap14, a plastid localized non-intrinsic ABC protein in *Arabidopsis thaliana* results in the over-accumulation of transition metals and in aberrant chloroplast structures. Plant Cell Environ. *33*, 1029–1038.

Shingles, R., North, M., and McCarty, R.E. (2002). Ferrous ion transport across chloroplast inner envelope membranes. Plant Physiol. *128*, 1022–1030.

Shockey, J.M., Fulda, M.S., and Browse, J. (2003). *Arabidopsis* contains a large superfamily of acyl-activating enzymes. Phylogenetic and biochemical analysis reveals a new class of acyl-coenzyme A synthetases. Plant Physiol. *132*, 1065–1076.

Smith, P.M.C., and Atkins, V.A. (2002). Purine biosynthesis. Big in cell division, even bigger in nitrogen assimilation. Plant Physiol. *128*, 793–802.

Soll, J., Bölter, B., Wagner, R., and Hinnah, S.C. (2000). Response: The chloroplast outer envelope: a molecular sieve? Trends Plant Sci. *5*, 137–138.

Soll, J., and Schleiff, E. (2004). Protein import into chloroplasts. Nat. Rev. Mol. Cell Biol. *5*, 198–208.

Somerville, C.R. (2001). An early *Arabidopsis* demonstration. Resolving a few issues concerning photorespiration. Plant Physiol. *125*, 20–24.

Song, X.M., Forsgren, A., and Janson, H. (1998). Glycerol-3-phosphate transport in *Haemophilus influenzae*: cloning, sequencing, and transcription analysis of the *glpT* gene. Gene *215*, 381–388.

Song, C.P., Guo, Y., Qiu, Q., Lambert, G., Galbraith, D.W., Jagendorf, A., and Zhu, J.K. (2004). A probable $Na^+(K^+)/H^+$ exchanger on the chloroplast envelope functions in pH homeostasis and chloroplast development in *Arabidopsis thaliana*. Proc. Natl. Acad. Sci. U.S.A. *101*, 10211–10216.

Stael, S., Wurzinger, B., Mair, A., Mehlmer, N., Vothknecht, U.C., and Teige, M. (2012). Plant organellar calcium signalling: an emerging field. J. Exp. Bot. *63*, 1525–1542.

Stepien, P., and Johnson, G.N. (2009). Contrasting responses of photosynthesis to salt stress in the glycophyte *Arabidopsis* and the halophyte Thellungiella: Role of the plastid terminal oxidase as an alternative electron sink. Plant Physiol. *149*, 1154–1165.

Stork, S., Lau, J., Moog, D., and Maier, U.G. (2013). Three old and a new one: protein import into red-algal-derived plastids surrounded by four membranes. Protoplasma *250*, 1013–1023.

Strawn, M.A., Marr, S.K., Inoue, K., Inada, N., Zubieta, C., and Wildermuth, M.C. (2007). *Arabidopsis* isochorismate synthase functional in pathogen-induced salicylate biosynthesis exhibits properties consistent with a role in diverse stress responses. J. Biol. Chem. *282*, 5919–5933.

Subtil, A., Collingro, A., and Horn, M. (2014). Tracing the primordial Chlamydiae: extinct parasites of plants? Trends Plant Sci. *19*, 36–43.

Takabayashi, A., Kadoya, R., Kuwano, M., Kurihara, K., Ito, H., Tanaka, R., and Tanaka, A. (2013). Protein co-migration database (PCoM-DB) for *Arabidopsis* thylakoids and *Synechocystis* cells. SpringerPlus 2, 148.

Tanz, S.K., Castleden, I., Hooper, C.M., Vacher, M., Small, I., and Millar, H.A. (2013). SUBA3: a database for integrating experimentation and prediction to define the

SUBcellular location of proteins in *Arabidopsis*. Nucl. Acids Res. *41*, D1185-D1191.

Tapken, W., Ravet, K., and Pilon, M. (2012). Plastocyanin controls the stabilization of the thylakoid Cu-transporting P-type ATPase PAA2/HMA8 in response to low copper in *Arabidopsis*. J. Biol. Chem. *287*, 18544-18550.

Tarantino, D., Morandini, P., Ramirez, L., Soave, C., and Murgia, I. (2011). Identification of an *Arabidopsis* mitoferrinlike carrier protein involved in Fe metabolism. Plant Physiol. Biochem. *49*, 520–529.

Teng, Y.S., Su, Y., Chen, L.J., Lee, Y.J., Hwang, I., and Li, H. (2006). Tic21 is an essential translocon component for protein translocation across the chloroplast inner envelope membrane. Plant Cell *18*, 2247–2257.

Tester, M., and Blatt, M.R. (1989). Direct measurement of K^+ channels in thylakoid membranes by incorporation of vesicles into planar lipid bilayers. Plant Physiol. *91*, 249–252.

Thuswaldner, S., Lagerstedt, J.O., Rojas-Stuetz, M., Bouhidel, K., Der, C., Leborgne-Castel, N., Mishra, A., Marty, F., Schoefs, B., Adamska, I., Persson, B.L., and Spetea, C. (2007). Identification, expression, and functional analyses of a thylakoid ATP/ADP carrier from *Arabidopsis*. J. Biol. Chem. *282*, 8848–8859.

Tiwari, M., Sharma, D., and Trivedi, P.K. (2014). Artificial microRNA mediated gene silencing in plants: progress and perspectives. Plant Mol. Biol. *86*, 1–18.

Tomizioli, M., Lazar, C., Brugiere, S., Burger, T., Salvi, D., Gatto, L., Moyet, L., Breckels, L.M., Hesse, A.M., Lilley, K.S., Seigneurin-Berny, D., Finazzi, G., Rolland, N., and Ferro, M. (2014). Deciphering thylakoid sub-compartments using a mass spectrometry-based approach. Mol. Cell. Prot. *13*, 2147–2167.

Toth, S.Z., Schansker, G., and Garab, G. (2013). The physiological roles and metabolism of ascorbate in chloroplasts. Physiol. Plant. *148*, 161–175.

Tsukaguchi, H., Tokui, T., Mackenzie, B., Berger, U.V., Chen, X.Z., Wang, Y., Brubaker, R.F., and Hediger, M.A. (1999). A family of mammalian Na^+-dependent L-ascorbic acid transporters. Nature *399*, 70–75.

Tyra, H., Linka, M., Weber, A.P.M., and Bhattacharya, D. (2007). Host origin of plastid solute transporters in the first photosynthetic eukaryotes. Genome Biol. *8*, R212.

Vera, J.C., Rivas, C.I., Fischbarg, J., and Golde, D.W. (1993). Mammalian facilitative hexose transporters mediate the transport of dehydroascorbic acid. Nature *364*, 79–82.

Versaw, W.K., and Harrison, M.J. (2002). A chloroplast phosphate transporter, PHT2;1, influences allocation of phosphate within the plant and phosphate-starvation responses. Plant Cell *14*, 1751-1766.

Villarejo, A., Buren, S., Larsson, S., Dejardin, A., Monne, M., Rudhe, C., Karlsson, J., Jansson, S., Lerouge, P., Rolland, N., von Heijne, G., Grebe, M., Bako, L., and Samuelsson, G. (2005). Evidence for a protein transported through the secretory pathway *en route* to the higher plant chloroplast. Nature Cell Biol. *7*, 1224–1231.

von Berlepsch, S., Kunz, H.H., Brodesser, S., Fink, P., Marin, K., Flügge, U.I., and Gierth, M. (2012). The acyl-acyl carrier protein synthetase from *Synechocystis* sp. PCC 6803 mediates fatty acid import. Plant Physiol. *159*, 606–617.

Wang, X., Berkowitz, G.A., and Peters, J.S. (1993). K^+-conducting ion channel of the chloroplast inner envelope: functional reconstitution into liposomes. Proc. Natl. Acad. Sci. U.S.A. *90*, 4981–4985.

Weber, A.P.M. (2004). Solute transporters as connecting elements between cytosol and plastid stroma. Curr. Opin. Plant Biol. *7*, 247–253.

Weber, A.P.M., and Fischer, K. (2007). Making the connections – the crucial role of metabolite transporters at the interface between chloroplast and cytosol. FEBS Lett. *581*, 2215–2222.

Weber, A.P.M., and Fischer, K. (2009). The role of metabolite transporters in integrating chloroplasts with the metabolic network of plant cells. In Plant Cell Monographs, Vol. 13, The Chloroplast – Interactions with the Environment, A.S. Sandelius and H. Aronsson, eds. (Heidelberg, Germany: Springer), pp. 159–179.

Weber, A.P.M., Linka, M., and Bhattacharya, D. (2006). Single, ancient origin of a plastid metabolite translocator family in Plantae from an endomembrane-derived ancestor. Eukaryot. Cell *5*, 609–612.

Weber, A., Menzlaff, E., Arbinger, B., Gutensohn, M., Eckerskorn, C., and Flügge, U.I. (1995). The 2-oxoglutarate/malate translocator of chloroplast envelope membranes: Molecular cloning of a transporter containing a 12-helix motif and expression of the functional protein in yeast cells. Biochemistry *34*, 2621–2627.

Weber, A.P.M., Schwacke, R., and Flügge, U.I. (2005). Solute transporters of the plastid envelope membrane. Annu. Rev. Plant Biol. *56*, 133–164.

Wentz, J.B. (1926). Heritable characters of maize: XXVI-Concave. J. Hered. *17*, 327–329.

Werdan, K., Heldt, H.W., and Milovancev, M. (1975). The role of pH in the regulation of carbon fixation in the chloroplast stroma. Studies on CO_2 fixation in the light and dark. Biochim. Biophys. Acta *396*, 276–292.

Whatley, J.M., McLean, B., and Juniper, P.E. (1991). Continuity of chloroplast and endoplasmic reticulum membranes in *Phaseolus vulgaris*. New Phytol. *117*, 209–217.

Wheeler, G., Jones, M.A., and Smirnoff, N. (1998). The biosynthetic pathway of vitamin C in higher plants. Nature *93*, 365–368.

Wildermuth, M.C., Dewdey, J., Wu, G., and Ausubel, F.M. (2001). Isochorismate mutase is required to synthesize salicylic acid for plant defense. Nature *414*, 562–565.

Witz, S., Jung, B., Fürst, S., and Möhlmann, T. (2012). *De novo* pyrimidine nucleotide synthesis mainly occurs outside of plastids, but a previously undiscovered nucleobase importer provides substrates for the essential salvage pathway in *Arabidopsis*. Plant Cell *24*, 1549–1559.

Witz, S., Panwar, P., Schober, M., Deppe, J., Pasha, F.A., Lemieux, M.J., and Möhlmann, T. (2014). Structure–function relationship of a plant NCS1 member – homology modeling and mutagenesis identified residues critical for substrate specificity of PLUTO, a nucleobase transporter from *Arabidopsis*. PLOS One *9*, e91343.

Wolucka, B.A., and van Montagu, M. (2003). GDP-mannose 30,50-epimerase forms GDP-L-gulose, a putative intermediate for the *de novo* biosynthesis of vitamin C in plants. J. Biol. Chem. *278*, 47483–47490.

Wooding, F.B.P., and Northcote, D.H. (1965). Association of the Endoplasmic Reticulum and the plastids in Acer and Pinus. Am. J. Bot. *52*, 526–531.

Wu, W., and Berkowitz, G.A. (1992). Stromal pH and photosynthesis are affected by electroneutral K^+ and H^+ exchange through chloroplast envelope ion channels. Plant Physiol. *98*, 666–672.

Xing, H.L., Dong, L., Wang, Z.P., Zhang, H.Y., Han, C.Y., Liu, B., Wang, X.C., and Chen, Q.J. (2014). A CRISPR/Cas9 toolkit for multiplex genome editing in plants. BMC Plant Biol. *14*, Article 327.

Yamaguchi, M., Takechi, K., Myouga, F., Imura, S., Sato, H., Takio, S., Shinozaki, K., and Takano, H. (2012). Loss of the plastid envelope protein AtLrgB causes spontaneous chlorotic cell death in *Arabidopsis thaliana*. Plant Cell Physiol. 53, 125–134.

Yamasaki, K., Motomura, Y., Yagi, Y., Nomura, H., Kiguchi, S., Nakai, M., and Shiina, T. (2013). Chloroplast envelope localization of EDS5, an essential factor for salicylic acid biosynthesis in *Arabidopsis thaliana*. Plant Signal Behav. *8*, e23603.

Yang, Y., Jin, H., Chen, Y., Lin, W., Wang, C., Chen, Z., Han, N., Bian, H., Zhu, M., and Wang, J. (2012). A chloroplast envelope membrane protein containing a putative LrgB domain related to the control of bacterial death and lysis is required for chloroplast development in *Arabidopsis thaliana*. New Phytol. *193*, 81–95.

Yoon, H.S., Hackett, J.D., Pinto, G., and Bhattacharya, D. (2002). The single, ancient origin of chromist plastids. Proc. Natl. Acad. Sci. U.S.A. *99*, 15507–15512.

Yoon, H.S., Hackett, J.D., Ciniglia, C., Pinto, G., and Bhattacharya, D. (2004). A molecular timeline for the origin of photosynthetic eukaryotes. Mol. Biol. Evol. *21*, 809–818.

Zanetti, M., Teardo, E., La Rocca, N., Zulkifli, L., Checchetto, V., Shijuku, T., Sato, Y., Giacometti, G.M., Uozumi, N., Bergantino, E., and Szabo, I. (2010). A novel potassium channel in photosynthetic cyanobacteria. PLOS One 5, e10118.

Zechmann, B., Stumpe, M., and Mauch, F. (2011). Immunocytochemical determination of the subcellular distribution of ascorbate in plants. Planta 233, 1–12.

Zhang, H.W., Zhu, H.F., Pan, Y.J., Yu, Y.X., Luan, S., and Li, L.G. (2014). A DTX/MATE-type transporter facilitates abscisic acid efflux and modulates ABA sensitivity and drought tolerance in *Arabidopsis*. Mol. Plant 7, 1522–1532.

Zrenner, R., Stitt, M., Sonnewald, U., and Boldt, R. (2006). Pyrimidine and purine biosynthesis and degradation in plants. Ann. Rev. Plant Biol. 57, 805–836.

Zybailov, B., Rutschow, H., Friso, G., Rudella, A., Emanuelsson, O., Sun, Q., and van Wijk, K.J. (2008). Sorting signals, N-terminal modifications and abundance of the chloroplast proteome. Plos One 3, e1994.

The GreenCut – Functions and Relationships of Proteins Conserved in Green Lineage Organisms

11

Tyler M. Wittkopp, Shai Saroussi, Wenqiang Yang and Arthur R. Grossman

Abstract

The most recent GreenCut (GreenCut2) represents a collection of 597 proteins in *Chlamydomonas* and other green lineage photosynthetic organisms, but not in non-photosynthetic (heterotrophic) organisms. GreenCut2 proteins have diverse functions, with many involved in chloroplast processes including photosynthesis and various biosynthetic pathways. GreenCut2 proteins also have roles in assembly and maintenance of chloroplasts, as well as in plant and algal regulatory processes. This chapter focuses on how GreenCut2 bioinformatic analyses were performed, discusses new insights into GreenCut2 proteins that have recently been characterized, suggests new ways to group these proteins based on their known or inferred biological functions, and re-examines potential functions of some GreenCut2 'unknowns' for which there is now experimental information.

Introduction: brief history of photosynthesis and the study of chloroplasts

The characteristics of photosynthetic organisms have been examined for centuries. Many features of plants and algae were appreciated well before chloroplasts were discovered and their complex nature understood. Aristotle suggested that plants obtain nutrients from the earth and water, while in 1774 Priestley discovered that gases released by plants and those exhaled by animals differed in their ability to maintain the burning of a candle in a closed container (Priestley, 1774). In 1783, Lavoiser named this active component of air *oxygen* (Greek: 'acid former') (reviewed in Govindjee and Krogmann, 2004). Subsequently, Ingenhousz suggested that green plants absorb carbon dioxide (CO_2) and release oxygen (O_2) during the day (Ingenhousz, 1779).

During the nineteenth century, scientists continued to examine morphological and physiological/biochemical features of plants. In 1818, Pelletier and Caventou isolated the green pigment in leaves and named it *chlorophyll* (Greek: 'green leaf') (Pelletier and Caventou, 1818), while in 1845 Mayer hypothesized that plants transform energy from sunlight into chemical energy (Mayer, 1845). A few decades later, Boussingault made the first accurate measurements of gas exchange, which eventually led to the equation describing photosynthesis: $6\ CO_2 + 6\ H_2O + \text{solar energy} \rightarrow C_6H_{12}O_6 + 6\ O_2$ (Boussingault, 1864). During the late 1800s, instruments such as microscopes and prisms allowed scientists to probe photosynthetic organisms in novel ways. Schimper was among the first to observe chloroplast division in algae (Schimper, 1883), which set the stage for developing the endosymbiotic theory of eukaryotic cell evolution and the origins of chloroplasts (see 'Evolutionary Perspective', below). At around the same time, Engelmann discovered that the most effective light wavelengths for O_2 production were in the blue and red regions of the visible spectrum (Engelmann, 1882).

The twentieth century brought major advances to our understanding of photosynthetic processes. In 1906, Tswett developed chromatographic methods for separating leaf pigments (Tswett, 1906), while a few decades later Warburg and Negelein developed spectrophotometry (Warburg and Negelein, 1922). In the 1930s, van Niel demonstrated that some (anoxygenic) bacteria could assimilate

CO_2 in the light without evolving O_2 (van Niel, 1931), while Hill showed that isolated chloroplasts could split water (produce O_2) in a light-dependent process that did not require CO_2 (Hill, 1937, 1939). In 1945, Calvin, Benson and Bassham began their work using radioactive $^{14}CO_2$ to determine the biochemical pathway by which CO_2 is assimilated and converted into carbohydrates, and it was ultimately shown that ribulose 1,5-bisphosphate carboxylase/oxygenase (Rubisco) catalyses the formation of two molecules of 3-phosphoglycerate (3-PGA) from CO_2 and ribulose 1,5-bisphosphate (Calvin *et al.*, 1950). In 1951, Burk and Warburg proposed that photosynthetic energy capture and conversion to fixed carbon involved two processes; one process a light-dependent liberation of O_2, and the other a set of light-independent reactions ('dark reactions') that use the products formed in the light to fix or reduce CO_2 (Burk and Warburg, 1951). In the mid-1950s, Frenkel observed the formation of adenosine triphosphate (ATP) from adenosine diphosphate (ADP) by isolated chloroplast membranes (Frenkel, 1954), after which Arnon formulated the principle of photophosphorylation (Arnon 1954a,b). Shortly thereafter, Arnon's group showed that chloroplast grana membranes are the sight of light-dependent ATP and NADPH formation, while light-independent CO_2 assimilation occurs in the stroma (Allen *et al.*, 1955, 1958). Finally, the same group discovered cyclic photophosphorylation and proposed that it was a mechanism to provide additional ATP (above the level synthesized as a consequence of linear electron flow) needed for CO_2 fixation (Arnon *et al.*, 1958).

The 1950s were noteworthy for the pioneering work of Emerson, Kok, Duysens and others that demonstrated the existence of two integral membrane photochemical reaction centres [photosystems I and II (PSI and PSII, respectively)] in thylakoid membranes (Emerson and Lewis, 1943; Emerson *et al.*, 1957; Emerson and Chalmers, 1958), while Hill and Bendall synthesized this information into the concept of the Z-Scheme of electron transport, in which the two photosystems operate in series (Kok, 1959; Hill and Bendall, 1960; Duysens *et al.*, 1961; Duysens and Amesz, 1962). In the late 1950s, work in Arnon's laboratory led to the discovery of ferredoxin and the idea that ATP formation was coupled to light-driven electron transfer from water to $NADP^+$ (Tagawa and Arnon, 1962). Finally, work by Mitchell, Jagendorf and Uribe in the 1960s showed that electron transport generates a proton gradient across thylakoid membranes, which ultimately drives the activity of the ATP synthase (Mitchell, 1961a,b; Jagendorf and Uribe, 1966). Additionally, Joliot and Kok in the latter part of the 1960s gained key insights into the sequential nature of photosynthetic H_2O oxidation (Joliot *et al.*, 1969; Kok *et al.*, 1970). During this period there were also advances in genetic and molecular analyses, with the discovery of plastid DNA by Ris and Plaut (1962).

The work described above laid the foundation for further analysis of photosynthetic electron transport and the fixation of CO_2 by algae and plants, and also led to the initiation of genetic and molecular technologies for dissecting photosynthetic processes. In the 1950s, Sager's group understood the power of using *Chlamydomonas reinhardtii*, a unicellular, haploid green alga with defined genetics, for dissecting biological processes through tetrad analyses (Sager and Ishida, 1963), while Levine and colleagues realized that since photosynthesis is not obligate for *Chlamydomonas* (*Chlamydomonas* can grow on acetate in the dark), this organism could serve as a model for analysing photosynthesis through the generation of mutants. Indeed, Levine's group was the first to isolate and analyse *Chlamydomonas* mutants defective for photosynthesis (Levine, 1960; Ebersold *et al.*, 1962), and to use these mutants to delineate the order of specific electron carriers in the photosynthetic electron transport chain (Levine, 1969). *Chlamydomonas* has continued to be used as a model system for elucidating various aspects of photosynthesis ranging from the components of reaction centres (reviewed in Wollman *et al.*, 1999) to an understanding of state transitions (reviewed in Wollman, 2001; Minagawa, 2011; Rochaix, 2014), non-photochemical quenching (reviewed in Müller *et al.*, 2001; Niyogi and Truong, 2013), protein transport into chloroplasts (reviewed in Li and Chiu, 2010), chlororespiration (reviewed in Peltier and Cournac, 2002; Nawrocki *et al.*, 2015) and other regulatory processes (reviewed in Eberhard *et al.*, 2008; Grossman *et al.*, 2010).

The GreenCut: proteins conserved among green lineage organisms

With an increasing ability to perform large-scale genome sequencing, analyses of gene expression and comparative genomics, new strategies are being developed to explore photosynthetic processes. Within the last 10 years the *Chlamydomonas* and *Arabidopsis* genomes have been sequenced, annotated and characterized in several ways. Comparative genomic analyses using *Chlamydomonas, Arabidopsis* and other fully sequenced genomes has led to the identification of a set of proteins (the GreenCut) that are conserved among green lineage organisms and absent from (or highly diverged in) heterotrophic organisms; these proteins also may be present in other photosynthetic organisms that are not included in the green lineage (Merchant *et al.*, 2007). The GreenCut was originally identified based on the establishment of a complete set of proteins encoded by the *Chlamydomonas* nuclear genome, as predicted from gene models. These proteins were used to identify orthologous proteins present in other photosynthetic eukaryotes of the green lineage or Viridiplantae (included in the group were the green alga *Ostreococcus tauri*, the angiosperm *Arabidopsis thaliana*, and the moss *Physcomitrella patens*), but absent in non-photosynthetic organisms (Merchant *et al.*, 2007). Originally, the GreenCut included 349 proteins, with many known to have roles in chloroplast processes including photosynthetic electron transport, carbon fixation, and scavenging of reactive oxygen species. This group also contained proteins involved in chloroplast and thylakoid membrane biogenesis, redox homeostasis, and other metabolic processes. Of the original 349 *Chlamydomonas* GreenCut proteins, 214 had no known function, although many contained domains or other signatures from which a function/activity could be inferred (Merchant *et al.*, 2007).

An additional level of analysis was performed to determine which GreenCut protein families were also present in the diatoms *Thalassiosira pseudonana* and *Phaeodactylum tricornutum* (Merchant *et al.*, 2007). This DiatomCut (CGLD – Conserved in Green Lineage and Diatoms) includes 150 of the original 349 GreenCut proteins. A similar analysis identified GreenCut proteins in the red alga *Cyanidioschyzon merolae*, designated the PlantCut (CPL – Conserved in Plant Lineage; 117 of the 349 GreenCut proteins) (Merchant *et al.*, 2007). Finally, the PlastidCut (CPLD – Conserved in Plant Lineage and Diatoms) included GreenCut proteins that were shared between the DiatomCut and the PlantCut, representing an inventory of proteins conserved in plastid-containing organisms (Merchant *et al.*, 2007).

There were several issues with the way in which the original GreenCut analysis was performed, especially as new information became available and better gene models were developed. First, the number of full genomes used in the analysis was limited. At the time, only a handful of plant and algal genomes were available. Second, the requirement that a GreenCut protein must be present in *Chlamydomonas, Arabidopsis, Physcomitrella,* and *Ostreococcus* (in addition to being absent in non-photosynthetic organisms) limited the output (Karpowicz *et al.*, 2011). *Ostreococcus tauri* is a highly specialized green alga with a very small genome (12.6 Mb) (Derelle *et al.*, 2006); many gene families have a reduced membership and many pathways and proteins were optimized to its niche in the marine environment (Derelle *et al.*, 2006). Therefore, a number of nucleus-encoded proteins present in *Chlamydomonas, Arabidopsis,* and *Physcomitrella* were excluded from the initial GreenCut collection because they were absent or highly diverged in *Ostreococcus*.

The above-mentioned issues were addressed, and with improved sets of gene models a revised GreenCut (GreenCut2) that included 597 *Chlamydomonas* proteins was generated (Karpowicz *et al.*, 2011). The GreenCut2 inventory includes the *Chlamydomonas* nucleus-encoded proteins that have orthologues in *Arabidopsis thaliana, Physcomitrella patens, Oryza sativa, Populus trichocarpa,* and one or more of three *Ostreococcus* species (*O. tauri, O. lucimarinus,* or *Ostreococcus* sp. RCC809) for which there is full genome sequence information, and that are absent from or have become highly diverged in heterotrophic (non-photosynthetic) organisms (Karpowicz *et al.*, 2011). Similar to the original analysis (Merchant *et al.*, 2007), orthologues of many GreenCut2 proteins are also present in cyanobacteria and other non-green lineage organisms (e.g. diatoms and red algae). Furthermore, ~70% of these proteins are predicted to be targeted to chloroplasts where they may function in various biological processes ranging from photosynthetic and metabolic functions to chloroplast biogenesis/assembly and the regulation of photosynthetic processes. Some GreenCut2 proteins were either demonstrated or predicted to localize to other cellular compartments, including the nucleus (~8%), cytosol, secretory system, and mitochondrion (~14%) (Karpowicz *et al.*, 2011).

Many GreenCut2 proteins either lack informative domains or await validation of potential function (designated unknowns). Functions/activities of the unknowns are sometimes inferred from studies of gene expression, subcellular localization, proteomics, and/or mutant analysis in model photosynthetic organisms. A major assumption in the use of GreenCut2 information is that the function of a member protein of an orthologue group is conserved among all organisms within the green lineage. However, because of the long evolutionary distances, the functions of putative orthologues among the different organisms may have diverged. Indeed, even a limited number of changes in a coding sequence may result in major alterations in protein function (Ikeda *et al.*, 1997; Broun *et al.*, 1998). Divergent functionalities may reflect niche specialization (e.g. marine versus freshwater systems), the evolution of distinct developmental and morphological features (e.g. single celled alga versus multicellular plant), divergent reproductive strategies (e.g. sexual versus asexual) and the expansion of gene families coupled to the evolution of new functionalities. Furthermore, two closely related proteins that have the same or similar activities may operate in distinct physiological contexts. Such proteins may be part of different metabolic pathways or networks, may be differentially regulated with respect to tissue type or environmental conditions, may be localized to different subcellular compartments or have a role under a set of environmental conditions that may vary among organisms. Therefore, the loss of a specific GreenCut protein in one organism will not always result in the same phenotype as the analogous loss in another organism.

Evolutionary perspective

The endosymbiosis theory offers an explanation for the origin of eukaryotic cells from prokaryotes. The theory suggests that eukaryotic organelles such as mitochondria and plastids arose from the engulfment of bacterium/cyanobacterium by an ancestral eukaryote (Mereschkowsky, 1910), although this is likely a gross oversimplification of a long and complex process. The once free-living bacterium lost the ability to grow independently of the host (endosymbiont) and a significant portion of its genome was either lost or transferred to the host nucleus. Plastids in eukaryotic photosynthetic organisms arose from one or more engulfment events (reviewed in Keeling, 2010). The uptake of a cyanobacterium by a heterotrophic eukaryote represents the initial endosymbiotic event that eventually gave rise to different lineages harbouring a primary plastid, which include the Rhodophytes, Glaucophytes, and Chlorophytes. Over evolutionary time, the Chlorophytes evolved into both non-vascular (Bryophytes) and vascular (Traceophytes) land plants, which dominate the terrestrial environment. Secondary endosymbiotic events also occurred and involved the transfer of primary plastids from red or green algae to other lineages (e.g. diatoms).

The establishment of a plastid from an endosymbiont involved many processes, including the transfer of genetic material from the endosymbiont to the genome of the host organism. This phenomenon, referred to as endosymbiotic gene transfer, afforded the host greater control over gene expression and the activities of the symbiont, but also facilitated the evolution of regulatory and signalling factors to coordinate expression of nuclear and plastid encoded proteins that function together (e.g. part of the same protein complex) within plastids. In plants and algae, plastids are predicted to contain several thousand proteins, while the chloroplast genomes of these organisms often contain only ~100 genes.

GreenCut2 functional categories

The categorization of GreenCut2 proteins is based on their validated or predicted functions. In many cases, these proteins are binned into groups encompassing broad biological processes, with little detail concerning their precise functions. For example, the GreenCut2 'Protein' category contains a range of polypeptides known or predicted to function as proteases, tRNA synthetases, and chaperones, as well as factors involved in transcriptional regulation,

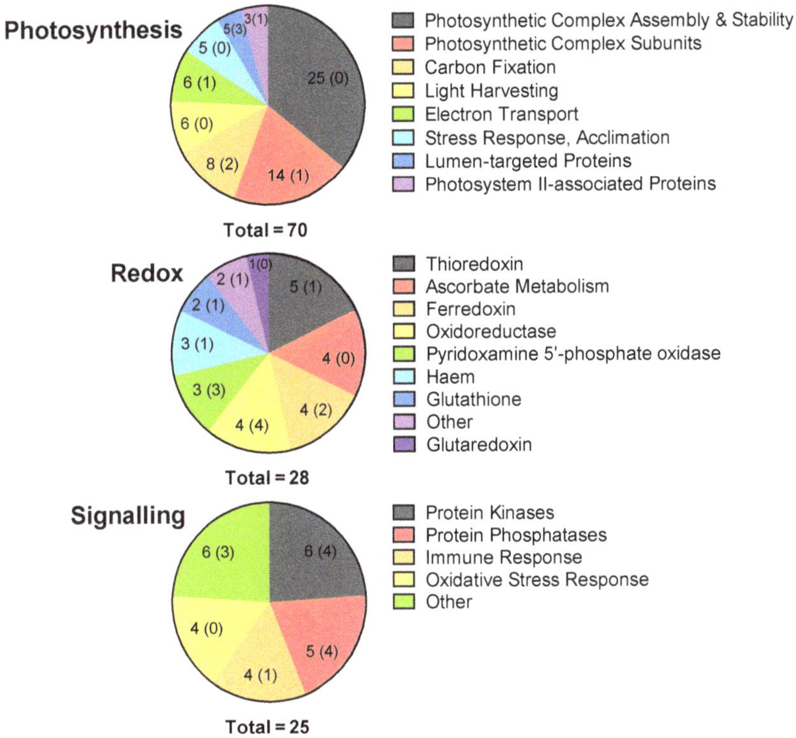

Figure 11.1 Sub-categorization of GreenCut2 proteins in the 'Photosynthesis', 'Redox', and 'Signalling' categories. The GreenCut2 categories were subdivided into groups based upon additional levels of detail regarding their biological functions. Numbers within a pie slice indicate the total number of proteins within the sub-category and the number of proteins of unknown function (in parentheses). For specific proteins in each category, refer to Tables 11.1–11.3.

translation initiation, and post-translational modification. In other instances, proteins span multiple functional categories, making their assignments to specific groups arbitrary. For example, the STT7/STN7 kinase, which is involved in state transitions, could be placed in the categories 'Protein', 'Signalling' or 'Photosynthesis'. To achieve a clearer sense of the function of a specific GreenCut2 protein, each of the classification categories used can be subdivided into groups that provide more detail with respect to potential function. We decided to examine three categories of GreenCut2 proteins, 'Photosynthesis', 'Redox' and 'Signalling', in more detail (Fig. 11.1). A comprehensive regrouping of all GreenCut2 proteins is beyond the scope of this review, so we chose three diverse groups with some functional overlap. The 'Photosynthesis' category contains a large number of proteins with known functions, while many proteins in the 'Redox' and 'Signalling' categories have no assigned function, although they may be involved in regulatory events, an area that we are currently exploring in more detail.

Photosynthesis

The 597 *Chlamydomonas* proteins of the GreenCut2 inventory have been divided into 14 functional categories, including those with vague descriptions ('Other', 'No Prediction', 'Uninformative'). Of these 14 categories, the 'Photosynthesis' category consisted of 62 proteins, of which only 3 were of unknown function. Since 2011, when the GreenCut2 inventory was published, a number of proteins have been characterized. Some of these proteins previously belonged to other groups, but because of new evidence that supports a role in photosynthesis, they are now considered 'known' proteins in the category 'Photosynthesis'. Furthermore, the 'Photosynthesis' category itself can be further divided into subcategories based on specific processes to which each protein can be assigned (Fig. 11.1 and Table 11.1).

Table 11.1 Subcategories of GreenCut2 proteins belonging to the 'Photosynthesis' category

Category	Cre Name	Cre *v*5.5 ID	Ath Name	TAIR ID	Known	Function
Photosynthetic Complex Subunits	PETC	Cre11.g467689	PETC	At4g03280	☑	Rieske iron–sulfur cluster subunit of cytochrome b_6f complex
	PSAD	Cre05.g238332	PSAD-1	At4g02770	☑	Stromal ridge subunit D of PSI
			PSAD-2	At1g03130	☑	
	PSAE	Cre10.g420350	PSAE-2	At2g20260	☑	Stromal ridge subunit E of PSI
	PSAF	Cre09.g412100	PSAF	At1g31330	☑	Subunit F of PSI
	PSAG	Cre12.g560950	PSAG	At1g55670	☑	Subunit G of PSI
	PSAH	Cre07.g330250	PSAH-1	At3g16140	☑	Subunit H of PSI
			PSAH-2	At1g52230	☑	
	PSAK	Cre17.g724300	PSAK	At1g30380	☑	Subunit K of PSI
	PSAL	Cre12.g486300	PSAL	At4g12800	☑	Subunit L of PSI
	PSAO	Cre07.g334550	PSAO	At1g08380	☑	Subunit O of PSI
	PSBW	Not found*	PSBW	At2g30570	☑	Possible subunit W of PSII
	PSB28	Cre10.g440450	PSB28	At4g28660	☐	Similar to PSBW subunit of PSII
	PSBO	Cre09.g396213	PSBO-1	At5g66570	☑	Oxygen-evolving enhancer 1
	PSBP	Cre12.g550850	PSBP-1	At1g06680	☑	Oxygen-evolving enhancer 2
			PSBP-2	At2g30790	☑	
	PSBQ	Cre08.g372450	PSBQ-1	At4g21280	☑	Oxygen-evolving enhancer 3
			PSBQ-2	At4g05180	☑	
	ATPG	Cre11.g481450	PDE334	At4g32260	☑	ATP synthase subunit II
Photosynthetic complex assembly and stability	CGL102	Cre02.g142146	HCF244	At4g35250	☑	PSII assembly
	CPLD10	Cre13.g578650	HCF173	At1g16720	☑	PSII assembly
	CGL54	Cre02.g073850	LPA19	At1g05385	☑	PSII assembly
	EFG10	Cre03.g165000	LPA1	At5g08650	☑	PSII assembly
	CPLD47	Cre07.g329000	PAM68	At4g19100	☑	PSII assembly
	CPLD28	Cre03.g184550	LPA3	At1g73060	☑	PSII assembly
	CPLD41	Cre12.g493150	LTO1	At4g35760	☑	PSII assembly
	RBD1	Cre07.g315150	RBD1	At1g54500	☑	PSII assembly
	TEF5	Cre09.g411200	PSB33	At1g71500	☑	PSII assembly
	TEF30	Cre01.g031100	MET1	At1g55480	☑	PSII supercomplex formation/repair
	PSBP3	Cre12.g509050	PPL1	At3g55330	☑	PSII assembly/repair
	CCB1	Cre16.g662150	CCB1	At3g26710	☑	Cytochrome b_6f complex assembly
	CCB2	Cre12.g537850	CCB2	At5g52110	☑	Cytochrome b_6f complex assembly
	CCB3	Cre01.g052400	CCB3	At5g36120	☑	Cytochrome b_6f complex assembly
	CCB4	Cre08.g382300	CCB4	At1g59840	☑	Cytochrome b_6f complex assembly
	CCS1	Cre13.g575000	na	At1g49380	☑	Cytochrome b_6f complex assembly
	CCS5	Cre17.g702150	HCF164	At4g37200	☑	Cytochrome b_6f complex assembly

Table 11.1 Continued

Category	Cre Name	Cre *v*5.5 ID	Ath Name	TAIR ID	Known	Function
	CPLD38	Cre01.g000850	DAC	At3g17930	☑	Cytochrome b_6f complex assembly
	CGL71	Cre12.g524300	PYG7	At1g22700	☑	PSI assembly
	CGL59	Cre06.g280650	Y3IP1	At5g44650	☑	PSI assembly
	PSBP4	Cre08.g362900	PPD1	At4g15510	☑	PSI assembly
	ZNJ1	Cre11.g475850	PSA2	At2g34860	☑	PSI assembly
	CGL160	Cre01.g049600	CGL160	At2g31040	☑	ATP synthase assembly
	SRP43	Cre04.g231026	CPSRP43	At2g47450	☑	Chlorophyll *a/b* light-harvesting complex assembly
	CGL129	Cre05.g233950	CURT1A	At4g01150	☑	Thylakoid membrane curvature, structural organization
Light harvesting	LHCA1	Cre06.g283050	LHCA1	At3g54890	☑	Light-harvesting protein of PSI
	LHCA3	Cre11.g467573	LHCA3	At1g61520	☑	Light-harvesting chlorophyll *a/b* protein of PSI, type III
	LHCA4	Cre10.g452050	LHCA4	At3g47470	☑	Light-harvesting protein of PSI
	LHCB4	Cre17.g720250	LHCB4.3	At2g40100	☑	Minor LHCII protein CP29
	LHCB5	Cre16.g673650	LHCB5	At4g10340	☑	Minor LHCII protein CP26
	LHCB7	Cre02.g110750	LHCB7	At1g76570	☑	Chlorophyll a/b binding protein
Electron transport	PETF	Cre14.g626700	Fd3	At2g27510	☑	Ferredoxin
	FNR1	Cre11.g476750	RFNR1	At4g05390	☑	Ferredoxin-NADP reductase
			RFNR2	At1g30510	☑	
	PCY1	Cre03.g182551	PETE1	At1g76100	☑	Plastocyanin
	CYC4	Cre16.g670950	CYTC6A	At5g45040	☐	Cytochrome c family member
	PGR5	Cre05.g242400	PGR5	At2g05620	☑	Proton gradient regulation 5 involved in cyclic electron transport
	PGRL1	Cre07.g340200	PGRL1A	At4g22890	☑	PGR5-like protein involved in cyclic electron transport
			PGRL1B	At4g11960	☑	
Carbon fixation	RBCS1	Cre02.g120100	RBCS1A	At1g67090	☑	Rubisco small subunit 1
			RBCS1B	At5g38430	☑	
	RBCS2	Cre02.g120150	RBCS2B	At5g38420	☑	Rubisco small subunit 2
	RCA1	Cre05.g229300	RCA	At2g39730	☑	Rubisco activase
	RCA2	Cre17.g718950	na	At1g73110	☐	Rubisco activase-like protein
	RBCMT1	Cre16.g551350	LSMT-L	At1g14030	☑	Rubisco large subunit N-methyltransferase
	PRK1	Cre12.g554800	PRK	At1g32060	☑	Phosphoribulokinase
	RPI2	Cre07.g314600	na	At5g44520	☐	Ribose-5-phosphate isomerase-related protein
	CGL41	Cre01.g030350	RBCX1	At4g04330	☑	RbcX domain-containing protein with chaperone-like function assisting in the folding of Rubisco
Stress response, acclimation	STT7	Cre02.g120250	STN7	At1g68830	☑	Protein kinase required for state transitions
	APE1	Cre16.g665250	APE1	At5g38660	☑	Protein involved in acclimation of photosynthesis to environment

Table 11.1 Continued

Category	Cre Name	Cre v5.5 ID	Ath Name	TAIR ID	Known	Function
	ELI3	Cre09.g394325	ELIP2	At4g14690	☑	Early light-inducible protein possibly involved in the regulation of chlorophyll biosynthesis
	ELI7	Cre03.g199535	LIL3:1	At4g17600	☑	LHC-like protein required for Chlorophyll and tocopherol biosynthesis
	HLIP	Cre02.g109950	HLIP	At5g02120	☑	Single helix LHC-like protein induced by high light
Photosystem II-associated proteins	PSBS1	Cre01.g016600	PSBS	At1g44575	☑	Chloroplast PSII-associated protein; luminal pH sensor in *Arabidopsis*
	PSBS2	Cre01.g016750	na	At1g44575	☐	Chloroplast PSII-associated protein
	PSBY2	Cre10.g452100	PSBY	At1g67740	☐	Ycf32-related protein of PSII
Lumen-targeted	CGL158	Cre10.g446350	PPD3	At1g76450	☐	PSBP domain protein
PSBP-related	CGL30	Cre03.g198950	PPD4	At1g77090	☐	PSBP domain protein
Proteins	PSBP9	Cre06.g280150	PPD6	At3g56650	☐	PSBP domain protein

☑ Function known or inferred; ☐ Function unknown; na: not available.

Subunits of photosynthetic complexes

In eukaryotes, the major photosynthetic complexes are PSI, PSII, the cytochrome b_6f complex, and the chloroplast ATP synthase. The GreenCut2 contains a number of nucleus-encoded subunits of these major photosynthetic complexes. For the cytochrome b_6f complex, among the four major subunits of the complex, only PETC[1] [Cre11.g467689; At4g03280], or the Rieske iron–sulfur cluster subunit, is present in the GreenCut2 inventory; it is the only major subunit encoded in the nuclear genome (reviewed in Wollman *et al.*, 1999).

PSI subunits that are included in the GreenCut2 are the stromal ridge subunits PSAD [Cre05.g238332; At1g03130, At4g02770], PSAE [Cre10.g420350; At2g20260], but not PsaC because it is chloroplast encoded. Several other PSI intrinsic membrane proteins are also conserved in the green lineage, including PSAF [Cre09.g412100; At1g31330], PSAG [Cre12.g560950; At1g55670], PSAH [Cre07.g330250; At3g16140, At1g52230], PSAK [Cre17.g724300; At1g30380], PSAL [Cre12.g486300; At4g12800], and PSAO [Cre07.g334550; At1g08380].

In contrast to PSI subunits, the majority of PSII subunits are encoded by chloroplast genes. PSII subunits included in the GreenCut2 are PSBW and the related PSB28 protein [Cre10.g440450; At4g28660]. PSBW was originally identified in spinach as a small (6.1 kDa), nuclear-encoded subunit of PSII (Schröder *et al.*, 1988; Lorković *et al.*, 1995). In *Arabidopsis*, a knockdown of PSBW [At2g30570] prevented stabilization of the dimeric PSII complex, which led to a reduction in the total number of PSII complexes per cell (Shi *et al.*, 2000). A separate *Arabidopsis* TDNA knockout line failed to accumulate PSII–LHCII supercomplexes, suggesting an important structural role for PSBW (Garcia-Cerdán *et al.*, 2011). In *Chlamydomonas*, PSBW was also shown to be a subunit of PSII (Bishop *et al.*, 2003). Interestingly, a gene model for PSBW is not given in the most recent *Chlamydomonas* genome browser (Phytozome *10.2*, *Chlamydomonas v5.5*).

PSB28 is one of the few GreenCut2 proteins in the 'Photosynthesis' category for which there is little experimental information concerning functionality. It is annotated as a PSBW-like protein and is localized to thylakoid membranes based

1 In keeping with conventional nomenclature described by Dutcher and Harris (1998), proteins encoded by nuclear genes are designated with uppercase letters (e.g. PETC), while proteins encoded by chloroplast genes have only the first letter capitalized (and often the last letter if there is one) with none of the letters italicized (e.g. PetA). Proteins encoded by chloroplast genes are not included in the GreenCut2.

on *Arabidopsis* proteomic studies (Peltier *et al.*, 2004). In cyanobacteria, Psb28 may function in the biogenesis of the PSII core antenna protein CP47 and may also have a role in regulating chlorophyll (Chl) synthesis (Dobáková *et al.*, 2009).

The oxygen-evolving complex (OEC) is integral to PSII activity. The three major subunits of this complex, all included in GreenCut2, are PSBO [Cre09.g396213; At5g66570], PSBP [Cre12.g550850; At1g06680, At2g30790], and PSBQ [Cre08.g372450; At4g05180, At4g21280].

Finally, ATPG [Cre11.g481450; AtPDE334; At4g32260] represents the only GreenCut2 protein that functions as a subunit of the chloroplast ATP synthase complex. ATPG (or subunit II) is one of two peripheral stalk proteins that associate with both the catalytic CF_1 portion and the membrane-bound CF_0 proton channel (reviewed in Wollman *et al.*, 1999). It is critical for both accumulation and function of the chloroplast ATP synthase. This was demonstrated in an *Arabidopsis* T-DNA insertion mutant; the *atpG* mutant was unable to grow photoautotrophically (Kong *et al.*, 2013). When grown on sucrose, the mutant also displayed disorganized thylakoid membrane structure, as well as altered accumulation of the CF_1–CF_0 complex (Kong *et al.*, 2013).

Assembly and stability

The assembly of photosynthetic complexes within the chloroplast requires coordinated expression of subunits encoded by both the nucleus and chloroplast genomes. Several proteins have recently been identified as having roles in the assembly and/or stability of such complexes. At least two GreenCut2 proteins are known to have a post-transcriptional role in PSII assembly. HCF244 [At4g35250; CrCGL102; Cre02.g142146] and HCF173 [At1g16720; CrCPLD10; Cre13.g578650] are members of the atypical short-chain dehydrogenase/reductase family of proteins that have evolved functions in translation initiation of *psbA* mRNA (Link *et al.*, 2012). HCF173 has an additional role in stabilizing *psbA* transcripts and is speculated to be an RNA-binding factor that assists in delivery of *psbA* transcripts to ribosomes (Schult *et al.*, 2007).

LPA19 [At1g05385; CrCGL54; Cre02.g073850] is a homologue of PSB27, a PSII-associated protein thought to have a critical role in the early stages of PSII assembly (Wei *et al.*, 2010). In accord with this function, *Arabidopsis* LPA19 was demonstrated to interact with both mature and precursor D1 protein (PsbA), potentially facilitating the processing of the D1 precursor prior to integration into PSII reaction centres (Wei *et al.*, 2010). The function of the LPA19 orthologue in *Chlamydomonas* has not been determined. A number of other PSII assembly factors have been identified. Several of these factors interact with D1 during *de novo* PSII biogenesis, including LPA1 [At5g08650; CrEFG10; Cre03.g165000], which is a tetratricopeptide (TPR) repeat-containing protein previously placed in the 'Protein' category of the GreenCut2. LPA1 is an integral thylakoid membrane chaperone that appears to facilitate efficient PSII assembly through interactions with D1 (Peng *et al.*, 2006). An additional PSII assembly factor, PAM68 [At4g19100; CrCPLD47; Cre07.g329000], interacts with several PSII reaction centre subunits (D1, D2, CP43, CP47) and with other assembly factors including LPA1 (Armbruster *et al.*, 2010). An *Arabidopsis pam68* mutant exhibited reduced PSII content, high levels of Chl fluorescence and defects in the formation of PSII dimers and supercomplexes (Armbruster *et al.*, 2010).

LPA3 [At1g73060; CrCPLD28; Cre03.g184550] was previously placed in the 'Uninformative' category of the GreenCut2. However, this protein is believed to participate in assembly of CP43 within PSII, and is therefore better placed in the 'Photosynthesis' category (Cai *et al.*, 2010). An *Arabidopsis* mutant lacking LPA3 exhibited reduced levels of several PSII subunits, including the core proteins D1, D2, CP47 and CP43 (Cai *et al.*, 2010). The mating-based split ubiquitin system (mb-SUS) demonstrated an interaction between LPA3 and CP43, but not D1, D2, or CP47 (Cai *et al.*, 2010). Additionally, LPA3 was shown to interact with LPA2 and Alb3, two non-GreenCut2 assembly factors that participate in CP43 integration into PSII (Ma *et al.*, 2007).

LTO1 [At4g35760; CrCPLD41; Cre12.g493150], or Lumen Thiol Oxidoreductase1, functions in disulfide bond formation within the chloroplast lumen (Karamoko *et al.*, 2011). An *Arabidopsis lto1* mutant displayed severe photoautotrophic growth defects including limited linear electron transport, which was correlated to a significant loss of PSII subunits (Karamoko *et al.*, 2011). LTO1, suggested to associate with the PSBO subunit of OEC based on yeast two–hybrid interactions, had an *in vitro* capacity to introduce a disulfide bond in PSBO (Karamoko *et al.*, 2011). Together, these results highlight a crucial role for thiol-disulfide chemistry in the assembly/stability of thylakoid membrane photosynthetic complexes. Little is known about the function of the LTO1 orthologue in other green lineage organisms.

RBD1 [Cre07.g315150; At1g54500], an iron–sulfur containing rubredoxin, is essential for the proper accumulation of PSII in a diverse set of oxygenic photoautotrophic organisms including cyanobacteria, green algae, and plants (Calderon *et al.*, 2013). While the specific function of this protein has yet to be determined, some evidence suggests that it could function in electron transport, oxygen tolerance, or in promoting photosynthetic complex assembly or stability (Calderon *et al.*, 2013). RBD1 was moved from the 'Redox' category to the 'Photosynthesis' category.

Recently, *Arabidopsis* PSB33 [At1g71500; CrTEF5; Cre09.g411200] was shown to have a role in the organization of PSII–LHCII supercomplexes (Fristedt *et al.*, 2014a). A *psb33* mutant was originally identified in the 'Chloroplast 2010: A Database for Large-Scale Phenotypic Screening of *Arabidopsis* Mutants' project and was found to have altered Chl fluorescence after several hours of photoinhibiting light (Ajjawi *et al.*, 2010). PSB33 is an integral thylakoid membrane protein thought to be a component of PSII based on interactions with D1, D2, CP43 and LHCII (Fristedt *et al.*, 2014a). A *psb33* mutant displays reduced levels of LHCII-containing PSII–LHCII supercomplexes and a lower capacity for non-photochemical quenching under high light conditions, all of which suggest that PSB33 has a critical role in the structural organization of the photosynthetic apparatus (especially PSII) in response to changes in environmental conditions (Fristedt *et al.*, 2014a).

Another GreenCut2 protein MET1 [At1g55480], previously placed in the 'Signalling' category, was recently determined to have a role in PSII supercomplex formation in *Arabidopsis* (Bhuiyan *et al.*, 2015). MET1 contains a TPR domain and is localized to thylakoid membranes where it may associate with the PSII core components CP43 and CP47, and is especially important for PSII repair (Bhuiyan *et al.*, 2015). MET1, also called ZKT because of its PDZ, K-box, and TPR domains, is suggested to be regulated by phosphorylation during the *Arabidopsis* wounding response (Ishikawa *et al.*, 2005).

Finally, a member of the PSBP-like (PPL) family of proteins, PPL1 [At3g55330; CrPSBP3; Cre12.g509050] is likely localized to the thylakoid lumen where it assumes a role in the repair of photodamaged PSII (Ishihara *et al.*, 2007). *PPL1* mRNA is co-expressed with other genes up-regulated in response to high light, and high light grown *Arabidopsis ppl1* mutant plants have compromised PSII activity as well as accelerated turnover of the D1 PSII reaction centre protein (Ishihara *et al.*, 2007). Additional members of the PSBP family of proteins are described below.

A number of GreenCut2 proteins have functional roles in the assembly and stability of the cytochrome b_6f complex. One defining feature that distinguishes the photosynthetic cytochrome b_6f complex from its mitochondrial counterpart, the ubihydroquinone:cytochrome *c* oxidoreductase (bc_1 complex), is a *c*' haem that is covalently bound to cytochrome b_6. The GreenCut2 includes four CCB proteins [cofactor assembly, complex C (b_6f), subunit B (PetB)]. These proteins catalyse the covalent attachment of *c*' haem to cytochrome b_6 during protein maturation. CCB1 [Cre16.g662150], CCB2 [Cre12.g537850], CCB3 [Cre01.g052400], and CCB4 [Cre08.g382300] were each able to functionally restore a mutant strain with a defect attributed to the lack of haem binding to cytochrome b_6 (Kuras *et al.*, 1997; Kuras *et al.*, 2007). All four *Chlamydomonas* CCBs are localized in chloroplasts and contain at least one transmembrane domain; they are thought to function on the stromal side of thylakoid membranes (Kuras *et al.*, 2007). In *Arabidopsis*, the orthologous proteins CCB1 [At3g26710], CCB2 [At5g52110], CCB3 [At5g36120] and CCB4 [At1g59840] also localize to chloroplasts, with CCB1, CCB2, and CCB4 necessary for *c*' haem binding to cytochrome b_6 (Lezhneva *et al.*, 2008).

The cytochrome *f* subunit of the cytochrome b_6f complex also contains covalently bonded haem. In *Chlamydomonas*, four nuclear genes (*CCS1*-*CCS4*) and one chloroplast gene (*ccsA*) encode proteins required for haem attachment to apocytochrome *f* (Xie *et al.*, 1998). CCS1 [Cre13.g575000; At1g49380] was originally shown to complement a *Chlamydomonas* mutant (*abf3*) that exhibited defective accumulation of the cytochrome b_6f complex (Inoue *et al.*, 1997). CCS1 was previously placed in the 'Other' category of the GreenCut2. More recently, an additional GreenCut2 protein, CCS5 [Cre17.g702150], was shown to have a similar function in *Chlamydomonas* (Gabilly *et al.*, 2010). The orthologue of CCS5 in *Arabidopsis*, HCF164 [At4g37200], is a membrane-anchored thioredoxin-like protein that accepts electrons from the stromal *m*-type thioredoxin and is capable of donating electrons to the PSAN subunit of PSI as well as to the cytochrome *f* and Rieske iron–sulfur proteins of the cytochrome b_6f complex (Motohashi and Hisabori, 2006). Like HCF164 in *Arabidopsis*, which participates in chloroplast redox chemistry, CCS5 in *Chlamydomonas* physically interacts with apocytochrome *f* and is capable of reducing its haem-binding site (Gabilly *et al.*, 2010). The disulfide reductase activity of CCS5/HCF164 is a critical prerequisite to haem ligation and necessary for normal accumulation of the cytochrome b_6f complex (Motohashi and Hisabori, 2006; Gabilly *et al.*, 2010).

CPLD38 [Cre01.g000850; AtDAC; At3g17930] was previously placed in the 'No Prediction' category of the GreenCut2 because it lacked any informative domain or motif. However, recent publications suggest that this protein is essential for proper accumulation of the cytochrome b_6f complex in both *Chlamydomonas* (Heinnickel *et al.*, 2013) and *Arabidopsis* (Xiao *et al.*, 2012). The CPLD38/DAC polypeptide is localized to thylakoid membranes, has two predicted transmembrane domains, and contains a hydrophilic C-terminal region that protrudes into the chloroplast stroma (Xiao *et al.*, 2012; Heinnickel *et al.*, 2013). While there is some evidence that in *Arabidopsis* DAC interacts with the PetD subunit of the cytochrome b_6f complex (Xiao *et al.*, 2012), the precise mechanism(s) of CPLD38/DAC function is not known.

GreenCut2 proteins required for the proper assembly and/or stability of PSI include PYG7, Y3IP1, PPD1, and PSA2. In *Arabidopsis*, PYG7 [At1g22700; CrCGL71; Cre12.g524300] is critical for assembly of PSI subunits into a functional complex. The absence of PSI in the *pale yellow green7–1* (*pyg7*) mutant is manifest as severe growth defects, leaf discoloration, and irregular chloroplast ultrastructure (Stöckel *et al.*, 2006). PYG7 is a thylakoid membrane TPR protein that co-purifies with PSI and is thought to function post-translationally during the early stages of PSI assembly (Stöckel *et al.*, 2006). A similar TPR-containing PSI assembly factor is encoded by *ycf3*, a gene conserved in the chloroplast genome of algae and vascular plants (Boudreau, *et al.*, 1997). Y3IP1 [At5g44650; CrCGL59; Cre06.g280650] was identified as a nucleus-encoded protein that interacts with Ycf3 (Albus *et al.*, 2010). A knockdown of this thylakoid-membrane GreenCut2 protein caused a significant reduction in the level of PSI in both *Arabidopsis* and tobacco (*Nicotiana tabacum*) (Albus *et al.*, 2010).

PPD1 is one of a number of PsbP-related proteins, several of which are included in GreenCut2 (see 'Lumen-targeted PSBP-related proteins' below). In *Arabidopsis*, PPD1 [At4g15510; CrPSBP4; Cre08.g362900] is required for the formation of stable PSI (Liu *et al.*, 2012). Similar to the *pyg7* mutant, the *ppd1* mutant does not exhibit defects in transcription or translation of PSI subunits, and is thus thought to act at the post-translational level (Liu *et al.*, 2012). PPD1 was demonstrated to interact with the luminal loops of PsaA and PsaB, and potentially assist in folding and membrane integration of these reaction centre proteins during assembly (Liu *et al.*, 2012).

Recently, a GreenCut2 protein belonging to the 'Protein' category and predicted to contain a DNAJ-like Zinc-finger domain was shown to be involved in PSI assembly (Fristedt *et al.*, 2014b). This protein, PSA2 [At2g34860; CrZNJ1, Cre11.g475850], was identified from a forward genetic screen in maize (*Zea mays*), but much of the characterization was performed with an *Arabidopsis psa2* mutant (Fristedt *et al.*, 2014b). PSA2 in *Arabidopsis* localizes to the thylakoid lumen and is part of a multimeric ~250 kDa complex containing the PSI subunit PSAG (Fristedt *et al.*, 2014b). This interaction appears to be specific to PSAG as other PSI subunits were not detected in this large protein complex. PSA2 also has disulfide reductase activity, suggesting that it might participate in thiol transactions during the assembly of PSI (Fristedt *et al.*, 2014b).

Recent work shows that CGL160 [At2g31040; Cre01.g049600] functions in assembly of the chloroplast ATP synthase (cpATPase) in *Arabidopsis* (Rühle *et al.*, 2014). CGL160 is integral to thylakoid membranes, interacts with the c-subunit of the cpATPase, and is speculated to function specifically in the incorporation of this subunit into the complex (Rühle *et al.*, 2014). Not surprisingly, the absence of CGL160 results in acidification of the thylakoid lumen, which triggers an increase in the xanthophyll cycle and energy-dependent non-photochemical quenching (Rühle *et al.*, 2014).

SRP43 [Cre04.g231026; At2g47450] is unique among GreenCut2 proteins in that it is the only protein necessary for proper accumulation of light harvesting complex proteins (Kirst *et al.*, 2012). The *tla3* (for truncated light harvesting antenna size3) mutant of *Chlamydomonas* has both reduced Chl per cell and an elevated Chl *a*/*b* ratio relative to its parental strain (Kirst *et al.*, 2012). The reduced Chl antenna size in this mutant correlates with the need for much higher light fluence rates to saturate photosynthesis (measured as O_2 evolution) (Kirst *et al.*, 2012). The disrupted gene in the *tla3* mutant [Cre04.g231026 or g5047] encodes a protein with similarity to SRP43 of the signal recognition particle, is localized to chloroplasts, and participates in assembly of peripheral components of the light harvesting antenna (Kirst *et al.*, 2012). The exact molecular mechanism by which SRP43 facilitates integration of proteins into thylakoid membranes is not known.

The GreenCut2 likely contains many additional proteins involved in chloroplast assembly and organization of the photosynthetic apparatus, although their functions are only beginning to be elucidated. One example of such a protein is CURT1A [At4g01150; CrCGL129; Cre05.g233950;], a member of a family of proteins enriched in the thylakoid grana margins that is thought to play a role in inducing membrane curvature (Armbruster *et al.*, 2013a).

Light harvesting proteins

Chlorophyll-binding light harvesting antenna proteins associate with PSI and PSII in green lineage organisms, but each organism possesses a unique combination of such light harvesting complex (LHC) proteins. A number of these LHC proteins are similar in *Chlamydomonas* and *Arabidopsis* (reviewed in Rochaix, 2014). In *Chlamydomonas*, there are 12 LHCII proteins, all of which are encoded by the nuclear genome, while *Arabidopsis* has 11 LHCII proteins, with several that exist as different isoforms. Only a few LHCII and LHCI proteins are included in the GreenCut2. These are the minor LHCII proteins LHCB4 [CP29; Cre17.g720250; At2g40100] and LHCB5 [CP26; Cre16.g673650; At4g10340], as well as the rarely expressed LHCB7 [Cre02.g110750; At1g76570] (Klimmek *et al.*, 2006). *Chlamydomonas* and *Arabidopsis* also differ in the number of PSI-associated LHCI proteins; *Chlamydomonas* has nine and *Arabidopsis* six. LHCI proteins included in the GreenCut2 are LHCA1 [Cre06.g283050; At3g54890], LHCA3 [Cre11.g467573; At1g61520], and LHCA4 [Cre10.g452050; At3g47470]. Many of the LHC proteins are not in the GreenCut2 because they do not exhibit a mutual best BLAST hit with an LHC in at least one green lineage organism used to generate the GreenCut2; this has been a limitation in the strategy used to generate GreenCut2 membership, but can be at least partially remedied by bringing in all protein family paralogues or generating membership using similarity cut-offs.

Electron transport

The photosynthetic electron transport chain comprises both membrane-bound, hydrophobic, and soluble, hydrophilic electron carriers. The GreenCut2 includes several conserved proteins that participate in electron transport reactions, including the PSI electron acceptor ferredoxin 1 [PETF; CrFDX1, Cre14.g626700; AtFd1, At2g27510], ferredoxin-$NADP^+$ reductase (FNR; see below), plastocyanin [PCY1; Cre03.g182551; At1g76100], and a member of the cytochrome *c* family (CYTC6A; see below). PGR5 [Cre05.g242400; At2g05620] and PGRL1 [Cre07.g340200; At4g11960; At4g22890] have received attention for their roles in cyclic electron transport around PSI and their association with photosynthetic supercomplexes (Munekage *et al.*, 2002; DalCorso *et al.*, 2008; Iwai *et al.*, 2010; Suorsa *et al.*, 2012; Hertle *et al.*, 2013; Johnson *et al.*, 2014).

In *Chlamydomonas*, the GreenCut2 contains a single ferredoxin-$NADP^+$ oxidoreductase protein [FNR1; Cre11.g476750]. There is some evidence suggesting that this protein exists as both a soluble, stromal protein and thylakoid membrane-associated protein, and that the difference in location facilitates the partitioning of electrons between different acceptors and donors (Yacoby *et al.*, 2011; Peden *et al.*, 2013). In *Chlamydomonas*, FNR1 catalyses the

FDX1-dependent reduction of $NADP^+$ (Peden *et al.*, 2013). FNR1 may also interact with and reduce FDX2 (Peden *et al.*, 2013). There are two paralogues of FNR1 in *Arabidopsis*, which are designated RFNR1 [At4g05390] and RFNR2 [At1g30510]. Both paralogues are root-type FNRs (Hanke *et al.*, 2005) that are present in non-photosynthetic plastids in *Arabidopsis* and other vascular plants. The root-type FNRs are significantly less efficient at reducing $NADP^+$ than their leaf counterparts (LFNRs) (Hanke *et al.*, 2005), and have specialized roles in utilization (oxidation) of NADPH for bioassimilation and biosynthetic processes rather than in $NADP^+$ reduction. *Chlamydomonas*, which contains a single photosynthetic plastid, may use FNR1 for both anabolic (NADPH-oxidizing) and catabolic ($NADP^+$-reducing) processes depending on the redox environment of the chloroplast and the energetic requirements of the cell.

CYTC6A [At5g45040; CrCYC4; Cre16.g670950] is related to cytochrome c_6, which in many cyanobacteria and algae accepts electrons from cytochrome *f* (of the cytochrome b_6f complex) and transfers them to oxidized P700 ($P700^+$, reaction centre of PSI) (Hervás *et al.*, 2003). Some algae have both the Cu-containing plastocyanin and the Fe-containing cytochrome c_6. Plastocyanin is the predominant soluble protein that transfers electrons to PSI under normal growth conditions, while under Cu-limitation, cytochrome c_6 can functionally replace plastocyanin (Wood, 1978). Cytochrome c_6 has not been identified in vascular plants, but the related protein CYTC6A in *Arabidopsis* is suggested to oxidize dithiol/disulfide groups of thylakoid lumen proteins (Marcaida *et al.*, 2006). Likewise, a specific role for CYC4 in *Chlamydomonas* has not been determined, leaving these orthologous proteins among the few GreenCut2 'Photosynthesis' proteins that still have unknown functions.

Carbon fixation

Photosynthesis has often been divided into the 'light' reactions, those involving solar energy capture and conversion into chemical energy, and the 'dark' reactions, those involving the use of that chemical energy for the fixation of inorganic carbon into metabolites and storage molecules (i.e. carbohydrates). Since the so-called 'dark' reactions also occur in the light, they are now collectively referred to as the 'carbon fixation' reactions. A major enzyme involved in carbon fixation, Rubisco, is comprised of two subunits. The large subunit, RbcL, is encoded on the chloroplast genome in *Chlamydomonas* and other green lineage organisms, and thus does not fit the criteria of the GreenCut2. The small subunit, RBCS, however, exists as multiple isoforms including RBCS1 [Cre02.g120100; At1g67090, At5g38420, At5g38430], and RBCS2 [Cre02.g120150; At1g67090, At5g38410, At5g38420, At5g38430]. Several proteins are involved in the regulation of Rubisco, including Rubisco activase [RCA1; Cre05.g229300; At2g39730], a protein similar to Rubisco activase [RCA2; Cre17.g718950; At1g73110)] and Rubisco N-methyltransferase [RBCMT1; Cre16.g661350; At1g14030]. Another known component of the Calvin–Benson-Bassham cycle and in the GreenCut2, phosphoribulokinase [PRK1; Cre12.g554800; At1g32060], catalyses the conversion of D-ribulose 5-phosphate to D-ribulose 1,5-bisphosphate. Finally, this group also includes a protein related to ribose-5-phosphate isomerase [RPI2; Cre07.g314600; At5g44520].

Homologues of the RBCMT1 methyltransferase, such as PsLSMT, which is involved in methylation of the Rubisco large subunit (RbcL) in pea (*Pisum sativum*), are encoded on all plant genomes, although RbcL methylation is not conserved (Mininno *et al.*, 2012). In *Arabidopsis*, the AtLSMT orthologue [At1g14030] trimethylates 1,6-bisphosphate aldolase rather than RbcL (Mininno *et al.*, 2012). There is still little known about the role of protein (specifically lysine) methylation in photosynthetic carbon fixation and chloroplast carbon metabolism.

Stress response, acclimation

A number of GreenCut2 proteins function in acclimation responses to stress or changes in environmental conditions. These include the protein kinase STT7 [Cre02.g120250; AtSTN7, At1g68830] (Depège *et al.*, 2003; Bellafiore *et al.*, 2005), which is involved in phosphorylation of mobile LHCII proteins during state transitions, as well as APE1 [Cre16.g665250; At5g38660], an Arabidsopsis protein of unknown function that is necessary for acclimation to high light (Walters *et al.*, 2003). Other GreenCut2 proteins involved in acclimatory processes include several members of the light harvesting complex (LHC) protein superfamily: the early light-inducible proteins (ELIPs) and a high light-inducible protein (HLIP).

ELIPs were first identified in pea as nucleus-encoded thylakoid membrane proteins that were rapidly and transiently expressed upon transfer of plants between different light conditions (Meyer and Kloppstech, 1984). It was later realized that ELIPs have transmembrane domains (TMDs) similar to those of LHC proteins (Grimm *et al.*, 1989); both LHCs and ELIPs have three transmembrane α-helices. However, these proteins differ in that ELIPs are only expressed transiently in developing thylakoid membranes or during exposure of the plant to high light conditions (Grimm *et al.*, 1989). In *Chlamydomonas*, there are a number of ELIPs, but only ELI3 [Cre09.g394325; AtELIP2, At4g14690] and ELI7 [Cre03.g199535] were identified in the most recent GreenCut2 analysis. ELI7, which appears to be orthologous to *Arabidopsis* LIL3:1 [At4g17600], showed a decrease in the abundance of its mRNA following transfer of the cells to high light (Teramoto *et al.*, 2004). This is in contrast to other ELIPs whose mRNA levels show a rapid, transient increase upon transfer to high light. *Arabidopsis* has a second LIL3 (LIL3:2), although only LIL3:1 is included in the GreenCut2. An *Arabidopsis* double mutant lacking both LIL3 isoforms does not synthesize mature phytylated Chl molecules and is defective in α-tocopherol synthesis (Tanaka *et al.*, 2010). LIL3 was shown to interact with geranylgeranyl reductase (GGR), an enzyme required for Chl, tocopherol and phyloquinone biosynthesis, and it was suggested that LIL3 functions to stabilize GGR (Tanaka *et al.*, 2010).

Surprisingly, the physiological functions of many ELIPs are still largely unknown. In *Arabidopsis*, overexpression of ELIP2 [At4g14690] caused a drastic decrease in Chl content with a corresponding reduction in PSI and PSII (although those complexes that were present were correctly assembled) (Tzvetkova-Chevolleau *et al.*, 2007). ELIP2 may control Chl levels so as to prevent accumulation of reactive Chl intermediates during the assembly of the photosynthetic apparatus, and especially when light levels are high (Tzvetkova-Chevolleau *et al.*, 2007).

HLIPs contain a single transmembrane α-helix and are thought to be the ancestral precursor of the LHC and ELIP families, which collectively comprise one of the largest and most defining protein superfamilies of the green lineage (Montané and Kloppstech, 2000). HLIP [Cre02.g109950; AtOHP, At5g02120] was originally identified in cyanobacteria (Dolganov *et al.*, 1995; Teramoto *et al.*, 2004). The exact function of HLIP is not well understood, but some have suggested that it may be a pigment carrier (binding Chl or carotenoids), while others assign it a potential role in energy dissipation (photoprotection) (Jansson *et al.*, 2000; He *et al.*, 2001; Xu *et al.*, 2002; Havaux *et al.*, 2005). Similar to LHC and ELIP mRNA levels, it does appear that there is diurnal variation in expression of HLIP (Jansson *et al.*, 2000).

PSII-associated proteins

There are at least three GreenCut2 proteins that are predicted to be associated with PSII, including PSBS [At1g44575; CrPSBS1; Cre01.g016600;] and the *Chlamydomonas* co-orthologue PSBS2 [Cre01.g016750], as well as PSBY1 [Cre10.g452100; At1g57740]. PSBS was first identified in spinach as a PSII subunit belonging to the LHC protein superfamily (Kim *et al.*, 1992), but differing from most LHCs in that it contains four transmembrane α-helices. In *Arabidopsis*, PSBS was shown to be critical for non-photochemical quenching (NPQ) (Li *et al.*, 2000) and to specifically function in the feedback de-excitation (qE) component of NPQ by acting as a sensor of luminal pH (Li *et al.*, 2000). Indeed, there is a stoichiometric correlation between qE and the amount of PSBS protein (Li *et al.*, 2000, 2002). The *Chlamydomonas* orthologues of PSBS (PSBS1 and PBSB2) do not likely function as pH sensors, nor are they thought to have a role in NPQ under a range of laboratory conditions (Bonente *et al.*, 2008). Instead, LHCSR3 [Cre08.g367500], which is not included in GreenCut2, is critical for the qE component of NPQ (Peers *et al.*, 2009). Unlike PSBS in vascular plants, *Chlamydomonas* LHCSR3 binds Chl and xanthophylls, and is itself both a pH sensor and energy quencher (Bonente *et al.*, 2010).

Lumen-targeted PSBP-related proteins

The GreenCut2 also includes several proteins related to PSBP, the luminal oxygen evolving enhancer2 (OEE2) protein associated with PSII. PSBP proteins are divided into the PSBP-like

(PPL) proteins and the PSBP domain (PPD) proteins, with the former having higher similarity to PSBP than the latter. *Arabidopsis* contains 2 PPLs (PPL1–2) and 7 PPDs (PPD1–7) (reviewed in Ifuku *et al.*, 2010). *Chlamydomonas* has orthologues for all of these proteins except PPL2. The PSBP protein family is diverse among photosynthetic organisms, but most family members, including PPL1, and PPD1–6, are conserved in GreenCut2 organisms. PPL1 and PPD1 function in recovery of PSII from photodamage and in the biogenesis of PSI, respectively (Ishihara *et al.*, 2007; Liu *et al.*, 2012). Since PPD5 [At5g11450; CrPSBP6; Cre07.g32820] has a potential role in strigolactone biosynthesis and *Arabidopsis* development (Roose *et al.*, 2011), we propose that it be moved from the GreenCut2 'Photosynthesis' category to 'Other'. *Arabidopsis* PPD2 [At2g28605; CrPSBP2; Cre16.g678851] should not be confused with a pyruvate phosphate dikinase [At4g15530] of the same name in *Chlamydomonas*. PSBP2 may be involved in singlet oxygen ($^1O_2^*$) sensing in *Chlamydomonas*, and therefore this protein has been moved to the 'Signalling' category (see 'Signalling' category below). Finally, the functions of PPD3 [At1g76450; CrCGL158; Cre10.g446350], PPD4 [At1g77090; CrCGL30; Cre03.198950], and PPD6 [At3g56650; CrPSBP9; Cre06.g280150] are not known. Given the diversity of this family of proteins, it is likely that the uncharacterized PPDs have key regulatory roles associated with the luminal surface of the thylakoid membranes.

Redox

In the GreenCut2 (Karpowicz *et al.*, 2011), the 'Redox' category contains 32 proteins, including proteins that participate in electron transfer reactions (ferredoxins), redox regulation (thioredoxins), and sensing and responding to redox cues. In contrast to the 'Photosynthesis' category, the 'Redox' category is much smaller, but contains a larger proportion of proteins of unknown function. In fact, only half (16) of the GreenCut2 'Redox' proteins were previously regarded as having a known or inferred function. A number of the unknown 'Redox' proteins have been characterized over the past several years. Based on updated information regarding their functions, we have moved four of these proteins to the 'Photosynthesis' category. These include LTO1 (CPLD41), PSB33 (TEF5), RBD1, and CCS5 (HCF164), all of which have been shown to be necessary for proper assembly or stability of photosynthetic protein–pigment complexes (see 'Photosynthesis' category above). Of the 28 proteins remaining in the 'Redox' category, 13 are still uncharacterized, including four oxidoreductases and three pyridoxamine 5′-phosphate oxidases (Fig. 11.1 and Table 11.2).

Ascorbate metabolism

Ascorbate, or vitamin C, is an important, versatile metabolite in plants and algae. Ascorbate has critical functions in growth and cellular development, cell wall synthesis, and modulating levels of signalling molecules (reviewed in Tóth *et al.*, 2013). In chloroplasts, ascorbate functions as a major redox buffer that scavenges reactive oxygen species (ROS), as well as a redox regulator of gene expression and enzyme activity (reviewed in Foyer and Noctor, 2011).

SAPX [At4g08930], or stromal ascorbate peroxidase, is one of eight ascorbate peroxidases in *Arabidopsis* (three are chloroplast-localized) (Kangasjärvi *et al.*, 2008). A mutant devoid of SAPX showed increased levels of photooxidative stress, including bleaching during the early greening process, although in mature leaves the loss of SAPX appears to be compensated for by the thylakoid-bound ascorbate peroxidase, APX (TAPX) (Kangasjärvi *et al.*, 2008). In *Chlamydomonas*, *APX1* transcript levels increase in response to $^1O_2^*$ in photoautotrophically grown cells (Ledford *et al.*, 2007). APX6 [At4g32320], one of three cytosolic ascorbate peroxidases in *Arabidopsis*, functions in protecting desiccating and germinating seeds from oxidative damage and may have a role in ROS signalling (Chen *et al.*, 2014).

In *Arabidopsis*, there is evidence that MDAR1 [At3g52880] functions as an NADH-dependent monodehydroascorbate reductase within peroxisomes (Lisenbee *et al.*, 2005). In *Chlamydomonas*, *MDAR1* transcript and protein [Cre17.g712100] levels increase when the cells are Fe-limited, part of the cellular response to increased hydrogen peroxide (H_2O_2) levels resulting from compromised PSI function (Urzica *et al.*, 2012).

GDP-mannose 3′,5′-epimerase, or GME1 [At5g28840], also appears to function in ascorbate synthesis. *Arabidopsis* GME1 catalyses the $NADP^+$-dependent epimerization of GDP-D-mannose to

Table 11.2 Subcategories of GreenCut2 proteins belonging to the 'Redox' category

Category	Cre Name	Cre v5.5 ID	Ath Name	TAIR ID	Known	Function
Ascorbate Metabolism	APX1	Cre02.g087700	SAPX	At4g08390	☑	Ascorbate peroxidase
	APX2	Cre06.g285150	APX6	At4g32320	☑	Putative L-ascorbate peroxidase
	MDAR1	Cre17.g712100	MDAR1	At3g52880	☑	Monodehydroascorbate reductase involved in ascorbate recycling
	SNE1	Cre01.g019250	GME1	At5g28840	☑	Sugar nucleotide epimerase involved in ascorbate biosynthesis
Oxidoreductase	CPLD49	Cre16.g666050	na	At1g50450	☐	Saccharopine dehydrogenase-like protein
	PNO4	Cre17.g744997	PDE327	At4g30720	☐	FAD-dependent oxidoreductase
	FAO3	Cre02.g109800	na	At2g22650	☐	Conserved FAD-dependent oxidoreductase
	OXR1	Cre01.g000350	na	At1g15140	☐	Possible oxidoreductase
Glutaredoxin	GRX6	Cre01.g047800	GRXS16	At2g38270	☑	Glutaredoxin
Haem	SOUL2	Cre13.g566850	HBP2	At2g37970	☑	SOUL haem-binding domain
	SOUL4	Cre06.g304400	HBP5	At5g20140	☑	SOUL haem-binding domain
	SOUL5	Cre06.g292400	na	At2g46100	☐	SOUL haem-binding domain
Glutathione	CPLD58	Cre06.g261500	na	At5g03880	☐	Glutathione S-transferase-like protein
	GSH1	Cre02.g077100	GSH1	At4g23100	☑	Gamma-glutamylcysteine synthetase
Thioredoxin	TRX1	Cre01.g052250	THX	At1g50320	☑	Thioredoxin x
	TRX2	Cre03.g157800	ACHT1	At4g26160	☑	Atypical cysteine-, histidine-rich thioredoxin
	TRX11	Cre14.g624201	CDSP32	At1g76080	☑	Plastidic thioredoxin-like protein
	TRX10	Cre01.g027450	na	At3g53220	☐	Thioredoxin-like protein with unusual active site
	TRX13	Cre02.g142800	TRXZ	At3g06730	☑	Thioredoxin-related protein
Ferredoxin	FDX4	Cre07.g334800	FDC1	At4g14890	☑	Ferredoxin with possible function in glycolysis or the oxidative stress response
	FDX6	Cre03.g183850	FDC2	At1g32550	☐	Ferredoxin of unknown function
	FDX11	Cre06.g291650	na	At4g32590	☐	Atypical ferredoxin of unknown function
	FTR3	Cre03.g193950	na	At2g04700	☑	Ferredoxin thioredoxin reductase catalytic subunit
Pyridoxamine 5′-phosphate oxidase	PPO2	Cre02.g095100	na	At2g46580	☐	Pyridoxamine 5′-phosphate oxidase-like protein
	PPO3	Cre02.g098250	na	At3g03890	☐	Pyridoxamine 5′-phosphate oxidase-like protein
	PPO5	Cre12.g520200	na	At3g21140	☐	Pyridoxamine 5′-phosphate oxidase-like protein
Other	CPLD30	Cre12.g510400	ENH1	At5g17170	☑	Rubredoxin domain protein involved in oxidative stress response
	CPLD66	Cre14.g628500	na	At1g19690	☐	Possible NAD-dependent epimerase

☑ Function known or inferred; ☐ Function unknown; na: not available.

GDP-L-galactose, a reaction immediately upstream of the committed step for ascorbate biosynthesis (Wolucka and Van Montagu, 2005). Additionally, GME1 was shown to catalyse the conversion of GDP-D-mannose to GDP-L-gulose, which could serve as an ascorbate precursor (Wolucka and Van Montagu, 2005). The exact mechanism(s) controlling whether GME1 converts GDP-D-mannose to GDP-L-galactose or GDP-L-gulose is not known, but is thought to involve the redox state of the cell and perhaps the GDP-sugar demand for cell wall biosynthesis (Wolucka and Van Montagu, 2005). In *Chlamydomonas*, SNE1 [Cre01.g019250] is the GME1 orthologue. Its activity appears to be limited to epimerization of GDP-D-mannose to GDP-L-galactose, based on the absence of enzymes that convert GDP-L-gulose to L-gulono-1,4-lactone (the substrate for guluno-lactone dehydrogenase and the synthesis of ascorbate) (Urzica *et al.*, 2012).

Oxidoreductase

Within the GreenCut2 'Redox' category are several putative oxidoreductases that remain largely uncharacterized. One such protein, CPLD49 [Cre16.g666050], is predicted to contain a saccharopine dehydrogenase domain. Saccharopine dehydrogenase catalyses the condensation of α-aminoadipate-δ-semialdehyde and glutamate to saccharopine, which is then hydrolysed to form lysine and α-ketoglutarate (Broquist and Trupin, 1966). *Arabidopsis* has three saccharopine dehydrogenase/saccharopine dehydrogenase-like proteins that have distinct subcellular localizations: At4g33150 in the cytosol, At5g39410 in mitochondria, and At1g50450 in chloroplasts. Only the cytosolic protein has been experimentally demonstrated to have saccharopine dehydrogenase activity (Tang *et al.*, 1997). CPLD49, the *Chlamydomonas* orthologue of the chloroplast-localized At1g50450, may function as an allyl-alcohol:$NADP^+$ oxidoreductase (Cunningham and Gantt, 2011). A similar protein in *Adonis aestivalis* (summer pheasant's eye) participates in the conversion of β-carotene to astaxanthin (Cunningham and Gantt, 2011). These alternative annotations of CPLD49 make it difficult to suggest a function for this protein.

Chlamydomonas PNO4 [Cre17.g744997] is a predicted flavin adenine dinucleotide (FAD-dependent oxidoreductase. Its orthologue in *Arabidopsis*, PDE327 [At4g30720], is a chloroplast-localized protein speculated to have a role in efficient photosynthetic electron transport by ensuring accumulation of proper levels of PSI and PSII (Vlad *et al.*, 2010). Finally, there is no evidence as yet to suggest a function for FAO3 [Cre02.g109800; At2g22650] or OXR1 [Cre01.g000350; At1g15140].

Glutaredoxin

Glutaredoxins (GRXs) are small redox proteins that are reduced by glutathione and participate in thiol/disulfide exchange reactions. GRXs typically function at more positive redox potentials than thioredoxins (Aslund *et al.*, 1997). There is still limited information about glutaredoxin function in green lineage organisms. In *Arabidopsis*, the GreenCut2 protein GRXS16 [At2g38270; CrGRX6, Cre01.g047800] has a N-terminal domain with an endonuclease motif that can function in chloroplast DNA cleavage, and a C-terminal Grx module (Liu *et al.*, 2013) that potentially links redox conditions with nucleic acid metabolism.

Haem

SOUL/haem-binding proteins (HBPs) were first identified in animals in a screen for mRNAs specifically expressed in the retina (Zylka and Reppert, 1999). SOUL/HBPs were shown to bind haem and were suggested to function in light signalling (Zylka and Reppert, 1999). Five SOUL/HPBs have been identified in *Chlamydomonas*, but the specific biological function of these proteins is still unknown (Merchant *et al.*, 2007). The GreenCut2 contains three proteins with SOUL haem-binding domains. These include SOUL2 [Cre13.g566850; AtHBP2; At2g37970], SOUL4 [Cre06.g304400; AtHBP5; At5g20140], and SOUL5 [Cre06.g292400; At2g46100]. Similar to results with animals, *Chlamydomonas* SOUL3 [Cre16.g666550], a non-GreenCut2 protein, was identified in proteomic studies of the *Chlamydomonas* eyespot (Schmidt *et al.*, 2006) and shown to localize to pigment globules of the eyespot, bind haemin, and potentially be involved in regulating eyespot size, organization, and position within the cell (Schulze *et al.*, 2012). In *Arabidopsis*, there are six haem-binding proteins (HBP). The mRNA for the SOUL2 orthologue in *Arabidopsis*, HBP2 [At2g37970], primarily accumulates in root tissue, with the cytosol-localized HBP2 protein able to bind haem and other porphyrins (Takahashi *et al.*, 2008). Several HBP family members may function as tetrapyrrole carriers, delivering cytosolic haem to apoproteins (Takahashi *et al.*, 2008). HBP5 [At5g20140], the *Arabidopsis* SOUL4 orthologue, also binds haem and likely functions in chloroplasts where it interacts with the haem oxygenase HY1 [At1g69720; CreHMOX1; Cre10.g423500], a GreenCut2 protein in the 'Pigments' category (Lee *et al.*, 2012).

Glutathione

Glutathione, a γ-Glu-Cys-Gly tripeptide, is a thiol cofactor critical for maintaining redox balance and alleviating oxidative stress in photosynthetic organisms (Rouhier *et al.*, 2008). At least two GreenCut2 proteins are known or predicted to be involved in glutathione metabolism. CPLD58 [Cre06.g261500], which is related to glutathione S-transferase, and γ-glutamylcysteine synthetase GSH1 [Cre02.g077100; At4g23100], which is required for glutathione synthesis (May and Leaver, 1994). GSH1 was recently shown to have an indirect, yet critical role in the hypersensitive response (HR) of *Arabidopsis* to fungal infection (Hiruma *et al.*, 2013). In this case, the synthesis of glutathione may be necessary for the production of sulfur-containing tryptophan-derived compounds that inhibit the growth of invading pathogens (Hiruma *et al.*, 2013). This specific role for GSH1 is not likely conserved in all green lineage organisms (such as *Chlamydomonas*), but highlights critical functions for glutathione that extend beyond redox transactions.

Thioredoxin

Thioredoxins (TRXs) are a family of nearly ubiquitous redox proteins that share a characteristic structural motif, are enzymatically reduced by NADPH or ferredoxins, and participate in reactions involving thiol/disulfide exchange (Buchanan and Balmer, 2005). *Arabidopsis* has at least 20 different thioredoxins, with five different types (*f*, *m*, *x*, *y*, and *z*) located in the chloroplast. The most well characterized plant thioredoxins are thioredoxins *m* and *f*, which have been linked to light-dependent redox regulation of $NADP^+$-malate dehydrogenase and fructose 1,6-bisphosphatase, respectively (Buchanan, 1980; Jacquot *et al.*, 1997). Thioredoxin *x*, or THX [At2g37970], is a prokaryotic type thioredoxin that is phylogenetically distinct from other *Arabidopsis* thioredoxins (Mestres-Ortega and Meyer, 1999). Features of its transit peptide suggest that it is localized to both mitochondria and chloroplasts (Mestres-Ortega and Meyer, 1999), while at this point only the chloroplast localization has been experimentally confirmed (Zybailov *et al.*, 2008). *In vitro* experiments with recombinant THX from *Arabidopsis* revealed that this protein is an efficient substrate for 2-Cys peroxiredoxin, but not for $NADP^+$-malate dehydrogenase or fructose 1,6-bisphosphatase, suggesting that this isoform may be specifically involved in oxidative stress responses (Collin *et al.*, 2003). The function of this THX orthologue in *Chlamydomonas* [Cre01.g052250] is not known.

In *Arabidopsis*, the thylakoid-associated, atypical cysteine- and histidine-rich thioredoxin ACHT1 [At4g26160; TRX2; Cre03.g157800], which is also a GreenCut2 protein, has high specificity for oxidation by 2-Cys peroxiredoxin under moderate light intensities (Dangoor *et al.*, 2012). Similarly, CDSP32 [At1g76080; TRX11; Cre14.g624201], first identified as a chloroplast drought-induced protein of potato (Pruvot *et al.*, 1996), contains two thioredoxin domains and can reduce the BAS1 peroxiredoxin *in vitro*. These thioredoxins are likely involved in mitigating the effects of oxidative stress in chloroplasts (Pruvot *et al.*, 1996; Dangoor *et al.*, 2012).

The GreenCut2 protein TRXZ [At3g06730; CrTRX13; Cre02.g142800], or thioredoxin *z*, belongs to a distinct branch of the plastidial thioredoxins (Arsova *et al.*, 2010). A *trxZ* knockout line in *Arabidopsis* exhibits an albino phenotype owing to inhibition of normal chloroplast development (Arsova *et al.*, 2010). In the absence of TRXZ, there was a specific decrease in transcripts dependent upon the plastid-encoded RNA polymerase (PEP) for expression, a phenotype also observed when the levels of two fructokinase-like proteins (FLN1 and FLN2) were knocked down (Arsova *et al.*, 2010). TRXZ was shown to interact with both FLNs, suggesting that this TRX regulates PEP-dependent chloroplast transcription through light/dark redox control of the activities of the FLNs (Arsova *et al.*, 2010).

Functions of thioredoxin/thioredoxin-like proteins in green lineage organisms have been primarily inferred from work on *Arabidopsis* orthologues, except for TRX10 [Cre01.g027450; At3g53220].

The physiological role for this thioredoxin is not clear (Meyer *et al.*, 2006).

Ferredoxin

Ferredoxins are small, soluble electron carriers containing a characteristic iron–sulfur cluster that mediates redox exchange in a range of biochemical reactions. The most studied ferredoxin is PETF, or Fd1 in *Arabidopsis* (called Fd1 or FDX1 in *Chlamydomonas*), which accepts electrons from PSI and donates them to a range of downstream acceptor molecules. The functions of most ferredoxins are known or inferred from their ability to reduce substrates *in vitro*, although ferredoxins can have multiple redox partners. Three members of the ferredoxin superfamily are included in the 'Redox' category of GreenCut2; these are FDX4 [Cre07.g334800], FDX6 [Cre03.g183850], and FDX11 [Cre06.g291650; At4g32590]. The *Chlamydomonas* genome encodes six 'typical' plant type ferredoxins (PETF, FDX2–6) that vary in their redox potentials, expression profiles, and substrate specificities (Terauchi *et al.*, 2009). *Chlamydomonas* also contains four 'atypical' ferredoxins (FDX7–10), of which FDX8 is synonymous with what has been designated FDX11.

Recently, a yeast two-hybrid screen was performed to identify targets of *Chlamydomonas* ferredoxins (Peden *et al.*, 2013). FDX4 was shown to interact with the glycolytic enzymes phosphofructokinase (PFK2) and glyceraldehyde-3-phosphate dehydrogenase (GAP3), as well as with peroxiredoxin (PRX1), suggesting that this isoform may have diverse functions (Peden *et al.*, 2013). In *Arabidopsis*, the orthologue of FDX4 is designated FDC1 [At4g14890]. FDCs are related to ferredoxins but have extended C terminal regions (Voss *et al.*, 2010). *FDC1* mRNA and the corresponding FDC1 protein increase when the cells experience PSI acceptor side limitation (knockout or knockdown of *Arabidopsis* Fd2) (Voss *et al.*, 2010). The chloroplast-localized FDC1 can accept electrons from PSI, but because its redox potential is significantly more positive (~−280 mV) than the ferredoxin that is the dominant primary PSI electron acceptor (~−430 mV), it would not be able to support reduction of $NADP^+$ (Voss *et al.*, 2010). Instead, FDC1 likely acts as an alternative electron acceptor under conditions in which photosynthetic ferredoxins (e.g. Fd1 and Fd2 in *Arabidopsis*) are reduced and downstream acceptors (such as $NADP^+$) are limiting (Voss *et al.*, 2010). The ability of FDC1 to serve as a source of electrons for other reactions in the cell has not been explored.

In yeast two-hybrid studies of *Chlamydomonas* FDXs, FDX6 had the fewest interacting partners, and of those that did interact, half were ribosomal proteins. These findings make it difficult to hypothesize a function for this ferredoxin (Peden *et al.*, 2013). It was speculated to have a role in the acclimation of cells to Fe limitation since its transcript increased ~10-fold when the cells were Fe deprived; however the FDX6 protein level does not appear to increase during Fe deprivation (Terauchi *et al.*, 2009). In *Arabidopsis*, the orthologue of FDX6, FDC2 [At1g32550], exhibits an increase in mRNA abundance when the photosynthetic ferredoxin Fd2 is knocked down (Voss *et al.*, 2010). The exact function of FDC2 is not known.

Interaction among redox proteins is critical for maintaining redox homeostasis in chloroplasts. The ferredoxin-thioredoxin system, which requires the activity of ferredoxin-thioredoxin reductase (FTR), plays a critical role in redox regulation of many chloroplast enzymes, especially in response to light (Buchanan and Balmer, 2005). The FTR enzyme has two subunits; a catalytic subunit with a 4Fe–4S cluster, and an enzymatically inactive subunit that may have a structural role (Schürmann, 2002). *Chlamydomonas* FTR is comprised of the FTR3 [Cre03.g193950] catalytic subunit, which is a GreenCut2 protein, and a variable subunit [product of Cre16.g687294] that is not part of GreenCut2 (Huppe *et al.*, 1990).

Pyridoxamine 5′-phosphate oxidase

The three pyridoxamine 5′-phosphate oxidase-like proteins in the GreenCut2 are PPO2 [Cre02.g095100; At2g46580], PPO3 [Cre02.g098250; At3g03890] and PPO5 [Cre12.g520200; At3g21140]. Pyridoxine/pyridoxamine phosphate oxidases catalyse the synthesis of pyridoxal phosphate (PLP) from pyridoxine 5′-phosphate (PNP) or pyridoxamine 5′-phosphate (PMP) (Sang *et al.*, 2007). PLP, the active form of vitamin B_6, is an essential coenzyme used in many biosynthetic pathways, including those for amino acids, lipids, and polysaccharides (Drewke and Leistner, 2001).

Structural derivatives of vitamin B_6 (called vitamers), including PMP and PNP, possess antioxidant properties (Jain and Lim, 2001). In *Arabidopsis*, the non-GreenCut2 protein pyridoxine/pyridoxamine phosphate oxidase (PPOX) was shown to catalyse PLP synthesis *in vitro* (Sang *et al.*, 2007). Additionally, *PPOX* transcript levels showed differential responses to stress (up-regulated in response to light, heat, abscisic acid and ethylene, and down-regulated in response to drought and salt), and a *ppox* mutant is sensitive to high light (Sang *et al.*, 2011). While little is known about the function of PPO2, PPO3 and PPO5 in green lineage organisms, it is possible that these proteins modulate the synthesis of vitamin B_6 and related vitamers to meet the metabolic and/or protective requirements of the cell. Finally, bioinformatic analyses suggest that PPO2 is localized to mitochondria, while PPO3 and PPO5 may function in chloroplasts (Karpowicz *et al.*, 2011).

Other

Two additional GreenCut2 proteins belonging to the 'Redox' category are ENH1 and CPLD66. ENH1 [At4g17170; CrCPLD30; Cre12.g510400] is a chloroplast-localized, rubredoxin domain-containing protein that functions in scavenging ROS produced during salt stress (Zhu *et al.*, 2007). CPLD66 [Cre14.g628500; At1g19690] is an unknown protein with a NAD-dependent epimerase domain.

Signalling

The GreenCut2 'Signalling' category includes proteins that sense extracellular and intracellular conditions and initiate signal transduction cascades. Many of these proteins are known or predicted to localize to chloroplasts, although several likely function in the cytosol. At the time of the initial publication, the GreenCut2 'Signalling' category contained 26 proteins, with 10 having known or inferred functions. This category contains protein kinases and phosphatases and other regulators associated with acclimation or stress responses. We have removed MPA6 from the 'Signalling' category because it is likely the same protein as CPL3 [Cre03.g185200]. The remaining 25 proteins in the 'Signalling' category have been divided into groups based on their predicted functions (Fig. 11.1 and Table 11.3).

Protein kinases

The GreenCut2 contains a number of proteins demonstrated or predicted to function as protein kinases. These include three ABC1-related serine/threonine kinases (AKC2, AKC3, AKC4), a family of proteins named for ABC1 (activity of bc_1 complex) of *Saccharomyces cerevisiae*, which has a critical role in mitochondrial electron transport (Bousquet *et al.*, 1991). The functions of AKC2 [Cre05.g301700; At4g24810; At5g50330] and AKC3 [Cre04.g215400; At3g24190] are not known, while the function of AKC4 [Cre13.g570350] has been inferred from characterization of the OSA1 orthologue in *Arabidopsis* [At5g64940] (see 'Oxidative Stress Response' below).

In *Chlamydomonas*, CDPK2 [Cre10.g466350] is annotated as a calcium/calmodulin-dependent protein kinase, although this has not been experimentally validated. *Arabidopsis* OST1 [At4g33950] (also referred to as SRK2E), the orthologue of CDPK2, was identified in a screen that used infrared thermography to isolate mutants defective in abscisic acid (ABA)-induced stomatal closure, a response that minimizes water loss through transpiration during drought (Mustilli *et al.*, 2002). OST1 functions downstream of the initial ABA signal but upstream of ROS production, both of which precede stomatal closure (Mustilli *et al.*, 2002). In contrast to the prediction that *Chlamydomonas* CDPK2 is a Ca^{2+}-dependent enzyme, *Arabidopsis* OST1 was experimentally shown to possess Ca^{2+}-independent protein kinase activity, although its specific protein substrates are not known (Mustilli *et al.*, 2002). OST1 protein kinase activity can be activated by the application of exogenous ABA or low humidity conditions (Yoshida *et al.*, 2002).

Chlamydomonas CDPK6 [Cre02.g114750] is also a predicted Ca^{2+}-dependent protein kinase, however, unlike CDPK2, it contains an EF hand domain characteristic of Ca^{2+}-binding proteins. In *Arabidopsis*, the CDPK6 orthologue, designated CPK34 [At5g19360], along with a functionally redundant 'co-orthologue' CPK17, participate in pollen tube tip growth (Myers *et al.*, 2009). A CPK34–YFP fusion protein localizes to plasma membranes and growing pollen tubes (Myers *et al.*, 2009). Mutants disrupted for both CPK17 and CPK34 exhibited a drastic reduction in pollen transmission (Myers *et al.*, 2009). This near sterile phenotype could be rescued by complementation

Table 11.3 Subcategories of GreenCut2 proteins belonging to the 'Signalling' category

Category	Cre Name	Cre v5.5 ID	Ath Name	TAIR ID	Known	Function
Protein Kinases	AKC2	Cre06.g307100	na	At4g24810	☐	ABC1-related serine/threonine kinase
			na	At5g50330	☐	
	AKC3	Cre04.g215400	na	At3g24190	☐	ABC1-related serine/threonine kinase
	CDPK2	Cre10.g466350	OST1	At4g33950	☑	Protein kinase involved in hyperosmotic and saline stress signalling
	CDPK6	Cre02.g114750	CPK34	At5g19360	☑	Calcium-dependent protein kinase involved in pollen tube tip growth
	MAPK2	Cre08.g385050	MPK20	At2g42880	☐	Mitogen-activated protein kinase
	CSK4	Cre16.g666200	na	At3g03940	☐	Protein kinase family protein
			na	At5g18190	☐	
Protein phosphatases	PP2C4	Cre01.g030200	na	At1g79630	☐	Protein phosphatase 2C family protein
			na	At1g16220	☐	
	PP2C5	Cre17.g715700	VTC3	At2g40860	☑	Dual function protein kinase::protein phosphatase involved in regulating ascorbate pool
	PP2C6	Cre03.g201400	na	At1g68410	☐	Protein phosphatase 2C family protein
	CPL3	Cre03.g185200	SLP1	At1g07010	☐	*Shewenella*-like protein phosphatase
	MPA14	Cre17.g718800	SLP2	At1g18480	☐	*Shewenella*-like protein phosphatase
Immune response	CPLD33	Cre03.g145207	RPH1	At2g48070	☑	Immune response protein of unknown function
	CGL44	Cre12.g532600	RabGAP22	At5g53570	☑	Rab-GTPase-TBC domain protein involved in innate immunity
	CGL91	Cre01.g015500	SS52	At4g34150	☑	C2 Calcium-dependent lipid-binding (CaLB) domain protein with possible signalling role in the hypersensitive response
	MLO3	Cre08.g358575	MLO1	At4g02600	☐	Transient receptor potential ion channel protein possibly involved in biotic signalling
Oxidative stress	AKC4	Cre13.g570350	OSA1	At5g64940	☑	ABC1-related serine/threonine kinase
Response	CGLD21	Cre03.g163500	EX1	At4g33630	☑	Mediator of singlet oxygen-induced programmed cell death
	PSBP2	Cre16.g678851	PPD2	At2g28605	☑	Possible chloroplast sensor of singlet oxygen
	CGL151	Cre09.g416500	MBS1	At3g02790	☑	Zinc-finger protein required for induction of singlet oxygen-dependent gene expression
Other	CGL8	Cre02.g111650	SPX1	At5g20150	☑	SPX-domain containing protein
			SPX2	At2g26660	☑	

Category	Cre Name	Cre v5.5 ID	Ath Name	TAIR ID	Known	Function
	TEF2	Cre12.g497300	CAS	At5g23060	☑	Rhodanese-like Ca-sensing receptor
	RDP4	Cre05.g247450	na	At3g59780	☐	Putative rhodanese-like protein
	CGL92	Cre03.g208721	CRSH	At3g17470	☑	Calcium-activated RelA/spot homologue involved in fertilization
	CPLD68	Cre13.g607900	na	At2g38780	☐	Possible signal transduction protein interacting with CGL80
	CGL80	Cre02.g108200	na	At1g08530	☐	Possible signal transduction protein interacting with CPLD68

☑ Function known or inferred; ☐ Function unknown; na: not available.

with CPK34-YFP, but not when the kinase or Ca^{2+}-activating EF hand domains were disrupted (Myers *et al.*, 2009).

These findings with respect to CDPK2 (*Chlamydomonas*) and CDPK6 (*Arabidopsis*) suggest that even if proteins exhibit an orthologous relationship they may perform different functions. Clearly, CDPK2 cannot play a role in ABA-mediated stomatal closure in *Chlamydomonas* (and in other unicellular photosynthetic organisms). However, there is some evidence that exposure of *Chlamydomonas* to exogenous ABA can increase tolerance to oxidative stress, perhaps by regulating expression of genes involved in scavenging ROS (Yoshida *et al.*, 2003). The conserved function of CDPK2/OST1 may be to serve as a signalling molecule that links the sensing of a particular stress to appropriate cellular responses. Likewise, while CDPK6 of *Chlamydomonas* cannot be involved in pollen tube growth, it may be a sensor of intracellular Ca^{2+} levels, relaying the information through its protein kinase domain. Hence, it is likely that CDPK2 and CDPK6 have the same or very similar biological functions in divergent organisms such as *Chlamydomonas* and *Arabidopsis*, but that the specific pathways in which these proteins operate are different.

Chlamydomonas MAPK2 [Cre08.g385050] is a member of the mitogen-activated protein kinase family. MAPK cascades involve activation (phosphorylation of specific threonine and tyrosine residues) by MAPK kinases (MAPKKs), which are themselves phosphorylated by MAPKK kinases (MAPKKKs) (reviewed in Schaeffer and Weber, 1999). MAPK cascades are involved in a wide range of signal transduction pathways in eukaryotes, with many associated with sensing an external stimulus or environmental perturbation and transmitting that information to the synthetic machinery critical for launching a cellular response. Plants have at least 20 MAPKs that can be classified according to the amino acid motif TxY, which is phosphorylated by MAPKKs (Ichimura *et al.*, 2002). Several MAPKs containing a TEY motif are important for the responses of plants to both environmental and hormonal conditions, and may also play a role in cell division (reviewed in Ichimura *et al.*, 2002). A different subgroup of MAPKs have a TDY motif and lack a common docking (CD) domain found in TEY motif-containing MAPKs (Ichimura *et al.*, 2002). *Chlamydomonas* MAPK2 and its *Arabidopsis* orthologue MPK20 [At2g42880] are examples of TDY domain-containing MAPKs; the specific functions of these kinases are not known.

Finally, CSK4 [Cre16.g666200] is a predicted protein kinase with homology to MUT9p [Cre06.g281100], a GreenCut2 protein belonging to the 'Nucleic Acid' category. *Chlamydomonas* MUT9p is a serine/threonine protein kinase that phosphorylates histones H3 and H2A and is critical for heritable epigenetic silencing (Casas-Mollano *et al.*, 2008).

Protein phosphatases

The GreenCut2 includes several protein phosphatases that likely function in sensing and signalling. These can be placed into two groups: the PP2C phosphatases and *Shewenella*-like protein phosphatases. The PP2C-type protein phosphatase family is very large in *Arabidopsis* (at least 76 members), with several members shown to function in signal transduction and development (Schweighofer *et al.*, 2004). The functions of PP2C4 [Cre01.g030200; At1g79630; At1g16220] and PP2C6 [Cre03.g201400; At1g68410] are not known. Interestingly, VTC3 [At2g40860], the *Arabidopsis* orthologue of *Chlamydomonas* PP2C5 [Cre17.g715700], is a dual function protein kinase::protein phosphatase originally identified in a screen for ascorbic acid-deficient mutants (Conklin *et al.*, 2000). Loss of VTC3 activity results in the reduction in the level of the ascorbic acid (AsA) pool, and an inability of *Arabidopsis* to regulate levels of AsA in response to light and heat (Conklin *et al.*, 2013). It is still unclear how VTC3 controls AsA biosynthesis (Conklin *et al.*, 2013).

The GreenCut2 also contains two *Shewenella*-like (bacterial) phosphatases, CPL3 [Cre03.g185200] and MPA14 [Cre17.g718800; SLP2; At1g18480] (Andreeva and Kutuzov, 2004) (although they are not enough like the *Shewenella* proteins to be excluded from the GreenCut). While little is known about the function of these proteins in *Chlamydomonas*, the *Arabidopsis* CPL3 orthologue, SLP1 [At1g07010], has been experimentally localized to chloroplasts, while its homologue SLP2 [At1g18480], the orthologue of MPA14, is in the cytosol (Uhrig and Moorhead, 2011). Both SLP1 and SLP2 are insensitive to okadaic acid and microcystin, two classic serine/threonine phosphatase inhibitors (Uhrig and Moorhead, 2011). Finally, SLP1 exhibits Mn^{2+}/Fe^{3+}-dependent phosphatase activity while SLP2 displayed high phosphatase activity in the presence of Mg^{2+} and Zn^{2+} (Uhrig and Moorhead, 2011). The specific functions of SLP1 and SLP2 are not understood, but differences in their subcellular localization and enzymatic properties suggest that they have non-overlapping functions.

Immune response

At least two GreenCut2 proteins belonging to the 'Signalling' category, RPH1 and RabGAP22, have been shown to function in immune responses in *Arabidopsis*. In plants, pathogen infection is often associated with the production of ROS (oxidative burst) followed by controlled death of cells at the infection site (hypersensitive response). An *Arabidopsis rph1* [At2g48070] mutant displayed increased susceptibility to the water mould pathogen *Phytophthora brassicae*, although it was still able to mount a hypersensitive response (Belhaj *et al.*, 2009). The *rph1* mutant exhibited reduced production of ROS and an inability to activate genes involved in host defence, suggesting that RPH1 is critical for a proper immune response (Belhaj *et al.*, 2009).

RabGAP22 [At5g53570] also functions in immune response reactions in *Arabidopsis*. RabGAP22 is a Rab GTPase-activating protein that functions in host defence against infections by the soil-borne fungal pathogen *Verticillium longisporum* (Roos *et al.*, 2014). RabGAP22 appears to also participate in brassinolide (BL) signalling based reduction of *RabGAP22* transcript levels in a *bak1* mutant (defective in brassinosteroid signalling and defence), and is important for eliciting a decrease in the proliferation of fungal hyphae upon BL pretreatment (Roux *et al.*, 2011; Roos *et al.*, 2014).

In *Chlamydomonas*, orthologues of RPH1 and RabGAP22 are CPLD33 [Cre03.g145207] and CGL44 [Cre12.g532600], respectively. While *Chlamydomonas* may encounter pathogenic organisms in its native soil environment, it is unlikely to mount the same types of immune responses that are observed in multicellular plants. For example, the oxidative burst ascribed to the functioning of RPH1 is not likely to occur in *Chlamydomonas*, where such a response is not confined to a specific tissue/cell type. Therefore, the evolutionary conservation of this GreenCut2 protein suggests that it may have a more general function, such as sensing particular physiological changes that may be associated with biotic or abiotic challenges. Alternatively, *Arabidopsis* RPH1 and *Chlamydomonas* CPLD33 could have the same enzymatic or regulatory roles, but operate in different biological contexts (e.g. different metabolic pathways or cellular responses).

The GreenCut2 'Signalling' category also contains a protein, designated SS52 [At4g34150] in *Arabidopsis*, involved in an immune response to viral infection. This protein is localized to the plasma membrane via an N-terminal C2 targeting domain, where it is believed to negatively regulate the plant hypersensitivity response (HR) (Sakamoto *et al.*, 2009). Interestingly, *SS52* mRNA levels are induced in response to both viral infection and other abiotic stresses such as wounding, but mechanistic details concerning its specific function are limited (Sakamoto *et al.*, 2009).

Oxidative stress response

Oxidative stress is a term used to describe a physiological imbalance between the production and detoxification of ROS. An inherent and inevitable consequence of photosynthesis is ROS generation, which largely results from the absorption of excitation energy beyond the level that can be used for photochemistry (electron transport and carbon fixation). A number of photoprotective mechanisms have been described including feedback de-excitation (Li *et al.*, 2000), state transitions (reviewed in Wollman, 2001; Minagawa, 2011; Rochaix, 2014), photoinhibition (reviewed in Müller *et al.*, 2001), and alternative electron transport (reviewed in Peltier *et al.*, 2010). Additional mechanisms specifically involved in ROS detoxification include non-enzymatic scavenging by cellular redox buffers (e.g. ascorbic acid, glutathione, tocopherol, and carotenoids) as well as enzymatic scavenging by proteins such as superoxide dismutase, ascorbate peroxidase, glutathione peroxidase and catalase (reviewed in Apel and Hirt, 2004).

There are several GreenCut2 proteins in the 'Signalling' category that are associated with oxidative stress responses. OSA1 [At5g64940] is a member the ABC1-like serine/threonine kinase protein family (described above). Disruption of chloroplast-localized OSA1 resulted in reduced Chl and xanthophyll content and a higher Chl *a*:*b* ratio, although photosynthetic electron transport was not compromised (Manara *et al.*, 2013). In addition, the *osa1* mutant exhibited reduced levels of the Fe-containing subunits of the cytochrome b_6f complex (PETC or the Rieske protein, and cytochrome *f*) and increased susceptibility to oxidative stress (Jasinski *et al.*, 2008; Manara *et al.*, 2013). The decreased accumulation of Fe-containing photosynthetic proteins and increased levels of ferritin (Fe-storage proteins) suggest that OSA1 is involved in Fe homeostasis within chloroplasts (Manara *et al.*, 2013). Ferritins are also associated with ROS production (Buchanan-Wollaston and Ainsworth, 1997) and the *osa1* mutant was shown to be more sensitive to H_2O_2 and maintained elevated levels of transcripts encoding oxidative stress response proteins (e.g. ascorbate peroxidase) under non-stress conditions (Manara *et al.*, 2013). Together, these results suggest that OSA1 functions at the interface of Fe metabolism and the acclimation of cells to oxidative stress.

EX1 [At4g33630], or EXECUTOR1, was selected as a second-site mutation that suppresses the conditional *Arabidopsis flu* mutant, which is defective in plastid-initiated stress ($^1O_2^*$) signalling (Wagner *et al.*, 2004; Lee *et al.*, 2007). Inactivation of the *EX1* gene attenuates the up-regulation of $^1O_2^*$-induced nuclear genes that are involved in scavenging ROS (Wagner *et al.*, 2004). Elimination of a second plastid-localized protein, EX2, in the *ex1/flu* background, caused near complete suppression of all $^1O_2^*$-responsive genes, suggesting that retrograde signalling in response to oxidative stress requires the function of both chloroplast localized EXECUTOR proteins (Lee *et al.*, 2007).

A *Chlamydomonas* $^1O_2^*$-sensitive reporter strain was generated by fusing the glutathione peroxidase (*GPX5*) promoter to the coding region of arylsulfatase 2 (*ARS2*) and introducing the construct into wild-type cells; this strain was then mutagenized and mutants responding abnormally to $^1O_2^*$ based on arylsulfatase induction were identified (Brzezowski *et al.*, 2012). A mutant disrupted in GreenCut2 protein PSBP2 [Cre16.g678851] failed to up-regulate the *GPX5–ARS2* fusion in response to increased levels of $^1O_2^*$ (Brzezowski *et al.*, 2012). PSBP2 is a member of the PSBP-related protein family (described in the 'Photosynthesis' section above), and the *psbp2* mutant does not respond to either endogenous $^1O_2^*$ (originating from PSII) or exogenous $^1O_2^*$ (generated by exposing cells to neutral red) (Brzezowski *et al.*, 2012). Although the specific function of PSBP2 is unknown, it is proposed to be involved in chloroplast-to-nucleus retrograde signalling in response to oxidative stress.

Finally, MBS1 [At3g02790; CrCGL151; Cre09.g416500], a GreenCut2 protein previously placed in the 'No Prediction' category, was also shown to have a significant role in cellular responses to $^1O_2^*$. MBS1 is a small zinc finger protein required for the induction of $^1O_2^*$-responsive genes (Shao *et al.*, 2013). *Arabidopsis mbs* (METHYLENE BLUE SENSITIVITY) mutants, including *mbs1*, are hypersensitive to oxidative stress, while *Arabidopsis* lines overexpressing MBS1 are more tolerant to high light-induced photooxidative stress (Shao *et al.*, 2013). MBS1 changes its cytosolic distribution by associating with RNA granules (thought to sequester untranslated mRNAs) upon $^1O_2^*$ stress, suggesting a role for MBS1 in regulating the stability or translation of mRNAs involved in the oxidative stress response (Shao *et al.*, 2013).

Other

Arabidopsis SPX1 [At5g20150] and its co-orthologue SPX2 [At2g26660] were recently shown to have a critical role in phosphate (Pi) sensing and regulation of PHR1 activity, a transcription factor involved in the Pi starvation response (Puga *et al.*, 2014). An *spx1spx2* double mutant exhibited increased PHR1 activity under Pi-replete conditions, with little impact on PHR1 activity in Pi-deprived cells (Puga *et al.*, 2014). Furthermore, SPX1 and PHR1 were shown to interact *in planta* in a Pi-dependent manner; the association of PHR1 with SPX1 competitively inhibited binding of PHR1 to its recognition sequence on the DNA (Puga *et al.*, 2014). The function of the *Chlamydomonas* orthologue of SPX1/2, CGL8 [Cre02.g111650], is not known, although it may also have role in Pi sensing.

The GreenCut2 'Signalling' category also contains a Ca^{2+}-sensing protein critical for a number of biological processes in eukaryotic photoautotrophs. In *Arabidopsis*, the chloroplast-localized Ca^{2+}-sensor protein CAS [At5g.23060] is involved in regulating stomatal closure in response to extracellular Ca^{2+} levels (Nomura *et al.*, 2008; Weinl *et al.*, 2008). CAS also plays a pivotal role in chloroplast-mediated immune signalling (Nomura *et al.*, 2012). In *Chlamydomonas*, CAS [TEF2; Cre12.g497300] localizes to thylakoid membranes where it impacts expression of *LHCSR3* under high light conditions (Petroutsos *et al.*, 2011). An RNAi knockdown of *Chlamydomonas* CAS caused diminished activity of PSII and severe light sensitivity, but these phenotypes could be rescued by increasing the Ca^{2+} concentration in the growth medium (Petroutsos *et al.*, 2011). CAS was also demonstrated to form a complex with PGRL1 (see 'Photosynthesis' category above) and the anaerobic response factor ANR1, which together are involved in regulating cyclic electron flow (Terashima *et al.*, 2012). These results highlight the conserved, pleotropic significance of Ca^{2+} and CAS in photosynthetic organisms.

In *Arabidopsis* and *Chlamydomonas*, the C-terminal region of CAS contains a rhodanese-like domain. Another GreenCut2 'Signalling' protein in *Chlamydomonas* with a similar domain is RDP4 [Cre05.g247450]. The functions of RDP4, and that of its orthologue in *Arabidopsis* [At3g49780], have still not been explored.

The bacterial stringent response regulates changes in gene expression through the production of guanosine 5′-diphosphate 3′-diphosphate (ppGpp), which is also known as the stringent factor (reviewed in Braeken *et al.*, 2006). The stringent response has been well studied and is critical for acclimation to a wide range of environmental stimuli. It was first shown to be conserved in eukaryotes based on work done on the *Chlamydomonas* RSH protein (encoded by Cre03.g159800), which is chloroplast-localized and involved in stringent factor production (Kasai *et al.*, 2002). In plants, the stringent response also occurs in chloroplasts, is critical for plant reproduction and involves the chloroplast-localized CRSH [At3g17440] (Masuda *et al.*, 2008). CRSH, a homologue of bacterial ppGpp synthase, has Ca^{2+}-dependent ppGpp synthase activity and is necessary to attain normal silique size and seed production (Masuda *et al.*, 2008). In *Chlamydomonas*, the orthologue of CRSH is CGL92 [Cre03.g208721], which has been suggested to have a role in the synthesis of stringent factor, although its exact physiological function is still to be determined.

The unknowns

Of the 597 GreenCut2 proteins (Karpowicz *et al.*, 2011), 311 were categorized as 'unknowns'. Of these 311 unknowns, 105 were placed into either the 'No Prediction' or 'Uninformative' categories because they did not contain a domain that suggested a function. Various strategies have been employed to discover the functions of unknown proteins, including the generation and analysis of mutants. *Chlamydomonas* and *Arabidopsis* are excellent model organisms with well-established molecular and genetic tools and fully sequenced genomes, and are also amenable to both forward and reverse genetic screens. The sequencing era has enabled large transcriptomic data sets to be obtained for both wild-type and specific mutant strains under a variety of different growth conditions. Co-expression analyses have also proven useful for determining the functions of unknown or poorly studied proteins. Finally, for many GreenCut2 proteins with orthologues in cyanobacteria, the analysis of operons can provide hints about cellular processes in which the unknowns participate.

No prediction and uninformative

We sought to re-examine the functions of the 105 proteins in the 'No Prediction'/'Uninformative' categories. For a number of these proteins, we found information from the literature or from BLAST searches (NCBI BLAST, http://blast.ncbi.nlm.nih.gov/Blast.cgi), which includes the presence of protein domains, that now makes it possible to place these proteins into more specific categories (Table 11.4). The largest group of newly annotated proteins is enriched for factors associated with the assembly/stability of photosynthetic complexes. This includes Y3IP1 (CGL59), required for PSI biogenesis (Albus *et al.*, 2010); DAC (CPLD38), necessary for cytochrome b_6f complex

Table 11.4 GreenCut2 proteins moved from 'No Prediction' and 'Uninformative' to new categories

Previous category	Cre Name	Cre v5.5 ID	New category	Known	Function
No prediction	CGL136	Cre01.g000800	Transport	☐	Protein with homology to mitochondrial import inner membrane translocase subunit Tim-10
	CGL41	Cre01.g030350	Photosynthesis – Calvin cycle	☐	RbcX domain-containing protein with possible chaperone-like function assisting in the folding of Rubisco
	CGL129	Cre05.g233950	Photosynthesis – complex assembly	☑	Orthologue of CURT1A, involved in curvature of thylakoid membrane
	CGL59	Cre06.g280650	Photosynthesis – complex assembly	☑	Ycf3-interacting protein1 involved in photosystem I assembly
	CGL111	Cre03.g174840	Lipids	☐	Lipid-binding START protein
	CGL127	Cre01.g032550	Nucleic acid	☐	Possible DNA- or chromatin-binding domain
	CGL151	Cre09.g416500	Signalling	☑	Zinc-finger protein MBS1 required for induction of singlet oxygen-dependent gene expression.
	CGL57	Cre16.g694703	Protein	☐	Mitochondrial inner membrane protease-related protein
	PYK6	Cre10.g426292	Other	☐	Pyruvate kinase. Mg ion-domain
	CPLD38	Cre01.g000850	Photosynthesis – complex assembly	☑	Protein necessary for the accumulation of the cytochrome b6f complex
	CPLD24	Cre10.g435850	Cofactor	☐	Possible 2-dehydropantoate 2-reductase involved in pantothenate (vitamin B5) synthesis
Uninformative	CGL6	Cre10.g419700	Transport	☐	Possible Mg transporter
	CGL149	Cre07.g340650	Other	☐	Homologue of ER membrane protein complex subunit 7
	CGLD27	Cre05.g237050	Other	☐	Conserved protein of unknown function with possible metabolic role under iron deficiency
	CPLD28	Cre03.g184550	Photosynthesis – complex assembly	☑	Protein necessary for the accumulation of photosystem II
	TPR1	Cre02.g092550	Protein	☑	Plastid outer envelope chaperone receptor
	SUR6	Cre02.g082300	Nucleic acid	☐	Possible nucleic acid binding protein
	FAP173	Cre16.g690879	Other	☑	Hypersensitive response induced protein
	RTN1	Cre16.g689087	Other	☑	Reticulon involved in ER membrane-shaping

☑ Function known or inferred; ☐ Function unknown.

accumulation (Xiao *et al.*, 2012; Heinnickel *et al.*, 2013); and LPA3 (CPLD28), a PSII assembly factor (Cai *et al.*, 2010). Additionally, a role for CURT1A (CGL129) in thylakoid membrane curvature was recently proposed (Armbruster *et al.*, 2013a). Finally, RBCX1 (CGL41) was moved from the 'No Predictions' category based on its possible role as a Rubisco chaperone.

Several other GreenCut2 proteins have been characterized since the initial publication. This includes MBS1 (CGL151), which was moved to the 'Signalling' category because of its involvement in modulating gene expression in response to $^1O_2{}^*$ (Shao *et al.*, 2013). Additionally, TPR1 was demonstrated to function as a chloroplast outer envelope chaperone receptor, and was moved to the 'Protein' category (von Loeffelholz *et al.*, 2011). Finally, two proteins moved from the 'Uninformative' category to the 'Other' category are HIR1 (CrFAP173) which is involved in the hypersensitive response in *Arabidopsis* (Qi *et al.*, 2011), and RTN1 which is a reticulon protein involved in shaping endoplasmic reticulum membranes (Sparkes *et al.*, 2010).

Analysis of interaction networks

Biological network analysis is an informatics approach that helps ascribe proteins to specific biological processes/functions as well as to identify 'systems level' connections among different proteins. Based on transcriptomic data for *Arabidopsis*, it is possible to assemble interaction networks that cluster genes (and their encoded proteins) according to their specific functions, associated processes, mechanisms and pathways, and localization to specific subcellular compartments. Such network analysis fosters the development of data-driven hypotheses about potential functions of unknown or uncharacterized GreenCut2 proteins.

PGR5 and PGRL1 are nuclear-encoded proteins localized to the chloroplast. They are essential components of cyclic electron flow, a process thought to balance the ATP/NADPH ratio in photosynthetic organisms (DalCorso *et al.*, 2008). An interaction analysis using STRING (http://string-db.org; Franceschini *et al.*, 2013) suggests that PGR5 and PGRL1 are part of a larger network of co-expressed genes that may be necessary for specific acclimation processes. Of the 44 and 16 genes co-expressed

A

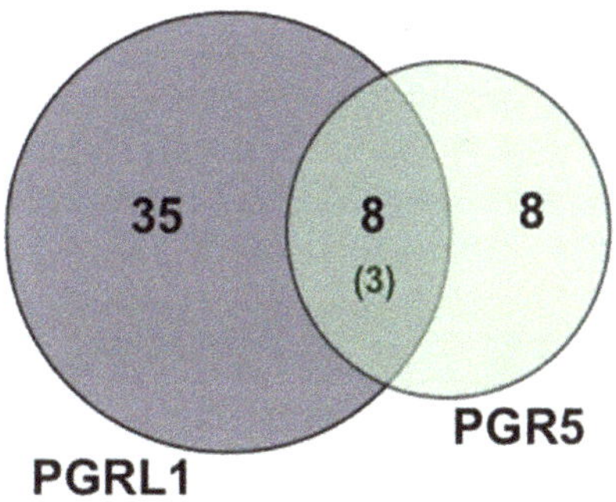

B

A. thaliana Locus	*C. reinhardtii* Locus	Description
At1g44575 (PSBS)	Cre01.g016600 (PSBS1) Cre01.g016750 (PSBS2)	Chloroplast photosystem II-associated 22 kDa protein
At4g03280 (PETC)	Cre11.g467689 (PETC)	Cytochrome b_6f complex subunit
At1g55480 (MET1)	Cre01.g031100 (TEF30)	Photosystem II assembly factor
At1g23740 (AOR)	Cre06.g297400 Cre16.g677950 (QOR1) Cre16.g674050 (ADH12)	Alkenal/one oxidoreductase
At1g68010 (HPR)	Cre06.g295450 (HPR1)	Hydroxypyruvate reductase
At1g09340 (CRB)	Cre10.g435800 Cre10.g440050	NAD-dependent epimerase/dehydratase
At3g55800 (SBPASE)	Cre03.g185550 (SBP1)	Sedoheptulose-1,7-bisphophatase
At1g75690 (LQY1)	na	Protein involved in repair of PSII

Figure 11.2 Interaction (co-expression) analysis of PGR5 and PGRL1, proteins associated with cyclic electron flow. Interactions were analysed using STRING (http://string-db.org; Franceschini *et al.*, 2013), based on the *Arabidopsis thaliana* genome. Parameters were restricted to co-expression, co-occurrence and experimental evidence with an output cut-off set to 0.8 (high probability). Following analysis, *Arabidopsis* proteins were used to identify *Chlamydomonas reinhardtii* orthologues with their corresponding annotations. (A) The numbers within each circle are the number of proteins that interact (or are co-expressed) with PGRL1 (grey), PGR5 (green) and the number of the proteins that show mutual interaction with both PGR5 and PGRL1 (overlapping grey-green). The number of GreenCut2 proteins shared between PGRL1 and PGR5 is shown in parentheses. (B) List of interacting proteins shared by PRGL1 and PGR5 along with their proposed functions. GreenCut2 proteins are highlighted in green.

with *PGRL1* and *PGR5*, respectively, eight were mutually co-expressed. Of these, three (PSBS, TEF30, and PETC) are GreenCut2 proteins (Fig. 11.2). In *Arabidopsis*, PSBS is involved in the activation of energy-dependent non-photochemical quenching (see 'Photosynthesis' category above), a process that redirects energy away from PSII (Li *et al.*, 2000). *Chlamydomonas* has 2 PSBS homologues, PSBS1 and PSBS2, but they do not appear to be required for NPQ (Bonente *et al.*, 2008), and therefore it is difficult to deduce a conserved function based on mutual co-expression of these genes, although they might facilitate NPQ under specific environmental conditions. PETC is a nucleus-encoded core subunit of the cytochrome b_6f complex, a complex which is critical for cyclic electron flow. TEF30 is conserved within the green lineage and also present in diatoms. Based on this analysis, TEF30 may participate in cyclic electron flow. The *Arabidopsis* orthologue of TEF30, ZKT, was hypothesized to serve as a molecular adaptor associated with wounding responses (see 'Photosynthesis' category above). A STRING interaction network analysis using ZKT as the query gene identified other known components of cyclic electron transport including components of the cytochrome b_6f and NAD(P)H-dehydrogenase (NDH) complexes. Recently, ZKT (also referred to as MET1) was demonstrated to act as a PSII assembly factor and to have an important role under fluctuating light (Bhuiyan *et al.*, 2015). Interestingly, split ubiquitin experiments showed strong interactions between *Arabidopsis* MET1 and PSII core subunits CP43 and CP47 and a weak interaction with PGR5 (Bhuiyan *et al.*, 2015). This work highlights the utility of network/co-expression analyses for generating hypotheses about protein functions and their interactions.

A similar strategy was successfully used to dissect the function of *Arabidopsis* GreenCut2 thylakoid proteins PAM68 and its PAM68-like (PAM68L) homologue (Armbruster *et al.*, 2010, 2013b). In this example, co-expression analysis helped delineate the different functions of the PAM68 homologues. PAM68, a conserved protein in plants, green algae and cyanobacteria, facilitates early stages of PSII assembly (Armbruster *et al.*, 2010). PAM68L, which is not conserved throughout the green linage, was found to function as a biogenesis factor for the chloroplast NAD(P)H dehydrogenase (NDH) complex (Armbruster *et al.*, 2013b).

Cyanobacterial operon analysis

The GreenCut2 analysis (Karpowicz *et al.*, 2011) also explored relationships between GreenCut2 proteins and their homologues in cyanobacteria. Comparative analyses reveal that ~30% of GreenCut2 proteins are of prokaryotic origin, suggesting that they may be fundamental to photosynthesis and other related metabolic pathways. A comparable number of GreenCut2 proteins are thought to have evolved after the primary endosymbiotic event leading to the establishment of photosynthetic eukaryotes. The remaining GreenCut2 proteins (40%) have a bacterial homologue in at least one cyanobacterial genome. Most proteins that are evolutionarily conserved from cyanobacteria to vascular plants have a known or inferred function, although ~30% of those still have no assigned specific function.

The Integrated Microbial Genomes (IMG, www.jgi.doe.h.img.gov) contains 226 sequenced cyanobacterial genomes from organisms that grow in a range of ecological niches and habitats. IMG informatics tools can help predict the function(s) of unknown GreenCut2 proteins. Some of these tools reveal conserved operons and gene neighbourhoods that may help place an unknown protein in a specific biological process and suggest ways to experimentally dissect the function of that protein.

PCD4 [Cre14.g616250, previously termed CGL9] is an unknown GreenCut2 protein that was placed in the 'Other' category (Karpowicz *et al.*, 2011). PCD4 in *Chlamydomonas* is predicted to localize to chloroplasts based on PredAlgo (http://giavap-genomes.ibpc.fr/predalgo; Tardif *et al.*, 2012). BLAST analyses (NCBI BLAST, http://blast.ncbi.nlm.nih.gov/Blast.cgi) show that PCD4 is conserved in cyanobacteria and may have a role in photosynthetic regulation. It is homologous to two proteins encoded on the *Synech*ocystis sp. PCC 7407 genome; the one with the most homology (E value: 6 E-32) is a polyketide cyclase/dehydrogenase that is encoded by a gene found in a putative operon with a gene encoding a ζ-carotene desaturase (Fig. 11.3). Bidirectional best BLAST hits suggest that this sequence is also conserved in other cyanobacteria, however, in some organisms ζ-carotene desaturase is annotated as 7,8-carotene desaturase (Fig. 11.3). In several organisms, the PCD4-related gene neighbourhood also contains a transcriptional repressor, although the proximity and orientation with respect to the carotene desaturase gene is variable. Several hypothetical proteins are also encoded in the gene neighbourhoods, one of which is consistently positioned immediately downstream of the gene encoding the cyanobacterial PCD4 orthologue (Fig. 11.3).

Carotene desaturase is an enzyme that functions in carotenoid biosynthesis. The ζ-carotene (also known as 7,7′,8,8′-carotene) and 7,8-carotene desaturases catalyse the formation of lycopene. Based on our analysis, it is difficult to determine whether the PDC4 orthologue in cyanobacteria is functionally involved in the same pathway as carotene desaturases. *Chlamydomonas* PDC4, along with the cyanobacterial orthologues, do not share any similarity with downstream enzymes such as lycopene cyclase. Moreover, in contrast to lycopene cyclase, *Chlamydomonas* PCD4 does not contain a predicted transmembrane domain, but could still function in a reaction that is downstream of ζ-carotene desaturation. Interestingly, interaction analysis of *Chlamydomonas* PCD4 suggests co-expression with a carotenoid cleavage dioxygenase enzyme, which cleaves β-carotene to form signalling molecules (Harrison and Bugg, 2014).

Another example of the use of operon associations to explore gene function involved the GreenCut2 protein CPLD38 (Heinnickel *et al.*, 2013). CPLD38 is localized to chloroplasts and was suggested to be involved in chlororespiration because the gene encoding the CPLD38 orthologue in cyanobacteria was found to be in close proximity (probably in the same operon) to a gene encoding *ndhL*, a component of the NAD(P)H dehydrogenase (NDH) complex that catalyses the reduction of the PQ pool during chlororespiration (Heinnickel *et al.*, 2013).

In summary, analysis of cyanobacterial operon structure is a powerful resource for exploring functions of unknown or poorly characterized GreenCut2 proteins. This process requires an understanding of biological pathways, and is enhanced when coupled to other bioinformatic approaches

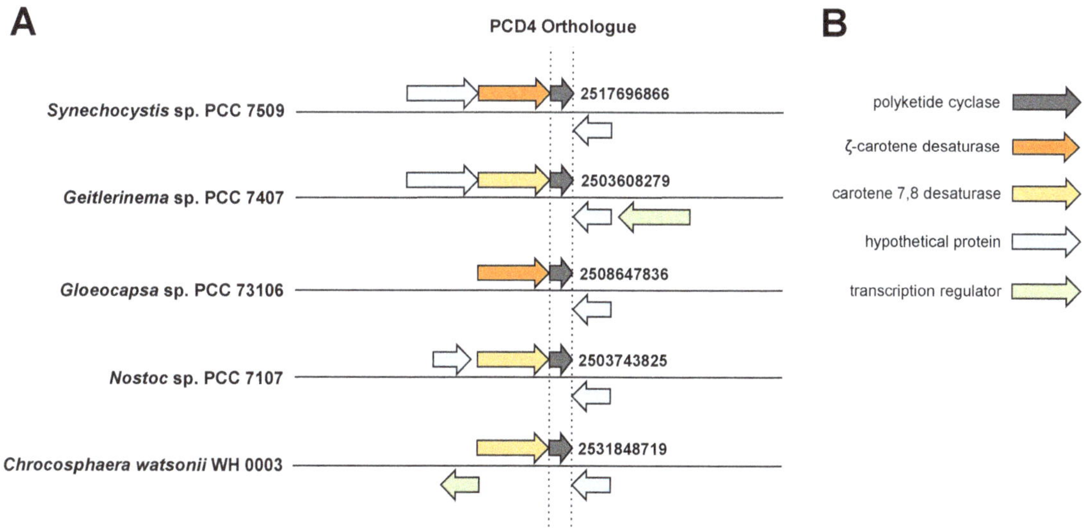

Figure 11.3 Analysis of operon structures for the gene encoding the PCD4 orthologue in cyanobacteria. *Chlamydomonas reinhardtii* PCD4 protein (polyketide cyclase) was BLASTED against the database of non-redundant cyanobacterial genomes (NCBI BLAST, http://blast.ncbi.nlm.nih.gov/Blast.cgi). The gene in *Synechocystis* sp. PCC 7407 encoding the PCD4 orthologue was localized in the genome using the IMG database (IMG, www.jgi.doe.h.img.gov); in the figure it is shown with its neighbouring genes (genes potentially in the same operon). The gene is annotated as encoding a polyketide cyclase/dehydrogenase and lipid transport protein (E value: 6 E-32). (A) The *Synechocystis* sp. PCC 7407 gene neighbourhood is presented (upper operon) along with conserved gene neighbourhoods from 4 different organisms. The gene upstream of the polyketide cyclase gene encodes a carotene desaturase. (B) Descriptors of the functions of the encoded proteins in each operon.

such as subcellular localization, hydrophobicity, domain annotation, and interaction analysis.

Perspectives – beyond the GreenCut2

Our understanding of chloroplasts and the biological processes occurring within photosynthetic organelles is advancing at a rapid rate. In 2007, when the *Chlamydomonas* genome was published, 214 (61%) out of the 349 original GreenCut proteins lacked an experimentally validated function. In 2011, the GreenCut2 inventory was expanded to 597 proteins, of which 311 (52%) were unknown. In the past 5 years, a significant number of these poorly understood proteins have been characterized to varying extents. Many have been shown to have critical roles in photosynthesis, redox homeostasis, and signalling in response to intracellular and environmental cues. An increasing number of proteins are also being categorized as 'regulatory' based on their roles in assembly and maintenance of chloroplasts, as well as in fine-tuning plastid metabolism.

From an evolutionary perspective, the GreenCut2 is an inventory of proteins that is present in green lineage organisms, although many in the inventory are also present in other photosynthetic organisms (e.g. diatoms, red algae). In a number of cases, the functions of specific GreenCut2 proteins are highly conserved (e.g. subunits of photosynthetic protein–pigment complexes), while in other cases they have evolved organism- or niche-specific roles (e.g. proteins involved in immune responses). The GreenCut2 is also a useful bioinformatics resource for evaluating functional relationships among proteins. A number of the remaining unknown proteins in the GreenCut2 are members of large protein families (e.g. PSBP-like proteins, oxidoreductases, and protein phosphatases) that await functional assignments. Continued efforts are under way to predict the functions of these and other unknowns of the GreenCut2.

As we move beyond the genomics era, several important questions remain: (i) How do GreenCut2 proteins interact with one another (and with other chloroplast- or nucleus-encoded proteins)? (ii) How do GreenCut2 proteins facilitate communication between chloroplasts and other organelles in eukaryotes? (iii) How can the vast assemblage of knowledge available from GreenCut2 analyses be leveraged in a productive manner? In other words, are there synthetic biology, agricultural, or other applications that can benefit from understanding the physiological roles of GreenCut2 proteins with respect to photosynthesis and other chloroplast functions as well as extrachloroplastic, green lineage-specific functions?

Acknowledgements

This work was supported by NSF Grant MCB 0951094 awarded to A.R.G., NSF MCB-1146621 (awarded to A.R.G. and Martin Jonikas) and the Department of Energy, Grant No. DE-FG02-12ER16338. S.S. is sponsored by the Vaadia –BARD Postdoctoral Fellowship, Grant No. FI-485-2013.

References

Ajjawi, I., Lu, Y., Savage, L.J., Bell, S.M., and Last, R.L. (2010). Large-scale reverse genetics in *Arabidopsis*: case studies from the Chloroplast 2010 project. Plant Physiol. *152*, 529–540.

Albus, C.A., Ruf, S., Schöttler, M.A., Lein, W., Kehr, J., and Bock, R. (2010). Y3IP1, a nucleus-encoded thylakoid protein, cooperates with the plastid-encoded Ycf3 protein in photosystem I assembly of tobacco and *Arabidopsis*. Plant Cell. *22*, 2838–2855.

Allen, M.B., Arnon, D.I., Capindale, J.B., Whatley, F.R., and Durham, I.J. (1955). Photosynthesis by isolated chloroplasts. III. Evidence for complete photosynthesis. J. Am. Chem. Soc. *77*, 4149–4155.

Allen, M.B., Whatley, F.R., and Arnon, D.I. (1958). Photosynthesis by isolated chloroplasts. VI. Rates of conversion of light into chemical energy in photosynthetic phosphorylation. Biochim. Biophys. Acta. *27*, 16–23.

Andreeva, A.V., and Kutuzov, M.A. (2004). Wide-spread presence of 'bacterial-like' PPP phosphatases in eukaryotes. BMC Evol. Biol. *4*, 47–61.

Apel, K., and Hirt, H. (2004). Reactive oxygen species: metabolism, oxidative stress, and signal transduction. Annu. Rev. Plant Biol. *55*, 373–399.

Armbruster, U., Zühlke, J., Rengstl, B., Kreller, R., Makarenko, E., Rühle, T., Schünemann, D., Jahns, P., Weisshaar, B., Nickelsen, J., and Leister, D. (2010). The *Arabidopsis* thylakoid protein PAM68 is required for efficient D1 biogenesis and photosystem II assembly. Plant Cell. *22*, 3439–3460.

Armbruster, U., Labs, M., Pribil, M., Viola, S., Xu, W., Scharfenberg, M., Hertle, A.P., Rojahn, U., Jensen, P.E., Rappaport, F., Joliot, P., Dörmann, P., Wanner, G., and Leister, D. (2013a). *Arabidopsis* CURVATURE THYLAKOID1 proteins modify thylakoid architecture by inducing membrane curvature. Plant Cell. *25*, 2661–2678.

Armbruster, U., Rühle, T., Kreller, R., Strotbek, C., Zühlke, J., Tadini, L., Blunder, T., Hertle, A.P., Qi, Y., Rengstl, B., Nickelsen, J., Frank, W., and Leister, D. (2013b). The photosynthesis affected mutant68-like protein evolved from a PSII assembly factor to mediate assembly of the chloroplast NAD(P)H dehydrogenase complex in *Arabidopsis*. Plant Cell. *25*, 3926–3943.

Arnon, D.I., Allen, M.B., and Whatley, F.R. (1954a). Photosynthesis by isolated chloroplasts. Nature. *174*, 394–396.

Arnon, D.I., Whatley, F.R., and Allen, M.B. (1954b). Photosynthesis by isolated chloroplasts. II. Photosynthetic phosphorylation, the conversion of light energy into phosphate bond energy. J. Am. Chem. Soc. *76*, 6324–6329.

Arnon, D.I., Whatley, F.R., and Allen, M.B. (1958). Assimilatory power in photosynthesis: photosynthetic phosphorylation by isolated chloroplasts is coupled with TPN reduction. Science. *127*, 1026–1034.

Arsova, B., Hoja, U., Wimmelbacher, M., Greiner, E., Ustün, S., Meizer, M., Petersen, K., Lein, W., and Börnke, F. (2010). Plastidial thioredoxin z interacts with two fructokinase-like proteins in a thiol-dependent manner: evidence for an essential role in chloroplast development in *Arabidopsis* and *Nicotiana benthamiana*. Plant Cell. *22*, 1498–1515.

Aslund, F., Berndt, K.D., and Holmgren, A. (1997). Redox potentials of glutaredoxins and other thiol-disulfide oxidoreductases of the thioredoxin superfamily determined by direct protein-protein redox equilibria. J. Biol. Chem. *272*, 30780–30786.

Belhaj, K., Lin, B., and Mauch, F. (2009). The chloroplast protein RPH1 plays a role in the immune response of *Arabidopsis* to *Phytophthora brassicae*. Plant J. *58*, 287–298.

Bellafiore, S., Barneche, F., Peltier, G., and Rochaix, J.D. (2005). State transitions and light adaptation require chloroplast thylakoid protein kinase STN7. Nature. *433*, 892–895.

Bhuiyan, N.H., Friso, G., Poliakov, A., Ponnala, L., and van Wijk, K.J. (2015). MET1 is a thylakoid-associated TPR protein involved in photosystem II supercomplex formation and repair in *Arabidopsis*. Plant Cell. *27*, 262–285.

Bishop, C.L., Purton, S., and Nugent, J.H.A. (2003). Molecular analysis of the *Chlamydomonas* nuclear gene encoding PsbW and demonstration that PsbW is a subunit of photosystem II, but not photosystem I. Plant Mol. Biol. *52*, 285–289.

Bonente, G., Passarini, F., Cazzaniga, S., Mancone, C., Buia, M.C., Tripodi, M., Bassi, R., and Caffarri, S. (2008). The occurrence of the *psbS* gene product in *Chlamydomonas reinhardtii* and in other photosynthetic organisms and its correlation with energy quenching. Photochem. Photobiol. *84*, 1359–1370.

Bonente, G., Ballottari, M., Truong, T.B., Morosinotto, T., Ahn, T.K., Fleming, G.R., Niyogi, K.K., and Bassi, R. (2010). Analysis of LhcSR3, a protein essential for feedback de-excitation in the green alga *Chlamydomonas reinhardtii*. PLoS Biol. *9*, e1000577.

Boudreau, E., Takahashi, Y., Lemieux, C., Turmel, M., and Rochaix, J.D. (1997). The chloroplast *ycf3* and *ycf4* open reading frames of *Chlamydomonas reinhardtii* are required for the accumulation of the photosystem I complex. EMBO J. *16*, 6095–6104.

Bousquet, I., Dujardin, G., and Slonimski, P.P. (1991). ABC1, a novel yeast nuclear gene has a dual function in mitochondria: it suppresses a cytochrome b mRNA translation defect and is essential for electron transfer in the bc 1 complex. EMBO J. *10*, 2023–2031.

Boussingault, J.B. (1864). De la végétation dans l'obscurité. Ann. Sci. Nat. (Paris) *1*, 314–324.

Braeken, K., Moris, M., Daniels, R., Vanderleyden, J., and Michiels, J. (2006). New horizons for (p)ppG*pp.* in bacterial and plant physiology. Trends Microbiol. *14*, 45–54.

Broquist, H.P., and Trupin, J.S. (1966). Amino acid metabolism. Annu. Rev. Biochem. *35*, 231–274.

Broun, P., Shanklin, J., Whittle, E., and Somerville, C. (1998). Catalytic plasticity of fatty acid modification enzymes underlying chemical diversity of plant lipids. Science. *282*, 1315–1317.

Brzezowski, P., Wilson, K.E., and Gray, G.R. (2012). The PSBP2 protein of *Chlamydomonas reinhardtii* is required for singlet oxygen-dependent signaling. Planta. *236*, 1289–1303.

Buchanan, B.B. (1980). Role of light in the regulation of chloroplast enzymes. Annu. Rev. Plant Physiol. *31*, 341–374.

Buchanan, B.B., and Balmer, Y. (2005). Redox regulation: a broadening horizon. Annu. Rev. Plant Biol. *56*, 187–220.

Buchanan-Wollaston, V., and Ainsworth, C. (1997). Leaf senescence is *Brassica napus*: cloning of senescence-related genes by subtractive hybridisation. Plant Mol. Biol. *33*, 821–834.

Burk, D., and Warburg, O. (1951). Ein-Quanten-Reaktion und Kreisprozeß der Energie bei der Photosynthese. Z. Naturforsch. *6b*, 12.

Cai, W., Ma, J., Chi, W., Zou, M., Guo, J., Lu, C., and Zhang, L. (2010). Cooperation of LPA3 and LPA2 is essential for photosystem II assembly in *Arabidopsis*. Plant Physiol. *154*, 109–120.

Calderon, R.H., Garcia-Cerdán, J.G., Malnoë, A., Cook, R., Russell, J.J., Gaw, C., Dent, R.M., de Vitry, C., and Niyogi, K.K. (2013). A conserved rubredoxin is necessary for photosystem II accumulation in diverse oxygenic photoautotrophs. J. Biol. Chem. *288*, 26688–26696.

Calvin, M., Bassham, J.A., and Benson, A.A. (1950). Chemical transformations in photosynthesis. Fed. Proc. *9*, 524–534.

Casas-Mollano, J.A., Jeong, B., Xu, J., Moriyama, H., and Cerutti, H. (2008). The MUT9p kinase phosphorylates histone H3 threonine 3 and is necessary for heritable epigenetic silencing in *Chlamydomonas*. Proc. Natl. Acad. Sci. U.S.A. *105*, 6486–6491.

Chen, C., Letnik, I., Hacham, Y., Dobrev, P., Ben-Daniel, B.H., Vanková, R., Amir, R., and Miller, G. (2014). Ascorbate Peroxidase 6 protects *Arabidopsis* desiccating and germinating seeds from stress and mediates cross talk between reactive oxygen species, abscisic acid, and auxin. Plant Physiol. *166*, 370–383.

Collin, V., Issakidis-Bourquet, E., Marchand, C., Hirasawa, M., Lancelin, J.M., Knaff, D.B., and Miginiac-Maslow, M. (2003). The *Arabidopsis* plastidial thioredoxins: new functions and new insights into specificity. J. Biol. Chem. *278*, 23747–23752.

Conklin, P.L., Saracco, S.A., Norris, S.R., and Last, R.L. (2000). Identification of ascorbic acid-deficient *Arabidopsis thaliana* mutants. Genetics *154*, 847–856.

Conklin, P.L., DePaolo, D., Wintle, B., Schatz, C., and Buckenmeyer, G. (2013). Identification of *Arabidopsis* VTC3 as a putative and unique dual function protein kinase::protein phosphatase involved in the regulation of the ascorbic acid pool in plants. J. Exp. Bot. *64*, 2793–2804.

Cunningham, F.X., and Gantt, E. (2011). Elucidation of the pathway to astaxanthin in the flowers of *Adonis aestivalis*. Plant Cell. *23*, 3055–3069.

DalCorso, G., Pesaresi, P., Masiero, S., Aseeva, E., Schünemann, D., Finazzi, G., Joliot, P., Barbato, R., and Leister, D. (2008). A complex containing PGRL1 and PGR5 is involved in the switch between linear and cyclic electron flow in *Arabidopsis*. Cell. *132*, 273–285.

Dangoor, I., Peled-Zehavi, H., Wittenberg, G., and Danon, A. (2012). A chloroplast light-regulated oxidative sensor for moderate light intensity in *Arabidopsis*. Plant Cell. *24*, 1894–1906.

Depège, N., Bellafiore, S., and Rochaix, J.D. (2003). Role of chloroplast protein kinase Stt7 in LHCII phosphorylation and state transitions in *Chlamydomonas*. Science. *299*, 1572–1575.

Derelle, E., Ferraz, C., Rombauts, S., Rouzé, P., Worden, A.Z., Robbens, S., Partensky, F., Degroeve, S., Echetnié, S., Cooke, R., Saeys, Y., Wuyts, J., Jabbari, K., Bowler, C., Panaud, O., Piégu, B., Ball, S.G., Ral, J.P., Bouget, F.Y., Piganeau, G., De Baets, B., Picard, A., Delseny, M., Demaille, J., Van de Peer, Y., and Moreau, H. (2006). Genome analysis of the smallest free-living eukaryote *Ostreococcus tauri* unveils many unique features. Proc. Natl. Acad. Sci. U.S.A. *103*, 11647–11652.

Dobáková, M., Sobotka, R., Tichy, M., and Komenda, J. (2009). Psb28 protein is involved in the biogenesis of the photosystem II inner antenna CP47 (PsbB) in the cyanobacterium *Synechocystis* sp. PCC 6803. Plant Physiol. *149*, 1076–1086.

Dolganov, N.A.M., Bhaya, D., and Grossman, A.R. (1995). Cyanobacterial protein with similarity to the chlorophyll a/b binding proteins of higher plants: evolution and regulation. Proc. Natl. Acad. Sci. U.S.A. *92*, 636–640.

Drewke, C., and Leistner, E. (2001). Biosynthesis of vitamin B_6 and structurally related derivatives. *In* G Litwack, T Begley, eds, Vitamins and Hormones. Advances in Research and Applications, Vol 61. Academic Press, San Diego, *pp.* 121–125.

Dutcher, S.K., and Harris, E. (1998). *Chlamydomonas reinhardtii. In* Genetic nomenclature guide. Trends in Genetics. Elsevier Science Ltd., Cambridge, United Kingdom, *pp.* S18-S19.

Duysens, L.N.M., Amesz, J., and Kamp, B.M. (1961). Two photochemical systems in photosynthesis. Nature. *190*, 510–511.

Duysens, L.N.M., and Amesz, J. (1962). Function and identification of two photochemical systems in photosynthesis. Biochim. Biophys. Acta. *64*, 243–260.

Eberhard, S., Finazzi, G., and Wollman, F.A. (2008). The dynamics of photosynthesis. Annu. Rev. Genet. *42*, 463–515.

Ebersold, W.T., Levine, R.P., Levine, E.E., and Olmsted, M.A. (1962). Linkage maps of *Chlamydomonas reinhardi*. Genetics. *47*, 531–543.

Emerson, R., and Lewis, C. (1943). The dependence of the quantum yield of Chlorella photosynthesis on wavelength of light. Am. J. Bot. *30*, 165–178.

Emerson, R., Chalmers, R.V., and Cederstrand, C. (1957). Some factors influencing the long-wave limit of photosynthesis. Proc. Natl. Acad. Sci. U.S.A. *43*, 133–143.

Emerson, R., and Chalmers, R.V. (1958). Speculations concerning the function and phylogenetic significance of the accessory pigments in algae. Phycol. Soc. Am. News. Bull. *11*, 51–56.

Engelmann, T.W. (1882). Über Sauerstoffausscheidung von Pflanzenzellen im Mikrospectrum. Bot. Z. *40*, 419–426.

Foyer, C.H., and Noctor, G. (2011). Ascorbate and glutathione: the heart of the redox hub. Plant Physiol. *155*, 2–18.

Franceschini, A., Szklarczyk, D., Frankild, S., Kuhn, M., Simonovic, M., Roth, A., Lin, J., Minguez, P., Bork, P., von Mering, C., and Jensen, L.J. (2013). STRING v9.1: protein–protein interaction networks, with increased coverage and integration. Nucleic Acids Res. *41*, D808–815.

Frenkel, A. (1954). Light induced phosphorylation by cell-free preparations of photosynthetic bacteria. J. Am. Chem. Soc. *76*, 5568–5570.

Fristedt, R., Herdean, A., Blaby-Haas, C.E., Marnedov, F., Merchant, S.S., Last, R.L., and Lundin, B. (2014a). PSB33, a protein conserved in the plastid lineage, is associated with the chloroplast thylakoid membrane and provides stability to photosystem II supercomplexes in *Arabidopsis*. Plant. Physiol. *167*, 481–492.

Fristedt, R., Williams-Carrier, R., Merchant, S.S., and Barkan, A. (2014b). A thylakoid membrane protein harboring a DnaJ-type zinc finger domain is required for photosystem I accumulation in plants. J. Biol. Chem. *289*, 30657–30667.

Gabilly, S.T., Dreyfuss, B.W., Karamoko, M., Corvest, V., Kropat, J., Page, M.D., Merchant, S.S., and Hamel, P.P. (2010). CCS5, a thioredoxin-like protein involved in the assembly of plastid c-type cytochromes. J. Biol. Chem. *285*, 29738–29749.

Garcia-Cerdán, J.G., Kovács, L., Tóth, T., Kereïche, S., Aseeva, E., Boekema, E.J., Mamedov, F., Funk, C., and Schröder, W.P. (2011). The PsbW protein stabilizes the supramolecular organization of photosystem II in higher plants. Plant J. *65*, 368–381.

Govindjee, and Krogmann, D. (2004). Discoveries in oxygenic photosynthesis (1727–2003): a perspective. Photosynth. Res. *80*, 15–57.

Grimm, B., Kruse, E., and Kloppstech, K. (1989). Transiently expressed early light-inducible thylakoid proteins share transmembrane domains with light-harvesting chlorophyll binding proteins. Plant Mol. Biol. *13*, 583–593.

Grossman, A.R., Karpowicz, S.J., Heinnickel, M., Dewez, D., Hamel, B., Dent, R., Niyogi, K.K., Johnson, X., Alric, J., Wollman, F.A., Li, H., and Merchant, S.S. (2010). Phylogenomic analysis of the *Chlamydomonas* genome unmasks proteins potentially involved in photosynthetic function and regulation. Photosynth. Res. *106*, 3–17.

Hanke, G.T., Okutani, S., Satomi, Y., Takao, T., Suzuki, A., and Hase, T. (2005). Multiple iso-proteins of FNR in *Arabidopsis*: evidence for different contributions to chloroplast function and nitrogen assimilation. Plant Cell Environ. *28*, 1146–1157.

Harrison, P.J., and Bugg, T.D.H. (2014). Enzymology of the carotenoid cleavage dioxygenases: Reaction mechanisms, inhibition and biochemical roles. Arch. Biochem. Biophys. *544*, 105–111.

Havaux, M., Guedeney, G., Hagemann, M., Yeremenko, N., Matthijs, H.C.P., and Jeanjean, R. (2005). The chlorophyll-binding protein IsiA is inducible by high light and protects the cyanobacterium *Synechocystis* PCC6803 from photooxidative stress. FEBS Lett. *579*, 2289–2293.

He, G., Dolganov, N., Björkman, O., and Grossman, A.R. (2001). The high light-inducible polypeptides in *Synechocystis* PCC 6803. Expression and function in high light. J. Biol. Chem. *276*, 306–314.

Heinnickel, M.L., Alric, J., Wittkopp, T., Yang, W., Catalanotti, C., Dent, R., Niyogi, K.K., Wollman, F.A., and Grossman, A.R. (2013). Novel thylakoid membrane GreenCut protein CPLD38 impacts accumulation of the cytochrome b_6f complex and associated regulatory processes. J. Biol. Chem. *288*, 7024–7036.

Hertle, A.P., Blunder, T., Wunder, T., Pesaresi, P., Pribil, M., Armbruster, U., and Leister, D. (2013). PGRL1 is the elusive ferredoxin-plastoquinone reductase in photosynthetic cyclic electron flow. Mol. Cell *49*, 511–523.

Hervás, M., Navarro, J.A., and De la Rosa, M.A. (2003). Electron transfer between membrane complexes and soluble proteins in photosynthesis. Acc. Chem. Res. *36*, 798–805.

Hill, R. (1937). Oxygen evolution by isolated chloroplasts. Nature *139*, 881–882.

Hill, R. (1939). Oxygen production by isolated chloroplasts. Proc. R. Soc. Lond. Ser. B. *127*, 192–210.

Hill, R., and Bendall, F. (1960). Function of the cytochrome components in chloroplasts: a working hypothesis. Nature *186*, 136–137.

Hiruma, K., Fukunaga, S., Bednarek, P., Piślewska-Bednarek, M., Watanabe, S., Narusaka, Y., Shirasu, K., and Takano, Y. (2013). Glutathione and tryptophan metabolism are required for *Arabidopsis* immunity during the hypersensitive response to hemibiotrophs. Proc. Natl. Acad. Sci. U.S.A. *110*, 9589–9594.

Huppe, H.C., de Lamotte-Guéry, F., Jacquot, J.P., and Buchanan, B.B. (1990). The ferredoxin-thioredoxin system of a green alga, *Chlamydomonas reinhardtii*. Planta *180*, 341–351.

Ichimura, K., Shinozaki, K., Tena, G., Sheen, J., Henry, Y., Champion, A., Kreis, M., Zhang, S., Hirt, H., and Wilson, C. (2002). Mitogen-activated protein kinase cascades in plants: a new nomenclature. Trends Plant Sci. *7*, 301–308.

Ifuku, K., Ishihara, S., and Sato, F. (2010). Molecular functions of oxygen–evolving complex family proteins in photosynthetic electron flow. J. Integr. Plant Biol. *52*, 723–734.

Ikeda, Y., Tanaka, T., and Noguchi, T. (1997). Conversion of non-allosteric pyruvate kinase isozyme into an allosteric enzyme by a single amino acid substitution. J. Biol. Chem. *272*, 20495–20501.

Ingenhousz, J. (1779). Experiments upon vegetables, discovering their great power of purifying the common air in the sun-shine and of injuring it in the shade and at night. (London). pp. xiii-xxxviii.

Inoue, K., Dreyfuss, B.W., Kindle, K.L., Stern, D.B., Mechant, S., and Sodeinde, O.A. (1997). *Ccs1*, a nuclear gene required for the post-translational assembly of chloroplast c-type cytochromes. J. Biol. Chem. *272*, 31747–31754.

Ishihara, S., Takabayashi, A., Ido, K., Endo, T., Ifuku, K., and Sato, F. (2007). Distinct functions for the two PsbP-like proteins PPL1 and PPL2 in the chloroplast thylakoid lumen of *Arabidopsis*. Plant Physiol. *145*, 668–679.

Ishikawa, A., Tanaka, H., Kato, C., Iwasaki, Y., and Asahi, T. (2005). Molecular characterization of the ZKT gene encoding a protein with PDZ, K-Box, and TPR motifs in *Arabidopsis*. Biosci. Biotechnol. Biochem. *69*, 972–978.

Iwai, M., Takizawa, K., Tokutsu, R., Okamuro, A., Takahashi, Y., and Minagawa, J. (2010). Isolation of the elusive supercomplex that drives cyclic electron flow in photosynthesis. Nature. *464*, 1210–1213.

Jacquot, J.P., Lancelin, J.M., and Meyer, Y. (1997). Thioredoxin: structure and function in plant cells. New Phytol. *136*, 543–570.

Jagendorf, A.T., and Uribe, E. (1966). ATP formation caused by acid–base transition of spinach chloroplasts. Proc. Natl. Acad. Sci. U.S.A. *55*, 170–177.

Jain, S.K., and Lim, G. (2001). Pyridoxine and pyridoxamine inhibits superoxide radicals and prevents lipid peroxidation, protein glycosylation, and ($Na^+ + K^+$)-ATPase activity reduction in high glucose-treated human erythrocytes. Free Radic. Biol. Med. *30*, 232–237.

Jansson, S., Andersson, J., Kim, S.J., and Jackowski, G. (2000). An *Arabidopsis thaliana* protein homologous to cyanobacterial high-light-inducible proteins. Plant Mol. Biol. *42*, 345–351.

Jasinski, M., Sudre, D., Schansker, G., Schellenberg, M., Constant, S., Martinoia, E., and Bovet, L. (2008). AtOSA1, a member of the Abc1-Like family, as a new factor in cadmium and oxidative stress response. Plant Physiol. *147*, 719–731.

Johnson, X., Steinbeck, J., Dent, R.M., Takahashi, H., Richaud, P., Ozawa, S., Houille-Vernes, L., Petroutsos, D., Rappaport, F., Grossman, A.R., Niyogi, K.K., and Alric, J. (2014). Proton gradient regulation 5-mediated cyclic electron flow under ATP- or redox-limited conditions: a study of *ΔATpase pgr5* and *ΔrbcL pgr5* mutants in the green alga *Chlamydomonas reinhardtii*. Plant Physiol. *165*, 438–442.

Joliot, P., Barbieri, G., and Chabaud, R. (1969). Un nouveau modèle des centres photochimiques du systeme II. Photochem. Photobiol. *10*, 309–329.

Kangasjärvi, S., Lepistö, A., Hännikäinen, K., Piippo, M., Luomala, E.M., Aro, E.M., and Rintamäki, E. (2008). Diverse roles for chloroplast stromal and thylakoid-bound ascorbate peroxidase in plant stress responses. Biochem. J. *412*, 275–285.

Karamoko, M., Cline, S., Redding, K., Ruiz, N., and Hamel, P.P. (2011). Lumen thiol oxidoreductase1, a disulfide bond-forming catalyst, is required for the assembly of photosystem II in *Arabidopsis*. Plant Cell. *23*, 4462–4475.

Karpowicz, S.J., Prochnik, S.E., Grossman, A.R., and Merchant, S.S. (2011). The GreenCut2 resource, a phylogenomically derived inventory of proteins specific to the plant lineage. J. Biol. Chem. *286*, 21427–21439.

Kasai, K., Usami, S., Yamada, T., Endo, Y., Ochi, K., and Tozawa, Y. (2002). A RelA-SpoT homolog (Cr-RSH) identified in *Chlamydomonas reinhardtii* generates stringent factor *in vivo* and localizes to chloroplasts *in vitro*. Nucleic Acids Res. *30*, 4985–4992.

Keeling, P.J. (2010). The endosymbiotic origin, diversification and fate of plastids. Phil. Trans. R. Soc. B. *365*, 729–748.

Kim, S., Sandusky, P., Bowlby, N.R., Aebersold, R., Green, B.R., Vlahakis, S., Yocum, C.F., and Pichersky, E. (1992). Characterization of a spinach *psbS* cDNA encoding the 22 kDa protein of photosystem II. FEBS Lett. *314*, 67–71.

Kirst, H., Garcia-Cerdán, J.G., Zurbriggen, A., Ruehle, T., and Melis, A. (2012). Truncated photosystem chlorophyll antenna size in the green microalga *Chlamydomonas reinhardtii* upon deletion of the TLA-Cp*SRP43* gene. Plant Physiol. *160*, 2251–2260.

Klimmek, F., Sjödin, A., Noutsos, C., Leister, D., and Jansson, S. (2006). Abundantly and rarely expressed Lhc protein genes exhibit distinct patterns in plants. Plant Physiol. *140*, 793–804.

Kok, B. (1959). Light-induced absorption changes in photosynthetic organisms. II. A split-beam difference spectrophotometer. Plant Physiol. *34*, 184–192.

Kok, B., Forbush, B., and McGloin, M. (1970). Cooperation of charges in photosynthetic O2 evolution-I. A linear four step mechanism. Photochem. Photobiol. *11*, 457–475.

Kong, M.M., Wang, F.F., Yang, Z.N., and Mi, H.L. (2013). ATPG is required for the accumulation and function of the chloroplast ATP synthase in *Arabidopsis*. Chin. Sci. Bull. *26*, 3224–3232.

Kuras, R., de Vitry, C., Choquet, Y., Girard-Bascou, J., Culler, D., Büschlen, S., Merchant, S., and Wollman, F.A. (1997). Molecular genetic identification of a pathway for heme binding to cytochrome b_6. J. Biol. Chem. *272*, 32427–32435.

Kuras, R., Saint-Marcoux, D., Wollman, F.A., and de Vitry, C. (2007). A specific c-type cytochrome maturation system is required for oxygenic photosynthesis. Proc. Natl. Acad. Sci. U.S.A. *104*, 9906–9910.

Ledford, H.K., Chin, B.L., and Niyogi, K.K. (2007). Acclimation to singlet oxygen stress in *Chlamydomonas reinhardtii*. Euk. Cell *6*, 919–930.

Lee, H.J., Mochizuki, N., Masuda, T., and Buckhout, T.J. (2012). Disrupting the bimolecular binding of the heme-binding protein 5 (AtHBP5) to haem oxygenase (HY1) leads to oxidative stress in *Arabidopsis*. J. Exp. Bot. *63*, 5967–5978.

Lee, K.P., Kim, C., Landgraf, F., and Apel, K. (2007). EXECUTER1- and EXUCUTER2-dependent transfer of stress-related signals from the plastid to the nucleus of *Arabidopsis thaliana*. Proc. Natl. Acad. Sci. U.S.A. *104*, 10270–10275.

Levine, R.P. (1960). Genetic control of photosynthesis in *Chlamydomonas reinhardi*. Proc. Natl. Acad. Sci. U.S.A. *46*, 972–977.

Levine, R.P. (1969). The analysis of photosynthesis using mutant strains of algae and higher plants. Annu. Rev. Plant Physiol. *20*, 523–540.

Lezhneva, L., Kuras, R., Ephritikhine, G., and de Vitry, C. (2008). A novel pathway of cytochrome c biogenesis is involved in the assembly of the cytochrome b_6f complex in *Arabidopsis* chloroplasts. J. Biol. Chem. *283*, 24608–24616.

Li, H.M., and Chiu, C.C. (2010). Protein transport into chloroplasts. Annu. Rev. Plant Biol. *61*, 157–180.

Li, X.P., Björkman, O., Shih, C., Grossman, A.R., Rosenquist, M., Jansson, S., and Niyogi, K.K. (2000). A pigment-binding protein essential for regulation of photosynthetic light harvesting. Nature. *403*, 391–395.

Li, X.P., Gilmore, A.M., and Niyogi, K.K. (2002). Molecular and global time-resolved analysis of a *psbS* gene dosage effect on pH- and xanthophyll cycle-dependent nonphotochemical quenching in photosystem II. J. Biol. Chem. *277*, 33590–33597.

Link, S., Engelmann, K., Meierhoff, K., and Westhoff, P. (2012). The atypical short-chain dehydrogenase HCF173 and HCF244 are jointly involved in translational initiation of the *psbA* mRNA in *Arabidopsis*. Plant Physiol. *160*, 2202–2218.

Lisenbee, C.S., Lingard, M.J., and Trelease, R.N. (2005). *Arabidopsis* peroxisomes possess functionally redundant membrane and matrix isoforms of monodehydroascorbate reductase. Plant J. *43*, 900–914.

Liu, J., Yang, H., Lu, Q., Wen, X., Chen, F., Peng, L., Zhang, L., and Lu, C. (2012). PSBP-DOMAIN PROTEIN1, a nuclear-encoded thylakoid luminal protein, is essential for photosystem I assembly in *Arabidopsis*. Plant Cell. *24*, 4992–5006.

Liu, X., Liu, S., Feng, Y., Liu, J.Z., Chen, Y., Pham, K., Deng, H., Hirschi, K.D., Wang, X., and Cheng, N. (2013). Structural insights into the N-terminal GIY-YIG endonuclease activity of *Arabidopsis* glutaredoxin AtGRXS16 in chloroplasts. Proc. Natl. Acad. Sci. U.S.A. *110*, 9565–9570.

Lorković, Z.J., Schröder, W.P., Pakrasi, H.B., Irrgang, K.D., Herrmann, R.G., and Oelmüller, R. (1995). Molecular characterization of PsbW, a nuclear-encoded component of the photosystem II reaction center complex in spinach. Proc. Natl. Acad. Sci. U.S.A. *92*, 8930–8934.

Ma, J., Peng, L., Guo, J., Lu, Q., Lu, C., and Zhang, L. (2007). LPA2 is required for efficient assembly of photosystem II in *Arabidopsis thaliana*. Plant Cell. *19*, 1980–1993.

Manara, A., DalCorso, G., Leister, D., Jahns, P., Baldan, B., and Furini, A. (2013). AtSIA1 and AtOSA1: two Abc1 proteins involved in oxidative stress responses and iron distribution within chloroplasts. New Phytol. *201*, 452–465.

Marcaida, M.J., Schlarb-Ridley, B.G., Worrall, J.A.R., Wastl, J., Evens, T.J., Bendall, D.S., Luisi, B.F., and Howe, C.J. (2006). Structure of cytochrome c6A, a novel dithio-cytochrome of *Arabidopsis thaliana*, and its reactivity with plastocyanin: implications for function. J. Mol. Biol. *360*, 968–977.

Masuda, S., Mizusawa, K., Narisawa, T., Tozawa, Y., Ohta, H., and Takamiya, K. (2008). The bacterial stringent response, conserved in chloroplasts, controls plant fertilization. Plant Cell Physiol. *49*, 135–141.

May, M.J., and Leaver, C.J. (1994). *Arabidopsis thaliana* γ-glutamylcysteine synthetase is structurally unrelated to mammalian, yeast, and *Escherichia coli* homologs. Proc. Natl. Acad. Sci. U.S.A. *91*, 10059–10063.

Mayer, J.R. (1845). Die organische Bewegung in ihrem Zussamenhag mit dem Stoffwechsel: Ein Beitrag zur Naturkunde. Verlag der C. Drechsler'schen Buchhandlung, Heilbronn.

Merchant, S.S., Prochnik, S.E., Vallon, O., Harris, E.H., Karpowicz, S.J., Witman, G.B., *et al.* (2007). The *Chlamydomonas* genome reveals the evolution of key animal and plant functions. Science. *318*, 245–251.

Mereschkowsky, C. (1910). 'Theorie der zwei plasmaarten als grundlage der symbiogenesis, einer neuen lehre von der entstehung der organismen. Biol. Centralbl. *30*, 353-367.

Mestres-Ortega, D., and Meyer, Y. (1999). The *Arabidopsis thaliana* genome encodes at least four thioredoxins m and a new prokaryotic-like thioredoxin. Gene *240*, 307–316.

Meyer, G., and Kloppstech, K. (1984). A rapidly light-induced chloroplast protein with a high turnover coded for by pea nuclear DNA. Eur. J. Biochem. *138*, 201–207.

Meyer, Y., Riondet, C., Constans, L., Abdelgawwad, M.R., Reichheld, J.P., and Vignols, F. (2006). Evolution of redoxin genes in the green lineage. Photosynth. Res. *89*, 179–192.

Minagawa, J. (2011). State transitions – the molecular remodeling of photosynthetic supercomplexes that controls energy flow in the chloroplast. Biochem. Biophys. Acta. *1807*, 897–905.

Mininno, M., Brugière, S., Pautre, V., Gilgen, A., Ma, S., Ferro, M., Tardif, M., Alban, C., and Ravanei, S. (2012). Characterization of chloroplast fructose 1,6-bisphosphate aldolases as lysine-methylated proteins in plants. J. Biol. Chem. *287*, 21034–21044.

Mitchell, P. (1961a). Coupling of phosphorylation to electron and hydrogen transfer by a chemiosmotic type of mechanism. Nature. *191*, 144–148.

Mitchell, P. (1961b). Chemiosmotic Coupling in Oxidative and Photosynthetic Phosphorylation. Glynn Res, Bodmin, Cornwall, UK.

Montané, M.H., and Kloppstech, K. (2000). The family of light-harvesting related proteins (LHCs, ELIPs, HLIPs): was the harvesting of light their primary function? Gene *258*, 1–8.

Motohashi, K., and Hisabori, T. (2006). HCF164 receives reducing equivalents from stromal thioredoxin across the thylakoid membrane and mediates reduction of target proteins in the thylakoid lumen. J. Biol. Chem. *281*, 35039–35047.

Müller, P., Li, X.P., and Niyogi, K.K. (2001). Non-photochemical quenching. A response to excess light energy. Plant Physiol. *125*, 1558–1566.

Munekage, Y., Hojo, M., Meurer, J., Endo, T., Tasaka, M., and Shikanai, T. (2002). PGR5 is involved in cyclic electron flow around photosystem I and is essential for photoprotection in *Arabidopsis*. Cell. *110*, 361–371.

Mustilli, A.C., Merlot, S., Vavasseur, A., Fenzi, F., and Giraudat, J. (2002). *Arabidopsis* OST1 protein kinase mediates the regulation of stomatal aperture by abscisic acid and acts upstream of reactive oxygen species production. Plant Cell. *14*, 3089–3099.

Myers, C., Romanowsky, S.M., Barron, Y.D., Garg, S., Azuse, C.L., Curran, A., Davis, R.M., Hatton, J., Harmon, A.C., and Harper, J.F. (2009). Calcium-dependent protein kinases regulate polarized tip growth in pollen tubes. Plant J. *59*, 528–539.

Nawrocki, W.J., Tourasse, N.J., Taly, A., Rappaport, F., and Wollman, F.A. (2015). The plastid terminal oxidase: its elusive function points to multiple contributions to plastid physiology. Annu. Rev. Plant Biol. *66*, 49–74.

Niyogi, K.K., and Truong, T.B. (2013). Evolution of flexible non-photochemical quenching mechanisms that regulate light harvesting in oxygenic photosynthesis. Curr. Opin. Plant Biol. *16*, 307–314.

Nomura, H., Komori, T., Kobori, M., Nakahira, Y., and Shiina, T. (2008). Evidence for chloroplast control of external Ca^{2+}-induced cytosolic Ca^{2+} transients and stomatal closure. Plant J. *53*, 988–998.

Nomura, H., Komori, T., Uemura, S., Kanda, Y., Shimotani, K., Nakai, K., Furuichi, T., Takebayashi, K., Sugimoto, T., Sano, S., Suwastika, I.N., Fukusaki, E., Yoshioka, H., Nakahira, Y., and Shiina, T. (2012). Chloroplast-mediated activation of plant immune signaling in *Arabidopsis*. Nat. Commun. *3*, 926.

Peden, E.A., Boehm, M., Mulder, D.W., Davis, R., Old, W.M., King, P.W., Ghirardi, M.L., and Dubini, A. (2013). Identification of global ferredoxin interaction networks in *Chlamydomonas reinhardtii*. J. Biol. Chem. *288*, 35192–35209.

Peers, G., Truong, T.B., Ostendorf, E., Busch, A., Elrad, D., Grossman, A.R., Hippler, M., and Niyogi, K.K. (2009). An ancient light-harvesting protein is critical for the regulation of algal photosynthesis. Nature. *462*, 518–521.

Pelletier, J., and Caventou, J.B. (1818). Sur la matiere verte des feuilles. Ann. Chim. Phys. Ser. 2, 194–196.

Peltier, G., and Cournac, L. (2002). Chlororespiration. Annu. Rev. Plant Biol. 53, 523–550.

Peltier, G., Tolleter, D., Billon, E., and Cournac, L. (2010). Auxiliary electron transport pathways in chloroplasts of microalgae. Photosynth. Res. *106*, 19–31.

Peltier, J.B., Ytterberg, A.J., Sun, Q., and van Wijk, K.J. (2004). New functions of the thylakoid membrane proteome of *Arabidopsis thaliana* revealed by a simple, fast, and versatile fractionation strategy. J. Biol. Chem. *279*, 49367–49383.

Peng, L., Ma, J., Chi, W., Guo, J., Zhu, S., Lu, Q., Lu, C., and Zhang, L. (2006). LOW PSII ACCUMULATION1 is involved in efficient assembly of photosystem II in *Arabidopsis thaliana*. Plant Cell. *18*, 955–969.

Petroutsos, D., Busch, A., Janßen, I., Trompelt, K., Verena Bergner, S., Weinl, S., Holtkamp, M., Karst, U., Kudla, J., and Hippler, M. (2011). The chloroplast calcium sensor CAS is required for photoacclimation in *Chlamydomonas reinhardtii*. Plant Cell. 23, 2950–2963.

Priestly, J. (1774). Experiments and observations on different kinds of air. Retrieved 10 March 2015 from http://www.gutenberg.org/files/29734/29734-h/29734-h.htm.

Pruvot, G., Massimino, J., Peltier, G., and Rey, P. (1996). Effects of low temperature, high salinity and exogenous ABA on the synthesis of two chloroplast drought-induced proteins in *Solanum tuberosum*. Physiol. Plant. *97*, 123–131.

Puga, M.I., Mateos, I., Charukesi, R., Wang, Z., Franco-Zorrilla, J.M., de Lorenzo, L., Irigoyen, M.L., Masiero, S., Bustos, R., Rodriguez, J., Leyva, A., Rubio, V., Sommer, H., and Paz-Ares, J. (2014). SPX1 is a phosphate-dependent inhibitor of Phosphate Starvation Response 1 in *Arabidopsis*. Proc. Natl. Acad. Sci. U.S.A. *111*, 14947–14952.

Qi, Y., Tsuda, K., Nguyen, L.V., Wang, X., Lin, J., Murphy, A.S., Glazebrook, J., Thordai-Christensen, H., and Katagiri, F. (2011). Physical association of *Arabidopsis* hypersensitive induced reaction proteins (HIRs) with the immune receptor RPS2. J. Biol. Chem. *286*, 31297–31307.

Ris, H., and Plaut, W. (1962). Ultrastructure of DNA-containing areas in the chloroplasts of *Chlamydomonas*. J. Cell. Biol. *13*, 383–391.

Rochaix, J.D. (2014). Regulation and dynamics of the light-harvesting system. Annu. Rev. Plant Biol. *65*, 287–309.

Roos, J., Bejal, S., Olde, S., and Dixelius, C. (2014). RabGAP22 is required for defense to the vascular pathogen *Verticillium longisporum* and contributes to stomata immunity. PLoS One. *9*, e88187.

Roose, J.L., Frankel, L.K., and Bricker, T.M. (2011). Developmental defects in mutants of the PsbP domain protein 5 in *Arabidopsis thaliana*. PLoS One. *6*, e28624.

Rouhier, N., Lemaire, S.D., and Jacquot, J.P. (2008). The role of glutathione in photosynthetic organisms: emerging functions for glutaredoxins and glutathionylation. Annu. Rev. Plant Biol. *59*, 143–166.

Roux, M., Schwessinger, B., Albrecht, C., Chinchilla, D., Jones, A., Holton, N., Malinovsky, F.G., Tör, M., de Vries, S., and Zipfel, C. (2011). The *Arabidopsis* leucine-rich repeat receptor-like kinase BAK1/SERK3 and BKK1/SERK4 are required for innate immunity to hemibiotrophic and biotrophic pathogens. Plant Cell. *23*, 2440–2455.

Rühle, T., Razeghl, J.A., Vamvaka, E., Viola, S., Gandini, C., Kleine, T., Schünemann, D., Barbato, R., Jahns, P., and Leister, D. (2014). The *Arabidopsis* protein CONSERVED ONLY IN THE GREEN LINEAGE160 promotes the assembly of the membranous part of the chloroplast ATP synthase. Plant Physiol. *165*, 207–226.

Sager, R., and Ishida, M.R. (1963). Chloroplast DNA in *Chlamydomonas*. Proc. Natl. Acad. Sci. U.S.A. *50*, 725–730.

Sakamoto, M., Tomita, R., and Kobayashi, K. (2009). A protein containing an XYPPX repeat and a C2 domain is associated with virally induced hypersensitive cell death in plants. FEBS Lett. *583*, 2552–2556.

Sang, Y., Barbosa, J.M., Wu, H., Locy, R.D., and Singh, N.K. (2007). Identification of a pyridoxine (pyridoxamine) 5′-phosphate oxidase from *Arabidopsis thaliana*. FEBS Lett. *581*, 344–348.

Sang, Y., Locy, R.D., Goertzen, L.R., Rashotte, A.M., Si, Y., Kang, K., and Singh, N.K. (2011). Expression, *in vivo* localization and phylogenetic analysis of a pyridoxine 5′-phosphate oxidase in *Arabidopsis thaliana*. Plant Physiol. Biochem. *49*, 88–95.

Schaeffer, H.J., and Weber, M.J. (1999). Mitogen-activated protein kinases: specific messages from ubiquitous messengers. Mol. Cell. Biol. *19*, 2435–2444.

Schimper, A.F.W. (1883). Über die entwicklung der chlorophyllkörner und farbkörper. Bot. Zeitung. *41*, 105–14, 121–31, 137–46, 153–162.

Schmidt, M., Gessner, G., Luff, M., Heiland, I., Wagner, V., Kaminski, M., Geimer, S., Eitzinger, N., Reissenweber, T., Voytsekh, O., Fiedler, M., Mittag, M., and Kreimer, G. (2006). Proteomic analysis of the eyespot of *Chlamydomonas reinhardtii* provides novel insights into its components and tactic movements. Plant Cell. *18*, 1908–1930.

Schröder, W.P., Henrysson, T., and Åkerlund, H.E. (1988). Characterization of low molecular mass proteins of photosystem II by N-terminal sequencing. FEBS Lett. *235*, 289–292.

Schult, K., Meierhoff, K., Paradies, S., Töller, T., Wolff, P., and Westhoff, P. (2007). The nuclear-encoded factor HCF173 is involved in the initiation of translation of the *psbA* mRNA in *Arabidopsis thaliana*. Plant Cell. *19*, 1329–1346.

Schulze, T., Schreiber, S., Iliev, D., Boesger, J., Trippens, J., Kreimer, G., and Mittag, M. (2012). The heme-binding protein SOUL3 of *Chlamydomonas reinhardtii* influences size and position of the eyespot. Mol. Plant *6*, 931–944.

Schürmann, P. (2002). Ferredoxin-dependent thioredoxin reductase: a unique iron–sulfur protein. Methods Enzymol. *347*, 403–411.

Schweighofer, A., Hirt, H., and Meskiene, I. (2004). Plant PP2C phosphatases: emerging functions in stress signaling. TRENDS Plant Sci. *9*, 236–243.

Shao, N., Duan, G.Y., and Bock, R. (2013). A mediator of singlet oxygen responses in *Chlamydomonas reinhardtii* and *Arabidopsis* identified by a luciferase-based genetic screen in algal cells. Plant Cell. *25*, 4209–4226.

Shi, L.X., Lorković, Z.J., Oelmüller, R., and Schröder, W.P. (2000). The low molecular mass PsbW protein is involved in the stabilization of the dimeric photosystem II complex in *Arabidopsis thaliana*. J. Biol. Chem. *275*, 37945–37950.

Sparkes, I., Tolley, N., Aller, I., Svozil, J., Osterrieder, A., Botchway, S., Mueller, C., Frigerio, L., and Hawes, C. (2010). Five *Arabidopsis* reticulon isoforms share endoplasmic reticulum location, topology, and membrane-shaping properties. Plant Cell. *22*, 1333–1343.

Stöckel, J., Bennewitz, S., Hein, P., and Oelmüller, R. (2006). The evolutionarily conserved tetratrico peptide repeat protein pale yellow green7 is required for photosystem I accumulation in *Arabidopsis* and copurifies with the complex. Plant Physiol. *141*, 870–878.

Suorsa, M., Järvi, S., Grieco, M., Nurmi, M., Pietrzykowska, M., Rantala, M., Kangasjärvi, S., Paakkarinen, V., Tikkanen, M., Jannson, S., and Aro, E.M. (2012). PROTON GRADIENT REGULATION5 is essential for proper acclimation of *Arabidopsis* photosystem I to naturally and artificially fluctuating light conditions. Plant Cell. *24*, 2934–2948.

Tagawa, K., and Arnon, D.I. (1962). Ferredoxins as electron carriers in photosynthesis and in the biological production and consumption of hydrogen gas. Nature. *195*, 537–543.

Takahahsi, S., Ogawa, S., Inoue, K., and Masuda, T. (2008). Characterization of cytosolic tetrapyrrole-binding proteins in *Arabidopsis thaliana*. Photochem. Photobiol. Sci. *7*, 1216–1224.

Tanaka, R., Rothbart, M., Oka, S., Takabayashi, A., Takahashi, K., Shibata, M., Myouga, F., Motohashi, R., Shinozaki, K., Grimm, B., and Tanaka, A. (2010). LIL3, a light-harvesting-like protein, plays an essential role in chlorophyll and tocopherol biosynthesis. Proc. Natl. Acad. Sci. U.S.A. *107*, 16721–16725.

Tang, G., Miron, D., Zhu-Shimoni, J.X., and Galili, G. (1997). Regulation of lysine catabolism through lysine-ketoglutarate reductase and saccharopine dehydrogenase in *Arabidopsis*. Plant Cell. *9*, 1305–1316.

Tardif, M., Atteia, A., Specht, M., Cogne, G., Rolland, N., Brugiere, S., Hippler, M., Ferro, M., Bruley, C., Peltier, G., Vallon, O., and Cournac, L. (2012). PredAlgo: a new subcellular localization prediction tool dedicated to green algae. Mol. Biol. Evol. *29*, 3625–3639.

Teramoto, H., Itoh, T., and Ono, T. (2004). High-intensity-light-dependent and transient expression of new genes encoding distant relatives of light-harvesting chlorophyll-a/b proteins in *Chlamydomonas reinhardtii*. Plant Cell Physiol. *45*, 1221–1232.

Terashima, M., Petroutsos, D., Hüdig, M., Tolstygina, I., Trompelt, K., Gäbelein, P., Fufezan, C., Kudla, J., Weinl, S., Finazzi, G., and Hippler, M. (2012). Calcium-dependent regulation of cyclic photosynthetic electron transport by a CAS, ANR1, and PGRL1 complex. Proc. Natl. Acad. Sci. U.S.A. *109*, 17717–17722.

Terauchi, A.M., Lu, S.F., Zaffagnini, M., Tappa, S., Hirasawa, M., Tripathy, J.N., Knaff, D.B., Farmer, P.J., Lemaire, S.D., Hase, T., and Merchant, S.S. (2009). Pattern of expression and substrate specificity of chloroplast ferredoxins from *Chlamydomonas reinhardtii*. J. Biol. Chem. *284*, 25867–25878.

Tóth, S.Z., Schansker, G., and Garab, G. (2013). The physiological roles and metabolism of ascorbate in chloroplasts. Physiol. Plant. *148*, 161–175.

Tswett, M. (1906). Absorption Analyse und Chromatographische Methode. Anwendung auf die Chemie des Chlorophylls. Ber. Deut. Bot. Ges. *24*, 384–393.

Tzvetkova-Chevolleau, T., Franck, F., Alawady, A.E., Dall'Osto, L., Carrière, F., Bassi, R., Grimm, B., Nussaume, L., and Havaux, M. (2007). The light stress-induced protein ELIP2 is a regulator of chlorophyll synthesis in *Arabidopsis thaliana*. Plant J. *50*, 795–809.

Uhrig, R.G., and Moorhead, G.B. (2011). Two ancient bacterial-like PPP family phosphatases from *Arabidopsis* are highly conserved plant proteins that possess unique properties. Plant Physiol. *157*, 1778–1792.

Urzica, E.I., Adler, L.N., Page, M.D., Linster, C.L., Arbing, M.A., Casero, D., Pellegrini, M., Merchant, S.S., and Clarke, S.G. (2012). Impact of oxidative stress on ascorbate biosynthesis in *Chlamydomonas* via regulation of the *VTC2* gene encoding a GDP-L-galactose phosphorylase. J. Biol. Chem. *287*, 14234–14245.

Vlad, D., Rappaport, F., Simon, M., and Loudet, O. (2010). Gene transposition causing natural variation for growth in *Arabidopsis thaliana*. PLoS Genet. *6*, e1000945.

van Niel, C.B. (1931). On the morphology and physiology of the purple and the green bacteria. Arch. Mikrobiol. *3*, 1–12.

von Loeffelholz, O., Kriechbaumer, V., Ewan, R.A., Jonczyk, R., Lehmann, S., Young, J.C., and Abell, B.M. (2011). OEP61 is a chaperone receptor at the plastid outer envelope. Biochem. J. *438*, 143–153.

Voss, I., Goss, T., Murozuka, E., Altmann, B., McLean, K.J., Rigby, S.E.J., Munro, A.W., Scheibe, R., Hase, T., and Hanke, G.T. (2010). FdC1, a novel ferredoxin protein capable of alternative electron partitioning, increases in conditions of acceptor limitation at photosystem I. J. Biol. Chem. *286*, 50–59.

Wagner, D., Przybyla, D., op den Camp, R., Kim, C., Landgraf, F., Lee, K.P., Würsch, M., Laloi, C., Nater, M., Hideg, E., and Apel, K. (2004). The genetic basis of singlet oxygen-induced stress responses of *Arabidopsis thaliana*. Science. *306*, 1183–1185.

Walters, R.G., Shephard, F., Rogers, J.J.M., Rolfe, S.A., and Horton, P. (2003). Identification of mutants of *Arabidopsis* defective in acclimation of photosynthesis to the light environment. Plant Physiol. *131*, 472–481.

Warburg, O., and Negelein, E. (1922). Über den Energieumsatz bei der Kohlensäureassimilation. Zeit. Physikal. Chem. *102*, 235–266; Naturwissenschaften. *10*, 647–653.

Wei, L., Guo, J., Ouyang, M., Sun, X., Ma, J., Chi, W., Lu, C., and Zhang, L. (2010). LPA19, a Psb27 homolog in *Arabidopsis thaliana*, facilitates D1 protein precursor processing during PSII biogenesis. J. Biol. Chem. *285*, 21391–21398.

Weinl, S., Held, K., Schlücking, K., Steinhorst, L., Kuhlgert, S., Hippler, M., and Kudia, J. (2008). A plastid protein critical for Ca^{2+}-regulated stomatal responses. New Phytol. *179*, 675–686.

Wollman, F.A., Minai, L., and Nechushtai, R. (1999). The biogenesis and assembly of photosynthetic proteins in thylakoid membranes. Biochem. Biophys. Acta. *1411*, 21–85.

Wollman, F.A. (2001). State transitions reveal the dynamics and flexibility of the photosynthetic apparatus. EMBO J. *20*, 3623–2630.

Wolucka, B.A., and Van Montagu, M. (2005). GDP-mannose 3′5′-epimerase forms GDP-L-gulose, a putative intermediate for the *de novo* biosynthesis of vitamin C in plants. J. Biol. Chem. *278*, 47483–47490.

Wood, P.M. (1978). Interchangable copper and iron proteins in algal photosynthesis. Studies on plastocyanin and cytochrome c-552 in *Chlamydomonas*. Eur. J. Biochem. *87*, 9–19.

Xiao, J., Li, J., Ouyang, M., Yun, T., He, B., Ji, D., Ma, J., Chi, W., and Zhang, L. (2012). DAC is involved in the accumulation of the cytochrome b_6/f complex in *Arabidopsis*. Plant Physiol. *160*, 1911–1922.

Xie, Z., Culler, D., Dreyfuss, B.W., Kuras, R., Wollman, F.A., Girard-Bascou, J., and Merchant, S. (1998). Genetic analysis of chloroplast c-type cytochrome assembly in *Chlamydomonas reinhardtii*: one chloroplast locus and at least four nuclear loci are required for heme attachment. Genetics. *148*, 681–692.

Xu, H., Vavilin, D., Funk, C., and Vermaas, W. (2002). Small Cab-like proteins regulating tetrapyrrole biosynthesis in the cyanobacterium *Synechocystis* sp. PCC 6803. Plant Mol. Biol. *49*, 149–106.

Yacoby, I., Pochekailov, S., Toporik, H., Ghirardi, M.L., King, P.W., and Zhang, S. (2011). Photosynthetic electron partitioning between [FeFe]-hydrogenase and ferredoxin:NADP+-oxidoreductase (FNR) enzymes *in vitro*. Proc. Natl. Acad. Sci. U.S.A. *108*, 9396–9401.

Yoshida, K., Igarashi, E., Mukai, M., Hirata, K., and Miyamoto, K. (2003). Induction of tolerance to oxidative stress in the green alga, *Chlamydomonas reinhardtii*, by abscisic acid. Plant Cell Environ. *26*, 451–457.

Yoshida, R., Hobo, T., Ichimura, T., Takahashi, F., Aronso, J., Ecker, J.R., and Shinozaki, K. (2002). ABA-activated SnRK2 protein kinase is required for dehydration stress signaling in *Arabidopsis*. Plant Cell Physiol. *43*, 1473–1483.

Zhu, J., Fu, X., Koo, Y.D., Zhu, J.K., Jenney Jr, F.E., Adams, M.W.W., Zhu, Y., Shi, H., Yun, D.J., Hasegawa, P.M., and Bressan, R.A. (2007). An enhancer mutant of *Arabidopsis* salt overly sensitive 3 mediates both ion homeostasis and the oxidative stress response. Mol. Cell. Biol. *27*, 5214–5224.

Zybailov, B., Rutschow, H., Friso, G., Rudella, A., Emanuelsson, O., Sun, Q., and van Wijk, K.J. (2008). Sorting signals, N-terminal modifications and abundance of the chloroplast proteome. PLoS One. *3*, e1994.

Zylka, M.J., and Reppert, S.M. (1999). Discovery of a putative heme-binding protein family (SOUL/HBP) by two-tissue suppression subtractive hybridization and database searches. Mol. Brain Res. *74*, 175–181.

Index

Cysteine 6, 71, 187, 189, 194, 200–201, 256, 258

Dinoflagellate 140

Diurnal 187, 191, 195, 198, 201, 254

Eukaryotic 1, 3, 5–6, 8–9, 18, 25, 42, 59, 138, 145, 147, 191, 199, 209, 211–216, 218, 224, 241, 244, 265

Lutein 15–16, 30–33, 37, 46, 60, 137–138, 179, 182

M

N

Tocopherol 1, 3, 12–15, 17, 37, 136, 171–172, 174–175, 177–181, 183, 215, 248, 254, 264

W

X

Y

Z

www.ingramcontent.com/pod-product-compliance
Ingram Content Group UK Ltd.
Pitfield, Milton Keynes, MK11 3LW, UK
UKHW062000041026
15591UKWH00016B/1097

9 781910 190470